Achieving sustainable cultivation of barley

It is widely recognised that agriculture is a significant contributor to global warming and climate change. Agriculture needs to reduce its environmental impact and adapt to current climate change whilst still feeding a growing population, i.e. become more 'climate-smart'. Burleigh Dodds Science Publishing is playing its part in achieving this by bringing together key research on making the production of the world's most important crops and livestock products more sustainable.

Based on extensive research, our publications specifically target the challenge of climate-smart agriculture. In this way we are using 'smart publishing' to help achieve climate-smart agriculture.

Burleigh Dodds Science Publishing is an independent and innovative publisher delivering high quality customer-focused agricultural science content in both print and online formats for the academic and research communities. Our aim is to build a foundation of knowledge on which researchers can build to meet the challenge of climate-smart agriculture.

For more information about Burleigh Dodds Science Publishing simply call us on +44 (0) 1223 839365, email info@bdspublishing.com or alternatively please visit our website at www.bdspublishing.com.

Related titles:

Achieving sustainable cultivation of wheat Volume 1: Breeding, quality traits, pests and diseases
Print (ISBN 978-1-78676-016-6); Online (ISBN 978-1-78676-018-0, 978-1-78676-019-7)

Achieving sustainable cultivation of wheat Volume 2: Cultivation techniques
Print (ISBN 978-1-78676-020-3); Online (ISBN 978-1-78676-022-7, 978-1-78676-023-4)

Advances in breeding techniques for cereal crops
Print (ISBN 978-1-78676-244-3); Online (ISBN 978-1-78676-246-7, 978-1-78676-247-4)

Integrated disease management of wheat and barley
Print (ISBN 978-1-78676-216-0); Online (ISBN 978-1-78676-219-1, 978-1-78676-218-4)

Chapters are available individually from our online bookshop:
https://shop.bdspublishing.com

climate•SMART•publishing
IN AGRICULTURAL SCIENCE

BURLEIGH DODDS SERIES IN AGRICULTURAL SCIENCE
NUMBER 74

Achieving sustainable cultivation of barley

Edited by Professor Glen Fox, University of California-Davis, USA and The University of Queensland, Australia; and Professor Chengdao Li, Murdoch University, Australia

burleigh dodds
SCIENCE PUBLISHING

Published by Burleigh Dodds Science Publishing Limited
82 High Street, Sawston, Cambridge CB22 3HJ, UK
www.bdspublishing.com

Burleigh Dodds Science Publishing, 1518 Walnut Street, Suite 900, Philadelphia, PA 19102-3406, USA

First published 2020 by Burleigh Dodds Science Publishing Limited

Library of Congress Control Number: 2019951744

British Library Cataloguing in Publication Data
A catalogue record for this book is available from the British Library

ISBN 978-1-78676-308-2 (Print)
ISBN 978-1-78676-311-2 (PDF)
ISBN 978-1-78676-310-5 (ePub)
ISSN 2059-6936 (print)
ISSN 2059-6944 (online)

DOI 10.19103/AS.2019.0060

Typeset by Deanta Global Publishing Services, Dublin, Ireland

Contents

Series list xii
Introduction xviii

Part 1 Plant physiology and genetics

1 Advances in understanding of barley plant physiology: plant development and architecture 3
Andrea Visioni, International Center for Agricultural Research in the Dry Areas (ICARDA), Morocco

1 Introduction 3
2 Barley plant structure/morphology and growth habit 4
3 Molecular control of vegetative development 7
4 Molecular control of reproductive development 11
5 Implications for breeding 14
6 References 16

2 Advances in understanding barley plant physiology: responses to abiotic stress 23
Alessandro Tondelli, Cristina Crosatti, Stefano Delbono and Luigi Cattivelli, CREA Research Centre for Genomics and Bioinformatics, Italy

1 Introduction 23
2 Cold acclimation: a coordinated metabolic rearrangement leading to frost tolerance 25
3 New methodologies for dissecting an old phenotype: resilience to drought 28
4 Adaptation to soil salinity 33
5 Low nitrogen: a stress condition matching crop sustainability 36
6 Adaptation to environment: a key target for future breeding improvement 38
7 Acknowledgements 40
8 Where to look for further information 40
9 References 41

3 Advances in the understanding of barley plant physiology: factors determining grain development, composition, and chemistry 53
Ljudmilla Borisjuk, Hardy Rolletschek and Volodymyr Radchuk, Leibniz Institute of Plant Genetics and Crop Plant Research (IPK), Germany

1 Introduction 53
2 Spike growth and how it influences traits of the grain 54
3 Role of cell death in barley grain development 57
4 Sucrose allocation during the grain-filling stage 62
5 The use of starch in the developing caryopsis 67
6 Proteins and barley grain quality 71
7 Particularities of energy metabolism in barley grain 73
8 Functional orchestration of the barley grain 80
9 Conclusion 83
10 Acknowledgements 83
11 Where to look for further information 84
12 References 84

4 Exploring barley germplasm for yield improvement under sulphur-limiting environments 97
Tefera Tolera Angessa, Murdoch University, Australia; Kefei Chen, Curtin University, Australia; David Farleigh, Jenifer Bussanich and Lee-Anne McFawn, Department of Primary Industries and Regional Development-Western Australia, Australia; Kevin Whitfield, CSBP Limited, Australia; Brendon Weir, Mullewa, Australia; Steve Cosh, Department of Primary Industries and Regional Development-Western Australia, Australia; Achalu Chimdi, Gudeta Nepir Gurmu and Tadesse Kenea Amentae, Ambo University, Ethiopia; and Chengdao Li, Murdoch University, Australia

1 Introduction 97
2 The origins of barley 98
3 Genetic diversity in barley 99
4 Using genetic diversity in breeding 101
5 The role of sulphur in barley growth 102
6 Assessing the effects of sulphur nutrition on barley and wheat grain yield 104
7 The effects of sulphur on yield, quality and response to stress 109
8 Farming systems and sulphur nutrition 113
9 Genotypic differences in sulphur use 114
10 Conclusion 117
11 Acknowledgement 117
12 References 118

5 Mapping and exploiting the barley genome: techniques for mapping genes and relating them to desirable traits 123
Hélène Pidon and Nils Stein, Leibniz Institute of Plant Genetics and Crop Plant Research (IPK), Germany

1 Introduction 123
2 New possibilities for genetic mapping in the genomics era 124
3 Classical mapping strategies and their improvement in the genomics era 129
4 The association mapping boom 130
5 Multiparental populations: the perfect balance? 131
6 From an interval to the causal gene: from high-resolution mapping to gene cloning 132
7 Emerging mapping strategies: fast NGS-enabled technologies 133
8 Conservation of barley germplasm 138
9 Genetic and genomic resources of barley 139
10 Case study: from *rym4* to *rym11*, illustration of paradigm shift in disease resistance mapping and cloning 140
11 Conclusion and future trends 142
12 Acknowledgement 144
13 Where to look for further information 144
14 References 145

Part 2 Advances in breeding

6 Advanced designs for barley breeding experiments 159
Alison Kelly, Queensland Department of Agriculture and Fisheries and Queensland Alliance for Agriculture and Food Innovation (QAAFI), The University of Queensland, Australia; and Clayton Forknall, Queensland Department of Agriculture and Fisheries, Australia

1 Introduction 159
2 Background to experimental design of field trials 161
3 Designs for late-generation field trials 164
4 Designs for early-generation field trials 169
5 Incorporating a genetic relationship matrix 172
6 Multi-phase design for laboratory experiments 176
7 Conclusions 178
8 References 179

7 Advances in molecular breeding techniques for barley: genome-wide association studies (GWAS) 183
W. T. B. Thomas, James Hutton Institute, UK

1 Introduction 183
2 Progress in barley breeding 184

3 Mapping of malting quality and yield traits 187
4 Genome-wide association studies (GWAS) mapping in barley 188
5 Application of results from genome-wide association studies (GWAS) in barley improvement 192
6 Conclusion and future trends 195
7 Acknowledgements 197
8 References 197

8 Advances in molecular breeding techniques for barley: targeted induced local lesions in genomes (TILLING) 203
Serena Rosignoli and Silvio Salvi, University of Bologna, Italy

1 Introduction 203
2 Technical details on artificial mutagenesis and mutation discovery in TILLING 204
3 TILLING resources in barley 209
4 Current and future trends of barley TILLING 209
5 TILLING versus other reverse genetics tools in barley 212
6 Conclusion 214
7 Where to look for further information 214
8 References 215

Part 3 Cultivation techniques, pest and disease management

9 Advances in postharvest storage and handling of barley: methods to prevent or reduce mycotoxin contamination 227
Zhao Jin and Paul Schwarz, North Dakota State University, USA

1 Introduction 227
2 Postharvest handling and storage operations for barley 228
3 Mycoflora and mycotoxins in barley 237
4 Prevention or decontamination of mycotoxins in barley storage 248
5 Post-storage treatment of barley 252
6 Conclusion and future trends 255
7 Where to look for further information 256
8 References 257

10 Fungal diseases affecting barley 265
Robert S. Brueggeman, Shyam Solanki, Gazala Ameen and Karl Effertz, Washington State University, USA; Roshan Sharma Poudel, North Dakota State University, USA; and Aziz Karakaya, Ankara University, Turkey

1 Introduction 265
2 Understanding plant genetic resistance to fungal pathogens 267
3 Biotrophic foliar diseases: stem rust 270

4 Leaf rust 274
5 Stripe rust 276
6 Powdery mildew 278
7 Necrotrophic diseases: spot blotch 282
8 Net blotch 286
9 Ramularia leaf spot 291
10 Septoria speckled leaf blotch 294
11 Scald 297
12 Fusarium head blight 300
13 A seed-borne disease: barley stripe 304
14 Conclusion 304
15 References 305

11 Integrated disease management of barley 323
Adrian C. Newton, James Hutton Institute and SRUC, UK; and Henry E. Creissen, Neil D. Havis, and Fiona J. Burnett, SRUC, UK

1 Introduction 323
2 Barley production context: requirements and constraints 324
3 Diseases overview 326
4 Inoculum management: sources and epidemiological conditions 330
5 Varietal resistance 331
6 Crop protectants 334
7 Agronomy 339
8 IPM knowledge sources and tools 339
9 Uptake and communication of IPM 340
10 Farming systems, soil and research platforms 342
11 Conclusion and future trends 344
12 Acknowledgements 345
13 Where to look for further information 345
14 References 345

12 Integrated weed management in barley cultivation 353
Michael Widderick, Queensland Department of Agriculture and Fisheries, Australia

1 Introduction 353
2 Integrated Weed Management 354
3 Weed control tactics 356
4 IWM in practice 366
5 Examples of IWM in barley 367
6 Conclusion 368
7 Where to look for further information 368
8 References 369

Part 4 Quality

13 Developing barley crops for improved malt quality 377
Glen Fox, University of California-Davis, USA and The University of Queensland, Australia; and Reg Lance, Queensland Department of Agriculture and Fisheries, Australia

1 Introduction 377
2 Malting quality 379
3 Case study: modern varieties for twenty-first century brewing 388
4 A brief history of barley improvement in Australia 393
5 Requirements for successful programmes in malting quality improvement 394
6 Conclusion 396
7 Future trends 396
8 Where to look for further information 397
9 References 397

14 Developing barley crops for improved brewing quality 405
Søren Knudsen, Finn Lok and Ilka Braumann, Carlsberg Research Laboratory, Denmark

1 Introduction 405
2 Converting barley into beer 407
3 Breeding barley for the brewing process 408
4 Brewing traits related to the quality of the final product 413
5 Conclusion and future trends 416
6 Acknowledgements 418
7 Where to look for further information 418
8 References 419

15 Optimising the use of barley as an animal feed 427
David M. E. Poulsen, Queensland University of Technology, Australia

1 Introduction 427
2 What is 'feed barley'? 429
3 What do we want from 'feed barley'? 430
4 Optimising feed barley use 433
5 Understanding and optimising feed barley quality for different livestock species 442
6 Future trends and research opportunities 450
7 Conclusion 456
8 Where to look for further information 456
9 References 456

16 Nutritional and bioactive compounds in barley 467
Nancy Ames, Joanne Storsley, Lovemore Malunga and Sijo Joseph Thandapilly, Agriculture and Agri-Food Canada, Canada

1 Introduction 467
2 Key issues and challenges 468
3 Barley bioactives 470
4 Health benefits of barley foods 477
5 Enhancing barley bioactivity 482
6 Summary 484
7 Future trends 485
8 Where to look for further information 486
9 References 486

Index 497

Series list

Title	Series number
Achieving sustainable cultivation of maize - Vol 1 From improved varieties to local applications Edited by: Dr Dave Watson, CGIAR Maize Research Program Manager, CIMMYT, Mexico	001
Achieving sustainable cultivation of maize - Vol 2 Cultivation techniques, pest and disease control Edited by: Dr Dave Watson, CGIAR Maize Research Program Manager, CIMMYT, Mexico	002
Achieving sustainable cultivation of rice - Vol 1 Breeding for higher yield and quality Edited by: Prof. Takuji Sasaki, Tokyo University of Agriculture, Japan	003
Achieving sustainable cultivation of rice - Vol 2 Cultivation, pest and disease management Edited by: Prof. Takuji Sasaki, Tokyo University of Agriculture, Japan	004
Achieving sustainable cultivation of wheat - Vol 1 Breeding, quality traits, pests and diseases Edited by: Prof. Peter Langridge, The University of Adelaide, Australia	005
Achieving sustainable cultivation of wheat - Vol 2 Cultivation techniques Edited by: Prof. Peter Langridge, The University of Adelaide, Australia	006
Achieving sustainable cultivation of tomatoes Edited by: Dr Autar Mattoo, USDA-ARS, USA & Prof. Avtar Handa, Purdue University, USA	007
Achieving sustainable production of milk - Vol 1 Milk composition, genetics and breeding Edited by: Dr Nico van Belzen, International Dairy Federation (IDF), Belgium	008
Achieving sustainable production of milk - Vol 2 Safety, quality and sustainability Edited by: Dr Nico van Belzen, International Dairy Federation (IDF), Belgium	009
Achieving sustainable production of milk - Vol 3 Dairy herd management and welfare Edited by: Prof. John Webster, University of Bristol, UK	010
Ensuring safety and quality in the production of beef - Vol 1 Safety Edited by: Prof. Gary Acuff, Texas A&M University, USA & Prof. James Dickson, Iowa State University, USA	011
Ensuring safety and quality in the production of beef - Vol 2 Quality Edited by: Prof. Michael Dikeman, Kansas State University, USA	012
Achieving sustainable production of poultry meat - Vol 1 Safety, quality and sustainability Edited by: Prof. Steven C. Ricke, University of Arkansas, USA	013
Achieving sustainable production of poultry meat - Vol 2 Breeding and nutrition Edited by: Prof. Todd Applegate, University of Georgia, USA	014

Achieving sustainable production of poultry meat - Vol 3 015
Health and welfare
Edited by: Prof. Todd Applegate, University of Georgia, USA

Achieving sustainable production of eggs - Vol 1 016
Safety and quality
Edited by: Prof. Julie Roberts, University of New England, Australia

Achieving sustainable production of eggs - Vol 2 017
Animal welfare and sustainability
Edited by: Prof. Julie Roberts, University of New England, Australia

Achieving sustainable cultivation of apples 018
Edited by: Dr Kate Evans, Washington State University, USA

Integrated disease management of wheat and barley 019
Edited by: Prof. Richard Oliver, Curtin University, Australia

Achieving sustainable cultivation of cassava - Vol 1 020
Cultivation techniques
Edited by: Dr Clair Hershey, formerly International Center for Tropical Agriculture (CIAT), Colombia

Achieving sustainable cultivation of cassava - Vol 2 021
Genetics, breeding, pests and diseases
Edited by: Dr Clair Hershey, formerly International Center for Tropical Agriculture (CIAT), Colombia

Achieving sustainable production of sheep 022
Edited by: Prof. Johan Greyling, University of the Free State, South Africa

Achieving sustainable production of pig meat - Vol 1 023
Safety, quality and sustainability
Edited by: Prof. Alan Mathew, Purdue University, USA

Achieving sustainable production of pig meat - Vol 2 024
Animal breeding and nutrition
Edited by: Prof. Julian Wiseman, University of Nottingham, UK

Achieving sustainable production of pig meat - Vol 3 025
Animal health and welfare
Edited by: Prof. Julian Wiseman, University of Nottingham, UK

Achieving sustainable cultivation of potatoes - Vol 1 026
Breeding improved varieties
Edited by: Prof. Gefu Wang-Pruski, Dalhousie University, Canada

Achieving sustainable cultivation of oil palm - Vol 1 027
Introduction, breeding and cultivation techniques
Edited by: Prof. Alain Rival, Center for International Cooperation in Agricultural Research for Development (CIRAD), France

Achieving sustainable cultivation of oil palm - Vol 2 028
Diseases, pests, quality and sustainability
Edited by: Prof. Alain Rival, Center for International Cooperation in Agricultural Research for Development (CIRAD), France

Achieving sustainable cultivation of soybeans - Vol 1 029
Breeding and cultivation techniques
Edited by: Prof. Henry T. Nguyen, University of Missouri, USA

Achieving sustainable cultivation of soybeans - Vol 2 030
Diseases, pests, food and non-food uses
Edited by: Prof. Henry T. Nguyen, University of Missouri, USA

Achieving sustainable cultivation of sorghum - Vol 1 031
Genetics, breeding and production techniques
Edited by: Prof. William Rooney, Texas A&M University, USA

Achieving sustainable cultivation of sorghum - Vol 2 032
Sorghum utilization around the world
Edited by: Prof. William Rooney, Texas A&M University, USA

Achieving sustainable cultivation of potatoes - Vol 2 033
Production, storage and crop protection
Edited by: Dr Stuart Wale, Potato Dynamics Ltd, UK

Achieving sustainable cultivation of mangoes 034
Edited by: Prof. Víctor Galán Saúco, Instituto Canario de Investigaciones Agrarias (ICIA), Spain & Dr Ping Lu, Charles Darwin University, Australia

Achieving sustainable cultivation of grain legumes - Vol 1 035
Advances in breeding and cultivation techniques
Edited by: Dr Shoba Sivasankar et al., formerly International Crops Research Institute for the Semi-Arid Tropics (ICRISAT), India

Achieving sustainable cultivation of grain legumes - Vol 2 036
Improving cultivation of particular grain legumes
Edited by: Dr Shoba Sivasankar et al., formerly International Crops Research Institute for the Semi-Arid Tropics (ICRISAT), India

Achieving sustainable cultivation of sugarcane - Vol 1 037
Cultivation techniques, quality and sustainability
Edited by: Prof. Philippe Rott, University of Florida, USA

Achieving sustainable cultivation of sugarcane - Vol 2 038
Breeding, pests and diseases
Edited by: Prof. Philippe Rott, University of Florida, USA

Achieving sustainable cultivation of coffee 039
Edited by: Dr Philippe Lashermes, Institut de Recherche pour le Développement (IRD), France

Achieving sustainable cultivation of bananas - Vol 1 040
Cultivation techniques
Edited by: Prof. Gert H. J. Kema, Wageningen University and Research, The Netherlands & Prof. André Drenth, University of Queensland, Australia

Global Tea Science 041
Current status and future needs
Edited by: Dr V. S. Sharma, formerly UPASI Tea Research Institute, India & Dr M. T. Kumudini Gunasekare, Coordinating Secretariat for Science Technology and Innovation (COSTI), Sri Lanka

Integrated weed management 042
Edited by: Emeritus Prof. Rob Zimdahl, Colorado State University, USA

Achieving sustainable cultivation of cocoa 043
Edited by: Prof. Pathmanathan Umaharan, Cocoa Research Centre – The University of the West Indies, Trinidad and Tobago

Robotics and automation for improving agriculture 044
Edited by: Prof. John Billingsley, University of Southern Queensland, Australia

Water management for sustainable agriculture 045
Edited by: Prof. Theib Oweis, ICARDA, Jordan

Improving organic animal farming 046
Edited by: Dr Mette Vaarst, Aarhus University, Denmark & Dr Stephen Roderick, Duchy College, UK

Improving organic crop cultivation 047
Edited by: Prof. Ulrich Köpke, University of Bonn, Germany

Managing soil health for sustainable agriculture - Vol 1 048
Fundamentals
Edited by: Dr Don Reicosky, Soil Scientist Emeritus USDA-ARS and University of Minnesota, USA

Managing soil health for sustainable agriculture - Vol 2 049
Monitoring and management
Edited by: Dr Don Reicosky, Soil Scientist Emeritus USDA-ARS and University of Minnesota, USA

Rice insect pests and their management 050
E. A. Heinrichs, Francis E. Nwilene, Michael J. Stout, Buyung A. R. Hadi & Thais Freitas

Improving grassland and pasture management in temperate agriculture 051
Edited by: Prof. Athole Marshall & Dr Rosemary Collins, IBERS, Aberystwyth University, UK

Precision agriculture for sustainability 052
Edited by: Dr John Stafford, Silsoe Solutions, UK

Achieving sustainable cultivation of temperate zone tree fruit and berries – Vol 1 053
Physiology, genetics and cultivation
Edited by: Prof. Gregory A. Lang, Michigan State University, USA

Achieving sustainable cultivation of temperate zone tree fruit and berries – Vol 2 054
Case studies
Edited by: Prof. Gregory A. Lang, Michigan State University, USA

Agroforestry for sustainable agriculture 055
Edited by: Prof. María Rosa Mosquera-Losada, Universidade de Santiago de Compostela, Spain & Dr Ravi Prabhu, World Agroforestry Centre (ICRAF), Kenya

Achieving sustainable cultivation of tree nuts 056
Edited by: Prof. Ümit Serdar, Ondokuz Mayis University, Turkey & Emeritus Prof. Dennis Fulbright, Michigan State University, USA

Assessing the environmental impact of agriculture 057
Edited by: Prof. Bo P. Weidema, Aalborg University, Denmark

Critical issues in plant health: 50 years of research in African agriculture 058
Edited by: Dr Peter Neuenschwander and Dr Manuele Tamò, IITA, Benin

Achieving sustainable cultivation of vegetables 059
Edited by: Emeritus Prof. George Hochmuth, University of Florida, USA

Advances in breeding techniques for cereal crops 060
Edited by: Prof. Frank Ordon, Julius Kuhn Institute (JKI), Germany & Prof. Wolfgang Friedt, Justus-Liebig University of Giessen, Germany

Advances in Conservation Agriculture – Vol 1 061
Systems and Science
Edited by: Prof. Amir Kassam, University of Reading, UK and Moderator, Global Conservation Agriculture Community of Practice (CA-CoP), FAO, Rome, Italy

Advances in Conservation Agriculture – Vol 2 062
Practice and Benefits
Edited by: Prof. Amir Kassam, University of Reading, UK and Moderator, Global Conservation Agriculture Community of Practice (CA-CoP), FAO, Rome, Italy

Achieving sustainable greenhouse cultivation 063
Edited by: Prof. Leo Marcelis & Dr Ep Heuvelink, Wageningen University, The Netherlands

Achieving carbon-negative bioenergy systems from plant materials 064
Edited by: Dr Chris Saffron, Michigan State University, USA

Achieving sustainable cultivation of tropical fruits 065
Edited by: Prof. Elhadi M. Yahia, Universidad Autónoma de Querétaro, Mexico

Advances in postharvest management of horticultural produce 066
Edited by: Prof. Chris Watkins, Cornell University, USA

Pesticides and agriculture 067
Profit, politics and policy
Dave Watson

Integrated management of diseases and insect pests of tree fruit 068
Edited by: Prof. Xiangming Xu and Dr Michelle Fountain, NIAB-EMR, UK

Integrated management of insect pests: Current and future developments 069
Edited by: Emeritus Prof. Marcos Kogan, Oregon State University, USA & Emeritus Prof. E. A. Heinrichs, University of Nebraska-Lincoln, USA

Preventing food losses and waste to achieve food security and sustainability 070
Edited by: Prof. Elhadi M. Yahia, Universidad Autónoma de Querétaro, Mexico

Achieving sustainable management of boreal and temperate forests 071
Edited by: Dr John Stanturf, Estonian University of Life Sciences , Estonia

Advances in breeding of dairy cattle 072
Edited by: Prof. Julius van der Werf, University of New England, Australia & Prof. Jennie Pryce, Agriculture Victoria and La Trobe University, Australia

Improving gut health in poultry 073
Edited by: Prof. Steven C. Ricke, University of Arkansas, USA

Achieving sustainable cultivation of barley 074
Edited by: Prof. Glen Fox, University of California-Davis, USA and The University of Queensland, Australia & Prof. Chengdao Li, Murdoch University, Australia

Advances in crop modelling for a sustainable agriculture 075
Edited by: Emeritus Prof. Kenneth Boote, University of Florida, USA

Achieving sustainable crop nutrition 076
Edited by: Prof. Zed Rengel, University of Western Australia, Australia

Achieving sustainable urban agriculture 077
Edited by: Prof. Han Wiskerke, Wageningen University, The Netherlands

Climate change in agriculture 078
Edited by: Dr Delphine Deryng, McGill University, Canada

Advances in poultry genetics and genomics 079
Edited by: Prof. Samuel E. Aggrey, University of Georgia, USA; Prof. Huaijun Zhou, University of California-Davis, USA; Dr Michèle Tixier-Boichard, INRA, France; and Prof. Douglas D. Rhoads, University of Arkansas, USA

Achieving sustainable management of tropical forests 080
Edited by: Prof. Jürgen Blaser, Bern University of Life Sciences, Switzerland; and Dr Patrick D. Hardcastle, Forestry Development Specialist, UK

Improving the nutritional and nutraceutical properties of wheat and other cereals 081
Edited by: Prof. Trust Beta, University of Manitoba, Canada

Achieving sustainable cultivation of ornamental plants 082
Edited by: Emeritus Prof. Michael Reid, University of California-Davis, USA

Improving rumen function 083
Edited by: Dr Chris McSweeney, CSIRO, Australia; and Prof. Rod Mackie, University of Illinois, USA

Biostimulants for sustainable crop production 084
Edited by: Prof. Youssef Rouphael, University of Naples, Italy; Prof. Patrick du Jardin, University of Liège, Belgium; Prof. Stefania de Pascale, University of Naples, Italy; Prof. Giuseppe Colla, University of Tuscia, Italy; and Prof. Patrick Brown, University of California-Davis, USA

Improving data management and decision support systems in agriculture 085
Edited by: Dr Leisa Armstrong, Edith Cowan University, Australia

Achieving sustainable cultivation of bananas – Volume 2 086
Germplasm and genetic improvement
Edited by: Prof. Gert Kema, Wageningen University, The Netherlands; and Prof. Andrè Drenth, The University of Queensland, Australia

Reconciling agricultural production with biodiversity conservation 087
Edited by: Prof. Paolo Bàrberi and Dr Anna-Camilla Moonen, Scuola Superiore Sant'Anna, Pisa, Italy

Advances in postharvest management of cereals and grains 088
Edited by: Prof. Dirk Maier, Iowa State University

Biopesticides for sustainable agriculture 089
Edited by: Prof. Nick Birch, James Hutton Institute, UK; and Prof. Travis Glare, Lincoln University, New Zealand

Understanding and improving crop root function 090
Edited by: Emeritus Prof. Peter Gregory, University of Reading, UK

Understanding the behaviour and improving the welfare of chickens 091
Edited by: Prof. Christine Nicol, Royal Veterinary College – University of London, UK

Advances in measuring soil health for sustainable agriculture 092
Edited by: Prof. Wilfred Otten, Cranfield University, UK

Supporting smallholders in achieving sustainable agriculture 093
Edited by: Dr Dominik Klauser and Dr Michael Robinson, Syngenta Foundation for Sustainable Agriculture (SFSA), Switzerland

Advances in horticultural soilless culture 094
Edited by: Prof. Nazim Gruda, University of Bonn, Germany

Reducing greenhouse gas emissions from livestock production 095
Edited by: Dr Richard Baines, Royal Agricultural University, UK

Understanding the behaviour and improving the welfare of pigs 096
Edited by: Emerita Prof. Sandra Edwards, University of Newcastle, UK

Genome editing for precision crop breeding 097
Edited by: Dr Matthew Willman, Cornell University, USA

Introduction

This collection reviews advances in research on improving barley cultivation across the value chain. Part 1 reviews advances in understanding barley physiology in such areas as plant growth, grain development and plant response to abiotic stress. Chapters also review current developments in exploiting genetic diversity and mapping the barley genome. Building on this foundation, the second part of the book summarizes advances in breeding with chapters on breeding trial design as well as advances in molecular breeding techniques such as genome wide association studies (GWAS) and targeted induced lesions in genomes (TILLING). Part 3 looks further along the value chain at ways of optimizing cultivation practices. There are chapters on post-harvest storage as well as fungal diseases, weeds and integrated methods for their management. The final part of the book assesses current developments in optimising barley for particular end uses such as malting, brewing and animal feed as well as current research on the nutraceutical properties of barley.

Part 1 Plant physiology and genetics

Chapter 1 summarizes recent advances in understanding the genetics of barley development and architecture. In particular it discusses developments in understanding barley plant structure and morphology; molecular control of vegetative development; and molecular control of reproductive development. Finally, the chapter looks at the implications of these developments for breeding more resilient and productive varieties.

The next chapter addresses the importance of cold acclimation as a coordinated metabolic rearrangement leading to frost tolerance, before going on to consider new methodologies for understanding barley's resilience to drought. The chapter considers barley's adaptation to soil salinity, its resistance to low nitrogen, and the importance of environmental adaptation as a key target for future breeding improvement. The chapter concludes by looking ahead to future research trends in this area and gives detailed suggestions on further reading.

Chapter 3 highlights the progress in our understanding of barley grain, its functional architecture, and energy metabolism shaped by the constraints of internal hypoxia. The chapter provides a current view on the relevance of programmed cell death for grain development, and mechanisms regulating sugar intake. Finally, the chapter discusses the outcome of multiscale metabolic modelling studies and how they have advanced our understanding of grain physiology. This provides an insider's view on the life of the developing barley grain and raises new questions, which remain to be answered by applying the most advanced approaches, including nuclear magnetic resonance imaging.

Following that, Chapter 4 reviews genetic diversity in barley and its role in improving varieties, including adaptation to abiotic stresses. Sulphur is an essential macronutrient required in plants for normal growth and development. Its deficiency in agricultural soils reduces grain yield and grain quality traits. Studies conducted with barley and wheat varieties demonstrate substantial variations among crops and varieties in their response to application of different levels of sulphur. The chapter looks at factors affecting sulphur nutrition in barley and the potential role of genetic differences in breeding more resilient varieties.

Finally, Chapter 5 discusses recent changes in the field of genetic mapping that allow for new possibilities of mapping barley genomes. The chapter evaluates ways in which these strategies can be used to efficiently breed for traits that will improve the resistance of barley to various stresses as well as to meet the requirements of its several uses. This section includes a case study of the shift from rym4 to rym11. The chapter concludes by looking ahead to future research trends in this area and suggests further reading on the topic.

Part 2 Advances in breeding

Part 2 begins with a review of key developments in experimental design in barley breeding. After a brief history to set the scene, Chapter 6 covers the background of experimental design for field trials, highlighting the key principles that are still fundamental for modern comparative experiments, including model-based design. The following section explores the quantification of genetic relationships through either pedigree or molecular marker information. Finally, the chapter presents the principles of multi-phase experiments for testing material both in the field and in the laboratory. Three case studies are included to highlight non-standard experimental designs that should be in the toolkit of every agricultural scientist and which are essential for modern plant breeding programs.

Chapter 7 begins by summarizing progress in barley breeding and how the advent of molecular markers has seen development in genome mapping with relation to malting quality and yield traits. This chapter shows how genome wide association studies (GWAS) have highlighted contrasts between different breeding germplasm groups, revealing where crossing between groups can produce greater advances than continuing to cross within. GWAS can also be used as training populations for genomic selection but will remain a key R&D technology as it provides a route to candidate gene identification and hence to suitable sources of genetic diversity to maintain breeding progress. Integration of multi-environment GWAS with climatic variables is essential to breed for adaptation to climate change.

Chapter 8 introduces artificial mutagenesis and mutation discovery in TILLING, following a review of advances in TILLING resources in barley. The chapter evaluates the efficiency and TILLING versus other reverse genetics tools in barley. It also assesses the range of TILLING applications in barley. It gives special emphasis to developments in molecular screening approaches, and on opportunities for using TILLING in barley breeding.

Part 3 Cultivation techniques, pest and disease management

Part 3 opens with Chapter 9 which focuses on post-harvest storage and handling practices of barley grain and how these methods can be used to mitigate mycotoxin contamination. It also discusses management of insect pests in stored barley. The chapter goes on to review the various mycotoxins and fungi that are associated with barley, followed by the various post-storage treatments of feed and malting barley. It concludes by summarising how post-harvest storage is an important component in the sustainable production of barley and highlights potential areas for future research.

Chapter 10 reviews current research on the main fungal diseases affecting barley. It first reviews what we know about the mechanisms of barley genetic resistance to fungal pathogens. The chapter then focuses on the description of major fungal pathogens effecting barley production, new insights into their mechanisms of virulence and implications for achieving sustainable resistance to these important pathogens. The chapter reviews current knowledge about biotrophic foliar diseases: stem rust, leaf rust, stripe rust and powdery mildew. It then discusses necrotrophic diseases: spot blotch, net blotch, ramularia leaf spot, septoria speckled leaf blotch, scald and fusarium head blight. The chapter finally discusses barley stripe.

The following chapter, Chapter 11, looks at how integrated pest management (IPM) can be applied to barley production, considering the different disease threats, the tools available to counteract them and possible approaches to deploying them. The chapter evaluates varietal disease resistance, the range of crop protectants available and how agronomy can be used optimise protection. The chapter also discusses barriers to IPM use in practice. Finally, the chapter looks ahead to future research trends in this area. Chapter 12 examines the problem of weeds in barley and explains the application of integrated weed management (IWM) in barley cultivation. The chapter outlines weed control tactics and the practical implementation of IWM, focussing on specific examples of IWM in barley. Finally, the chapter provides detailed further reading on this issue.

Part 4 Quality

Chapter 13 introduces current challenges for improving malting barley. It goes on to review typical traits of malting quality, such as grain size, protein and germination. It also highlights the importance of malt extract obtained after the malting process, and reviews other important traits such as starch degrading enzymes, malt colour, grain hardness and other traits that are not routinely measured. A case study on modern varieties for twenty-first century brewing is also included. The chapter concludes by discussing requirements for successful programs in malting quality improvement and potential future trends in research.

Chapter 14 follows the application of barley throughout the brewing process. The chapter describes the different traits relevant during mashing, such as starch quality and heat stability of starch degrading enzymes; as well as traits during wort boiling, filtration, maturation as well as in the final product, putting special emphasis on barley-derived off-flavours. The chapter discusses breeding strategies to improve brewing quality. Finally, the chapter looks ahead to future research trends in this area.

Chapter 15 discusses the use of barley as feed for a range of livestock. The chapter reviews ways of optimising the use of barley for animal feed, from production and breeding through to the application of new technologies such as near infrared spectroscopy and molecular markers. The chapter then examines the specific grain quality and nutritional requirements of the major animal species routinely fed barley-based diets. The chapter concludes by assessing future research trends in optimising the use of feed barley.

The final chapter, Chapter 16, reviews the known and potential bioactive compounds in barley. Whole grain barley has been widely recognized as a valuable source of a number of biologically active compounds with unique health benefits. The great number of bioactive nutrients make barley an ideal raw material for the development of functional foods. This chapter discusses key issues and challenges currently faced by barley growers and manufacturers in producing high-quality products with health-promoting properties. It also reviews the known and potential bioactive compounds in barley, as well as research that has been carried out on barley and its health benefits. It concludes by discussing research that examines potential influences of barley bioactivity as well as future trends in research.

Part 1

Plant physiology and genetics

Chapter 1

Advances in understanding of barley plant physiology: plant development and architecture

Andrea Visioni, International Center for Agricultural Research in the Dry Areas (ICARDA), Morocco

1 Introduction
2 Barley plant structure/morphology and growth habit
3 Molecular control of vegetative development
4 Molecular control of reproductive development
5 Implications for breeding
6 References

1 Introduction

Crop plant improvement for food security in the face of population growth and climate change remains a key challenge for breeders. The intensive crop production achieved in recent years, which relies heavily on fertilizers, insecticides and fungicides, is not sustainable and is not a viable strategy for the future. Future crops need to be more resource-efficient and also able to adapt to their environment in a better way. Plant adaptation to environmental conditions can be enhanced by manipulating the architecture of agronomic traits with a consequent increase in grain yield.

The barley genome is large and complex but, at the same time, it offers a large reservoir of genes that can be exploited by breeders to develop new varieties with increased grain yield. The availability of barley mutants and the advances in genomics have led to an increased knowledge of the genetic factors controlling plant architecture that can be exploited by breeders. Epigenetics can also play an important role in generating further information that can be related to phenotypic response in adaptation to climate change. Ideotype breeding, proposed by Donald in 1968, is an alternative breeding strategy that aims at designing crops with optimal adaptation to target environments by combining a set of predefined traits. The recent advances in genetics and genomics could facilitate the obtainment of such ideotypes.

http://dx.doi.org/10.19103/AS.2019.0060.01

The on-going revolution in genomics also needs to be complemented by high-throughput phenotypic approaches to increase the rate of identification of trait-gene associations. Combining genetic knowledge of plant architecture with information about developmental processes is crucial for breeding varieties with increased adaptation and resilience to climate change, which are able to deliver competitive yields with lower inputs.

This chapter summarizes recent advances in understanding the genetics of barley development and architecture. In particular it discusses developments in understanding: (i) barley plant structure and morphology, (ii) molecular control of vegetative development and (iii) molecular control of reproductive development. Finally, the chapter looks at the implications of these developments for breeding more resilient varieties.

2 Barley plant structure/morphology and growth habit

The barley embryo is located dorsally near the basal mark of the seed. It is composed of four parts: scutellum, radicle, epicotyl and a nodal region between the epicotyl and radicle (MacLeod and Palmer 1966; Rossini et al. 2018). Vegetative development starts a few days after germination with the formation of the radicle from the apical-basal axis. The roots are then originated from the radicle. The epicotyl also starts to grow from the axis and initiates the shoot (Shaaf et al. 2019). The barley embryo contains shoot and root apical meristems (SAM and RAM, respectively). The SAM and the leaf primordia enclosed by the coleoptile are part of the epicotyl, while the RAM together with the radicle is part of the coleorhiza.

Both SAM and RAM are responsible for the architecture of the aerial and basal parts of the barley plant. SAM controls the development of the above-ground structures including the nodes, internodes, leaves, axillary meristems and the inflorescence (Sussex 1989; Babb and Muehlbauer 2003). The SAM structure has been described by Doring et al. (1999) as a system composed of two main and clonally distinct layers: L1 (*tunica*) and L2 (*corpus*). Its activity starts with the development of three or four-leaf primordia during embryogenesis. As reviewed by Rossini et al. (2014), lateral buds are located in the axil of the coleoptile and in the axil of the first leaf. Shoot architecture consists of units or modules called phytomers. The basic module of plant architecture, the phytomer, is made of an internode, a leaf and axillary bud (Sharman 1942). The basal region of the plant (also known as the crown) is composed of the first set of phytomers developed from the SAM. Their internodes do not elongate. In fact, internode elongation in barley occurs only after the transition from the vegetative to reproductive phase. The number of leaves produced, and consequently the number of basal internodes at the barley culm, depends on genetic and environmental factors. A part of those internodes will elongate after the transition from the vegetative to reproductive phase (Kirby and Appleyard 1987).

Tiller development in barley occurs as a result of the proliferation of axillary meristems (AXMs) located between the stem and the leaf/coleoptile. As reported by Hussien et al. (2014), the AXM produce an axillary bud that may produce a lateral shoot with a structure similar to the main culm. The AXM produces a stem cell accumulation in the leaf axil. Afterward, cells differentiate into primordial leaves that finally result in axillary buds. Tillers will finally originate only from subsets of the axillary buds while the rest will remain at the bud stage. Tillers are usually classified as primary tillers if they are originated from the main stem and as secondary tillers when they are derived from the primary tillers (Hussien et al. 2014). Tillers production is reiterative and further tillers (tertiary tillers) can originate from secondary tillers. Tiller outgrowth is controlled by a tangled network of hormonal and regulatory signals that leads to a high morphological diversity between genotypes and within the same genotypes (Kebrom et al. 2012; Shaaf et al. 2019). In barley and wheat the number of stems produced depends on genetic and environmental factors and, in general, winter genotypes produce more tillers (Kirby and Appleyard 1987). The transition phase between the vegetative stage and the reproductive stage is also under the control of genetic, environmental and hormonal factors.

Stem elongation begins when aerial internodes start to grow at the beginning of the reproductive phase. Elongation starts from basal internodes located at the bottom of the plant and continues toward the top with the elongation of internodes located in a more apical position (Briggs 1978). The process is complete when the last internode (called a peduncle) completes its elongation. Usually the internodes of a mature barley plant are longer than those immediately below. The only exception in some cases is the peduncle (Briggs 1978). A meristematic zone located in the base of the internode ensures the restoration of a vertical stem orientation in the case of lodging. The meristem is capable of asymmetric growth and delayed lignification (Briggs 1978). The inner layer of the internode is composed of a cavity pith surrounded by the parenchyma that contains additional vascular bundles and by a ring of sclerenchyma. The outer part is composed of alternate vertical files and sclerenchyma fibers supporting vascular bundles surrounded by a silicified epidermis (Briggs 1978).

Leaves originate from a ring of founder cells recruited on the SAM flank (Bossinger et al. 1992; Shaaf et al. 2019). The leaf primordium is located in the insertion disk at the node (Sharman 1942). A barley leaf has a strap-like appearance and is divided into a distal blade and proximal sheath that wraps around the culm and supports the blade (Rossini et al. 2018). The blade is the major photosynthetic organ of the plant. The ligular region, composed of a ligule and two auricles, separates the blade from the sheath (Rossini et al.

2018). Cell division ensures the differentiation and the growth of the leaf. The expansion pattern occurs in a basipetal wave, from the tip to the base, with the sheath cells still dividing when the blade cells are fully differentiated (Sylvester et al. 1990; Kołodziejek et al. 2006). Cell division undergoes both longitudinal and transverse divisions to support leaf growth in width and length (Sylvester and Smith 2009). The final leaf size (an important factor for photosynthesis efficiency) and leaf shape depends on the spatial and temporal coordination of these processes (Shaaf et al. 2019). Leaf orientation and angle also plays an important role in photosynthesis efficiency and is determined by the lamina joint that connects the blade to the sheath (Shaaf et al. 2019).

Inflorescence development starts with the transition of the SAM to an inflorescence meristem through the so-called double ridge phase. During this phase the apex is elongated about 1 mm. Each inflorescence primordium is composed of a leaf primordium and a lateral meristem (Kirby and Appleyard 1987). The lateral meristem then becomes the main growing point from where three spikelet triplet meristem (one central and two laterals) will originate, while the leaf primordium will not develop further (Bossinger et al. 1992; Rossini et al. 2014). The three-spikelet meristem stage is also called the triple-mound stage where each spikelet meristem will evolve into a floral meristem. Afterward, two outer glumes primordia originate from the floral meristems to give rise to the floral organ primordia. The final step of floral development is the sequential differentiation of glume, lemma and stamen primordium in the mature spikelet (Rossini et al. 2014). A spikelet located in the central region of inflorescence axis will develop earlier than basal and apical spikelets. The final number of spikelets is defined by the apex that continues to initiate new spikelet meristems until the awn primordium stage. At this stage the spike layout and spikelet structure are completed (Kirby and Appleyard 1987). The fully developed barley spike is composed of floral units (spikelets) located on the floral stem. Each spikelet consists of a floret and two subtending bracts called the outer glume. The spikelets are usually organized in triplets at each rachis node.

The spikelet axis also bears the so-called lemma, an abaxial floral bract that encloses single florets and carries the awn. Several authors have reported that this complex can be considered to be a reduced vegetative leaf where the lemma corresponds to the leaf blade while the awn is considered as the sheath (Dahlgren et al. 1985; Clifford 1988; Pozzi et al. 2000; Rossini et al. 2014). Spikelet fertility is different between two- and six-rowed barley varieties. In the first, each triplet contains only one fertile spikelet while, in the second, all three are fertile. Lateral spikelets develop slower than the central spikelet and this may result in the development of very rudimentary and sterile structures that represent the typical two-rowed barley structure. The six-rowed barley spike structure is the result of a recessive mutation at the Vrs1 gene locus. Vrs1 locus promote the development of the lateral spikelets that become morphologically indistinguishable from the central spikelet, resulting in the typical six-rowed barley spike (Kirby and Appleyard 1987; Komatsuda et al. 2007).

3 Molecular control of vegetative development

The pre-anthesis developmental phase in barley consists of the vegetative and early and late reproductive stages. All of these three phases impact grain yield. During the vegetative and early reproductive phases in particular, the number of tillers, plant height, biomass accumulation and leaf angle and area can have a direct impact on grain yield.

Tillering is one of the most important and critical traits for improving grain yield in temperate cereals (Sreenivasulu and Schnurbusch 2012; Jia et al. 2011). Cereals such as barley are able to increase grain yield through an increased number of tillers (Evers and Vos 2013). On the other hand, as reviewed by Shaaf et al. (2019), tillering potential needs to be carefully balanced to avoid a number of problems (Peng et al. 1994; Kennedy et al. 2017; Tripathi et al. 2003; Kuczynska et al. 2013; Mew 1991):

1 an excessive number of tillers will result in unfertile spikes;
2 unfertile tillers divert resources from developing spikes;
3 tillers can have negative effects on other traits related to biomass accumulation; and
4 a crowded canopy can foster the spread of disease.

Numerous mutants have been identified and characterized recently. Many of them have been used to develop near-isogenic (NI) lines in the cv. Bowman collection (Druka et al. 2011). These have been classified into:

1 mutants which fail to develop auxiliary buds (single culms mutants);
2 mutants with weak auxiliary bud development;
3 mutants with low tillering reduction; and
4 mutants producing high tiller number.

A comprehensive and more detailed review of available barley tillering mutants can be found in a recent publication from Shaaf et al. (2019). Tillering is considered as a complex trait that is strongly influenced by environmental and growing conditions such as light intensity and water availability and in which phytohormones also play a crucial role (Kebrom et al. 2013; Alqudah et al. 2016). As reviewed by several authors, phytohormones regulate bud growth through a very complex pathway in which different phytohormones interact (Evers and Vos 2013; Kebrom et al. 2013). Recent evidence suggests that sugars might also be involved in the branching process through the regulation of phytohormonal gene expression (Barbier et al. 2015).

Several QTL and GWAS studies in the past decade have suggested a correlation between tillering and other important regulatory traits such as vernalization and photoperiod sensitivity, flowering time and plant height (Abeledo et al. 2004; Borràs et al. 2009: Alqudah and Schnurbusch 2013; Alqudah et al. 2016). Several authors reported that Vrn-H1, Vrn-H2 and Ppd-H1 have a significant effect on tiller production. In particular, it was reported that tiller number increased in genotypes with a strong vernalization requirement and reduced photoperiod sensitivity (Karsai et al. 1999; von Korff et al. 2006; Wang et al. 2010). On the basis of these studies, it was suggested that those three genes may regulate tillering indirectly by controlling flowering time (Corbesier et al. 2007; Tamaki et al. 2007).

The Vrs1 and Int-C genes are responsible for spike morphology and are also related to tillering. Vrs1 is a major gene controlling the row type of barley spike. Vrs1 encodes for the two-row spike while mutations in the gene result in the six-row spike. Alqudah and Schnurbusch (2014) found significant differences in tiller number associated with row number under different growing conditions. Furthermore pleiotropic effects on tiller number are associated with allelic status at the Vrs1 locus. Int-C is the barley homolog of maize TB1 (*TEOSINTE BRANCHED 1*). It inhibits bud outgrowth by regulating the GT1 gene (*GRASSY TILLERS* 1; Alqudah et al. 2016). Int-C has been identified as a major gene-controlling lateral spikelet development that also represses the number of tillers in barley early in development (Ramsay et al. 2011).

A more recent study revealed that there are five major row-type loci (Vrs1, Vrs2, Vrs3, Vrs4 and Vrs5) that can convert spikes from the two-rowed to six-rowed type (Zwirek et al. 2019). The recessive alleles at vrs4 and vrs5 also affect tillering. Genotypes carrying Vrs1 and paired vrs3, vrs4 or vrs5 recessive alleles showed increased spikelet fertlity and variation in tillers numbers. In the case of vrs3 in vrs4 background, Zwirek et al. (2019) reported loss of spikelet identity and determinacy, improved grain homogeneity and increased tillering, while vrs5 in the same background was reported to be associated to decreased tiller number and increased grain weight.

A further study by Alqudah et al. (2016) investigated QTL underlying natural variation in number of tillers per plant at different pre-anthesis stages, and differences in plant height at harvest based on differences in row type and photoperiod response in a spring barley collection. This work clearly demonstrated a link between tillering, heading date and plant height. The study from Alqudah et al. (2016) revealed that the major loci controlling tillering were Vrs1 and Ppd-H1. Accessions carrying the Ppd-H1 allele showed several QTL associated with an increased number of productive tillers. Two-rowed accessions showed a more complex genetic make-up of tillering than six-rowed accessions. The Vrs1 gene was seen to have a pleiotropic effect in increasing the number of tillers, probably as compensation for the reduced number of seed produced. Experimental evidence suggests that the Vrs1 gene also controls plant height. Accessions carrying this allele were taller than accession carrying the Vrs1 allele. The QTL detected for plant height seems to colocate with genes involved with flowering time regulators and sugar-related genes. Interestingly, QTL for tillering and plant height are co-localized in genomic regions harboring plant stature-related phytohormones and sugar-related genes.

Brassinosteroids, gibberellins and stringolactones are the three phytohormones that play a key role in determining plant height. Dwarf and semi-dwarf phenotypes can be the results of lower production or insensitivity to those phytohormones, which usually arise from disorders in their biosynthesis or signaling pathways (Marzec and Alqudah 2018). Semi-dwarf varieties were one of the key factors in the success of the Green Revolution. Shorter plants usually have increased stem sturdiness, lodging resistance, improved response to fertilizers and enhanced grain yield. These semi-dwarf varieties were achieved using mutations in gibberellin pathway genes and metabolism (Hedden 2003). Semi-dwarf wheat and rice varieties introduced during the Green Revolution have dramatically increased yield due to the repartitioning of assimilate from stems to grain production (Khush 2001).

Since the beginning of Green Revolution, reduced plant height has always been a priority for breeders, especially to increase lodging resistance. For this reason numerous loci involved in plant height have been identified through QTL mapping and GWAS (Salvi et al. 2013; Rossini et al. 2018). A huge number of mutants have been identified and classified into different categories on the basis of their phenotypic changes, pleiotropic characteristics that often accompany short culm phenotypic changes, parental background such as brachytic (brh), breviaristatum (ari), dense spike (dsp), erectoides (ert), semibrachytic (uzu), semi-dwarf (sdw) or slender dwarf (sld) (Franckowiak and Lundqvist 2013; Dockter et al. 2014). More information about barley mutants can be found in recent reviews (Rossini et al. 2014; Dockter and Hansson 2015). These mentioned pleiotropic effects have limited the use of semi-dwarfing genes in breeding programs due to the negative effects on phenotypes such as low vigor and reduced yield (Rossini et al. 2018).

Uzu is a semi-dwarf gene that seems to be the ortholog of Arabidopsis and rice BRASSINOSTEROID-INSENSITIVE1 (BRI1), encoding a brassinosteroid receptor (Li and Chory 1997; Chono et al. 2003). It is one of the few genes shaping plant height that have been successfully used in breeding. Accessions carrying the uzu1.a allele show sturdy culm, lodging resistance and tolerance to dense planting due to leaf erectness. Uzu1.a expression is sensitive to temperature, exhibiting a mild phenotype at 14°C and a stronger phenotype at 26°C as reported by Dockter et al. (2014). This sensitivity to heat is a limitation to the diffusion of this allele that is mainly restricted to East Asia. However recently new uzu alleles, less sensible to high temperature, have been identified and they could represent an alternative to uzu1.a (Dockter and Hansson 2015).

Two mutations that are more extensively used by breeders in Europe and North America and Australia include sdw1 and denso (Hellewell et al. 2000), both located on chromosome 3H and thought to be allelic (Hellewell et al. 2000; Jia et al. 2009). Beside mutants, many QTL and GWAS studies have identified loci and candidate genes controlling plant height, like the brassinosteroid biosynthesis gene DWARF4 (HvD4), that has not been described in other species, and HvCPD that encodes a protein involved in brassinosteroid biosynthesis (Alqudah et al. 2018; Dockter et al. 2014; Marzec and Alqudah 2018).

Leaf angle is regulated by phytohormones, among which brassinosteroids play a major role (Sakamoto et al. 2006a; Hartwig et al. 2011). Brassinosteroids have been reported to control many physiological processes like cell expansion, stomata development, photo-morphogenesis, plant height, grain size and stress response (as reviewed by Shaaf et al. 2019). The leaf angle is controlled at the lamina joint level by brassinosteroids. The phytohormone seems to promote cell proliferation on the adaxial side while it suppresses cell division on the abaxial side (Sun et al. 2015). Enlarged leaf angle is associated with increased brassinosteroid content or increased brassinosteroid signaling, while brassinosteroid-deficient mutants have erect leaves (Shaaf et al. 2019).

As mentioned before, the uzu1.a allele is associated with erect leaves, making genotypes carrying this allele suitable for high-density planting. Uzu1.a was the first gene cloned among the brassinosteroid mutants in barley. Gruszka et al. (2016) identified two semi-dwarf barley mutants for HvDWARF, the barley ortholog of rice OsDWARF. A mutation on the rice gene causes reduced plant height and erect leaves (Hong et al. 2002). Resequencing of the barley mutants showed that missense mutation in the coding sequence can potentially affect the conserved sequence of the protein. The mutants also showed a reduced transcription of HvBAk1, another component of the brassinosteroid signal pathway. In rice and Arabidopsis this is associated with changes in plant height, leaf erectness, grain morphological features and disease resistance (Li et al. 2009). HvDWARF4 has been identified as one of the genes from the brassinosteroid pathway controlling plant height in barley that could also play a role in controlling leaf angle as its rice ortholog OsDWARF4. Further work on its functional characterization may confirm the role of HvDWARF4 in determining leaf angle (Sakamoto et al. 2006b; Dockter et al. 2014).

Leaf area is also an important trait for breeding. The size and the shape of leaves and their position are closely related to photosynthesis, photosynthetic efficiency and with the amount of assimilates produced by plants (Jiang et al. 2015; Driever et al. 2014). The amount of assimilates produced by plants generally means more fertile tillers, more spikes and increased spikelets survival with direct consequences on grain yield (Marzec and Alqudah 2018).

A recent GWAS analysis identified nine genes involved in the genetic control of leaf area in barley, all involved in different phytohormone pathways (gibberellins, brassinosteroids and stringolactones), suggesting that these phytohormones might regulate the leaf area independently from other traits such as branching (Marzec and Alqudah 2018; Alqudah et al. 2016). Results showed the effect of the Ppd-H1 gene, while the less photoperiod-sensitive allele (ppd-H1) was also associated with increased leaf areas when associated with the Ppd-H1 allele. Other associations with leaf area were found for other genes related to heading time and sugar-related genes (Alqudah et al. 2018). Molecular regulation of leaf area development needs to be further investigated in order to identify genes and use these genes in breeding programs.

4 Molecular control of reproductive development

Because of its relationship with grain yield and grain number, understanding the genetic control of barley inflorescence development is of primary importance in plant breeding. Dozens of barley mutants with altered spike and spikelet morphology have been described and mapped, providing an ideal starting point for the genetic analysis of inflorescence development (Druka et al. 2011; Franckowiak and Lundqvist 2010; Sreenivasulu and Schnurbusch 2012; Rossini et al. 2014). Most of the differences in spike morphology in grasses are reflected by the extent of inflorescence branching (Koppulu and Schnurbusch 2019). Several studies suggest that branched inflorescence is the basis of the evolution of most evolved/derived inflorescences (Vegetti and Anton 1995; Endress 2010; Kellogg et al. 2013).

Mutants with non-canonical spike branching have been identified in barley. Recessive mutations at the branched1 (brc1) and compositum (com1 and com2) loci cause the development of branches from rachis nodes in the basal portion of the spike (Franckowiak and Lundqvist 2010; Druka et al. 2011). Wild-type COM1 and COM2 genes determine spikelet meristem characteristics and branching (Franckowiak and Lundqvist 2013; Koppulu and Schnurbusch 2019). Paired or supernumerary spikelet formation in barley seems to be under the control of the flo locus. The underlying genetic mechanism is not known but was recently elucidated in wheat (Boden et al. 2015; Dixon et al. 2018). The loss of function of the photoperiod-insensitive Ppd-D1 allele attenuates FT1 expression, thus promoting paired spikelet formation (Boden et al. 2015). Interestingly, TB1 (that also regulate the expression of FT1) promotes paired spikelet formation by interacting directly with FT1. TB1 competitively binds FT1, making it less available and delaying meristem maturation (Dixon et al. 2018).

A characteristic feature of the barley spike is the triplet spikelet meristem. In wild and two-rowed barley the lateral spikelets are sterile. Each spike node develops one central fertile grain-bearing spikelet and two lateral sterile spikelets. The characteristic two-row shape comes from the position of the two fertile central spikelets that are located at the rachis. Six-rowed barley shows three fertile spikelets that form six rows of grains (Zohary and Hopf 2000). Lundqvist and Lundqvist (1988) identified 11 loci that can modify lateral spikelet fertility and that can independently convert two-rowed barley into a six-rowed barley. The genes are vrs1, vrs2, vrs3, vrs4 and Int-C. In natural six-rowed barley the predominant genotype is vrs1.a that promotes lateral spikelet fertility. Vrs1 also affects leaf primordium size and leaf area (Thirulogachandar et al. 2017). Different combinations of vrs1 and Int-c alleles are found in barley germplasm, leading to various levels of lateral spikelet development and fertility (Ramsay et al. 2011). For example, Int-c.a is the less functional allele of HvTB1, the barley orthologous of the maize teosinte branched gene. It facilitates the growth of lateral spikelets together with the vrs1.a allele and contributes to increased grain size (Ramsay et al. 2011). On the other hand, the recessive int-c.b allele is commonly found in two-rowed (Vrs1) cultivars where it inhibits anther development in lateral florets, while in six-rowed (vrs1) cultivars it results in reduced lateral spikelet development (Rossini et al. 2014). It has been found that in several six-rowed barley accessions the mutation in Vrs1 is absent and this accession also does not show alterations in Vrs1 expression (Komatsuda et al. 2007; Youssef et al. 2017b), suggesting that the suppression of lateral spikelet fertility might also be controlled by other genes.

It has also been reported that intermediate spike morphology is under the control of Vrs3. Genotypes carrying the Vrs3 mutant allele show a two-rowed condition in the lower third of their spike. The weaker six-rowed genotype may indicate a specific developmental regulation for Vrs3 across the spike (Koppulu and Schnurbusch 2019).

Rachis length and the number of spikelets per spike units (spike density) are generally correlated traits. In fact, longer rachis correlates positively with longer internodes and also negatively with spike density. The allelic variation that controls rachis development results in a range of phenotypes, often associated with alterations in other traits (Rossini et al. 2014). As reviewed by Terzi et al. (2017) several loci are involved in the modulation of spike density, such as dense spike, zeocriton, lax spike and laxatum.

The laxatum mutant phenotype is characterized by long rachis internode, a large base of lemma awns and five anthers instead of the regular three (Civáň and Brown 2017; Jost et al. 2016). The spike density seems to be under the control of several major genes, for example, dense spike genes (dsp) which control rachis internode length, resulting in dense or compact spikes. The Dsp.ar and Dsp1 gene seems to be located on the same position on chromosome

7H close to the lks2 gene that produces short awns (Shahinnia et al. 2012; Taketa et al. 2011). As reviewed by Terzi et al. (2017), the transcription factor HvAP2 (ortholog of APETALA2), which is driven by microRNA 172, regulates the length of the critical developmental window required for the elongation of the inflorescence internodes which has a critical effect on spike density.

The awn, the extension of the lemma, is a photosynthetic organ that plays an important role in grain size and yield and also shows a wide natural variation in length and shape (Terzi et al. 2017). Mutants can show this variation. A study conducted by Yuo et al. (2012) on the short awn 2 (lks2) gene demonstrated that this mutation affects the awn's cell proliferation, resulting awns 50% shorter than the wild type. Natural recessive variants of the lks2 gene are widespread in Eastern Asia, possibly offering adaptation to high-precipitation conditions (Rossini et al. 2014). The so-called hooded phenotypes are another example of awns morphology variation, where a mutation causes the appearance of an extra flower of inverse polarity on the lemma. The mutant phenotype is caused by a 305-base pair duplication in intron 4 at the single dominant genetic locus Knox3 (Müller et al. 1995). Hooded barley cultivars represent an interesting option for feed use as the presence of awns is unsafe for animals (Blake et al. 2011).

In covered barleys, the typical barley is used for both animal feed and malting. A lipid layer is present and favors the adhesion of the hull to the caryopsis surface. Hulless or naked barley results from a mutation on the nud gene that regulates the deposition of lipids on the epidermis of the pericarp. The nud gene is a transcription factor of the ethylene response factor (ERF) family regulating the lipid biosynthesis pathway (Terzi et al. 2017; Taketa et al. 2008). Naked barley is becoming popular in several countries as a health food due to its nutritional properties (i.e. b-glucans). Naked barley does not need to be pearled, thus reducing production costs while also preserving its nutraceutical content.

Barley is an autogamous plant and its stigmas become receptive before anther extrusion. When the anthers become ready for pollination, the stigmas are able to capture sufficient pollen. Another feature of barley florets is cleistogamy that ensures self-pollination through floret morphology. Florets show smaller lodicules in comparison with the non-cleistogamous types. In typical barley florets, palea and lemma remain tightly closed throughout the period of pollen release. Lodicule development is under the control of the Cly1 gene, an APETALA2 transcription factor (Nair et al. 2010). A mutation on the recessive allele of Cly1 gene leads to an increased expression that produces lodicule reduction. In genotypes carrying the Cly1 dominant allele, the downregulation of the Cly1 gene expression promotes the growth and the swelling of the lodicules, making the plant chasmogamous (Rossini et al. 2014).

5 Implications for breeding

Barley is generally known as an adaptable crop with the ability to cope with both abiotic and biotic stresses. However, even though it could be considered as a more resilient 'climate change' crop, it is still susceptible to major abiotic and biotic stresses. To ensure food security, breeders need to continue to develop new varieties that are capable of adapting to climate change and lower inputs. Current knowledge of plant architectural traits can be exploited to achieve improved genotypes.

In the case of rice, grain yield increase through the manipulation of tiller numbers was achieved by the International Rice Research Institute (IRRI; www.irri.org) by reducing the number of unproductive tillers (Peng et al. 2008). The Chinese 'super rice' initiative also achieved a significant grain yield increase by exploiting natural variation available in rice germplasm, combining optimal plant architecture and number of tillers (Qian et al. 2016; Wenfu et al. 2007). In barley the genetic factors controlling tillering plasticity are still only partially understood. However, it has been shown that vernalization and flowering time can be manipulated to increase grain yield by optimizing the number of fertile tillers (Shaaf et al. 2019). New insights on genes controlling tillering at the pre-anthesis developmental phase can be found in a GWAS study by Alqudah et al. (2016). This identified a QTL on chromosome 2H shown to be associated with the production of productive tillers in six-row genotypes. The QTL overlaps with the HvDRM1 gene in wheat. The expression of the DRM1-like gene is associated with tiller bud dormancy in a tiller inhibition mutant (Kebrom et al. 2012). Alqudah et al. (2016) identified three further QTL on chromosomes 1H, 2H and 5H associated with productive tiller production. The QTL on 2H is located near the HvBRD locus, an important regulator of plant height in barley. Once validated, these QTLs might be useful for breeders to genetically manipulate the number of tillers in barley to increase grain yield. The same study also identified novel QTL for plant height and suggested the role of Vrs1 in controlling plant height in addition to lateral spikelet development and tillering.

Alqudah et al. (2016) also reported a QTL for plant height located in a putatively sugar-related chromosomal region on chromosome 3H, suggesting that sugars may also be involved in plant height regulation. A few genes controlling plant height have been used in breeding such as uzu1.a, sdw and denso. New uzu alleles which are less sensitive to temperature have been identified which might help to overcome the limitations of uzua.1 (Dockter and Hansson 2015). Recent GWAS studies have demonstrated the effect of loci-involved brassinosteroid biosynthesis (i.e. HvD4 and HvCPD) in regulating plant height. The link between tillering and plant height needs to be further investigated for a better understanding of the genetic relationship between the two traits that may then lead to the identification of useful genes for breeders.

Table 1 Table of genes involved in barley plant development and architecture

Gene abbreviation	Chromosome	Gene product annotation	References
Vrs-3	1H	Putative jumonji C-type (JMJC) H3K9me2/3 histone demethylase	Bull et al. (2017), van Esse et al. (2017)
Ppd-H1	2H	PPR pseudo response regulator	Turner et al. (2005)
Vrs-1 (HvHox1)	2H	Homeodomain-leucine zipper class I protein (HD-ZIP I)	Komatsuda et al. (2007)
HvAP2	2H	APETALA2 transcription factor	Houston et al. (2013)
Cly1	2H	Cleistogamy 1 gene	Wang et al. (2015)
HvDRM1	2H	DORMANCY-ASSOCIATED1	Alqudah et al. (2016)
HvBRD	2h	BRASSINOSTEROID-DEFICIENT DWARF2	Dockter et al. (2014), Alqudah et al. (2016)
Vrs-4 (HvRA2)	3H	RAMOSA2	Koppolu et al. (2013)
uzu (HvBR1)	3H	Brassinosteroid insensitive	Chono et al. (2003), Dockter et al. (2014)
uzu1.a (HvBR1)	3H	Brassinosteroid insensitive	Chono et al. (2003), Dockter et al. (2014)
sdw1/denso	3H	Gibberellin 20-oxidase gene (HvGA20ox2)	Kuczynska et al. (2013), Xu et al. (2017)
Vrn-H2 (HvZCCT-Hb)	4H	ZCCT gene family member	von Ziztewiz et al. (2005)
Vrs-5 or Int-C (HvTB1)	4H	TEOSINTE BRANCHED1	Ramsay et al. (2011)
HvD4	4H	Brassinosteroid biosynthesis gene DWARF4	Dockter et al. (2014), Alqudah et al. (2018)
Vrn-H1 (HvBM5A)	5H	HvBM5A; MADS-box	von Ziztewiz et al. (2005)
Vrs-2	5H	Homolog of the Arabidopsis SHORT INTERNODES gene	Youssef et al. (2017a)
HvCPD	5H	BRASSINOSTEROID C-23 HYDROXYLASE	Dockter et al. (2014)
FT1	7H	FLOWERING LOCUS T	Faure et al. (2007)
Iks2	7H	*NA*	Shahinnia et al. (2012), Taketa et al. (2011)
Dsp1	7H	*NA*	Shahinnia et al. (2012), Taketa et al. (2011)
Dsp.ar	7H	*NA*	Shahinnia et al. (2012), Taketa et al. (2011)
Nud	7H	Ethylene response factor (ERF) family	Taketa et al. (2008)

Brassinosteroids also play a key role in leaf angle and leaf area (Sakamoto et al. 2006a,b; Hartwig et al. 2011; Marzec and Alqudah 2018; Alqudah et al. 2016). The ortholog of OsDWARF4 in barley HvD4 has been reported as part of the brassinosteroid pathway and seems to be associated with leaf angle. This gene is also associated with plant height. The relationship between brassinosteroids, plant height and leaf angle has also been reported for other brassinosteroid mutants like uzu. Leaf area is controlled by phytohormones, sugar-related genes and photoperiod. The genetic factors controlling leaf area need to be better understood in order to exploit these factors in breeding. The number of characterized mutants available, combined with new genomics technology such as genome editing and targeted mutagenesis, may, subject to public approval, making it possible to create targeted mutations in breeding lines (Dockter et al. 2014) (Table 1).

6 References

Abeledo, L. G., Calderini, D. F. and Slafer, G. A. (2004), Leaf appearance, tillering and their coordination in old and modern barleys from Argentina. *Field Crop Res.* 86: 23–32. doi:10.1016/S0378-4290(03)00168-0.

Alqudah, A. M. and Schnurbusch, T. (2013), Awn primordium to tipping is the most decisive developmental phase for spikelet survival in barley. *Funct. Plant Biol.* 4: 424–36.

Alqudah, A. M. and Schnurbusch, T. (2014), Awn primordium to tipping is the most decisive developmental phase for spikelet survival in barley. *Funct. Plant Biol.* 41: 424–36. doi:10.1071/FP13248.

Alqudah, A. M., Koppolu, R., Wolde, G. M., Graner, A. and Schnurbusch, T. (2016), The genetic architecture of barley plant stature. *Front. Genet.* 7: 117.

Alqudah, A.M., Youssef, H.M., Graner, A. and Schnurbusch, T. (2018) Natural variation and genetic make-up of leaf blade area in spring barley. *Theor. Appl. Genet.* 131(4): 873–86. https://doi.org/10.1007/s00122-018-3053-2.

Babb, S. and Muehlbauer, G. (2003), Genetic and morphological characterization of the barley uniculm2 (cul2) mutant. *Theor. Appl. Genet.* 106: 846. https://doi.org/10.1007/s00122-002-1104-0.

Barbier, F. F., Lunn, J. E. and Beveridge, C. A. (2015), Ready, steady, go! A sugar hit starts the race to shoot branching. *Curr. Opin. Plant Biol.* 25: 39–45. doi:10.1016/j.pbi.2015.04.004.

Blake, T., Blake, V. C., Bowman, J. G. P. and Abdel-Haleem, H. (2011), Barley feed uses and quality improvement. In: Ullrich, S. E. (Ed.), *Barley: Production, Improvement, and Uses*. John Wiley-Blackwell, West Sussex, pp. 522–31.

Boden, S. A., Cavanagh, C., Cullis, B. R., Ramm, K., Greenwood, J., Jean Finnegan, E., Trevaskis, B. and Swain, S. M. (2015), Ppd-1 is a key regulator of inflorescence architecture and paired spikelet development in wheat. *Nat. Plants* 1: 14016.

Borràs, G., Romagosa, I., van Eeuwijk, F. and Slafer, G. (2009), Genetic variability in the duration of pre-heading phases and relationships with leaf appearance and tillering dynamics in a barley population. *Field Crop Res.* 113(2): 95–104.

Bossinger, G., Maddaloni, M., Motto, M. and Salamini, F. (1992), Formation and cell lineage patterns of the shoot apex of maize. *Plant J.* 2: 311–20.

Briggs, D. E. (1978), The morphology of barley; the vegetative phase. In: Briggs, D. E. (Ed.), *Barley*. Chapman and Hall, London, pp. 1–38. ISBN-13: 978-94-009-5717-6.

Bull, H., Casao, M., Zwirek, M., Flavell, A. J., Thomas, W. T. B., Guo, W., Zhang, R., Rapazote-Flores, P., Kyriakidis, S., Russell, J., Druka, A., McKim, S. M. and Waugh, R. 2017. Barley SIX-ROWED SPIKE3 encodes a putative Jumonji C-type H3K9me2/me3 demethylase that represses lateral spikelet fertility. *Nat. Commun.* 8, 936. doi:10.1038/s41467-017-00940-7.

Chono, M., Honda, I., Zeniya, H., Yoneyama, K., Saisho, D., Takeda, K., Takatsuto, S., Hoshino, T. and Watanabe, Y. (2003), A semidwarf phenotype of barley uzu results from a nucleotide substitution in the gene encoding a putative brassinosteroid receptor. *Plant Physiol.* 133: 1209–19.

Civáň, P. and Brown, T. A. (2017), A novel mutation conferring the non-brittle phenotype of cultivated barley. *New Phytol.* 214: 468–72.

Clifford, H. T. (1988), Spikelet and floral morphology. In: Soderstrom, T. R., Hiu, K. W., Campbell, C. S. and Barkworth, M. E. (Eds), *Grass Systematics and Evolution*. Smithsonian Institution Press, Washington DC, pp. 21–30.

Corbesier, L., Vincent, C., Jang, S., Fornara, F., Fan, Q., Searle, I., Giakountis, A., Farrona, S., Gissot, L., Turnbull. C. and Coupland, G. (2007), FT protein movement contributes to long-distance signaling in floral induction of Arabidopsis. *Science* 316(5827): 1030–3.

Dahlgren, R. H. T., Clifford, H. T. and Yeo, P. F. (1985), The families of the monocotyledons. In: *Structure, Evolution and Taxonomy*. Springer, New York, NY, 520pp.

Dixon, L. E., Greenwood, J. R., Bencivenga, S., Zhang, P., Cockram, J., Mellers, G., Ramm, K., Cavanagh, C., Swain, S. M. and Boden, S. A. (2018), TEOSINTE BRANCHED1 regulates inflorescence architecture and development in bread wheat (Triticum aestivum L.). *Plant Cell* 30: 563–81.

Dockter, C. and Hansson, M. (2015), Improving barley culm robustness for secured crop yield in a changing climate. *J. Exp. Bot.* 66: 3499–509.

Dockter, C., Gruszka, D., Braumann, I., Druka, A., Druka, I., Franckowiak, J., Gough, S. P., Janeczko, A., Kurowska, M., Lundqvist, J., Lundqvist, U., Marzec, M., Matyszczak, I., Muller, A. H., Oklestkova, J., Schulz, B., Zakhrabekova, S. and Hansson, M. (2014), Induced variations in brassinosteroid genes define barley height and sturdiness, and expand the green revolution genetic toolkit. *Plant Physiol.* 166: 1912–27.

Donald, C. M. (1968), The breeding of crop ideotypes. *Euphytica* 17: 385–403.

Doring, H. P., Lin, J., Uhrig, H. and Salamini. F. (1999), Clonal analysis of the development of the barley (Hordeumvulgare L.) leaf using periclinal chlorophyll chimeras. *Planta* 207(3): 335–42.

Driever, S. M., Lawson, T., Andralojc, P., Raines, C. A. and Parry, M. (2014), Natural variation in photosynthetic capacity, growth, and yield in 64 field-grown wheat genotypes. *J. Exp. Bot.* 65: 4959–73.

Druka, A., Franckowiak, J., Lundqvist, U., Bonar, N., Alexander, J., Houston, K., Radovic, S., Shahinnia, F., Vendramin, V., Morgante, M., Stein, N. and Waugh. R. (2011), Genetic dissection of barley morphology and development. *Plant Physiol.* 155: 617–27.

Endress, P. (2010), Disentangling confusions in inflorescence morphology: Patterns and diversity of reproductive shoot ramification in angiosperms. *J. Syst. Evol.* 48: 225–39.

Evers, J. B. and Vos, J. (2013), Modeling branching in cereals. *Front. Plant Sci.* 4: 399. doi:10.3389/fpls.2013.00399.

Faure, S., Higgins, J., Turner, A. and Laurie, D. A. (2007) The FLOWERING LOCUS-T-like family in barley (*Hordeum vulgare*). *Genetics* 176: 599–609.

Franckowiak, J. D. and Lundqvist, U. (2010) Descriptions of barley genetic stocks for 2010. *Barley Genet. Newsl.* 40: 45–177.

Franckowiak, J. D. and Lundqvist, U. (2013), Descriptions of barley genetic stocks. *Barley Genet. Newsl.* 43: 48–223.

Gruszka, D., Janeczko, A., Dziurka, M., Pociecha, E., Oklestkova, J. and Szarejko, I. (2016), Barley brassinosteroid mutants provide an insight into phytohormonal homeostasis in plant reaction to drought stress. *Front. Plant Sci.* 7: 1824.

Hartwig, T., Chuck, G. S., Fujioka, S., Klempien, A., Weizbauer, R., Potluri, D. P. V., Choe, S., Johal, G. S. and Schulz, B. (2011), Brassinosteroid control of sex determination in maize. *Proc. Natl. Acad. Sci.* 108: 19814–19.

Hedden, P. (2003), The genes of the Green Revolution. *Trends Genet.* 19: 5–9.

Hellewell, K. B., Rasmusson, D. C. and Gallo-Meagher, G. (2000), Enhancing yield in semidwarf barley. *Crop Sci.* 40: 352–58.

Hong, Z., Ueguchi-Tanaka, M., Shimizu-Sato, S., Inukai, Y., Fujioka, S., Shimada, Y., Takatsuto, S., Agetsuma, M., Yoshida, S., Watanabe, Y., Uozu, S., Kitano, H., Ashikari, M. and Matsuoka, M. (2002), Loss-of-function of a rice brassinosteroid biosynthetic enzyme, C-6 oxidase, prevents the organized arrangement and polar elongation of cells in the leaves and stem. *Plant J.* 32: 495–508.

Houston, K., McKim, S. M., Comadran, J., Bonar, N., Druka, I., Uzrek, N., Cirillo, E., Guzy-Wrobelska, J., Collins, N. C., Halpin, C., et al. (2013), Variation in the interaction between alleles of HvAPETALA2 and microRNA172 determines the density of grains on the barley inflorescence. *Proc. Natl. Acad. Sci. USA* 110: 16675–16680.

Hussien, A., Tavakol, E., Horner, D. S., Munoz-Amatriain, M., Muehlbauer, G. J. and Rossini, L. (2014), Genetics of tillering in rice and barley. *Plant Genome* 7(1): 1–20.

Jia, Q., Zhang, J., Westcott, S., Zhang, X. Q., Bellgard, M., Lance, R. and Li, C. (2009), GA-20 oxidase as a candidate for the semidwarf gene sdw1/denso in barley. *Funct. Integr. Genomics* 9: 255–62.

Jia, Q., Zhang, X. Q., Westcott, S., Broughton, S., Cakir, M., Yang, J., Lance, R. and Li, C. (2011), Expression level of a gibberellin 20-oxidase gene is associated with multiple agronomic and quality traits in barley. *Theor. Appl. Genet.* 122: 1451. https://doi.org/10.1007/s00122-011-1544-5.

Jiang, D., Fang, J., Lou, L., Zhao, J., Yuan, S., Yin, L., Sun, W., Peng, L., Guo, B. and Li, X. (2015), Characterization of a null allelic mutant of the rice NAL1 gene reveals its role in regulating cell division. *PLoS ONE* 10: e0118169.

Jost, M., Taketa, S., Mascher, M., Himmelbach, A., You, T., Shahinnia, F., Rutten, T., Druka, A., Schmutzer, T., Steuernagel, B., Beier, S., Taudien, S., Scholz, U., Morgante, M., Waugh, R. and Stein, N. (2016), A homolog of Blade-On-Petiole 1 and 2 (BOP1/2) controls internode length and homeotic changes of the barley inflorescence. *Plant Physiol.* 171: 1113–27.

Karsai, I., Eszaros, K. M., Szucs, P., Hayes, P. M., Lang, L. and Bedo, Z. (1999), Effects of loci determining photoperiod sensitivity (Ppd-H1) and vernalization response (Sh2) on agronomic traits in the 'Dicktoo' x 'Morex' barley mapping population. *Plant Breed* 118(5): 399–403.

Kebrom, T. H., Chandler, P. M., Swain, S. M., King, R. W., Richards, R. A. and Spielmeyer, W. (2012), Inhibition of tiller bud outgrowth in the tin mutant of wheat is associated with precocious internode development. *Plant Physiol.* 160: 308–18.

Kebrom, T. H., Spielmeyer, W. and Finnegan, E. J. (2013). Grasses provide new insights into regulation of shoot branching. *Trends Plant Sci.* 18, 41–48. doi: 10.1016/j.tplants.2012.07.001.

Kellogg, E., Camara, P., Rudall, P., Ladd, P., Malcomber, S., Whipple, C. and Doust, A. (2013), Early inflorescence development in the grasses (Poaceae). *Front. Plant Sci.* 4: 250.

Kennedy, S. P., Bingham, I. J. and Spink, J. H. (2017), Determinants of spring barley yield in a high-yield potential environment. *J. Agr. Sci.* 155: 60–80.

Khush, G. S. (2001), Green revolution: The way forward. *Nat. Rev. Genet.* 2: 815–22.

Kirby, E. J. M. and Appleyard, M. (1987), *Cereal Development Guide*, 2nd edn. Arable Unit, Stoneleigh, Warwickshire.

Kołodziejek, I., Wałęza, M. and Mostowska, A. (2006), Morphological, histochemical and ultrastructural indicators of maize and barley leaf senescence. *Biol. Plant* 50(4): 565–73.

Komatsuda, T., Pourkheirandish, M., He, C., Azhaguvel, P., Kanamori, H., Perovic, D., Stein, N., Graner, A., Wicker, T., Tagiri, A., Lundqvist, U., Fujimura, T., Matsuoka, M., Matsumoto, T. and Yano, M. (2007), Six-rowed barley originated from a mutation in a homeodomain-leucine zipper I-class homeobox gene. *Proc. Natl. Acad. Sci. USA* 104: 1424–9.

Koppolu, R., Anwar, N., Sakuma, S., Tagiri, A., Lundqvist, U., Pourkheirandish, M., Rutten, T., Seiler, C., Himmelbach, A., Ariyadasa, R., Youssef, H. M., Stein, N., Sreenivasulu, N., Komatsuda, T. and Schnurbusch, T. (2013), Six-rowed spike4 (Vrs4) controls spikelet determinacy and row-type in barley. *Proc. Natl. Acad. Sci. USA* 110: 13198–203.

Koppolu, R. and Schnurbusch, T. (2019), Developmental pathways for shaping spike inflorescence architecture in barley and wheat. *J. Integr. Plant Biol.* 61: 278–95.

Kuczynska, A., Surma, M., Adamski, T., Mikołajczak, K., Krystkowiak, K. and Ogrodowicz, P. (2013), Effects of the semidwarfing sdw1/denso gene in barley. *J. Appl. Genet.* 54: 381–90.

Li, J. and Chory, J. (1997), A putative leucine-rich repeat receptor kinase involved in brassinosteroid signal transduction. *Cell* 90: 929–38.

Li, D., Wang, L., Wang, M., Xu, Y. Y., Luo, W., Liu, Y. J., Xu, Z. H., Li, J. and Chong, K. (2009), Engineering OsBAK1 gene as a molecular tool to improve rice architecture for high yield. *Plant Biotechnol. J.* 7: 791–806.

Lundqvist, U. and Lundqvist, A. (1988), Induced intermedium mutants in barley: Origin, morphology and inheritance. *Hereditas* 108: 13–26.

MacLeod, A. M. and Palmer, G. H. (1966), The embryo of barley in relation to modification of the endosperm. *J. Inst. Brew.* 72(6): 580–9.

Marzec, M. and Alqudah, A. (2018), Key hormonal components regulate agronomically important traits in barley. *Int. J. Mol. Sci.* 19(3), 795. http://dx.doi.org/10.3390/ijms19030795.

Mew, T. W. (1991), Disease management in rice. In: Pimentel, D. (Ed.), *CRC Handbook of Pest Management in Agriculture*, 2nd edn. Vol. III, CRC Press Inc, Boca Raton, FL, pp. 279–99.

Müller, K. J., Romano, N., Gerstner, O., Garcia-Marotot, F., Pozzi, C., Salamini, F. and Rohde, W. (1995), The barley Hooded mutation caused by a duplication in a homeobox gene intron. *Nature* 374: 727–30.

Nair, S. K., Wang, N., Turuspekov, Y., Pourkheirandish, M., Sinsuwongwat, S., Chen, G., Sameri, M., Tagiri, A., Honda, I., Watanabe, Y., Kanamori, H., Wicker, T., Stein, N., Nagamura, Y., Matsumoto, T. and Komatsuda, T. (2010), Cleistogamous flowering in barley arises from the suppression of microRNA-guided HvAP2 mRNA cleavage. *Proc. Natl. Acad. Sci. USA* 107: 490–5.

Peng, S., Khush, G. S. and Cassman, K. G. (1994), Evolution of the new plant ideotype for increase yield potential. In: Cassman, K. G. (Ed.), *Breaking the Yield Barrier: Proceedings of a Workshop on Rice Yield Potential in Favorable Environments*. LosBanos, Philippine, pp. 5–20.

Peng, S. B., Khush, G. S., Virk, P., Tang, Q. Y., and Zou, Y. B. (2008). Progress in ideotype breeding to increase rice yield potential. Field Crops Res. 108, 32–38.

Pozzi, C., Faccioli, P., Terzi, V., Stanca, A. M., Cerioli, S., Castiglioni, P., Fink, R., Capone, R., Müller, K. J., Bossinger, G., Rohde, W. and Salamini, F. (2000), Genetics of mutations affecting the development of a barley floral bract. *Genetics* 154: 1335–46.

Qian, Q., Guo, L., Smith, S. M. and Li, J. 2016. Breeding high-yield superior quality hybrid super rice by rational design. *Natl. Sci. Rev.* 3(3): 283–24. https://doi.org/10.1093/nsr/nww006.

Ramsay, L., Comadran, J., Druka, A., Marshall, D. F., Thomas, W. T. B., MacAulay, M., MacKenzie, K., Simpson, C., Fuller, J., Bonar, N., Hayes, P. M., Lundqvist, U., Franckowiak, J. D., Close, T. J., Muehlbauer, G. J. and Waugh, R. (2011), INTERMEDIUM-C, a modifier of lateral spikelet fertility in barley, is an ortholog of the maize domestication gene TEOSINTE BRANCHED 1. *Nat. Genet.* 43: 169–72.

Rossini, L., Okagaki, R., Druka, A. and Muehlbauer, G. J. (2014), Shoot and Inflorescence Architecture. In: Kumlehn, J. and Stein, N. (Eds), *Biotechnological Approaches to Barley Improvement*. Biotechnology in Agriculture and Forestry (vol. 69). Springer, Berlin, Heidelberg.

Rossini, L., Muehlbauer, G. J., Okagaki, R., Salvi, S. and von Korff, M. (2018), Genetics of Whole Plant Morphology and Architecture. In: Stein, N. and Muehlbauer, G. (Eds), *The Barley Genome. Compendium of Plant Genomes*. Springer, Cham.

Sakamoto, T., Morinaka, Y., Ohnishi, T., Sunohara, H., Fujioka, S., Ueguchi Tanaka, M., Mizutani, M., Sakata, K., Takatsuto, S., Yoshida, S., Tanaka, H., Kitano, H. and Matsuoka, M. (2006a), Erect leaves caused by brassinosteroid deficiency Increase biomass production and grain yield in rice. *Nat. Biotechnol.* 24: 105–9. doi:10.1038/nbt1173.

Sakamoto, T., Sakakibara, H., Kojima, M., Yamamoto, Y., Nagasaki, H., Inukai, Y., Sato, Y. and Matsuoka, M. (2006b), Ectopic expression of KNOTTED1-like homeobox protein induces expression of cytokinin biosynthesis genes in rice. *Plant Physiol.* 142: 54–62.

Salvi, S., Porfiri, O. and Ceccarelli, S. (2013), Nazareno Strampelli, the 'Prophet' of the green revolution. *J. Agric. Sci.* 151(1): 1–5.

Shaaf, S., Bretani, G., Biswas, A., Fontana, I. M. and Rossini, L. (2019), Genetics of barley tiller and leaf development. *J. Integr. Plant Biol.* 61: 226–56.

Shahinnia, F., Druka, A., Franckowiak, J., Morgante, M., Waugh, R. and Stein, N. (2012), High resolution mapping of Dense spike-ar (dsp.ar) to the genetic centromere of barley chromosome 7H. *Theor. Appl. Genet.* 124: 373–84.

Sharman, B. C. (1942), Developmental anatomy of the shoot *Zea mays* L. *Ann. Bot.* 6(2): 245–82.

Sreenivasulu, N. and Schnurbusch, T. (2012), A genetic playground for enhancing grain number in cereals. *Trends Plant Sci.* 17: 91–101.doi: 10.1016/j.tplants.2011.11.003.

Sun, S., Chen, D., Li, X., Qiao, S., Shi, C., Li, C., Shen, H. and Wang, X. (2015), Brassinosteroid signaling regulates leaf erectness in Oryza sativa via the control of a specific U-type cyclin and cell proliferation. *Dev. Cell* 34: 220–8.

Sussex, I. M. (1989), Developmental programming of the shoot meristem. *Cell* 56: 225–9.

Sylvester, A. W. and Smith, L. G. (2009), Cell biology of maize leaf development. In: Bennetzen, J. L., Hake, S. C. (Eds), *Handbook of Maize: Its Biology*. Springer, New York, NY, pp. 179-203.

Sylvester, A. W., Cande, W. Z. and Freeling, M. (1990), Division and differentiation during normal and liguleless-1 maize leaf development. *Development* 110(3): 985–1000.

Taketa, S., Amano, S., Tsujino, Y., Sato, T., Saisho, D., Kakeda, K., Nomura, M., Suzuki, T., Matsumoto, T., Sato, K., Kanamori, H., Kawasaki, S. and Takeda, K. (2008), Barley grain with adhering hulls is controlled by an ERF family transcription factor gene regulating a lipid biosynthesis pathway. *Proc. Natl. Acad. Sci. USA* 105: 4062–7.

Taketa, S., Yuo, T., Sakurai, Y., Miyake, S. and Ichii, M. (2011), Molecular mapping of the short awn 2 (lks2) and dense spike 1 (dsp1) genes on barley chromosome 7H. *Breed Sci.* 61: 80.

Tamaki, S., Matsuo, S., Wong, H. L., Yokoi, S. and Shimamoto, K. (2007), Hd3a protein is a mobile flowering signal in rice. *Science* 316: 1033-6.

Terzi, V., Tumino, G., Pagani, D., Rizza, F., Ghizzoni, R., Morcia, C. and Stanca, A. M. (2017), Barley Developmental Mutants: The High Road to Understand the Cereal Spike Morphology. *Diversity* 9, 21.

Thirulogachandar, V., Alqudah, A. M., Koppolu, R., Rutten, T., Graner, A., Hensel, G., Kumlehn, J., Brautigam, A., Sreenivasulu, N., Schnurbusch, T. and Kuhlmann, M. (2017), Leaf primordium size specifies leaf width and vein number among row-type classes in barley. *Plant J.* 91: 601–12.

Tripathi, S. C., Sayre, K. D., Kaul, J. N. and Narang, R. S. (2003), Growth and morphology of spring wheat (*Triticum aestivum* L.) culms and their association with lodging: Effects of genotypes, N levels and ethephon. *Field Crop Res.* 84: 271–90.

Turner, A., Beales, J., Faure, S., Dunford, R. P. and Laurie, D. A. (2005) The pseudoresponse regulator Ppd-H1 provides adaptation to photoperiod in barley. *Science* 310: 1031–4.

van Esse, G. W., Walla, A., Finke, A., Koornneef, M., Pecinka, A. and von Korff, M. 2017. Six-Rowed Spike3 (VRS3) is a histone demethylase that controls lateral spikelet development in barley. *Plant Physiol,* 174: 2397-408.

Vegetti, A. and Anton, A. M. (1995), Some evolution trends in the inflorescence of Poaceae. *Flora* 190: 225–8.

von Korff, M., Wang, H., Léon, J. and Pillen, K. (2006), AB-QTL analysis in spring barley: II. Detection of favourable exotic alleles for agronomic traits introgressed from wild barley (*H. vulgare* ssp. spontaneum). *Theor. Appl. Genet.* 112(7): 1221–31.

Von Zitzewitz J., Szucs P., Dubcovsky J., Yan L., Francia E., Pecchioni N., Casas A., Chen T., Hayes P. and Skinner J. (2005), Molecular and structural characterization of barley vernalization genes. *Plant Mol. Biol.* 59: 449–67.

Wang, G., Schmalenbach, I., von Korff, M., Léon, J., Kilian, B., Rode, J. and Pillen, K. (2010), Association of barley photoperiod and vernalization genes with QTLs for flowering time and agronomic traits in a BC2DH population and a set of wild barley introgression lines. *Theor. Appl. Genet.* 120(8): 1559–74.

Wang, N., Ning, S., Wu, J., Tagiri, A. and Komatsuda, T. (2015) An epiallele at cly1 affects the expression of floret closing (cleistogamy) in barley. *Genetics* 199(1): 95–104. doi:10.1534/genetics.114.171652.

Wenfu, C., Zhengjin, X. and Longbu, Z. (2007), Theories and practices of breeding japonica rice for super high yield. *Sci. Agric. Sin.* 40: 869–74.

Xu, Y., Jia, Q., Zhou, G., Zhang, X. Q., Angessa, T., Broughton, S., Yan, G., Zhang, W. and Li, C. (2017) Characterization of the sdw1 semi-dwarf gene in barley. *BMC Plant Biol.* 17(1):11.

Youssef, H. M., Eggert, K., Koppolu, R., Alqudah, A. M., Poursarebani., N., Fazeli, A., Sakuma, S., Tagiri, A., Rutten, T., Govind, G., Lundqvist, U., Graner, A., Komatsuda, T., Sreenivasulu, N. and Schnurbusch, T. (2017a), VRS2 regulates hormone-mediated inflorescence patterning in barley. *Nat. Genet.* 49: 157–61.

Youssef, H. M., Mascher, M., Ayoub, M. A., Stein, N., Kilian, B. and Schnurbusch, T. (2017b), Natural diversity of inflorescence architecture traces cryptic domestication genes in barley (*Hordeum vulgare* L.). *Genet. Resour. Crop Evol.* 64: 843–53.

Yuo, T., Yamashita, Y., Kanamori, H., Matsumoto, T., Lundqvist, U., Sato, K., Ichii, M., Jobling, S. A. and Taketa, S. (2012), A SHORT INTERNODES (SHI) family transcription factor gene regulates awn elongation and pistil morphology in barley. *J. Exp. Bot.* 63: 5223–32.

Zohary, D. and Hopf, M. (2000), *Domestication of Plants in the Old World: The Origin and Spread of Cultivated Plants in West Asia, Europe and the Nile Valley*. Oxford University Press, Oxford.

Zwirek, M., Wough, R. and McKim, S. (2019), Interaction between row-type genes in barley controls meristem determinacy and reveals novel routes to improved grain. *New Phytol.* 221: 1950–65. doi: 10.1111/nph.15548.

Chapter 2

Advances in understanding barley plant physiology: responses to abiotic stress

Alessandro Tondelli, Cristina Crosatti, Stefano Delbono and Luigi Cattivelli, CREA Research Centre for Genomics and Bioinformatics, Italy

1 Introduction
2 Cold acclimation: a coordinated metabolic rearrangement leading to frost tolerance
3 New methodologies for dissecting an old phenotype: resilience to drought
4 Adaptation to soil salinity
5 Low nitrogen: a stress condition matching crop sustainability
6 Adaptation to environment: a key target for future breeding improvement
7 Acknowledgements
8 Where to look for further information
9 References

1 Introduction

The development and the expression of the yield potential of any crop plant depend on the environment and, given that very often the environmental conditions are far from optimal, on the resilience capacity that allow to face the multiplicity of different abiotic stress situations that every plant experiences during its life cycle. An abiotic stress can result from the shortage of an essential resource (i.e. nutrient), from the excess of a toxic substance or from climatic extremes. Occurrence, severity, timing and duration of stresses vary from location to location and from year to year. Furthermore, an abiotic stress seldom occurs alone, the plants often face growing conditions characterized by a combination of different physical stresses (Cattivelli et al., 2008).

The impact of abiotic stresses on barley is expected to increase in the future because of the ongoing climatic changes. The scientific evidences indicate that the warming of the climate system is unequivocal, as it is now evident from observations of increases in global average temperatures, widespread melting

http://dx.doi.org/10.19103/AS.2019.0060.02

of snow and ice and rising global average sea level (IPCC, 2013). Consequently, there are expectations of increase in the frequency and severity of extreme temperature events as well as of droughts, and by 2050 droughts are expected to cause water shortages which will have a significant impact on crops yield. A recent study has evaluated the effects of concurrent drought and heat extreme events projected under a range of future climate scenarios on barley. The authors have found that these extreme events may cause substantial decreases on barley yields worldwide with an estimation of yield losses ranging from 3% to 17% depending on the severity of the conditions. A trend that will impact on all the barley chain, and ultimately is expected to result in increased beer prices (Xie et al., 2018).

Drought is the main factor limiting crop production; furthermore, drought events are often associated to high temperatures which impose an additional level of stress to plants. Irrigation is the standard solution to alleviate drought, but it contributes to increasing soil salinization and the presence of excessive amounts of soluble salts, with sodium chloride being often the dominant one, represent an increasing emergence worldwide (Hanin et al., 2016). The expected climatic changes will also modify the annual temperature profile (e.g. less frost during winter, more heat stress during summer (IPCC, 2013)), which implies a consequent variation in the sowing date, growth habit or/and heading time (Marcinkowski and Piniewski, 2018). A clear example of the impact of climate changes on flowering time in barley is illustrated in Fig. 1, where the heading date of two varieties sown for 15 consecutive years in the same location (Fiorenzuola d'Arda, Italy; 44°55'40"N, 9°53'41"E) is reported. The trend of anticipation of flowering can be interpreted as an effect of higher winter temperatures.

Barley is well known for its outstanding capacity to grow over a wide range of latitudes (from Iceland to North Africa) and conditions (from cool climates to the limits of the desert), a capacity largely dependent on the mechanisms

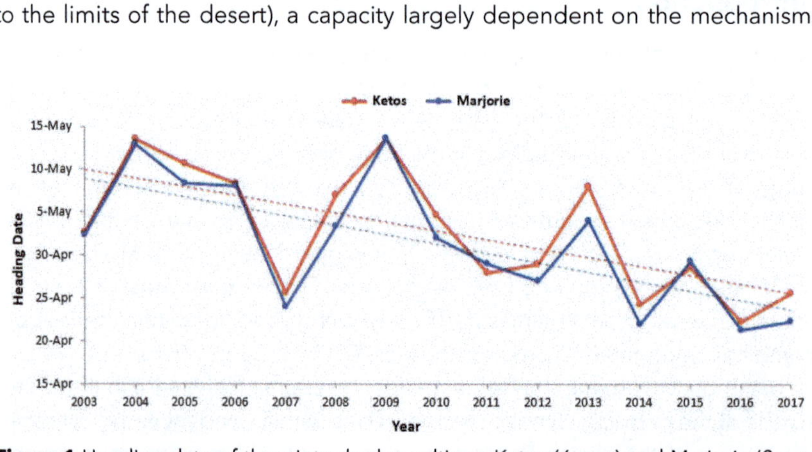

Figure 1 Heading date of the winter barley cultivars Ketos (6-row) and Marjorie (2-row) across 15 years of field trials (2003–17) in Fiorenzuola d'Arda, Italy (44°55'40"N, 9°53'41"E).

controlling flowering time in response to vernalization, photoperiod and temperature (Drosse et al., 2014). The synchronization of barley life cycle with the environmental climatic conditions is the first requirement to limit the impact of stress and assure yield. Spring cultivars avoid excessive frost and early flowering cultivars avoid terminal drought, and the combination of growth habit and response to photoperiod makes barley the most flexible crop in terms of adaptation to climates and climate changes (Dawson et al., 2015).

Besides stress avoidance, barley also has a significant stress resilience capacity generally based on adaptive mechanisms elicited when the plant is exposed to non-lethal unfavourable events. A network of hormones and transcription factors controls the response to stress and supports the expression of large sets of genes which, in turn, determine many physiological and metabolic modifications (for a review on plant molecular response to abiotic stress refer to Zhu, 2016). The level of stress tolerance achieved through these adaptive mechanisms is dependent on the specific genotype and, to some extent, on plant's growth stage (Cattivelli, et al., 2008). In barley exists a vast genetic diversity for tolerance in response to different stress conditions and the breeders are taking advantage from this diversity to select new varieties adapted to the different, sometime new, climatic conditions. In fact, fitting crops to the environment is a more sustainable strategy than modifying the environment to fit the crops.

2 Cold acclimation: a coordinated metabolic rearrangement leading to frost tolerance

The ability of temperate cereals such as barley and wheat to survive over winter (winter-hardness) is largely dependent on the degree of frost tolerance acquired at the vegetative stage. Maximum tolerance is reached by induction of a coordinated and complex genetic network that follows exposure to low but not freezing temperatures, a process known as cold acclimation or hardening (Rizza et al., 1994). Low temperature induces a general metabolic switch involving modifications of the membrane composition, increased activity of enzymes for sucrose synthesis, reactive oxygen species (ROS) scavenging and xanthophyll cycle, enhanced capacity for energy-dependent fluorescence quenching (Ensminger et al., 2006), accumulation of osmolytes (e.g. proline, glycinebetaine), soluble sugars, abscisic acid (ABA), induction of Cold-Regulated (*COR*) genes and accumulation of the corresponding COR proteins (Cattivelli and Bartels, 1990; Murelli et al., 1995; Crosatti et al., 1995). Furthermore, freezing also imposes a mechanical stress due to ice formation with impact on cell walls and cytoskeleton structure.

A range of methodologies have been developed for assessing the ability of barley plants to cold acclimate (Rizza et al., 1994; Prášil et al., 2007;

Cattivelli, 2014). Accurate evaluation of plant survival is usually carried out in fields during vegetative growth or under controlled conditions. Alternatively, rapid and simplified experimental systems, such as those based on chlorophyll fluorescence (Rizza et al., 2001), can be used for fast and precise monitoring of frost tolerance. Damage to cellular integrity, as in the case of freezing stress, leads to a decrease in the maximum quantum yield of the photosystem II, as it can be assessed by measuring the ratio of variable (Fv) to maximal (Fm) chlorophyll fluorescence in the dark-adapted state (Fv/Fm parameter; Rizza et al., 2001). In barley, it has been observed that these alternative methods are strongly correlated with both winter survival and measurement of regrowth after stress, when plants are analysed at the first leaf stage (Rizza et al., 2011).

The chloroplast acts as a sensor of the environmental changes integrating the light and low temperature into variations of the redox state, that triggers specific signalling pathways (Pinas Fernandez and Strand, 2008) leading to cold acclimation. The ability of many plant species to develop cold tolerance is associated with the presence of light and photosynthetic activity during low temperature growth. Energy is an essential factor to drive acclimation and sugar accumulation has important effects on plant freezing tolerance and resilience of photosynthesis under stress conditions (Gusta and Wisniewski, 2013; Dahal et al., 2012). Cold acclimation at low light intensity is much less effective than under normal light conditions; conversely elevated light intensity at normal temperatures may induce some cold tolerance (Janda et al., 2014). In barley, it has been demonstrated that light affects the expression of several *COR* genes (Crosatti et al., 1999) and that plants carrying a mutation preventing chloroplast development (*albino* and *xantha*) are completely frost susceptible as well as impaired in the expression of many *COR* genes (Dal Bosco et al., 2003; Svensson et al., 2006). These results, as well as data from *Arabidopsis* (Kindgren et al., 2015), demonstrate the relevance of a functional chloroplast for the cold acclimation process and further suggest that impaired plastid function could result in inhibition of protein synthesis at low temperature (Crosatti et al., 2013).

The molecular response to cold starts with the low temperature-dependent activation of ICE1, a MYC-like basic helix-loop-helix transcription factor (Skinner et al., 2006; Badawi et al., 2008) that induces the expression of the *C-repeat binding factor* (*CBF*) genes, whose products activate a set of *COR* genes (e.g. *COR14*), with a direct role in protecting the plant cells from frost damage allowing plants to acquire freezing tolerance (Dal Bosco et al., 2003; Skinner et al., 2005; Tondelli et al., 2011). Coherently, faster and/or higher levels of *COR* genes and proteins have been found associated to higher frost tolerance (Crosatti et al., 1995) and transgenic barley overexpressing *CBF* genes of wheat (*TaCBF14* and *TaCBF15*) showed enhanced cold acclimation and frost tolerance (Soltesz et al., 2013).

Extensive genetic analyses in barley have identified two main *Frost Resistance* loci accounting for a large proportion of the observed phenotypic diversity and repeatedly mapped 30 cm apart on the long arm of the chromosome 5H (Francia et al., 2004; Skinner et al., 2006; von Zitzewitz et al., 2011; Tondelli et al., 2014). The *Frost resistance-H2* (*Fr-H2*) locus overlaps with a cluster of *CBF* genes (Francia et al., 2007; Tondelli et al., 2011). The barley *CBF* locus is a gene cluster encompassing more than 11 members with some of them subject to copy number variations, therefore a relevant question concerns the identification of the specific *CBF* gene responsible for the extensive genetic variation observed in barley. Sequencing of genomic clones encompassing the *Fr-H2* locus in frost-hardy and frost-sensitive barleys suggested that an increase in the copy numbers of *HvCBF2* and *HvCBF4* might be the causal functional polymorphism underlying frost tolerance (Knox et al., 2010; Francia et al., 2016). Nevertheless, a candidate gene association mapping study suggests that allelic variation at *HvCBF14* is related to frost tolerance (Fricano et al., 2009). The *Frost resistance-H1* (*Fr-H1*) locus most probably represents a pleiotropic effect of *HvBM5A*, a MADS-box gene which is the candidate for the major vernalization gene *Vrn-H1* (Yan et al., 2003; von Zitzewitz et al., 2005). Temperatures that induce cold acclimation also satisfy the vernalization requirement of winter barley, allowing the switch from vegetative to the reproductive growth phase. It has been observed that levels of frost tolerance decrease after vernalization and the extent of *COR* gene expression is influenced by allelic variation at *Vrn-H1* with increased levels of *Vrn1* mRNAs associated to reduced *CBF* expression (Stockinger et al., 2007).

While cold acclimation is an on-off process, vernalization produces a memory in the plant cells. *Vrn-H1* is induced during exposure to low temperature, then, after vernalization, the stably high expression level of *Vrn-H1* in meristematic tissues promotes the transition from vegetative to reproductive phase (for recent reviews, see Xu and Chong, 2018). The memory of vernalization is due to epigenetic modifications in both the first intron and the promoter (Oliver et al., 2013). Before vernalization, the repressive histone modification mark H3K27me3 is deposited in *Vrn-H1* locus to inhibit gene expression, while during vernalization, the active marks H3K4me3 and H3K36me3 are gradually increased and H3K27me3 levels decreased to allow the expression of the gene (Oliver et al., 2013). Consequently, deletions in the 10 kb promoter of *Vrn-H1* can abolish the epigenetic control and induce flowering without vernalization (spring grow habit).

The interconnection between frost tolerance, vernalization requirement and photoperiod sensitivity has led to the definition of three different classes for barley, namely winter, facultative and spring. Winter and facultative types are both frost resistant, but they differ in sensitivity to vernalization. Rizza et al. (2011) have presented experimental evidences that support a faster acclimation capacity in facultative versus winter genotypes. Plants with facultative growth habit are more ready to react to changes in environmental factors assuring flowering under a wide range of climatic conditions, a trait associated with a high adaptation capacity. Interestingly, a positive allelic contribution to frost acclimation from a facultative barley line has been observed for the *Frost resistant-H3* (Fr-H3), a QTL mapped on the short arm of barley chromosome 1H (Fisk et al., 2013). Spring barleys are capable of flowering without vernalization (or with a minimal vernalization) and, on average, are less frost tolerant than winter or facultative cultivars. Nevertheless, spring barley can also cold acclimate, and a significant genetic diversity has been observed for frost tolerance within spring germplasm, mainly associated to allelic variation at the *Fr-H2* locus (Tondelli et al., 2014). In the context of global warming, a trend towards more mild winters and larger fluctuation of winter temperatures has been observed (Bellard et al., 2012). This climatic scenario will extend the cultivation of winter barley towards northern latitudes and support the use of facultative or spring cultivars in winter sowing in warmer regions.

3 New methodologies for dissecting an old phenotype: resilience to drought

Drought, which is when water availability is below that required for maximum crop yield, is the most relevant abiotic stress limiting both the ecological distribution of the species and the final grain yield of the crops. Drought avoidance (early flowering), tolerance (survival) and resilience (recovery) are the main strategies adopted to limit yield loss. Among small grain cereal, barley is considered the most adapted to drought, thanks to its avoidance and resilience capacities. In fact, barley is commonly grown in marginal lands where drought can occur at any moment during the life cycle, being particularly frequent during the terminal stages (Turner, 2004), when different components of grain yield can be largely influenced (Araus et al., 2002). According to different climate change scenarios, the impact of drought will increase in many barley production areas (Xie et al., 2018). For this reason, screening genetic resources to identify relevant physiological mechanisms to cope with water limitations, together with the responsible genes, is pivotal.

A crucial aspect is the identification of parameters capable to assess the degree of drought tolerance of a given genotype. Often, a single physiological parameter is not a reliable indicator of the yield performance in drought-prone environments. Rather, field evaluation over a range of environments should be used as the main indicator for drought tolerance (Voltas et al., 2005) and several indices have been proposed to describe yield performance in stressed versus non-stressed conditions (e.g. Finlay and Wilkinson, 1963; Fischer and Maurer, 1978) or in relations with the soil-plant-atmosphere interaction (water available versus evapotranspiration demand, Rizza et al., 2004).

From a physiological point of view, a plant experiences drought when evapotranspiration exceeds water uptake. In these conditions, a low leaf water potential leads to a decrease in leaf turgor, stomatal conductance and photosynthesis, thus limiting plant growth and yield. Depending on the timing, duration and intensity of the stress, the water deficit has different effects on the cellular processes, triggers different response mechanisms (e.g. the accumulation of osmolytes to maintain cell turgor pressure, a process known as osmotic adjustment, Moinuddin et al., 2005) and limits yield through different mechanisms (for recent reviews, see Tardieu et al., 2018; Kumar et al., 2018). In barley, a drought event during the vegetative growth phase impacts on plant biomass and on spike fertility, while a terminal drought, occurring during or after flowering, has more impact on grain filling (Cattivelli et al., 2008). At a molecular level, drought induces the expression of a large number of genes (for barley see Cantalapiedra et al., 2017) regulated through complex transcriptional networks with two main pathways known in most plant species: an ABA-dependent signalling pathway and an ABA-independent regulatory network mediated by dehydration-responsive element-binding (*DREB*)-type transcription factors (Kumar et al., 2018). The latter class of transcription factors triggers the expression of many drought-responsive genes, including the *LEA* (also known as *Dehydrin*) gene family (Choi et al., 1999), and have been suggested as a biotechnological target for drought improvement also in barley (Morran et al., 2011). It is remarkable that the overexpression of barley *HVA1*, a Group 3 LEA gene, was also able to confer better growth and higher water-use efficiency (WUE) to transgenic wheat plants (Sivamani et al., 2000).

Since drought impacts the whole plant growth and developmental programme, many different physiological traits relevant for response to drought conditions have been suggested (Kishor et al., 2014; Cattivelli et al., 2008), although, only in a few instances, the genetic inheritance of these complex traits has been dissected, up to the identification of the molecular mechanisms and of the genes involved. Recently, Muzammil et al. (2018) have demonstrated that an ancestral allele of the *pyrroline-5-carboxylate synthase* (*P5cs1*) gene from a wild barley accession is majorly responsible of quantitative, drought-inducible proline accumulation under water shortage. Higher proline accumulation was then related to higher tissue water status and efficiency of photochemistry under drought conditions. Besides proline, many phenolic and terpenoid compounds significantly changed their level because of drought stress with a role as antioxidants, regulators of gene expression and modulators of protein function during drought (Piasecka et al., 2017).

A typical symptom of drought during grain filling is an accelerated leaf senescence, which, in turn, limits the photosynthetic tissues and reduces crop yield. Consequently, a delay in leaf senescence has often been seen as a drought-tolerance mechanism and in many crops a 'stay-green' phenotype (i.e. the ability to retain green areas in the grain during filling) has been reported. Functional stay-green have a prolonged capacity to accumulate assimilates to harvested tissues (Thomas and Howarth, 2000). In crops such as sorghum, maize, rice and wheat, there are many data supporting the positive effect of stay-green phenotypes to increase yield under water-limited conditions (Kamara et al., 2003; Harris et al., 2007; Rong et al., 2013; Christopher et al., 2016). Cytokines are the key hormone controlling leaf senescence and the expression of isopentenyltransferase, the rate-limiting step in cytokinin biosynthesis under the control of senescence-associated promoters, leads to stay-green phenotypes (Gan and Amasino, 1995). A wheat stay-green mutant isolated after chemical mutagenesis has been associated with altered cytokinin metabolism (Wang et al., 2016). Since the stay-green phenotype is defined as the capacity of retaining green leaves longer than others, the precise assessment of this trait is a relevant question. A method has recently been proposed to assess the quantitative components of stay-green traits in field conditions (Christopher et al., 2014). The dynamics of canopy senescence are estimated based on NDVI measurements taken during the whole senescence period and fitted to a logistical model standardized to thermal time with respect to anthesis for each genotype, enabling the comparison of data both within and across environments (Christopher et al., 2016). In barley, selection for yield per se, even in water-limited environments, may have led to some degree of stay-green phenotype; nevertheless, very few works describe stay-green barleys. There are reports on QTL analyses describing changes in leaf senescence in response to drought stress (Gregersen et al., 2008), but only one study presents a QTL analysis for stay-green in barley grown under simulated heat- and water-stress conditions (Gous et al., 2016). This work identified several QTLs controlling a stay-green phenotype, although more data are needed to demonstrate the impact of this trait on yield under field drought conditions.

A 'steep, cheap and deep' root system (Lynch, 2013) was observed in Australian barley varieties from drought-prone environments and carrying a spring allele at the major vernalization-requirement gene, *HvVRN1* (Voss-Fels et al., 2018). The same allele induces early flowering and terminal drought escape. A TILLING mutant of the *HvCBP20* gene encoding for a small subunit of the cap-binding complex was associated to a lower stomatal conductance and a higher relative water content with respect to the wild type under drought stress (Daszkowska-Golec et al., 2017). Through a combination of transcriptome and phenotypic analyses, the gene was proposed as a regulatory master coordinating an early and more efficient activation of different mechanisms able to cope with water shortage, such as leaf rolling and change in epidermal cell pattern to limit transpiration, stomatal closure, ROS detoxification, but without severe phenotype changes. A second mutant in the *HvERA1* gene coding for the β-subunit of farnesyltransferase showed a better photosynthetic efficiency under drought stress compared to the wild type, together with enhanced ABA sensitivity during germination, suggesting a possible regulatory role in the modulation of ABA and ethylene signalling under drought stress (Daszkowska-Golec et al., 2018). Plant perception and stomatal response to water balance is mediated by a signalling network that has been explored in depth in *Arabidopsis* (Merilo et al., 2013; Yoshida et al., 2014), but not yet in barley. Knowledge on phytohormonal crosstalks could be used to alter the drought sensitivity of barley and help breeding for resilience to multiple stresses related to water scarcity. Loci with superior performance under drought conditions can be identified in barley wild relatives. Indeed, a barley line carrying a genomic introgression from *Hordeum spontaneum* showed an improved growth under drought stress at the juvenile phase, a behaviour subsequently confirmed under field conditions (Honsdorf et al., 2014). Despite the dependence on different genetic backgrounds and the still unexplored possibility of undetected epistatic effects, all these genes may be considered potential targets of breeding efforts towards the development of drought-resilient crops.

Recent progresses in plant phenomics are providing high-resolution and high-throughput platforms for screening large and still unexplored genetic resources, thus reducing the gap between plant physiology and genetics (Pratap et al., 2015). By using a non-destructive imaging analysis system under semi-controlled greenhouse conditions, Chen et al. (2014) monitored the phenotype of different barley cultivars in response to a drought stress period during the vegetative stage and after re-watering. The derived models were further improved to better focus on drought stress symptoms and to quantify differences in the accumulation of vegetative biomass in barley genotypes grown under different watering regimes. An earlier termination of biomass accumulation during stress was observed in old cultivars with respect to modern ones (Neumann et al., 2015), suggesting that modern breeding has positively impacted on drought tolerance.

The need for efficient 'functional' or 'physiological' phenotyping, for example by means of gravimetric system with weighing lysimeters, was recently claimed to continuously measure quantitative physiological traits (i.e. water-use efficiency, the relative measure of the amount of carbon fixed into biomass or grains versus the amount of water transpired) on the whole plant environment (from soil to atmosphere), in different genotypes simultaneously (Halperin et al., 2017; Chaka Gosa et al., 2018). This allows the measurement of growth rate and transpiration of any plant before, during and after a period of water deprivation period and it may overcome some of the negative factors of phenomic platforms that mainly aim at measuring biomass. By this system, the authors were able to use transpiration rate to define the point at which plant growth (hence productivity) is hampered by limitation in soil water content. They observed different risk-management strategies in wild barley accessions: samples collected from water-stable environments were able to take more risk in regulating their water balance with respect to accessions from sites with higher rainfall uncertainty, but recovered more slowly once irrigation restarted (Galkin et al., 2018). Plants might in fact have evolved different mechanisms of water usage depending on the duration of water shortage between two rainfall events, and the probability that a dry spell would last more than expected (Peleg et al., 2008). Screening cultivars released by breeding across years or aiming at different target environments and final end-uses would be extremely useful for understanding the relationships between barley resilience and yield potential. In fact, wild barleys need strategies to minimize the risk of failure to set seed, while breeders have generally selected for yield potential under the expected water availability for the target region. Timely stomatal closure conserves water but will eventually lead to carbon starvation and reduce yield; therefore, the most useful plant may not be the most resistant one, but on the contrary the one that better uses water when available and is still able to cope with a short period of water scarcity (Dalal et al., 2017).

In recent years, reliable high-throughput methods for crop phenotyping have become available and imaging techniques can be used to monitor plant health under real field conditions throughout the whole life cycle. For example, hyperspectral data combined with a learning algorithm performed better than well-established vegetation indices in the recognition of early water stress in barley plants grown under a rain-out shelter (Römer et al., 2012). To this extent, it is essential to identify traits that at the canopy level and over time integrate the ability of the crop in using water resources, thus helping in the selection of advanced breeding lines for drought-prone areas.

Because of lower plant evapotranspiration and high air temperature observed under prolonged periods of water scarcity, under field conditions plants frequently experience drought in combination with heat stress (Lawas et al., 2018). Heat stress per se causes a reduction of the photosynthetic activity, accelerates grain filling and anticipates plant senescence. This decline affects the contribution of available assimilates to the grain causing a reduction of kernel weight (Wardlaw and Wrigley, 1994). The catalytic activity of enzymes is often reduced as soon as the temperature exceeds the optimal value. In seeds, high temperatures have a negative impact on the many enzymes of the starch biosynthesis (ADP-glucose pyrophosphorylase, branching enzyme, granule-bound starch synthase, soluble starch synthase) that, together with the limited supply of carbon from photosynthetic tissues, results in a reduced accumulation of starch during heat stress (Wallwork et al., 1998). At a cellular level, heat stress triggers multiple signal transduction pathways that lead to an enhanced thermotolerance. The Heat Shock Factors are the key transcriptional regulators of heat response, capable of activating a large set of heat shock proteins which act as chaperones in protein folding to preserve protein stability and functionality during stress (Kotak et al., 2007). It is remarkable that in barley allelic diversity in a specific heat shock protein gene (HSP17.8) has been found associated to grains per spike and thousand kernel weight (Xia et al., 2013).

When heat comes together with drought the impact is even more pronounced; nevertheless, few studies have investigated combined drought and heat stress in barley, thus limiting the knowledge of molecular mechanisms underlying the tolerance to multiple stress. A proteomic approach undertaken by Rollins et al. (2013) identified barley leaf proteins specifically accumulated under a combination of heat and drought stress and linked to different cellular functions. Most interestingly, some of these proteins were differentially regulated when comparing genotypes with different plant growth and photosynthetic performance in response to the stresses. Furthermore, a metabolomic study comparing European breeding lines and landraces from Mediterranean drought-prone environments has suggested that stress combination led to a higher accumulation of different metabolites associated with antioxidant defence with respect to drought stress alone in both germplasm groups (Templer et al., 2017).

In barley, drought or heat stress during grain filling, besides the negative impact on yield, also deteriorates malting quality. Malting quality is associated with the kernel composition (i.e. low protein and low beta-glucan content), high beta-amylase, a key factor for starch degradation closely related to malt diastatic power and high malt extract (i.e. the amount of soluble carbohydrate in the malt). Most evidences suggest that heat and/or drought stress during the terminal phase limits carbohydrate accumulation and kernel weight leading to high protein concentration and reduced malt extract. Nevertheless, significant variations among cultivars have been reported for malting profile in response to stress, suggesting that genetic diversity can be used to protect quality from abiotic stresses (Wu et al., 2017; Mahalingam, 2017; Mahalingam and Bregitzer, 2019).

4 Adaptation to soil salinity

A consistent amount of agricultural land worldwide is affected by soil salinity, mainly because of salt accumulation in drought areas, human activities (such as land clearing or irrigation without right drainage) and increasing use of marginal lands (Hanin et al., 2016). This is already limiting agricultural production, but it is expected that climate change will further negatively affect soil salinity (Dagar et al., 2016). Under high salt conditions, an increase in the osmotic potential of soil water is observed, that results in the reduction of available water for the plants. In addition, the higher concentration of some ions (i.e. Na^+) in plant tissues may have a toxic effect and impair physiological activities such as photosynthesis and transpiration, up to the necrosis of the tissues in the most severe instances. Finally, the antagonism between ions, such as Na^+ versus K^+ and Ca^{++}, and Cl^- versus NO_{3-} results in ion stress imbalance (Tavakkoli et al., 2011; Munns and Tester, 2008; Negrão et al., 2017). Overall, these negative effects of salt stress translate in slower growth rates, lower tillering and a heavy reduction of crop productivity.

Halophytes plants can grow with concentration of salt greater than that of seawater, thanks to the activation of different physiological responses, that jointly contribute to the adaptation to saline soils (Shabala, 2013). However, most of the cultivated crop plants are glycophytes and are very sensitive to this kind of stress. Being able to tolerate up to 250 mM NaCl, barley is one of the most salt-tolerant crops, a particularly important feature for salinity-affected arid and semi-arid agro-environments in the world. Barley is also a valuable species for studying the underlying physiological mechanisms and their genetic inheritance (Maas and Hoffman, 1977; Munns and Tester, 2008). Several studies have shown a wide genetic variability in salinity stress tolerance in barley, and genes/QTLs controlling the trait have been discovered in both cultivated and wild accessions (Saade et al., 2018 for a recent review; Hazzouri et al., 2018; Shen et al. 2018; Zhu et al, 2015; Xue et al., 2017). For example, introgression lines carrying a *Hordeum spontaneum* allele at a locus on chromosome 2H yielded 30% more under saline conditions in replicated field trials with respect to lines with the allele from the recurrent parent (Saade et al., 2016). Tolerance at the reproductive phase under field conditions is probably the most valuable for breeders; nevertheless, the effect of salt stress, hence the resistance loci involved, may change depending to the growth stage (Mano and Takeda, 1997), with germination and seedling stages being the most sensitive ones.

According to Roy et al. (2014), the mechanisms of plant tolerance to salt group into three main categories: osmotic tolerance (ion-independent), ion exclusion and tissue tolerance. Many genes are supposed to contribute to these complex mechanisms (for recent reviews, see Ismail and Horie, 2017; Hanin et al., 2016). As for barley, Zhu et al. (2017) recently showed that under mild salt stress (200 mM NaCl), the concentration of Na^+ in the xylem increased less over time in tolerant barley genotypes with respect to the sensitive ones. Under these conditions, this would allow the shoot to rapidly adjust the osmotic potential in the first days after stress onset, but at the same time to avoid the subsequent accumulation of Na^+ in the leaf tissue, where it disturbs the photosynthesis. By studying ion fluxes and expression levels of involved genes in response to exogenous H_2O_2 and ABA, the same authors proposed a signalling pathway of xylem ion loading under moderate saline concentrations. Similarly, contrasting barley accessions selected from a large germplasm screening showed no differences in Na^+ root uptake under salt stress, but a higher sodium accumulation in the flag leaf of sensitive genotypes. In contrast, a higher sodium sequestration from xylem was detected in leaf sheaths of tolerant genotypes, suggesting a lower rate of transport from roots to shoots. Most interestingly, this was coupled with a decrease in the expression levels of the *High-Affinity K^+ Transporter HKT1;5*, the candidate gene underlying a major QTL in the same germplasm panel (Hazzouri et al., 2018). Conversely, when the barley Na^+ and K^+ transporter *HKT2;1* gene was constitutively overexpressed, an increased tolerance to mild salt stress was observed, together with higher Na^+ uptake, concentration in the xylem sap and translocation to leaves (Mian et al., 2011). Members of this gene family are interesting targets for genetic manipulation when their expression is modulated according to specific cell types and in a stress-inducible mode (Roy et al., 2014).

Salt tolerance in shoot tissues is also a component of the overall salt tolerance since it ensures the plant to sustain a higher photosynthetic activity and, hence, grain yield (Adem et al., 2014). Recently, tissue tolerance of different barley cultivars was screened through a method based on excised leaves instead of whole plants, showing a positive correlation between chlorophyll content and the overall tolerance to salinity. The higher tissue tolerance of some genotypes was most probably due to a more efficient compartmentalization of Na^+ in mesophyll cell's vacuole with respect to salt sensitive barleys, where an excess of toxic sodium was observed in chloroplasts instead (Wu et al., 2015a). Differences in the ability of K^+ retention, in the mesophyll cells, hence maintenance of cytosolic K^+ homeostasis, were also observed in the same genetic materials (Wu et al., 2015b), suggesting the simultaneous contribution of more mechanisms in determining tissue tolerance to salt. There are evidences in barley that cytosolic K^+ homeostasis might depend on plant's scavenging and detoxification from stress-induced hydroxyl radical and ROS affecting plasma membrane transporters (Adem et al., 2014). A lower amount of H_2O_2 and a lower oxidative stress, together with the maintenance of steady-state levels of potassium, were observed in roots and shoots of a salt-tolerant barley mutant with respect to the wild plant (Kiani et al., 2017).

As already pointed out (Rajendran et al., 2009), different mechanisms can be adopted by different genotypes to overcome the negative effects of salt stress. More attention should now be paid on how the integration and combination of these mechanisms contribute to higher salt stress tolerance, and how pyramiding the related genes may help in breeding barley cultivars more adapted to unfavourable conditions. High-throughput phenotyping methods have been already adopted to study the differential resistance to salt stress in *Arabidopsis* and rice (Awlia et al., 2016; Al-Tamimi et al., 2016), and it is expected they will significantly contribute in clarifying the physiological, genetic and molecular bases of barley higher resilience. This will ultimately help breeders in deciding which mechanisms can be manipulated to select barley cultivars adapted to saline soils (Roy et al., 2014). Finally, useful genes can also be found in the wild halophyte sea barley (*Hordeum marinum*) growing in coastal areas around the Mediterranean basin and Middle East (Bothmer et al., 1995). The physiological and molecular bases of sea barley's remarkable salt stress tolerance have been recently dissected (Huang et al., 2018; Saoudi et al., 2019), and amphiploid lines generated through hybridization with wheat already showed improved tolerance to salt stress (Munns et al., 2011).

5 Low nitrogen: a stress condition matching crop sustainability

Crop productivity relies heavily on nitrogen (N) fertilization, N being an essential macronutrient limiting the growth and development of plants in agriculture. Its extensive use has, however, contributed to nitrate leaching from fields, mineralization and emissions of the greenhouse gas nitrous oxide (Perchlik and Tegeder, 2017), thereby selecting plants capable to cope with low N is becoming an essential target in barley breeding. Insufficient N nutrition results in stunted growth, yellow leaves, reduced tillering and significant yield reductions (Robson and Snowball, 1986). Nitrogen-use efficiency (NUE) consists of two components: N-uptake efficiency (NUpE), defined as the efficiency of absorption/uptake of N supplied to the soil, and N-utilization efficiency (NUtE), which is the efficiency of assimilation and remobilization of plant N to ultimately produce grains (Moll et al., 1982). The contribution of the two components to NUE has been found to vary depending on conditions. Under low or moderate N availability, NUtE is more relevant (Beatty and Good, 2011; Bingham et al., 2012), whereas at higher N availability, NUtE and NUpE would contribute equally to NUE (Foulkes et al., 2009; Beatty and Good, 2011). Although the results are not entirely consistent, these reports showed that genotype differences in NUE under low N conditions were more likely related to differences in N uptake, while differences under higher N were largely due to differences in NUtE.

Different experimental strategies have been applied to identify the genes controlling NUE: (1) standard QTL mapping, (2) analysis of candidate genes, (3) generation of mutants and (4) transcriptome analysis. Genetic analyses of NUE in barley have described a complex genetic situation, where many QTLs control key agronomic traits associated with NUE, but with most QTLs showing a strong interaction with the environment depending on N levels and years (Kindu et al., 2014). Notably, several QTLs controlling N-related traits were mapped in the proximity of the *denso* locus, a main factor controlling plant height in barley (Kindu et al., 2014).

In wheat and barley, there are consisting evidences for a specific locus with impact on both NUE and GPC (grain protein content). The wheat *Gpc-B1* locus located on the chromosome 6BS and encoding for a NAC transcription factor (*TtNAM-B1*) accelerates leaf senescence with pleiotropic effects on N reallocation resulting in increased grain protein, NUE and N allocation within the plant (Uauy et al., 2006). In the orthologous position on the barley genome, a QTL for GPC was mapped, and the peak of the QTL overlapped the barley *TtNAM-B1* orthologous gene, making the barley NAC transcription factor the best candidate for QTL explaining GPC in barley (Distelfeld et al., 2008; Sharma et al., 2018).

Although no specific evidences are available for barley, there are many reports in cereals suggesting that two loci coding for glutamine synthetase (GS) and glutamate synthase (GOGAT) are the key players in the determination of NUE. Several QTLs for agronomic traits related to NUE and yield have been mapped to the chromosomal regions containing GS and GOGAT genes in wheat and rice (Han et al., 2016). Physical mapping, sequencing, annotation and candidate gene validation of an NUE metaQTL on wheat chromosome 3B highlighted the glutamate synthase as a locus evolutionarily conserved across maize, sorghum, rice and *Brachypodium* genomes that affect NUE (Quraishi et al., 2011). The whole body of evidence suggests that GS may be a valuable target for increasing NUE. Furthermore, in wheat, the genes controlling photoperiod sensitivity (*Ppd*), growth habit (*Vrn*) and the semidwarf gene *Rht* were also found highly correlated with NUE-related traits (Quraishi et al., 2011) and given the strong conservation of these genes in wheat and barley a similar result could be suggested for barley (Sharma et al., 2018). A positive breeding effect for NUE and other plant N traits was documented in a study describing an extensive germplasm panel made of 72 landraces and 123 cultivars released since 1910 grown at two different N levels in southern Finland (Rajala et al., 2017); nevertheless, this result could reflect changes in the main plant morphological and phenological traits.

Barley mutants with improved NUE have been described by Gao et al. (2018). The authors have generated a population of barley lines by microspore mutagenesis, the population was then field tested at low and high N level using the number of productive tillers as the main screening index. When the selected mutants were tested for NUE, some of them showed improved NUpE under both high and low N conditions; nevertheless, no attempt to genetically map the mutations is reported.

When the molecular response of barley exposed to low N level was analysed in two Tibetan wild barley genotypes differing for NUE, many low N tolerance-associated genes belonged to amino acid metabolism, starch and sucrose metabolism, transporter and antioxidant activities. The higher expression of nitrate transporters and energy-saving assimilation pattern was highlighted as the typical feature of the transcriptomic response on the genotype with higher NUE (Quan et al., 2016).

6 Adaptation to environment: a key target for future breeding improvement

Climate changes and progressively more limited input of resources is now rapidly altering the environment and generate an increasing demand for a more sustainable agriculture production with limited environmental impact. In this context, the selection of a new generation of varieties resilient to abiotic stresses and capable of a more efficient use of available resources is urgent. The understanding of the physiological mechanisms and of the genetic bases controlling the resilience of barley to abiotic stresses is an important issue; nevertheless, it is evident that there is a gap between the amount of knowledge accumulated on abiotic stress response/resistance and the genetic progress achieved by breeding. As highlighted in this work, there are examples of mutants, transgenic lines and introgression lines showing improvements for some traits associated to stress response, but there are few examples of new varieties with a significant improved abiotic stress tolerance and superior yield performance in field conditions. The first attempt to improve varieties in response to specific conditions was based on the selection of genotypes that maximize a specific stress response mechanism, but very often this approach led to plants that performed well only under the target stress conditions, a situation that is unlikely to happen with the same intensity every year or in different location of the same region/environment (Fig. 2a). For instance, a good level of earliness as well as an increased mobilization of the vegetative reserves from stems to ears (Blum, 1988) are effective breeding strategies for enhancing yield stability in Mediterranean environments where barley is often exposed to terminal drought, since they allow the synchronization of the crop cycle with the most favourable environmental conditions. Nevertheless, the timing of drought events varies among years and locations (and eventually may also not happen) and the optimal degree of earliness or the convenience of an early mobilization of stem reserves differ consequently.

The studies on barley genetic progress highlighted that some yield improvement have been achieved even in drought-prone environments (Rizza et al., 2004), as well as a significant increase in frost tolerance was found by comparing moderns versus old spring barleys (Tondelli et al., 2014). Nevertheless, these results are mainly due to a selection for high yield that has, unconsciously, carried on some improvements in stress tolerance. There are few examples where an understanding of the physiology and the genetics of putative important drought-related traits has led to improved yields. For instance, in wheat, carbon isotope discrimination as a surrogate for water-use efficiency in drought-prone environments was successfully applied to select drought-tolerant cultivars (Rebetzke et al., 2002). Osmotic adjustment was suggested as an effective selection criterion for drought tolerance, when

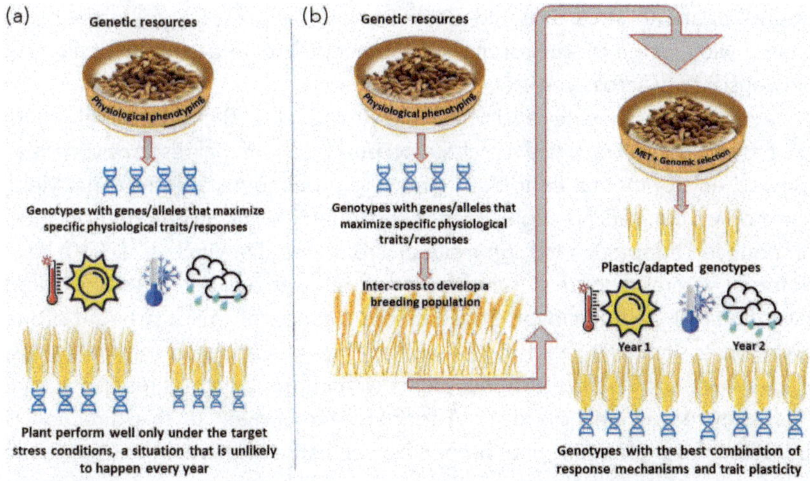

Figure 2 Schematic representation of breeding strategies for selecting stress-tolerant genotypes. (a) Traditional approach based on physiological knowledge focussing on the selection of genotype that maximizes the expression of one or few response mechanisms leading to genotypes adapted to specific stress situations. (b) Selection strategy where physiological knowledge are integrated in a genomic selection process capable of identifying genotypes with the best combination of response mechanisms and trait plasticity. MET = Multi-Environment Trials.

water deficit occurs during the reproductive growth stage (Moinuddin et al., 2005).

New phenotyping platforms coupled with the advanced genome knowledge are now offering a new opportunity for breeding. Advanced phenotyping allows to integrate the contribution of morphological (including root system architecture), metabolic and physiological changes in a global response evaluated as biomass or yield under controlled and reproducible conditions. This allows an easy association with molecular markers and loci (Araus and Cairns, 2014). This approach promotes a deeper understanding of the genetic bases of stress tolerance, although the transfer of the results to field remain, still, a challenge.

Overall, the selection for the high expression of specific stress-related traits did not capitalize the great work carried out for the understanding of stress tolerance, suggesting that probably this approach is not sufficient. To overcome this bottleneck, Dalal et al. (2017) have suggested to select for plasticity of yield-related quantitative physiological traits (e.g. photosynthesis rate or stomatal conductance), assuming that trait plasticity, more than the maximum expression of the trait, allows the plants to adapt their physiology to environmental variations and optimize yield performance in a wide range of environmental conditions. An increased plasticity in traits controlling the response to abiotic stresses could result in a resilient plant capable to perform well in a wide range of environments, a concept known as yield stability (i.e. yield consistency across years and environments).

Several studies in barley and wheat (e.g. Rizza et al., 2004; De Vita et al., 2010) have proven that old cultivars are characterized by a minimal responsiveness to improved environmental conditions, showing an almost stable nominal yield, while modern cultivars are highly responsive to fertility improvements and show a pronounced adaptation to high-input environments. De Vita et al. (2010) have also found several modern durum wheat cultivars with a high-yield stability and a high nominal yield (yield potential) across a range of (mild) drought-prone environments, indicating that the breeding strategies adopted during the last decades have contributed to select plants with improved stress resilience and, perhaps, increased trait plasticity. A further development in this direction is expected through the combination of crop simulation modelling and genomic prediction, two technologies designed to improve selection accuracy limiting the confounding effects of the environment (Bassi et al., 2016; Crossa et al., 2017).

A 50-year-long experience in the analysis of abiotic stress response and many attempts to use this knowledge to improve varieties has demonstrated that, with a few exceptions, no single trait is sufficient to guarantee the yield under the different stress conditions that a plant experiences in different years. We suggest that new breeding models resulting from the combination of physiological concepts (response mechanisms and plasticity of physiological traits) with genomic selection should be used to identify genotypes with the best combination of response mechanisms and high trait plasticity (Fig. 2b). Genotypes expressing different responsive mechanisms could be inter-crossed and subject to recurrent selection with selection based on molecular markers (genomic selection).

7 Acknowledgements

This work was supported by the FACCE-JPI projects BARISTA (Advanced tools for breeding BARley for Intensive and SusTainable Agriculture under climate change scenarios) and BarPLUS (Modifying canopy architecture and photosynthesis to maximize Barley biomass and yield for different end uses).

8 Where to look for further information

A website dedicated to abiotic stress tolerance: http://www.plantstress.com/.

A comprehensive book on the barley genome, including a review on genomic studies of abiotic stress tolerance: Stein, N. and Muehlbauer, G. J. (Eds). 2018. *The Barley Genome*. Springer International Publishing. ISBN: 978-3-319-92527-1.

A book describing the most relevant implications of crop physiology in plant breeding: Sadras, V. O. and Calderini, D. F. (Eds). 2015. *Crop Physiology Applications for Genetic Improvement, Agronomy and Farming Systems* (2nd edn.). Academic Press. ISBN: 978-0-12-417104-6.

9 References

Adem, G. D., Roy, S. J., Zhou, M., Bowman, J. P. and Shabala, S. 2014. Evaluating contribution of ionic, osmotic and oxidative stress components towards salinity tolerance in barley. *BMC Plant Biol.* 14(1), 113. doi:10.1186/1471-2229-14-113.

Al-Tamimi, N., Brien, C., Oakey, H., Berger, B., Saade, S., Ho, Y. S., Schmöckel, S. M., Tester, M. and Negrão, S. 2016. Salinity tolerance loci revealed in rice using high-throughput non-invasive phenotyping. *Nat. Commun.* 7, 13342. doi:10.1038/ncomms13342.

Araus, J. L. and Cairns, J. E. 2014. Field high-throughput phenotyping: the new crop breeding frontier. *Trends Plant Sci.* 19(1), 52–61. doi:10.1016/j.tplants.2013.09.008.

Araus, J. L., Slafer, G. A., Reynolds, M. P. and Royo, C. 2002. Plant breeding and drought in C3 cereals: what should we breed for? *Ann. Bot.* 89(7), 925–40. doi:10.1093/aob/mcf049.

Awlia, M., Nigro, A., Fajkus, J., Schmoeckel, S. M., Negrão, S., Santelia, D., Trtílek, M., Tester, M., Julkowska, M. M. and Panzarová, K. 2016. High-throughput non-destructive phenotyping of traits that contribute to salinity tolerance in *Arabidopsis thaliana*. *Front. Plant Sci.* 7, 1414. doi:10.3389/fpls.2016.01414.

Badawi, M., Reddy, Y. V., Agharbaoui, Z., Tominaga, Y., Danyluk, J., Sarhan, F. and Houde, M. 2008. Structure and functional analysis of wheat *ICE* (inducer of CBF expression) genes. *Plant Cell Physiol.* 49(8), 1237–49. doi:10.1093/pcp/pcn100.

Bassi, F. M., Bentley, A. R., Charmet, G., Ortiz, R. and Crossa, J. 2016. Breeding schemes for the implementation of genomic selection in wheat (*Triticum* spp.). *Plant Sci.* 242, 23–36. doi:10.1016/j.plantsci.2015.08.021.

Beatty, P. H. and Good, A. G. 2011. Plant science. Future prospects for cereals that fix nitrogen. *Science* 333(6041), 416–17. doi:10.1126/science.1209467.

Bellard, C., Bertelsmeler, C., Leadley, P., Thuiller, W. and Courchamp, F. 2012. Impacts of climate change on the future of biodiversity. *Ecol. Lett.* 15(4), 365–77. doi:10.1111/j.1461-0248.2011.01736.x.

Bingham, I. J., Karley, A. J., White, P. J., Thomas, W. T. B. and Russell, J. R. 2012. Analysis of improvements in nitrogen use efficiency associated with 75 years of spring barley breeding. *Eur. J. Agron.* 42, 49–58. doi:10.1016/j.eja.2011.10.003.

Blum, A. 1988. Improving wheat grain filling under stress by stem reserve mobilisation. *Euphytica* 100, 77–83.

Bothmer, R., Jacobsen, N., Baden, C., Jørgensen, R. B. and Linde-Laursen, I. 1995. *An Ecogeographical Study of the Genus Hordeum* (2nd edn.). Systematic and Ecogeographic Studies on Crop Genepools 7. International Plant Genetic Resources Institute, Rome, Italy.

Cantalapiedra, C. P., García-Pereira, M. J., Gracia, M. P., Igartua, E. A., Casas, A. M. and Contreras-Moreira, B. 2017. Large differences in gene expression responses to drought and heat stress between elite barley cultivar Scarlett and a Spanish landrace. *Front. Plant Sci.* 8, 1–23. doi:10.3389/fpls.2017.00647.

Cattivelli, L. 2014. Conducting field trials for frost tolerance breeding in cereals. In: Hincha, D. and Zuther, E. (Eds), *Plant Cold Acclimation. Methods in Molecular Biology (Methods and Protocols)*. Humana Press, New York, NY, pp. 25–33.

Cattivelli, L. and Bartels, D. 1990. Molecular cloning and characterization of cold-regulated genes in barley. *Plant Physiol.* 93(4), 1504–10. doi:10.1104/pp.93.4.1504.

Cattivelli, L., Rizza, F., Badeck, F. W., Mazzucotelli, E., Mastrangelo, A. M., Francia, E., Marè, C., Tondelli, A. and Stanca, A. M. 2008. Drought tolerance improvement in crop plants: an integrated view from breeding to genomics. *Field Crops Res.* 105(1–2), 1–14. doi:10.1016/j.fcr.2007.07.004.

Chaka Gosa, S., Lupo, Y. and Moshelion, M. 2018. Quantitative and comparative analysis of whole-plant performance for functional physiological traits phenotyping: new tools to support pre-breeding and plant stress physiology studies. *Plant Sci.* 282, 49–59. doi:10.1016/j.plantsci.2018.05.008.

Chen, D., Neumann, K., Friedel, S., Kilian, B., Chen, M., Altmann, T. and Klukas, C. 2014. Dissecting the phenotypic components of crop plant growth and drought responses based on high-throughput image analysis. *Plant Cell* 26(12), 4636–55. doi:10.1105/tpc.114.129601.

Choi, D.-W., Zhu, B. and Close, T. J. 1999. The barley (*Hordeum vulgare* L.) dehydrin multigene family: sequences, allele types, chromosome assignments, and expression characteristics of 11 *Dhn* genes of cv Dicktoo. *Theor. Appl. Genet.* 98(8), 1234–47. doi:10.1007/s001220051189.

Christopher, J. T., Veyradier, M., Borrell, A. K., Harvey, G., Fletcher, S. and Chenu, K. 2014. Phenotyping novel stay-green traits to capture genetic variation in senescence dynamics. *Funct. Plant Biol.* 41(11), 1035. doi:10.1071/FP14052.

Christopher, J. T., Christopher, M. J., Borrell, A. K., Fletcher, S. and Chenu, K. 2016. Stay-green traits to improve wheat adaptation in well-watered and water-limited environments. *J. Exp. Bot.* 67(17), 5159–72. doi:10.1093/jxb/erw276.

Crosatti, C., Soncini, C., Stanca, A. M. and Cattivelli, L. 1995. The accumulation of a cold-regulated chloroplastic protein is light-dependent. *Planta* 196(3), 458–63. doi:10.1007/BF00203644.

Crosatti, C., Polverino de Laureto, P., Bassi, R. and Cattivelli, L. 1999. The interaction between cold and light controls the expression of the cold-regulated barley gene cor14b and the accumulation of the corresponding protein. *Plant Physiol.* 119(2), 671–80. doi:10.1104/pp.119.2.671.

Crosatti, C., Rizza, F., Badeck, F. W., Mazzucotelli, E. and Cattivelli, L. 2013. Harden the chloroplast to protect the plant. *Physiol. Plant.* 147(1), 55–63. doi:10.1111/j.1399-3054.2012.01689.x.

Crossa, J., Pérez-Rodríguez, P., Cuevas, J., Montesinos-López, O., Jarquín, D., de los Campos, G., Burgueño, J., González-Camacho, J. M., Pérez-Elizalde, S., Beyene, Y., et al. 2017. Genomic selection in plant breeding: methods, models, and perspectives. *Trends Plant Sci.* 22(11), 961–75. doi:10.1016/j.tplants.2017.08.011.

Dagar, J. C., Sharma, P. C., Chaudhari, S. K., Jat, H. S. and Ahamad, S. 2016. Climate change vis-à-vis saline agriculture: impact and adaptation strategies. In: Dagar, J. C., Sharma, P. C., Sharma, D. K. and Singh, A. K. (Eds), *Innovative Saline Agriculture*. Springer, Dordrecht, pp. 5–53.

Dahal, K., Kaine, K., Gadapati, W., Webb, E., Savitch, L. V., Singh, J., Sharma, P., Sarhan, F., Longstaffe, F. J., Grodzinski, B., et al. 2012. The effects of phenotypic plasticity on photosynthetic performance in winter rye, winter wheat and Brassica napus. *Physiol. Plant.* 144(2), 169–88. doi:10.1111/j.1399-3054.2011.01513.x.

Dalal, A., Attia, Z. and Moshelion, M. 2017. To produce or to survive: how plastic is your crop stress physiology? *Front. Plant Sci.* 8, 2067. doi:10.3389/fpls.2017.02067.

Dal Bosco, C., Busconi, M., Govoni, C., Baldi, P., Stanca, A. M., Crosatti, C., Bassi, R. and Cattivelli, L. 2003. Cor gene expression in barley mutants affected in chloroplast development and photosynthetic electron transport. *Plant Physiol.* 131(2), 793–802. doi:10.1104/pp.014530.

Daszkowska-Golec, A., Skubacz, A., Marzec, M., Slota, M., Kurowska, M., Gajecka, M., Gajewska, P., Plociniczak, T., Sitko, K., Pacak, A., et al. 2017. Mutation in *HvCBP20* (Cap Binding Protein 20) adapts barley to drought stress at phenotypic and transcriptomic levels. *Front. Plant. Sci.* 8, 942. doi:10.3389/fpls.2017.00942.

Daszkowska-Golec, A., Skubacz, A., Sitko, K., Słota, M., Kurowska, M. and Szarejko, I. 2018. Mutation in barley *era1* (enhanced response to aba1) gene confers better photosynthesis efficiency in response to drought as revealed by transcriptomic and physiological analysis. *Environ. Exp. Bot.* 148, 12–26. doi:10.1016/j.envexpbot.2018.01.003.

Dawson, I. K., Russell, J., Powell, W., Steffenson, B., Thomas, W. T. B. and Waugh, R. 2015. Barley: a translational model for adaptation to climate change. *New Phytol.* 206(3), 913–31. doi:10.1111/nph.13266.

De Vita, P., Mastrangelo, A. M., Matteu, L., Mazzucotelli, E., Virzì, N., Palumbo, M., Lo Storto, M., Rizza, F. and Cattivelli, L. 2010. Effect of genetic improvement on yield stability in selected durum wheat genotypes. *Field Crops Res.* 119, 68–77.

Distelfeld, A., Korol, A., Dubcovsky, J., Uauy, C., Blake, T. and Fahima, T. 2008. Colinearity between the barley grain protein content (*GPC*) QTL on chromosome arm 6HS and the wheat *Gpc-B1* region. *Mol. Breed.* 22(1), 25–38. doi:10.1007/s11032-007-9153-3.

Drosse, B., Campoli, C., Mulki, A. and von Korff, M. 2014. Genetic control of reproductive development. In: Kumlehn, J. and Stein, N. (Eds), *Biotechnological Approaches to Barley Improvement. Biotechnology in Agriculture and Forestry* (vol. 69). Springer, Berlin and Heidelberg.

Ensminger, I., Busch, F. and Huner, N. P. A. 2006. Photostasis and cold acclimation: sensing low temperature through photosynthesis. *Physiol. Plant.* 126(1), 28–44. doi:10.1111/j.1399-3054.2006.00627.x.

Finlay, K. W. and Wilkinson, G. N. 1963. The analysis of adaptation in a plant breeding programme. *Aust. J. Agric. Res.* 14(6), 742–54. doi:10.1071/AR9630742.

Fischer, R. A. and Maurer, R. 1978. Drought resistance in spring wheat cultivars. 1. Grain yield response. *Aust. J. Agric. Res.* 29(5), 897–912. doi:10.1071/AR9780897.

Fisk, S. P., Cuesta-Marcos, A., Cistue, L., Russell, J., Smith, K. P., Baenziger, S., Bedo, Z., Corey, A., Filichkin, T., Karsai, I., et al. 2013. FR-H3: a new QTL to assist in the development of fall-sown barley with superior low temperature tolerance. *Theor. Appl. Genet.* 126(2), 335–47. doi:10.1007/s00122-012-1982-8.

Foulkes, M. J., Hawkesford, M. J., Barraclough, P. B., Holdsworth, M. J., Kerr, S., Kightley, S. and Shewry, P. R. 2009. Identifying traits to improve the nitrogen economy of wheat: recent advances and future prospects. *Field Crops Res.* 114(3), 329–42. doi:10.1016/j.fcr.2009.09.005.

Francia, E., Rizza, F., Cattivelli, L., Stanca., A. M., Galiba, G., Toth, B., Hayes, P. M., Skinner, J. S. and Pecchioni, N. 2004. Two loci on chromosome 5h determine low-temperature tolerance in a 'Nure' (winter) × 'Tremois' (spring) barley map. *Theor. Appl. Genet.* 108(4), 670–80. doi:10.1007/s00122-003-1468-9.

Francia, E., Barabaschi, D.,Tondelli, A., Laidò, G., Rizza, F., Stanca, A. M., Busconi, M., Fogher, C., Stockinger, E. J. and Pecchioni, N. 2007. Fine mapping of a *HvCBF* gene cluster at the frost resistance locus *Fr-H2* in barley. *Theor. Appl. Genet.* 115(8), 1083–91. doi:10.1007/s00122-007-0634-x.

Francia, E., Morcia, C., Pasquariello, M., Mazzamurro, V., Milc, J. A., Rizza, F., Terzi, V. and Pecchioni, N. 2016. Copy number variation at the *HvCBF4-HvCBF2* genomic segment is a major component of frost resistance in barley. *Plant Mol. Biol.* 92(1–2), 161–75. doi:10.1007/s11103-016-0505-4.

Fricano, A., Rizza, F., Faccioli, P., Pagani, D., Pavan, P., Stella, A., Rossini, L., Piffanelli, P. and Cattivelli, L. 2009. Genetic variants of *HvCBF14* are statistically associated with frost tolerance in a European germplasm collection of *Hordeum vulgare*. *Theor. Appl. Genet.* 119(7), 1335–48. doi:10.1007/s00122-009-1138-7.

Galkin, E., Dalal, A., Evenko, A., Fridman, E., Kan, I., Wallach, R. and Moshelion, M. 2018. Risk-management strategies and transpiration rates of wild barley in uncertain environments. *Physiol. Plant.* 164(4), 412–28, doi:10.1111/ppl.12814.

Gan, S. and Amasino, R. M. 1995. Inhibition of leaf senescence by autoregulated production of cytokinin. *Science* 270(5244), 1986–8. doi:10.1126/science.270.5244.1986.

Gao, R., Guo, G., Fang, C., Huang, S., Chen, J., Lu, R., Huang, J., Fan, X. and Liu, C. 2018. Rapid generation of barley mutant lines with high nitrogen uptake efficiency by microspore mutagenesis and field screening. *Front. Plant Sci.* 9, 450. doi:10.3389/fpls.2018.00450.

Gous, P. W., Hickey, L., Christopher, J. T., Franckowiak, J. and Fox, G. P. 2016. Discovery of QTL for stay-green and heat-stress in barley (*Hordeum vulgare*) grown under simulated abiotic stress conditions. *Euphytica* 207(2), 305–17. doi:10.1007/s10681-015-1542-9.

Gregersen, P. L., Holm, P. B. and Krupinska, K. 2008. Leaf senescence and nutrient remobilisation in barley and wheat. *Plant Biol. Stuttg.* 10, 37–49. doi:10.1111/j.1438-8677.2008.00114.x.

Gusta, L. V. and Wisniewski, M. 2013. Understanding plant cold hardiness: an opinion. *Physiol. Plant.* 147(1), 4–14. doi:10.1111/j.1399-3054.2012.01611.x.

Halperin, O., Gebremedhin, A., Wallach, R. and Moshelion, M. 2017. High-throughput physiological phenotyping and screening system for the characterization of plant-environment interactions. *Plant J.* 89(4), 839–50. doi:10.1111/tpj.13425.

Han, M., Wong, J., Su, T., Beatty, P. H. and Good, A. G. 2016. Identification of nitrogen use efficiency genes in barley: searching for QTLs controlling complex physiological traits. *Front. Plant Sci.* 7, 1587. doi:10.3389/fpls.2016.01587.

Hanin, M., Ebel, C., Ngom, M., Laplaze, L. and Masmoudi, K. 2016. New insights on plant salt tolerance mechanisms and their potential use for breeding. *Front. Plant Sci.* 7, 1787. doi:10.3389/fpls.2016.01787.

Harris, K., Subudhi, P. K., Borrell, A., Jordan, D., Rosenow, D., Nguyen, H., Klein, P., Klein, R. and Mullet, J. 2007. Sorghum stay-green QTL individually reduce post-flowering drought-induced leaf senescence. *J. Exp. Bot.* 58, 327.

Hazzouri, K. M., Khraiwesh, B., Amiri, K. M. A., Pauli, D., Blake, T., Shahid, M., Mullath, S. K., Nelson, D., Mansour, A. L., Salehi-Ashtiani, K., et al. 2018. Mapping of HKT1;5 gene in barley using GWAS approach and its implication in salt tolerance mechanism. *Front. Plant Sci.* 9, 156. doi:10.3389/fpls.2018.00156.

Honsdorf, N., March, T. J., Berger, B., Tester, M. and Pillen, K. 2014. High-throughput phenotyping to detect drought tolerance QTL in wild barley introgression lines. *PLoS ONE* 9(5), e97047. doi:10.1371/journal.pone.0097047.

Huang, L., Kuang, L., Li, X., Wu, L., Wu, D. and Zhang, G. 2018. Metabolomic and transcriptomic analyses reveal the reasons why *Hordeum marinum* has higher salt tolerance than *Hordeum vulgare*. *Env. Exp. Bot.* 156, 48–61. doi:10.1016/j.envexpbot.2018.08.019.

IPCC. 2013. Summary for policymakers. In: Stocker, T. F., Qin, D., Plattner, G.-K., Tignor, M., Allen, S. K., Doschung, J., Nauels, A., Xia, Y., Bex, V. and Midgley, P. M. (Eds), *Climate Change 2013: the Physical Science Basis. Contribution of Working Group I to the Fifth Assessment Report of the Intergovernmental Panel on Climate Change*. Cambridge University Press, Cambridge, UK and New York, NY, 1535pp.

Ismail, A. M. and Horie, T. 2017. Genomics, physiology, and molecular breeding approaches for improving salt tolerance. *Annu. Rev. Plant Biol.* 68(1), 405–34. doi:10.1146/annurev-arplant-042916-040936.

Janda, T., Majlath, I. and Szalai, G. 2014. Interaction of temperature and light in the development of freezing tolerance in plants. *J. Plant Growth Regul.* 33(2), 460–69. doi:10.1007/s00344-013-9381-1.

Kamara, A. Y., Menkir, A., Badu-Apraku, B. and Ibikunle, O. 2003. Reproductive and stay-green trait responses of maize hybrids, improved open-pollinated varieties and farmers' local varieties to terminal drought stress. *Maydica* 48, 29.

Kiani, D., Soltanloo, H., Ramezanpour, S. S., Nasrolahnezhad Qumi, A. A., Yamchi, A., Zaynali Nezhad, K. H. and Tavakol, E. 2017. A barley mutant with improved salt tolerance through ion homeostasis and ROS scavenging under salt stress. *Acta Physiol. Plant.* 39(3), 90. doi:10.1007/s11738-017-2359-z.

Kindgren, P., Dubreuil, C. and Strand, Å 2015. The recovery of plastid function is required for optimal response to low temperatures in *Arabidopsis*. *PLoS ONE* 10(9), e0138010. doi:10.1371/journal.pone.0138010.

Kindu, G. A., Tang, J., Yin, X. and Struik, P. C. 2014. Quantitative trait locus analysis of nitrogen use efficiency in barley (*Hordeum vulgare* L.). *Euphytica* 199(1–2), 207–21. doi:10.1007/s10681-014-1138-9.

Kishor, P. B. K., Rajesh, K., Reddy, P. S., Seiler, C. and Sreenivasulu, N. 2014. Drought stress tolerance mechanisms in barley and its relevance to cereals. In: Kumlehn, J. and Stein, N. (Eds), *Biotechnological Approaches to Barley Improvement. Biotechnology in Agriculture and Forestry* (vol. 69). Springer, Berlin and Heidelberg.

Knox, A. K., Dhillon, T., Cheng, H., Tondelli, A., Pecchioni, N. and Stockinger, E. J. 2010. CBF gene copy number variation at *Frost Resistance-2* is associated with levels of freezing tolerance in temperate-climate cereals. *Theor. Appl. Genet.* 121(1), 21–35. doi:10.1007/s00122-010-1288-7.

Kotak, S., Larkindale, J., Lee, U., von Koskull-Doring, P., Vierling, E. and Scharf, K. D. 2007. Complexity of the heat stress response in plants. *Curr. Opin. Plant Biol.* 10(3), 310–16. doi:10.1016/j.pbi.2007.04.011.

Kumar, S., Sachdeva, S., Bhat, K. V. and Vats, S. 2018. Plant responses to drought stress: physiological, biochemical and molecular basis. In: Vats, S. (Ed.), *Biotic and Abiotic Stress Tolerance in Plants*. Springer, Singapore.

Lawas, L. M. F., Zuther, E., Jagadish, S. K. and Hincha, D. K. 2018. Molecular mechanisms of combined heat and drought stress resilience in cereals. *Curr. Opin. Plant Biol.* 45, 212–7. doi:10.1016/j.pbi.2018.04.002.

Lynch, J. P. 2013. Steep, cheap and deep: an ideotype to optimize water and N acquisition by maize root systems. *Ann. Bot. Lond.* 112(2), 347–57. doi:10.1093/aob/mcs293.

Maas, E. V. and Hoffman, G. J. 1977. Crop salt tolerance - current assessment. *J. Irrig. Drain. Div.* 103, 115–34.

Mahalingam, R. 2017. Phenotypic, physiological and malt quality analyses of US barley varieties subjected to short periods of heat and drought stress. *J. Cereal Sci.* 76, 199–205. doi:10.1016/j.jcs.2017.06.007.

Mahalingam, R. and Bregitzer, P. 2019. Impact on physiology and malting quality of barley exposed to heat, drought and their combination during different growth stages under controlled environment. *Physiol. Plant.* 165(2), 277–89. doi:10.1111/ppl.12841.

Mano, Y. and Takeda, K. 1997. Mapping quantitative trait loci for salt tolerance at germination and the seedling stage in barley (*Hordeum vulgare* L.). *Euphytica* 94(3), 263–72. doi:10.1023/A:1002968207362.

Marcinkowski, P. and Piniewski, M. 2018. Climate change effect on sowing and harvest dates of spring barley and maize in Poland. *Int. Agrophys.* 32, 265–71.

Merilo, E., Laanemets, K., Hu, H., Xue, S., Jakobson, L., Tulva, I., Gonzalez-Guzman, M., Rodriguez, P. L., Schroeder, J. I., Brosche, M., et al. 2013. PYR/RCAR receptors contribute to ozone-, reduced air humidity-, darkness- and CO_2-induced stomatal regulation. *Plant Physiol.* 162(3), 1652–68. doi:10.1104/pp.113.220608.

Mian, A., Oomen, R. J. F. J., Isayenkov, S., Sentenac, H., Maathuis, F. J. M. and Very, A.-A. 2011. Over-expression of an Na^+- and K^+-permeable HKT transporter in barley improves salt tolerance. *Plant J.* 68(3), 468–79. doi:10.1111/j.1365-313X.2011.04701.x.

Moinuddin, A., Fischer, R. A., Sayre, K. D. and Reynolds, M. P. 2005. Osmotic adjustment in wheat in relation to grain yield under water deficit environments. *Agron. J.* 97(4), 1062–71. doi:10.2134/agronj2004.0152.

Moll, R. H., Kamprath, E. J. and Jackson, W. A. 1982. Analysis and interpretation of factors which contribute to efficiency of nitrogen utilization. *Agron. J.* 74(3), 562–64. doi:10.2134/agronj1982.00021962007400030037x.

Morran, S., Eini, O., Pyvovarenko, T., Parent, B., Singh, R., Ismagul, A., Eliby, S., Shirley, N., Langridge, P. and Lopato, S. 2011. Improvement of stress tolerance of wheat and barley by modulation of expression of DREB/CBF factors. *Plant Biotechnol. J.* 9(2), 230–49. doi:10.1111/j.1467-7652.2010.00547.x.

Munns, R. and Tester, M. 2008. Mechanisms of salinity tolerance. *Annu. Rev. Plant Biol.* 59(1), 651–81. doi:10.1146/annurev.arplant.59.032607.092911.

Munns, R., James, R. A., Islam, A. K. M. R. and Colmer, T. D. 2011. *Hordeum marinum*-wheat amphiploids maintain higher leaf K^+:Na^+ and suffer less leaf injury than wheat parents in saline conditions. *Plant Soil* 348(1–2), 365–77. doi:10.1007/s11104-011-0934-4.

Murelli, C., Rizza, F., Albini, F. M., Dulio, A., Terzi, V. and Cattivelli, L. 1995. Metabolic changes associated with cold-acclimation in contrasting cultivars of barley. *Physiol. Plant.* 94(1), 87–93. doi:10.1111/j.1399-3054.1995.tb00788.x.

Muzammil, S., Shrestha, A., Dadshani, S., Pillen, K., Siddique, S., Léon, J. and Naz, A. A. 2018. An ancestral allele of Pyrroline-5-carboxylate synthase1 promotes proline accumulation and drought adaptation in cultivated barley. *Plant Physiol.* 178(2), 771–82. doi:10.1104/pp.18.00169.

Negrão, S., Schmöckel, S. M. and Tester, M. 2017. Evaluating physiological responses of plants to salinity stress. *Ann. Bot.* 119(1), 1–11. doi:10.1093/aob/mcw191.

Neumann, K., Klukas, C., Friedel, S., Rischbeck, P., Chen, D., Entzian, A., Stein, N., Graner, A. and Kilian, B. 2015. Dissecting spatio-temporal biomass accumulation in barley under different water regimes using high-throughput image analysis. *Plant Cell Environ.* 38(10), 1980–96. doi:10.1111/pce.12516.

Oliver, S. N., Deng, W., Casao, M. C. and Trevaskis, B. 2013. Low temperatures induce rapid changes in chromatin state and transcript levels of the cereal *VERNALIZATION1* gene. *J. Exp. Bot.* 64(8), 2413–22. doi:10.1093/jxb/ert095.

Peleg, Z., Saranga, Y., Krugman, T., Abbo, S., Nevo, E. and Fahima, T. 2008. Allelic diversity associated with aridity gradient in wild emmer wheat populations. *Plant Cell Environ.* 31, 39–49. doi:10.1111/j.1365-3040.2007.01731.x.

Perchlik, M. and Tegeder, M. 2017. Improving plant nitrogen use efficiency through alteration of amino acid transport processes. *Plant Physiol.* 175(1), 235–47. doi:10.1104/pp.17.00608.

Piasecka, A., Sawikowska, A., Kuczyńska, A., Ogrodowicz, P., Mikołajczak, K., Krystkowiak, K., Gudyś, K., Guzy-Wróbelska, J., Krajewski, P. and Kachlicki, P. 2017. Drought-related secondary metabolites of barley (*Hordeum vulgare* L.) leaves and their metabolomic quantitative trait loci. *Plant J.* 89(5), 898–913. doi:10.1111/tpj.13430.

Pinas Fernandez, A. P. and Strand, Å 2008. Retrograde signaling and plant stress: plastid signals initiate cellular stress responses. *Curr. Opin. Plant Biol.* 11(5), 509–13. doi:10.1016/j.pbi.2008.06.002.

Prášil, I. T., Prasilova, P. and Marik, P. 2007. Comparative study of direct and indirect evaluations of frost tolerance in barley. *Field Crops Res.* 102(1), 1–8. doi:10.1016/j.fcr.2006.12.012.

Pratap, A., Tomar, R., Kumar, J., Pandey, V. R., Mehandi, S. and Katiyar, P. K. 2015. High-throughput plant phenotyping platforms. In: Kumar, J., Pratap, A. and Kumar, S. (Eds), *Phenomics in Crop Plants: Trends, Options and Limitations*. Springer, New Delhi.

Quan, X., Zeng, J., Ye, L., Chen, G., Han, Z., Shahm, J. M. and Zhang, G. 2016. Transcriptome profiling analysis for two Tibetan wild barley genotypes in responses to low nitrogen. *BMC Plant Biol.* 16(1), 30. doi:10.1186/s12870-016-0721-8.

Quraishi, U. M., Abrouk, M., Murat, F., Pont, C., Foucrier, S., Desmaizieres, G., Confolent, C., Rivière, N., Charmet, G., Paux, E., et al. 2011. Cross-genome map based dissection of a nitrogen use efficiency ortho-meta QTL in bread wheat unravels concerted cereal genome evolution. *Plant J.* 65(5), 745–56. doi:10.1111/j.1365-313X.2010.04461.x.

Rajala, A., Peltonen-Sainio, P., Jalli, M., Jauhiainen, L., Hannukkala, A., Tenhola-Roininen, T., Ramsay, L. and Manninen, O. 2017. One century of Nordic barley breeding: nitrogen use efficiency, agronomic traits and genetic diversity. *J. Agric. Sci.* 155(4), 582–98. doi:10.1017/S002185961600068X.

Rajendran, K., Tester, M. and Roy, S. J. 2009. Quantifying the three main components of salinity tolerance in cereals. *Plant Cell Environ.* 32(3), 237–49. doi:10.1111/j.1365-3040.2008.01916.x.

Rebetzke, G. J., Condon, A. G., Richards, R. A. and Farquhar, G. D. 2002. Selection for reduced carbon isotope discrimination increases aerial biomass and grain yield of rainfed bread wheat. *Crop Sci.* 42(3), 739–45. doi:10.2135/cropsci2002.0739.

Rizza, F., Crosatti, C., Stanca, A. M. and Cattivelli, L. 1994. Studies for assessing the influence of hardening on cold tolerance of barley genotypes. *Euphytica* 75(1–2), 131–38. doi:10.1007/BF00024540.

Rizza, F., Pagani, D., Stanca, A. M. and Cattivelli, L. 2001. Use of chlorophyll fluorescence to evaluate the cold acclimation and freezing tolerance of winter and spring oats. *Plant Breed.* 120, 389–96.

Rizza, F., Badeck, F. W., Cattivelli, L., Li Destri, O., Di Fonzo, N. and Stanca, A. M. 2004. Use of a water stress index to identify barley genotypes adapted to rainfed and irrigated conditions. *Crop Sci.* 44(6), 2127–37. doi:10.2135/cropsci2004.2127.

Rizza, F., Pagani, D., Gut, M., Prášil, I. T., Lago, C., Tondelli, A., Orrù, L., Mazzucotelli, E., Francia, E., Badeck, F.-W., et al. 2011. Diversity in the response to low temperature in representative barley genotypes cultivated in Europe. *Crop Sci.* 51(6), 2759–79. doi:10.2135/cropsci2011.01.0005.

Robson, A. D. and Snowball, K. 1986. Nutrient deficiency and toxicity symptoms. In: Reuter, D. J. and Robinson, J. B. (Eds), *Plant Analysis, an Interpretation Manual*. Inkata Press, Melbourne, pp. 13–9.

Rollins, J. A., Habte, E., Templer, S. E., Colby, T., Schmidt, J. and von Korff, M. 2013. Leaf proteome alterations in the context of physiological and morphological responses to drought and heat stress in barley (*Hordeum vulgare* L.). *J. Exp. Bot.* 64(11), 3201–12. doi:10.1093/jxb/ert158.

Römer, C., Wahabzada, M., Ballvora, A., Pinto, F., Rossini, M., Panigada, C., Behmann, J., Léon, J., Thurau, C. and Bauckhage, C. 2012. Early drought stress detection in cereals: simplex volume maximization for hyperspectral image analysis. *Funct. Plant Biol.* 39, 878–90.

Rong, H., Tang, Y., Zhang, H., Wua, P., Chen, Y., Li, M., Wua, G. and Jiang, H. 2013. The Stay-Green Rice like (SGRL) gene regulates chlorophyll degradation in rice. *J. Plant Physiol.* 170(15), 1367–73. doi:10.1016/j.jplph.2013.05.016.

Roy, S. J., Negrão, S. and Tester, M. 2014. Salt resistant crop plants. *Curr. Opin. Biotech.* 26, 115–24. doi:10.1016/j.copbio.2013.12.004.

Saade, S., Maurer, A., Shahid, M., Oakey, H., Schmöckel, S. M., Negrão, S., Pillen, K. and Tester, M. 2016. Yield-related salinity tolerance traits identified in a nested association mapping (NAM) population of wild barley. *Sci. Rep.* 6, 32586.

Saade, S., Negrão, S., Plett, D., Garnett, T. and Tester, M. 2018. Genomic and genetic studies of abiotic stress tolerance in barley. In: Stein, N. and Muehlbauer, G. (Eds), *The Barley Genome. Compendium of Plant Genomes*. Springer, Cham, pp. 259–86.

Saoudi, W., Badri, M., Taamalli, W., Zribi, O. T., Gandour, M. and Abdelly, C. 2019. Variability in response to salinity stress in natural Tunisian populations of *Hordeum marinum* subsp. marinum. *Plant Biol.* 21(1), 89–100. doi:10.1111/plb.12890.

Shabala, S. 2013. Learning from halophytes: physiological basis and strategies to improve abiotic stress tolerance in crops. *Ann. Bot.* 112(7), 1209–21. doi:10.1093/aob/mct205.

Sharma, R., Draicchio, F., Bull, H., Herzig, P., Maurer, A., Pillen, K., Thomas, W. T. B. and Flavell, A. J. 2018. Genome-wide association of yield traits in a nested association mapping population of barley reveals new gene diversity for future breeding. *J. Exp. Bot.* 69(16), 3811–22. doi:10.1093/jxb/ery178.

Shen, Q., Yu, J., Fu, L., Wu, L., Dai, F., Jiang, L., Wu, D. and Zhang, G. 2018. Ionomic, metabolomic and proteomic analyses reveal molecular mechanisms of root adaption to salt stress in Tibetan wild barley. *Plant Physiol. Biochem.* 123, 319–30. doi:10.1016/j.plaphy.2017.12.032.

Sivamani, E., Bahieldin, A., Wraith, J. M., Al-Niemi, T., Dyer, W. E., Ho, T.-D. and Qu, R. 2000. Improved biomass productivity and water use efficiency under water deficit conditions in transgenic wheat constitutively expressing the barley *HVA1* gene. *Plant Sci.* 155, 1-9.

Skinner, J. S., von Zitzewitz, J., Szucs, P., Marquez-Cedillo, L., Filichkin, T., Amundsen, K., Stockinger, E. J., Thomashow, M. F., Chen, T. H. and Hayes, P. M. 2005. Structural, functional, and phylogenetic characterization of a large CBF gene family in barley. *Plant Mol. Biol.* 59, 533-51.

Skinner, J. S., Szucs, P., von Zitzewitz, J., Marquez-Cedillo, L., Filichkin, T., Stockinger, E. J., Thomashow, M. F., Chen, T. H. H. and Hayes, P. M. 2006. Mapping of barley homologs to genes that regulate low temperature tolerance in *Arabidopsis*. *Theor. Appl. Genet.* 112, 832-42.

Soltesz, A., Smedley, M., Vashegyi, I., Galiba, G., Harwood, W. and Vaguifalvi, A. 2013. Transgenic barley lines prove the involvement of *TaCBF14* and *TaCBF15* in the cold acclimation process and in frost tolerance. *J. Exp. Bot.* 64, 1849-62.

Stockinger, E. J., Skinner, J. S., Gardner, K. G., Francia, E. and Pecchioni, N. 2007. Expression levels of barley *Cbf* genes at the *Frost resistance*-H2 locus are dependent upon alleles at *Fr-H1* and *Fr-H2*. *Plant J.* 51(2), 308-21. doi:10.1111/j.1365-313X.2007.0141.x.

Svensson, J. T., Crosatti, C., Campoli, C., Bassi, R., Stanca, A. M., Close, T. J. and Cattivelli, L. 2006. Analysis of cold acclimation in barley albina and xantha mutants reveals a key role of the chloroplast during plant adaptation to low temperature. *Plant Physiol.* 141, 257-70.

Tardieu, F., Simonneau, T. and Muller, B. 2018. The physiological basis of drought tolerance in crop plants: a scenario-dependent probabilistic approach. *Annu. Rev. Plant Biol.* 69(1), 733-59. doi:10.1146/annurev-arplant-042817-040218.

Tavakkoli, E., Fatehi, F., Coventry, S., Rengasamy, P. and McDonald, G. K. 2011. Additive effects of Na^+ and Cl^- ions on barley growth under salinity stress. *J. Exp. Bot.* 62, 2189-203.

Templer, S. E., Ammon, A., Pscheidt, D., Ciobotea, O., Schuy, C., McCollum, C., Sonnewald, U., Hanemann, A., Forster, J., Ordon, F., et al. 2017. Metabolite profiling of barley flag leaves under drought and combined heat and drought stress reveals metabolic QTLs for metabolites associated with antioxidant defense. *J. Exp. Bot.* 68(7), 1697-713. doi:10.1093/jxb/erx038.

Thomas, H. and Howarth, C. J. 2000. Five ways to stay green. *J. Exp. Bot.* 51 (Suppl. 1), 329-37. doi:10.1093/jexbot/51.suppl_1.329.

Tondelli, A., Francia, E., Barabaschi, D., Pasquariello, M. and Pecchioni, N. 2011. Inside the *CBF* locus in Poaceae. *Plant Sci.* 180(1), 39-45. doi:10.1016/j.plantsci.2010.08.012.

Tondelli, A., Pagani, D., Naseh Ghafoori, I. N., Rahimi, M., Ataei, R., Rizza, F., Flavell, A. J. and Cattivelli, L. 2014. Allelic variation at *Fr-H1/Vrn-H1* and *Fr-H2* loci is the main determinant of frost tolerance in spring barley. *Env. Exp. Bot.* 106, 148-55. doi:10.1016/j.envexpbot.2014.02.014.

Turner, N. C. 2004. Sustainable production of crops and pastures under drought in a Mediterranean environment. *Ann. Appl. Biol.* 144, 139-47.

Uauy, C., Distelfeld, A., Fahima, T., Blechl, A. and Dubcovsky, J. 2006. A *NAC* gene regulating senescence improves grain protein, zinc, and iron content in wheat. *Science* 314(5803), 1298-301. doi:10.1126/science.1133649.

Voltas, J., Lopez-Corcolesand, H. and Borras, G. 2005. Use of biplot analysis and factorial regression for the investigation of superior genotypes in multi-environment trials. *Eur. J. Agron.* 22(3), 309-24. doi:10.1016/j.eja.2004.04.005.

von Zitzewitz, J., Szucs, P., Dubcovsky, J., Yan, L., Francia, E., Pecchioni, N., Casas, A., Chen, T. H. H., Hayes, P. M. and Skinner, J. S. 2005. Molecular and structural characterization of barley vernalization genes. *Plant Mol. Biol.* 59(3), 449–67. doi:10.1007/s11103-005-0351-2.

von Zitzewitz, J., Cuesta-Marcos, A., Condon, F., Castro, A. J., Chao, S., Corey, A., Filichkin, T., Fisk, S. P., Gutierrez, L., Haggard, K., et al. 2011. The genetics of winterhardiness in barley: perspectives from genome-wide association mapping. *Plant Genome* 4(1), 76–91. doi:10.3835/plantgenome2010.12.0030.

Voss-Fels, K. P., Robinson, H., Mudge, S. R., Richard, C., Newman, S., Wittkop, B., Stahl, A., Friedt, W., Frisch, M., Gabur, I., et al. 2018. *VERNALIZATION1* modulates root system architecture in wheat and barley. *Mol. Plant* 11(1), 226–29. doi:10.1016/j.molp.2017.10.005.

Wallwork, M. A. B., Logue, S. J., MacLeod, L. C. and Jenner, C. F. 1998. Effect of high temperature during grain filling on starch synthesis in the developing barley grain. *Aust. J. Plant Physiol.* 25, 173–81.

Wang, W. Q., Hao, Q., Tian, F. X., Li, Q. and Wang, W. 2016. The stay-green phenotype of wheat mutant *tasg1* is associated with altered cytokinin metabolism. *Plant Cell Rep.* 35(3), 585–99. doi:10.1007/s00299-015-1905-7.

Wardlaw, I. F. and Wrigley, C. W. 1994. Heat tolerance in temperate cereals: an overview. *Aust. J. Plant Physiol.* 21, 695–703.

Wu, H., Shabala, L., Zhou, M., Stefano, G., Pandolfi, C., Mancuso, S. and Shabala, S. 2015a. Developing and validating a high-throughput assay for salinity tissue tolerance in wheat and barley. *Planta* 242(4), 847–57. doi:10.1007/s00425-015-2317-1.

Wu, H., Zhu, M., Shabala, L., Zhou, M. and Shabala, S. 2015b. K^+ retention in leaf mesophyll, an overlooked component of salinity tolerance mechanism: a case study for barley. *J. Integr. Plant Biol.* 57, 171–85.

Wu, X., Cai, K., Zhang, G. and Zeng, F. 2017. Metabolite profiling of barley grains subjected to water stress: to explain the genotypic difference in drought-induced impacts on malting quality. *Front. Plant Sci.* 8, 1547. doi:10.3389/fpls.2017.01547.

Xia, Y., Li, R., Ning, Z., Bai, G., Siddique, K. H. M., Yan, G., Baum, M., Varshney, R. K. and Guo, P. 2013. Single nucleotide polymorphisms in HSP17.8 and their association with agronomic traits in barley. *PLoS ONE* 8(2), e56816. doi:10.1371/journal.pone.0056816.

Xie, W., Xiong, W., Pan, J., Ali, T., Cui, Q., Guan, D., Meng, J., Mueller, N. D., Lin, E. and Davis, S. J. 2018. Decreases in global beer supply due to extreme drought and heat. *Nat. Plants* 4, 964–73.

Xu, S. and Chong, K. 2018. Remembering winter through vernalisation. *Nat. Plants* 4(12), 997–1009. doi:10.1038/s41477-018-0301-z.

Xue, W., Yan, J., Zhao, G., Jiang, Y., Cheng, J., Cattivelli, L. and Tondelli, A. 2017. A major QTL on chromosome 7HS controls the response of barley seedling to salt stress in the Nure × Tremois population. *BMC Genet.* 18(1), 79. doi:10.1186/s12863-017-0545-z.

Yan, L., Loukoianov, A., Tranquilli, G., Helguera, M., Fahima, T. and Dubcovsky, J. 2003. Positional cloning of the wheat vernalization gene *VRN1*. *Proc. Natl. Acad. Sci. U. S. A.* 100(10), 6263–68. doi:10.1073/pnas.0937399100.

Yoshida, H., Hirano, K., Sato, T., Mitsuda, N., Nomoto, M., Maeo, K., Koketsu, E., Mitani, R., Kawamura, M., Ishiguro, S., et al. 2014. DELLA protein functions as a transcriptional activator through the DNA binding of the indeterminate domain family proteins. *Proc. Natl. Acad. Sci. U. S. A.* 111(21), 7861–66. doi:10.1073/pnas.1321669111.

Zhu, J. K. 2016. Abiotic stress signalling and responses in plants. *Cell* 167(2), 313–24. doi:10.1016/j.cell.2016.08.029.

Zhu, M., Zhou, M. X., Shabala, L. and Shabala, S. 2015. Linking osmotic adjustment and stomatal characteristics with salinity stress tolerance in contrasting barley accessions. *Funct. Plant Biol.* 42(3), 252–63. doi:10.1071/FP14209.

Zhu, M., Zhou, M. X., Shabala, L. and Shabala, S. 2017. Physiological and molecular mechanisms mediating xylem Na^+ loading in barley in the context of salinity stress tolerance. *Plant Cell Environ.* 40(7), 1009-20. doi:10.1111/pce.12727.

Chapter 3

Advances in the understanding of barley plant physiology: factors determining grain development, composition, and chemistry

Ljudmilla Borisjuk, Hardy Rolletschek and Volodymyr Radchuk, Leibniz Institute of Plant Genetics and Crop Plant Research (IPK), Germany

1 Introduction
2 Spike growth and how it influences traits of the grain
3 Role of cell death in barley grain development
4 Sucrose allocation during the grain-filling stage
5 The use of starch in the developing caryopsis
6 Proteins and barley grain quality
7 Particularities of energy metabolism in barley grain
8 Functional orchestration of the barley grain
9 Conclusion
10 Acknowledgements
11 Where to look for further information
12 References

1 Introduction

A mature barley grain appears simple, but its structure is a unique cryptogram of evolution and human interference over millennia. The history of barley cultivation has revealed multiple factors that influence the development and composition of the mature grain; a large number of studies have characterized distinct events in grain development (Langridge, 2018); the recent sequencing of the barley genome has revealed the full set of molecular players behind these events (Mascher et al., 2017). The challenge is to uncover how the various factors interact in the living grain. The tailoring of crop seeds to meet human needs has turned out to be more challenging than what was perhaps expected. For this reason, we need to enhance our knowledge of grain metabolism and reconsider our strategies. Progress in genomics and phenomics, supported by

http://dx.doi.org/10.19103/AS.2019.0060.03

machine learning and artificial intelligence approaches, raises our expectations for solving agricultural problems. A clue for converting scientific output into value lies in the understanding of the physiological context and, eventually, the biology of the developing grain.

Barley's importance as a source of food and feed has meant that the structural, genetic, and molecular aspects of barley grain have been studied in great detail (Newton et al., 2011; Shewry and Ullrich, 2014). Recent progress has shed some light on internal processes by establishing topographical and imaging approaches that allow investigating the spatial arrangement of grain metabolism. Examples include MALDI-imaging (Peukert et al., 2014), infrared microspectroscopy (Guendel et al., 2018), respiration and oxygen mapping (Tschiersch et al., 2012), and X-ray imaging (Hughes et al., 2017) as well as a series of molecular tools applicable for barley (Harwood, 2019). For gaining a realistic picture of dynamic processes such as development, assimilate allocation, or cellular metabolism, noninvasive technologies have been implemented. These have been able to image and analyze the workings of a living seed. One such technology is nuclear magnetic resonance imaging (NMR/MRI), referred to within the field of clinical diagnostics as 'tomography' (Borisjuk et al., 2012). When combined with other experimental tools as well as with metabolic modeling approaches, a concept for the metabolic architecture of a living cereal grain was elaborated (Rolletschek et al., 2015). Perhaps the most exciting opportunity provided by NMR is its ability to view physiological processes *in vivo*, revealing their dynamics and spatial relationships. The data that can be generated in this way have the potential to greatly improve our understanding of how a seed–or a plant–lives.

This chapter highlights progress in the current understanding of barley grain physiology, gained by applying new technologies and exploring the spatial arrangement of grain metabolism and assimilating allocation within and among different seed tissues. The emerging roles of programed cell death (PCD) and hypoxia on grain metabolism are discussed, as well as the outcome of multiscale modeling studies on developing grain, to provide an insider's view on barley grain biology.

2 Spike growth and how it influences traits of the grain

Barley refers not only to the cereal *Hordeum vulgare* subsp. *vulgare*, but also more generally to the barley genus *Hordeum* (Triticeae tribe) that, apart from cultivated barley, comprises more than 30 wild grass species distributed in temperate and arid regions of the world (Blattner, 2018). Barley grains are produced on a structure called inflorescence that is a spike instead of the panicle that occurs in most other grasses. How does growing on a spike affect grain features?

2.1 Barley grain in botanical terms

The two-rowed and six-rowed barley are distinguished by the spike structure (Fig. 1) (Stanca et al., 2016). Row number is genetically determined by the VRS locus (Lundqvist et al., 1997), and loss-of-function alleles of any of the five genes (VRS1-5) cause complete-to-intermediate gain of lateral spikelet fertility. Current six-row cultivars contain naturally defective vrs1 and vrs5 alleles and produce grains that are smaller and more variable in size and shape than those of two-rowed cultivars. This effect may be associated with altered assimilate partitioning or decreased competition between the central and lateral spikelets (Sakuma et al., 2017). The alleles of the row-type architecture genes affect grain size by unknown mechanisms. Nonetheless, it appears feasible to manipulate

Figure 1 Structure of barley in spike. (a) two-row barley spike (left), where kernels tend to be symmetrical and of even size; six-row barley spike (right), where the two lateral rows of kernels are a little shorter, thinner, and slightly twisted; (b) enlarged view on spike with grains attached to the rachis at their base; some organs (lemma with long awns, glume surrounding pericarp and rachis) change from green to yellow during grain maturation. (c) naked grain (no hull) on the left; embryo region is arrowed; hulled grain on the right; hull (palea) is slightly wrinkled.

seed features by combining distinct alleles (Zwirek et al., 2019). For example, introgression of the mutant VRS3 allele into the cultivar 'Morex' (containing six-rowed VRS1 and VRS5 alleles) improves the uniformity of grain size (Bull et al., 2017).

In botanical terms, the barley grain is a dry fruit comprising only one seed. The seed is intimately fused with the fruit wall into a single unit (also called caryopsis). Additional external structures (rachilla, palea, and lemma) could also firmly adhere to the seed (Fig. 1). In the case where the palea and lemma form an additional hull of grain, the barley is referred to as covered (hulled). Barley, in which the palea and lemma do not adhere to the seed, is known as hulless (naked). This morphotype (covered vs. naked) is controlled by a single locus (nud, for nudum) (Franckowiack and Konishi, 1997). The corresponding gene encodes an ethylene response factor family transcription factor (TF) regulating a lipid biosynthesis pathway (Taketa et al., 2008). Its expression is strictly localized to the testa and leads to the secretion of a sticky, adhesive substance (cuticular lipid) on the surface of the pericarp of hulled grains. The clear differentiation of these two grain morphotypes is of agronomic relevance because it is important for the end use. Naked barley yields less than the hulled type and contains more protein (Sterna et al., 2017). Most cultivated barley types are hulled, whereas naked barley is an ancient food crop. Great potential exists to improve the yield based on manipulation of inflorescence (Sakuma, 2018; Schnurrbusch, 2019).

2.2 Temporal sequence of grain growth

Caryopsis is initiated after pollination of the flower; thus, the fertility of the flower and double fertilization are prerequisites for caryopsis development (Wilkinson et al., 2018). The zygote divides 20–24 h following fertilization (Engell, 1989), so that the developmental time sequence is usually measured in days after pollination (DAP). A tight correlation has been reported between the enlargement of the ovule and the embryo sac in the first days of development. Three main phases of development are described: (i) early or prestorage phase, starting at 0 DAP until the architecture of the caryopsis is completed; (ii) maturation phase (main storage), when biosynthesis and accumulation of storage product occurs; and (iii) desiccation phase (water loss), during which grain development is accomplished by acquisition of desiccation tolerance. The transition between the prestorage and storage phase is also considered as an intermediate phase of development (Sreenivasulu et al., 2010). Detailed descriptions of the cellular differentiation and tissue formation are provided elsewhere (Gubatz and Weschke, 2014; Olsen and Weschke, 2014). At the early stage the caryopsis consists of a tiny embryo sac and the tissues surrounding this are the pericarp (outermost layers) and integuments. The pericarp differentiates into two major tissue types, the inner green,

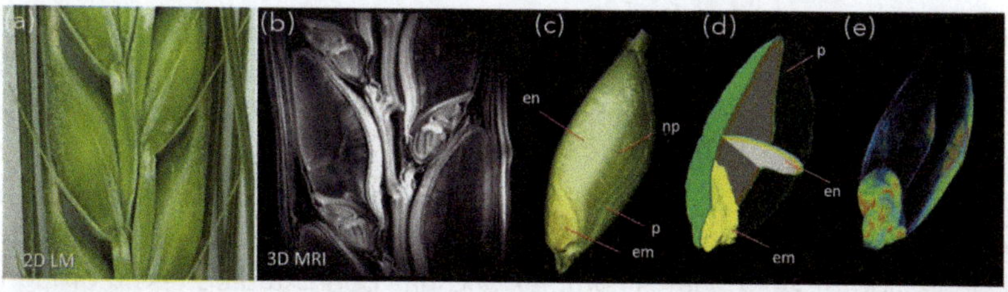

Figure 2 Noninvasive imaging of a living barley grain using magnetic resonance imaging (MRI). (a) fragment of a barley spike as pictured by light microscopy (LM), internal structures are not visible; (b) fragment of three-dimensional (3D) imaging of intact spike by MRI demonstrating the internal structure of grains and spike with spatial resolution of 35 µm; (c) cross-section through the grain; (d) 3D-model of grain structure based on MRI; (e) quantitative image of lipid distribution inside of living grain by MRI; lipids are mainly found in the embryo and the aleurone layer; lipid content is color-coded. Abbreviations: em, embryo; en, endosperm; np, nucellar projection; p, pericarp. For further details, see Borisjuk et al. (2012).

chloroplast-containing chlorenchyma and the outer nongreen cell layers. The seed coat is formed by two integuments underlaid by nucellus tissues, surrounding the embryo sac (Radchuk and Borisjuk, 2014). The major vein is embedded in the parenchyma tissue on the dorsal side of the caryopsis. The endosperm begins its development as the coenocytic tissue. Rapid nuclear divisions followed by cell wall formation lead to the cellularized endosperm; subsequently, this becomes the largest caryopsis organ (Fig. 2). The embryo undergoes complex tissue differentiation, but constitutes a minor part of the mature caryopsis.

3 Role of cell death in barley grain development

The shaping of the grain structure relies on cell proliferation and elimination events. While proliferation predetermines growth and biomass accumulation, cell death allows the removal and elimination of superfluous, irreversibly damaged, and/or potentially harmful cells. Together, these processes are required for forming the grain body. For a long time, the role of cell death was dismissed by plant biologists, but experimental evidence over the past decades has changed this stance. Here, we provide a current concept of cell death and perspective of how death integrates within the grain developmental process.

3.1 Arrangement of death events in grain

Cell death is defined as the irreversible degeneration of vital cellular functions (e.g. ATP production or redox homeostasis), culminating in the loss of cellular integrity, collapsed membrane permeability, and inevitable cellular fragmentation. Regulated cell death results from the activation of one or more signal transduction modules; hence, it can be modulated to a certain extent (e.g. genetically or pharmacologically). One specific form of regulated death is PCD, which occurs in strictly physiological scenarios and is a part of the developmental program (i.e. it is not associated with perturbations in inadequate stress response).

Despite seeming paradoxical, barley grain development is accompanied by cell death, which eliminates cells and reshapes tissues and organs. The coordination of cell death is complex and has been studied less thoroughly. While each grain organ allows its own developmental process, the regulation of cell death events is distinct in the pericarp, embryo, and endosperm (Fig. 3). Thereby, a large portion of maternal tissues is eliminated, providing sufficient space for growing the endosperm and embryo (Tran et al., 2014; Radchuk et al., 2011).

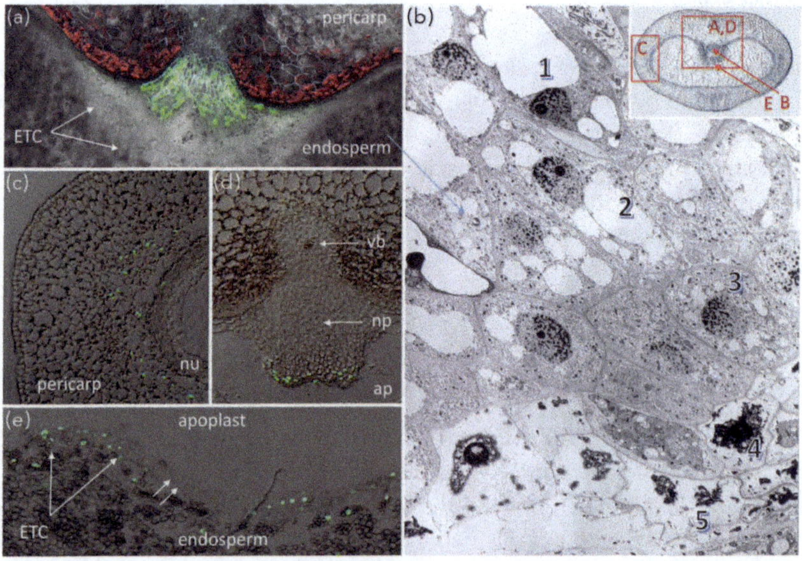

Figure 3 Pattern of programed cell death (PCD) during development of the barley grain. (a) expression of *Jekyll* gene within the nucellar projection as visualized by CLSM (green signal); chlorophyll (in red); endosperm transfer cells (ETC) are arrowed; (b) gradient of cell death evident in nucellar projection; scheme on the top right shows position of images A–E; representative cells are labeled by number from 1 to 4: (1) differentiated cells already formed a large central vacuole; nucleus begins to condensate; (2) nuclear condensation more pronounced; rupture of tonoplast leads to formation of differently sized vacuoles; (3) tonoplast rupture coincides with the rupture of nuclear envelope; (4) nuclear content is released into the cytoplasm; clearance of cytoplasm becomes obvious; (5) cell membrane ruptures; cell content is released in the apoplastic space; cell debris is seen; (c–e) TUNEL assay identifies degenerating nuclei (hallmark of PCD) in the pericarp (c), nucellar projection (d) and endosperm transfer cells (e). Abbreviations: ap-apoplastic space; ETC, endosperm transfer cells; np-nucellar projection; nu-nucellus; vb-vascular bundle.

In the nucellus, cell death progression coincides with cell proliferation and expansion of the endosperm. As early as 2 DAP, a TUNEL assay could identify degenerating nuclei in the surrounding endosperm cell layers. The contents of disintegrated dead cells were released in the apoplast and nourished the growing endosperm (Radchuk et al., 2011). The nucellus is digested completely (apart from the nuclear epidermis) within the dorsal region of the caryopsis. The fate of the nucellar cells adjacent to the vascular bundle (ventral region) is different (Fig. 3a and b). Within this region, cell division continues, new cells expand toward the inner part of grain, and they reshape their cell walls (acquisition of transfer cells function) shortly before dying. A gradient of cell differentiation (from proliferation to death), termed nuclear projection, became evident within this region (Radchuk et al., 2006). The content of disintegrating cells is released at the proximal end of nucellar projection, toward the endosperm transfer cells (ETCs) (Thiel et al., 2008). Hormone signaling and transcriptional networks regulate PCD initiation and progression (Sreenivasulu et al., 2006; Thiel et al., 2008; Julian et al., 2013).

In the pericarp, the first signs of PCD are detectable at ~4 DAP and extend to the entire tissue during 6–15 DAP (Fig. 3c). PCD in the pericarp begins from the innermost cell layer of the mesocarp (Radchuk et al., 2018; Tran et al., 2014). DNA fragmentation, a hallmark of PCD, is observed in the pericarp at 3–10 DAP. Disintegration of the lateral and dorsal pericarp sections coincides with growth in the longitudinal direction until 10–12 DAP. The ventral region of the pericarp and the chlorenchyma cell layers undergo gradual degeneration at a much later stage. The persistence of these particular tissues highlights their functional importance for grain physiology until maturation. Finally, the pericarp is reduced to the thin cuticle layer covering the grain (Radchuk et al., 2011).

In the barley endosperm, the cellular dynamics is controlled mainly by retinoblastoma-related proteins and cyclin-dependent kinase, as a general regulator of cereal endosperm cell division, endoreduplication, and death (Sabelli et al., 2013). The death interferes with growth during the grain-filling stage. The decline in storage activity after 16–18 DAP (Tran et al., 2014) is due to PCD occurring in the majority of cells. The ETCs begin disintegrating at the same time (Fig. 3e). Nutrient delivery is disturbed, and the endosperm activity switches from storage to maturation and desiccation. The cellular structure of the dying endosperm tissues remains intact, and the cellular content (mainly starch) is not disintegrated (nonlytic PCD; van Doorn et al., 2011). The endosperm of the mature grain consists mainly of dead tissue (starchy endosperm), with the exception of the thin aleurone layer.

The embryo forms two domains with distinct fates: the living part, which is the embryo proper, and the dying part, which is the suspensor (Peng and Sun, 2018). The suspensor cells connect the embryo to the surrounding tissues, and degenerate via the PCD. In the embryo proper, a number of TUNEL-positive nuclei were detected in the scutellum, indicating a significant tissue reorganization during embryo maturation (Tran et al., 2014). The functional role of PCD and the molecular mechanisms responsible for it in the late embryo require further investigation.

3.2 Cellular mechanisms and main actors in cell death

At the subcellular level, cell death in the pericarp, nucellus, and nucellar projection exhibits features of vacuolar PCD (van Doorn et al., 2011). Before the event, the differentiated cells form a large central vacuole, at which stage the nucleus already begins to condense. Cell disintegration becomes apparent with the rupture of the tonoplast. The nucleus can be still intact even in a condensed state. In certain cases, the tonoplast rupture coincides with the rupture of the nuclear envelope. The cytoplasm content then disintegrates, leading to cytoplasm clearance. Cell disintegration proceeds with the rupture of the nuclear envelope, releasing nuclear content into the cytoplasm. Finally, the cell membrane ruptures, releasing the digested cell content into the apoplastic space (Fig. 3b).

Despite considerable progress (Daneva et al., 2016), the mechanisms underlying the preparation, initiation, and execution of PCD in plants remain poorly understood. In animals, specific cysteine-dependent aspartate-directed proteases (caspases) are recognized as the main executors of PCD (Degterev et al., 2003). Caspase-like activities have also been detected in barley (Boren et al., 2006; Korthout et al., 2000). It has been speculated that caspase-like proteases are co-acting during plant PCD processes in a manner similar to that observed in animal cells (caspase cascade) (Degterev et al., 2003; Slee et al., 1999).

The coordinated caspase-like action during PCD in barley has been confirmed experimentally. Namely, different caspase-like activities were measured in separated tissues throughout whole grain development (Tran et al., 2014). A coordinated pattern of most analyzed caspase-like activities was evident in the pericarp and endosperm, and was correlated with PCD progression. Downregulation of the key (caspase-1-like activity) protease in transgenic grains leads to prompt downregulation of other caspases (Radchuk et al., 2018). Genes encoding caspases are absent in plant genomes (Vercammen et al., 2007); rather, other genes encoding proteases with equivalent features exist, for example, the vacuolar processing enzyme (VPE), also called legumain, with caspase-1-like activity (Hatsugai et al., 2015; Radchuk et al., 2017), and saspase and phytaspase with predominant caspase-6-like activity (Coffeen and Wolpert, 2004; Chichkova et al., 2010). Several proteases likely exhibit caspase-3-like activity, including the β1 subunit and possibly the β2 subunit of the 20*S* proteasome complex (Han et al., 2012), and cathepsin B (Ge et al., 2016).

The barley genome encodes eight *VPE* genes (Mascher et al., 2017; Julian et al., 2013) that are differentially expressed during development (Radchuk et al., 2011; Julian et al., 2013). The expression of *VPE1* was detected with increased abundance in the maturing endosperm (Radchuk et al., 2011), and coincides with caspase-1-like activity and nucleus fragmentation (Tran et al., 2014). Recombinant *VPE2a*, *VPE3*, and *VPE4* proteins exhibit legumain and caspase-1-like activities. Two of these (*VPE3* and *VPE4*) are redundantly expressed in the pericarp. *VPE4* expression is 50-fold higher than that of *VPE3* and its repression in transgenic barley plants decelerates pericarp disintegration (Radchuk et al., 2018). A group of vacuolar enzymes (*VPE2a*, *VPE2b*, and *VPE2d*) are specific for PCD in nucellar tissues. The nucellain gene, expressed exclusively in the nucellar tissues (Linnestad et al., 1998), also belongs to the *VPE2* subfamily and has been renamed as *VPE2a* (Radchuk et al., 2011). An increase in *VPE2a*–*VPE2d* gene expression coincides with elevation of caspase-1-like activity in the nucellus/nucellar projection (Tran et al., 2014). The *JEKYLL* gene, encoding a small cysteine-rich protein, regulates terminal cell differentiation in the nucellar tissues (Fig. 3a) (Radchuk et al., 2006, 2019).

3.3 Implications of cell death for grain development

PCD is an essential part of the developmental program. The mature barley grain mainly consists of dead tissues, with the exception of the tiny embryo and thin layer of aleurone. These tissues are sufficiently equipped to reutilize nutritional compounds from their environment. The most important function of PCD in barley grain physiology is (i) reshaping of the grain interior by targeted elimination of cells/tissues. Dying tissues, including the nucellus and pericarp, provide substantial room for the growing endosperm and embryo, required for establishing the new generation (Radchuk et al., 2011, 2018). Delayed cell elimination in the pericarp of transgenic *VPE4*-repressed grains causes a physical restraint on the growing endosperm: mature grains become smaller (Radchuk et al., 2018). (ii) Reutilization of cellular content accumulated in maternal tissues, required for nourishment of the low-sink organs. In particular, the early growth of the endosperm relies on nutrient supply from the degrading nucellus (Domínguez et al., 2001). (iii) The accumulation of storage compounds in the endosperm during the filling stage requires a significant assimilate supply from the maternal plant. We speculate that PCD contributes to the arrangement of nutrient supply to the main storage site in the endosperm. PCD occurs in the margins of the nucellar projection, facing the ETCs. *JEKYLL* expression in this region adds an additional control point in cell differentiation and the arrangement of the sucrose allocation route. Deceleration of PCD severely impairs starch accumulation in the endosperm and leads to the formation of irregular, small-sized grain (Radchuk et al., 2006). These results argue for the role of PCD in the maternal control of grain size in barley.

4 Sucrose allocation during the grain-filling stage

Sucrose is the major transport form for assimilates and is a source for growth and energy production in the growing grain. Sucrose metabolism is a gateway to diverse carbon use and sugar signaling (Ruan, 2014). It has been demonstrated that solute delivery to the seed occurs via the vascular system (phloem), but the postphloem routes of transport (within the seed) are likely distinct in various crops. Local (structural and environmental) conditions inside the grain can influence the partitioning of sucrose to starch, protein, and lipid. The investigation of sucrose allocation mechanisms was hampered for a long time due to the lack of appropriate technologies. In fact, barley is the only crop in which the sucrose allocation dynamics was visualized *in vivo* (Melkus et al., 2011). Here, we describe the pathway, sugar gradients, and key components of sucrose allocation within the grain.

4.1 Cellular sucrose allocation pathway within the grain

The vascular delivery route of nutrients ends in the pericarp. Therefore, the vascular bundle and entire transport route are hidden inside the seed, inaccessible for visual observation. Establishment of novel noninvasive technology for detection of sucrose in plants, based on high-field NMR, enabled functional imaging of sucrose allocation in the living seed for the first time (Melkus et al., 2011). The movement of ^{13}C-labeled sucrose was followed in real time and at submillimeter resolution. The findings demonstrated the main sucrose route inside the grain: the labeled sucrose first appears in the vascular bundle, clearly identifying the crease vein as its main entry point into the grain. The sucrose then moves via the nucellar projection toward the endosperm cavity, where it accumulates transiently. It enters the endosperm through the ETCs and gradually moves toward the periphery. Therefore, a sucrose concentration gradient forms, with high levels in the ETCs and low levels in peripheral endosperm regions. When the caryopses detach from the spike, the sucrose gradient gradually disappears. Expression of sucrose transporters (SUTs) was detected in different tissues (Weschke et al., 2000; Radchuk et al., 2017), but no sucrose influx into the endosperm was observed from either the dorsal pericarp or the nucellar epidermis during the main filling stage, only via the aforementioned route. In conclusion, the barley caryopsis utilizes a specialized sucrose delivery route during the grain-filling stage (Melkus et al., 2011); most, if not all, sucrose entering the endosperm during grain filling passes through this nucellar path (Fig. 4).

The impact of cellular nucellar projection arrangement on postphloem sucrose transport was demonstrated by comparing a series of transgenic barley lines. For example, *JEKYLL* RNAi repression caused aberrant cellular

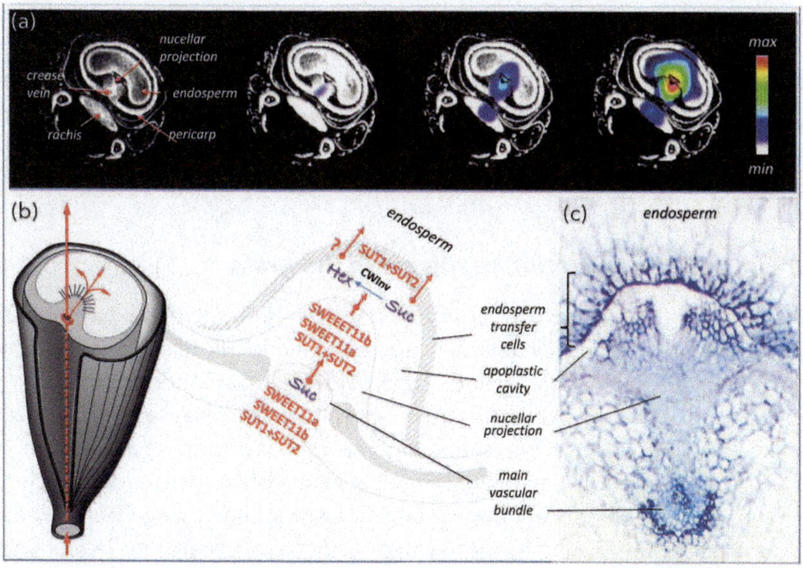

Figure 4 Sucrose allocation during onset of seed filling in barley. (a) spatiotemporal dynamic of the sucrose uptake as visualized by MRI; sucrose level is color-coded; for details see Melkus et al. (2011). (b) vascular bundle (thick red dot) and postphloem route (thin red arrows) of solute transfer into the endosperm; schematic representation of main sugar transporters along the allocation route; (c) localization of SUT1 expression by *in situ* hybridization, tissue specific expression is detectable (blue signal). Abbreviations: Suc-sucrose; Hex-hexoses; ?-unknown transporters; CWInv-cell wall-bound invertase.

differentiation of the nucellar projection (Radchuk et al., 2006). Caryopses with altered nucellar tissue structure displayed decreased sucrose flow toward the endosperm. In the caryopses of lines with 70% downregulation of *JEKYLL*, the flow rate dropped to 40% of the wild-type (WT) level. In caryopses where *JEKYLL* was downregulated by 80%, the flow was close to zero. In all transgenic lines, the endosperm remained underdeveloped and mature seeds were smaller (Melkus et al., 2011). Thus, sucrose release toward the endosperm via the nucellar projection provides an essential means for the maternal control of seed growth.

The structural and metabolic features of the endosperm influence sucrose allocation in the barley grain significantly. The role of ETCs in sucrose uptake was demonstrated by investigation of the barley mutant *seg8*, which has an impaired ability to form ETCs (Ramage and Crandall, 1981; Sreenivasulu et al., 2010). The ^{13}C-sucrose concentration gradient along the primary transport route was increased in the mutant grains, implying that the aberrant ETCs induce a higher flow resistance in the central endosperm, compared with the WT. The mature grain acquired a 'shrunken' phenotype. The importance of metabolic demand of the endosperm (termed 'sink activity') for sucrose allocation was verified by investigation of the barley mutant *Risø13*. Being morphologically normal, this mutant displays impaired starch synthesis, due to aberrant delivery of ADP-glucose from the cytosol to the plastid (Patron et al., 2004). Comparative analysis of sucrose allocation demonstrated an identical pattern of sucrose distribution in the mutant and the WT, but the flow rate driven by the decreased metabolic demand of the mutant endosperm was ~33% lower than that in the WT (Melkus et al., 2011).

4.2 Sucrose concentration gradients in the grain

Marked gradients in sucrose distribution across the caryopsis have been documented by recently established high-resolution FTIR microspectroscopy (Guendel et al., 2019). This technology provides a reliable tool for mapping sucrose concentrations along the grain postphloem pathway, at cellular-level resolution. FTIR maps confirmed that high sucrose levels are characteristic in tissues within the ventral part of the caryopsis, where phloem unloading occurs. In the pericarp, high concentrations were evident in the vascular bundle and surrounding tissue; levels decreased two- to fivefold toward the lateral and dorsal regions. High concentrations were also evident in the endosperm facing the vascular region of the pericarp, but the peak sucrose concentration was observed in the ETCs.

Within the endosperm, a steep sucrose gradient was formed between the ETCs and the central endosperm, with minimum levels at the periphery.

The FTIR images further revealed a slight drop in concentration from the vascular bundle toward the nucellar projection, while in the ETCs the sucrose level increased again (Fig. 5). This pattern is indicative of (membrane) barriers and/or transport events along the pathway, namely, passive unloading at the vein, but active loading (against the concentration gradient) at the transfer cells of nucellar projection and ETCs. From the ETCs into the storage tissue, most sucrose moves passively, resulting in the formation of a steep negative gradient toward the peripheral cell and aleurone layers.

4.3 Invoking of membrane transport events and active transporters

Allocation of sucrose between symplastically isolated grain organs requires membrane transport. Transporters are integral transmembrane proteins, facilitating diffusion or active transport across membranes. Both uptake (import) and efflux (export) transporters are important (Fig. 4b and c). Sucrose import into a sink cell relies primarily on plasma membrane–localized SUTs. Five *SUT* genes were identified in barley (Radchuk et al., 2017), based on full genome sequencing (Mascher et al., 2017). Of these, only *SUT1* and *SUT2* are highly expressed during grain development, indicating their importance for sugar allocation within barley grain (Radchuk et al., 2017). Whereas HvSUT1

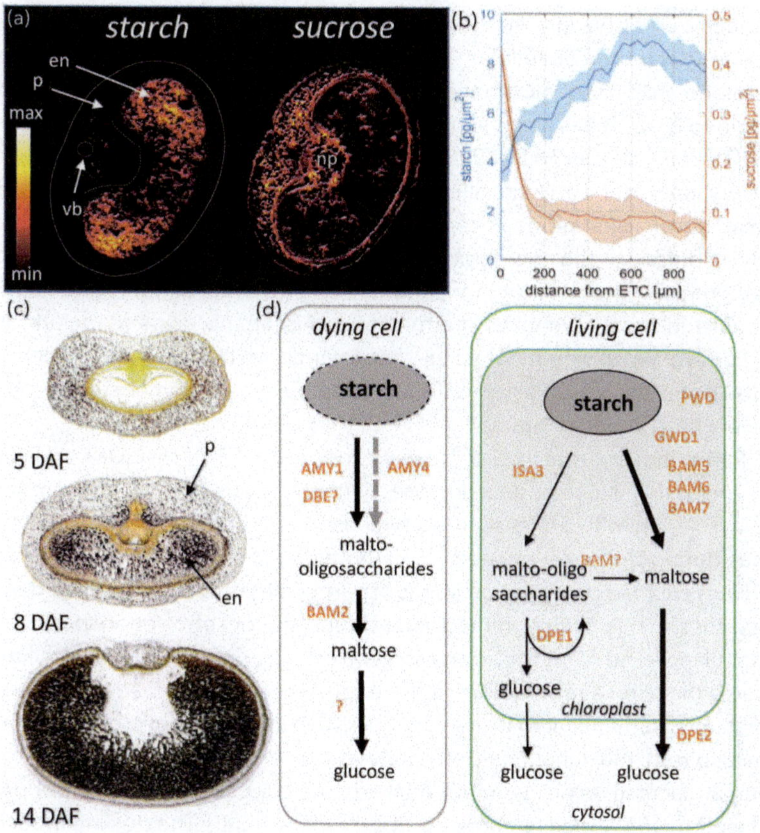

Figure 5 Sucrose gradients and starch dynamic in the barley grain. (a) gradients in sucrose and starch measured in cryo-sections through caryopsis at filling stage; (b) starch and sucrose gradients along a transect from endospermal transfer cells (ETC) toward endosperm periphery; data are given as mean and standard error (shaded area); (c) starch distribution pattern in developing caryopsis indicating shifts in allocation pattern; starch granules are visualized in black in median transverse sections of grain. (d) the proposed pathways of starch breakdown in distinct cell types of the pericarp. In cells undergoing programed cell death (left panel), starch degradation possibly occurs via AMY1 and AMY4 α-amylases. Linear malto-oligosaccharides released by the action of α-amylases and an unidentified starch debranching enzyme (DBE), providing substrates for the BAM2 β-amylase. Question mark indicates an unidentified enzyme converting maltose into glucose. In living pericarp cells (right panel), glucan-water dikinase (GWD1) and phosphoglucan, water dikinase (PWD) phosphorylate the surface of the starch granule, making it accessible for β-amylase action. BAM5, BAM6, and/or BAM7 β-amylases are likely involved in maltose production, acting either at the granule surface or on the linear malto-oligosaccharides. The action of isoamylase (ISA3) on the granule releases soluble malto-oligosaccharides. Short oligosaccharides can be metabolized by plastidial disproportionating enzyme (DPE1), liberating glucose, and larger malto-oligosaccharides for continued degradation. After transport into cytosol, maltose is converted into glucose by disproportionating enzyme (DPE2). Abbreviations: en, endosperm; np, nucellar projection; nu, nucellus; p, pericarp; Scalebar: 1mm.

was localized at plasma membranes, HvSUT2 was detected at the tonoplast. Despite different subcellular localizations, these genes are coordinately expressed in tissues along the assimilate transfer path from the vascular bundle to ETCs. It has been suggested that co-expressing plasma membrane *HvSUT1* and vacuolar *HvSUT2* controls the sucrose balance along the sucrose delivery path during grain filling. Downregulation of *HvSUT1* and *HvSUT2* caused severe alterations in carbohydrate metabolism, glycolysis-associated gene expression, the tricarboxylic acid (TCA) cycle, starch and amino acid (AA) synthesis, grain maturation, and abscisic acid signaling, together with a decrease in most glycolytic intermediates and amino acids (Radchuk et al., 2017). Overall, this highlights the importance of both the cell membrane *HvSUT1* and the tonoplast *HvSUT2* to sustain sucrose homeostasis along its delivery path in barley grains.

Rapid induction of *HvSUT1* expression in the ETCs (6–7 DAP) coincides with increasing sucrose and sucrose synthase levels (mRNA and enzymatic activity). Along with SUTs, sucrose synthase is one of the highest expressed genes during the onset of seed filling (Sreenivasulu et al., 2006). This enzyme mediates sucrose degradation in endosperm cells as a first step toward starch biosynthesis. This occurs immediately during the onset of endosperm starch accumulation and manifests in an elevation of sink activity. As starch is the major product of sucrose metabolism in the endosperm, the comparison of sucrose and starch gradients across the caryopsis (based on FTIR microspectrometry) is valuable in the functional interpretation of metabolite gradients (Fig. 5). Whereas sucrose levels generally decline from the ETCs toward the peripheral endosperm, the opposite pattern is observed for starch. The decline in sucrose and increase in starch is more pronounced toward the side lobes of the endosperm.

A unique family of proteins named Sugars Will Eventually be Exported Transporters (SWEET) was shown to facilitate cellular sugar efflux. The SWEET family consists of distinct transporter proteins capable of transferring diverse mono- and disaccharides (Chen et al., 2015), and even phytohormones (Kanno et al., 2016). Clade III of SWEETs is able to transport sucrose (Chen et al., 2015). Based on the full barley genome sequence, the family of transmembrane proteins mediating sugar efflux consists of 23 members (Mascher et al., 2017). An extension of the potentially sugar-transporting SWEET11, SWEET13, SWEET14, and SWEET15 subfamilies was observed in barley, with two or more genes for each subgroup compared with only a single orthologue in rice and Arabidopsis. SWEET11a and SWEET11b are highly expressed in maternal seed tissue, but differ in the distribution of expression domains (Mascher et al., 2017). Their transcripts were located in the nucellar projection, but in different cell layers. SWEET11a transcripts were detected in the main vascular bundle, and SWEET11b around it (Mascher et al., 2017). Divergent SWEET11a and SWEET11b expression patterns indicate neofunctionalization following gene duplication events. These proteins were identified as key candidates for sucrose release in the main vascular bundle and nucellar projection of the developing barley grain.

Current knowledge on the arrangement of SWEETs within the crease suggests a complex state in which these proteins are integrated within a network of transporters, enzymes, and pathways. Unraveling the contribution of SWEETs to the grain-filling process will require further experimentation, including the generation of appropriate genetic models (transgenic lines). Further studies will uncover the main features of SWEETs in barley, how the molecular and metabolic network of the barley grain responds to SWEET gene manipulation, and whether SWEETs could be relevant for barley improvement. This is required to achieve a mechanistic view on grain filling in the barley grain.

Different sucrose pathways are hypothesized for other crops: the specialized pedicel region and the basal endosperm transfer region were proposed as serving a major role in nutrient transfer in grains of sorghum (*Sorghum bicolor*) and maize (*Zea mays*) (Wang et al., 2012; Sosso et al., 2015). In rice (*Oryza sativa*), at least two pathways could be involved in the transport of nutrients toward the endosperm; one is similar to that in barley (nucellar projection-ETC pathway), and the other via the nucellar epidermis (Oparka and Gates, 1984). The main sucrose allocation route in numerous species from diverse Poaceae tribes was recently analyzed in relation to JEKYLL expression governing the establishment of the nuclear projection. The study suggests that the emergence of JEKYLL during evolution spurred innovations in the structure and physiology of the Triticeae grains, specifically by evolving a maternal/filial tissue conduit for transferring assimilates into the endosperm via the nucellar projection and ETCs (Radchuk et al., 2019).

5 The use of starch in the developing caryopsis

Starch represents the main form of carbon reserve in the endosperm. It has long been assumed that the capacity for starch synthesis influences final grain weight, drawing our attention to processes underlying starch storage in the endosperm. However, recent studies on various crops have revealed that starch in the endosperm can accumulate only to a physical limit, which is set much earlier during the development of maternal seed parts (Fahy et al., 2018; Radchuk et al., 2018). We focus here on the factors determining starch accumulation in the grain, and the orchestration of transient starch storage during development, which is of high relevance in grain physiology.

5.1 Physiological role of transient starch

In the barley grain, starch is synthesized in photosynthetic and non-photosynthetic plastids, and deposited in the form of differently sized starch granules. Starch is composed of two types of glucan polymers: unbranched amylose and highly branched amylopectin. At least four classes of enzyme activities have been identified as necessary for starch synthesis. ADP-glucose pyrophosphorylase (AGPase) synthesizes sugar-nucleotide precursor ADP-glucose. Granule-bound starch synthase (GBSS) is responsible for the synthesis of amylose and cannot be replaced by other synthases in this function (Patron et al., 2002). Different isoforms of soluble starch synthase (SS) are involved in specific steps of amylopectin synthesis (Morell et al., 2003; Li et al., 2011; Cuesta-Seijo et al., 2013). Starch-branching enzyme (SBE) and two classes of debranching enzyme isoamylase (ISA) and limit dextrinase (LDA) participate in shaping of final amylopectin structure (Burton et al., 2002; Regina et al., 2010). A concerted action of SS isoforms, branching and debranching enzymes, determines specificity and degree of amylopectin branching. The functional AGPase is a heterotetrameric enzyme made up of two large (AGP-L) and two small (AGP-S) subunits, each encoded by distinct genes (Emes et al., 2003). Two distinct AGP-S subunits exist, one is localized in the plastids and is the major subunit in leaves (Rösti et al., 2006), whereas the other is cytosolic and is a major contributor to starch synthesis exclusively in the endosperm (Johnson et al., 2003). In the barley pericarp, the same isoforms of enzymes as in leaves are expressed (*AGP-S1b, AGP-L2, SBE2a, GBSS1b*, and α-glucan-phosphorylase *PHO1*) (Radchuk et al., 2009). Expression of the respective genes coincides with starch accumulation in the (photosynthetically active) pericarp. Instant fixation of sugars as starch in the young pericarp is likely responsible for the establishment of an assimilate sink at early developmental stages (before endosperm differentiation). With the beginning of endosperm cellularization, the role of the pericarp as a sink tissue diminishes, and transient starch is progressively remobilized (Fig. 5c and d). Based on expression of genes encoding starch-degrading enzymes, two pathways/functions for starch degradation in the pericarp have been proposed. One supports cellular disintegration of tissues by PCD (Radchuk et al., 2018), and the other is associated with degradation of photosynthetically produced starch generated in the pericarp chlorenchyma layer (Radchuk et al., 2009).

In the barley endosperm, the enzymatic machinery required for starch synthesis is markedly more complex than that in the pericarp (Radchuk et al., 2009). Most AGPase activity (70–90%) here is cytosolic (Thorbjørnsen et al., 1996). However, cytosolic *AGP-S1a* and plastidial *AGP-S2* are simultaneously expressed in the endosperm during the main storage stage. Expression of these enzymes is characteristic of the most intense period of starch deposition, when the amount increases to ~70% of total grain weight. Currently, no evidence exists supporting significant starch turnover in the starchy endosperm during the filling period (Tomlinson and Denyer, 2003). Therefore, the detection of high β-amylase *BAM1* and *BAM2* expression together with high β-amylase activity in the endosperm starting from middle development was unexpected. Some starch breakdown occurs in ETCs and aleurone (Radchuk et al., 2009) and is required for functional differentiation of ETCs, allowing more efficient allocation of sugars to the endosperm. High β-amylase activity was detected in the endosperm fraction during grain maturation, clearly resulting from *BAM1* and *BAM2* expression. Similarly, *ISA1* and *LDA* transcript abundancies are increased in maturing grains (Radchuk et al., 2009). LDA and β-amylase enzymatic activities together with α-amylase activity play an essential role in starch degradation during grain germination and, therefore, are of primary importance for the malting process (Evans et al., 2009). The enzymes might be pre-synthesized during maturation, so that they can become rapidly involved in starch remobilization upon seed imbibition.

To summarize, starch accumulation in the barley grain is a dynamic process, including starch degradation as an important element of grain development. In the pericarp, starch synthesis and degradation share functional similarities with processes in the leaf, and involve similar sets of genes. Transient starch stores in the pericarp are likely to ensure the maintenance of sink strength in the young caryopsis, and fulfil a nourishment function for growing endosperm and embryo during later development. In the endosperm, starch storage is persistent, although some transient starch accumulation occurs in certain tissues, such as ETCs and aleurone.

5.2 Variability in starch storage and quality in mature grain

Great diversity exists in the chemical composition of starch, depending on the content of amylose, lipid, protein, ash, and phosphorus-containing compounds (Zhu, 2017). Amylose-to-amylopectin ratio is important for appearance, structure, and quality of food products and processing. The high amylose content determines the starch gelling and firmness, improves the product texture of starch, and turns it into a source of slowly digestible carbohydrate. The amylopectin is primarily responsible for the formation of crystalline granules, thickening of paste, and strong resistance to retrogradation (Wang et al., 2017). Amylopectin accounts for 75–90% of WT starches. Its fraction increases during grain development (9–24 DAP) (Källman et al., 2015), and is dependent on growth conditions (Gous et al., 2015). The amylose content can be effectively increased by either enhancing GBSS1 expression or eliminating SBEs, SSIIa, or other enzymes involved in amylopectin synthesis (Wang et al.,

2017). The first approach has only limited applications due to the limited non-reducing ends in amylose and the substrate competition between amylose and amylopectin. The second approach was proven as more promising. So, barley *sex6* mutant grains lacking starch synthase IIa activity accumulate about 50% amylose (Morell et al., 2003). While an almost complete SBEIIb inhibition causes only a minor change in amylose content (Regina et al., 2010), the amylose-only starch granules were obtained in barley grains after concerted suppression of all *SBE* genes (Carciofi et al., 2012). An increase of starch amylose is observed in the *amo1* mutant (Källman et al., 2015), and can reach up to 95% of total starch by suppressing SBEs ('amylose-only' genotype; Shaik et al., 2016). Amylopectin levels depend on the number and activity of enzymes involved (SS, SBEs, DBEs; Nakamura, 2015). The impact of genetic mutations on the amylopectin structure appears to be pleiotropic, and could depend on less well-studied protein-protein interactions, as well as phosphorylation processes in the granules during biosynthesis (Ahmed et al., 2015; Tetlow et al., 2015).

The large starch granules tend to have a higher amylose content than the small granules (Källman et al., 2015; Yangcheng et al., 2016). As the content of amylose and phosphate groups in starch increases during development (Borén et al., 2008), the structural features of starch also change. Phosphate groups in starch are important for its crystallinity (Carciofi et al., 2011), whereas a high amylose content make it amorphous (Waduge et al., 2006). Lipids form inclusion complexes with amylose, maintaining the granule integrity (Srichuwong and Jane, 2007). The size of starch granules further depends on the cultivar, and ranges between 1 μm and 24 μm (Jaiswal et al., 2014). Manipulation of biosynthetic enzymes substantially altered the size and shape of starch granules in the endosperm. SS mutation led to the formation of large granules (Sparla et al., 2014), whereas suppression of SBEs resulted in irregular and multilobular grains (Goldstein et al., 2016; Shaik et al., 2016). Environmental factors also played a role; for example, bigger grains were formed at higher temperature by the same cultivars (Yangcheng et al., 2016).

Factors such as granule size, amylose content, and amylopectin profile generally affect starch enzyme susceptibility (Asare et al., 2011; Ahmed et al., 2016). In the last decade, increased interest has been observed in using resistant starch (RS) for food products with health effects against diseases such as diabetes and obesity. Barley genotypes with 1–12% RS contents (Ahmed et al., 2016) have been selected to formulate such 'healthy' products. Notably, certain combinations of environmental factors can lead to the formation of amylose-lipid inclusion complexes in starch granules and thus provide grains with a decreased enzyme susceptibility (Stevnebø et al., 2006).

6 Proteins and barley grain quality

Barley grain contains relatively small concentrations of protein. Nevertheless, grain development, growth, and germination rely on protein functions, so eventually the protein present in the grain largely determines both grain quality and crop productivity (Witzel et al., 2010). Protein distribution within the different grain tissues corresponds to the functional role and physicochemical features of the tissues. In the next section, we discuss some aspects of protein accumulation in barley, which can be relevant for the grain's end-use value.

6.1 Specific properties and deposition

Mature barley grains consist of 7–20% protein on a dry-weight basis (Evers et al., 1999). The classical concept for protein classification, which was developed more than 100 years ago and efficiently used until today (Gubatz and Shewry, 2011), is based on protein extraction with various solvents. In addition, grain proteins can be distinguished based on their functions as storage, structural, metabolic, or protective proteins (Shewry and Halford, 2002). The main storage proteins are ethanol-soluble prolamins (hordeins), which account for 35–50% percent of the grain's total protein. They are diverse in structure and electrophoretic mobility, as detailed earlier (Finnie and Svensson, 2014). The albumins and globulins are water- and salt-soluble proteins, and they represent enzymes, metabolic regulators, and so forth. The structural proteins, so-called glutelins, can be extracted using detergents, acidic solutions, or alkali solutions. The various classes of seed storage proteins (SSP) make distinct contributions to the major grain compartments: storage globulins accumulate in both embryonic and aleuronic layers of endosperm, while hordeins accumulate in the starchy endosperm. The protein content, composition, and distribution varies depending on the environment, the grain's developmental state, and other factors (Finnie et al., 2002; Østergaard et al., 2002; Witzel et al., 2007, 2010; Bønsager et al., 2007; Cho and Rhee, 2009). For mostly unknown reasons, hordein accumulation in endosperm occurs along some tissue gradients. Thus, it is noteworthy that the spatial expression of hordein-encoding genes (most are present as multiple copies in the barley genome) does not necessarily reflect the spatial gradient in protein deposition. The syntheses of SSP are regulated at the transcriptional level by synergetic interactions between TFs and SSP gene promoters (Rubio-Somoza et al., 2006a,b; Yamamoto et al., 2006; Xi and Zheng, 2011). While most transcriptional regulators in cereal crops are activators, only a few repressors have been reported thus far (Boudet et al., 2019).

The relevance of the various protein fractions differs depending on the grain applications. In the brewing industry, hordein and glutelin are considered undesirable compounds, whereas a high protein level is generally desired if grain is used for feeding. The detailed compositional analyses via comparative proteomic profiling are important in both cases. For example, the influence of specific types of barley protein on malt and beer quality is complex and not fully understood (Fox, 2010). Mass spectrometry-based protein profiling is used to assess how genetic and environmental factors modify the proteins in grain. The study of Luo et al. (2019) shows that genetic factors were dominant for protein variation, although the environment also affected the protein composition. Matrix-assisted laser desorption/ionization time-of-flight mass spectrometry distinctly identified albumin, hordein, glutelin, and globulin as the most influenced proteins during the malting process in barley (Dai et al., 2014). When considering the nutritional quality of barley, the albumin-globulin fraction contains the highest lysine content (about 5–7%) compared to the storage protein fraction; thus, it is more nutritionally balanced. Slightly higher lysine content was found in hulled versus hull-less cultivars (Newman and Newman, 2008). In the context of a balanced diet, barley has considerable potential, but it is infrequently used (directly) for human nutrition. Given the high global rate of diet-related and cardiovascular diseases and the beneficial role of barley constituents (Topping and Morell, 2014), barley will potentially change from a solely low-value commodity to a high-value food ingredient.

The protein content and composition of barley represents characteristics featured in grain. Therefore, protein profiling is efficient for precise characterization of cultivars, identification of varieties, scoring specific features in population mapping, and providing a basis for selecting desired genotypes (Scobie and Jones, 2009; Mahalingam, 2017; Mock et al., 2018).

6.2 Regulation and manipulation of protein storage in barley

The final concentration of protein in grain depends on the availability of carbon and nitrogen compounds, whereas environmental factors influence the rate and duration of protein accumulation. In nature, variations in grain protein concentration induced by weather, water, and nitrogen availability, especially during the grain-filling period, could be larger than variations due to genotype, as was earlier demonstrated in wheat (Cooper et al., 2001).

AA serve as exclusive precursors for protein synthesis, and increased availability of AA is known to facilitate protein synthesis (Chope et al., 2014; Yu et al., 2017). This relationship essentially underlies the well-known promoting effect of N-fertilization on grain's protein content. It is therefore relevant to ask how AA are delivered to the developing cereal grain and metabolized therein. The allocation of AA has not yet been visually demonstrated, but collective evidence suggests that free AA are unloaded from the central vein, transported along the nucellar projection, and eventually loaded into the endosperm via a set of transporters localized in the transfer cells (Offler et al., 2003; Thiel et al., 2012). From there, diffusive (passive) allocation within the endosperm tissues is the most likely mechanism. Intense metabolic conversions can occur along the allocation route, as evidenced by transcriptome and biochemical data (Thiel et al., 2009; Rolletschek et al., 2011). The level of free AA in grain is high at the early developmental stages, but it strongly declines later on (Rolletschek et al., 2004; Mangelsen et al., 2010), probably reflecting the high activity of AA-demanding protein synthesis. There is evidence indicating that it is the availability of charged AAs (AA-tRNAs) that regulates the rate of protein synthesis, rather than the AA content itself (Fluitt et al., 2007; Gingold and Pilpel, 2011). The charging process is mediated by aminoacyl tRNA synthetases (ARS), which ligate the AA to their cognate tRNAs (Ibba and Söll, 2000). In turn, the transcription of ARS is closely regulated by the AA supply (Shan et al., 2016). Notably, the amino acid metabolism in developing grains is very intense; this also involves the incorporation of sugars as a carbon backbone into the AA molecule (Rolletschek et al., 2011). In this way, sucrose delivery can have a direct effect on AA metabolism and protein storage.

There are several ways to manipulate the content and composition of SSP in the mature cereal grain. The most obvious method involves enhancing or decreasing the expression of SSP-encoding genes (Lange et al., 2007; Hansen et al., 2007; Scossa et al., 2008; Pierucci et al., 2009). Substantial improvements in the nutritional quality of barley protein were reached by antisense technology, and high-lysine–containing plants also showed uncompromised yield and grain quality (Izydorczyk and Edney, 2017). However, such attempts have not been successful in changing the total protein content of the grain, as distinct storage proteins were compensated by some other SSP. The mechanisms underlying such protein re-balancing are still unknown, but they appear to be a widespread phenomenon. This phenomenon was also observed in non-cereal crops. Nonetheless, such approaches might be valuable when attempting to reduce the fractions of proteins involved in celiac disease. Another means to stimulate protein deposition and/or composition in the cereal grain relies on the modified expression of genes encoding either transporters for AA uptake and/or enzymes active in the AA metabolism representation. Such approaches can be effective (Hu et al., 2018), but they require careful target selection based on genomic and biochemical data.

7 Particularities of energy metabolism in barley grain

Growth and generation of storage compounds in grain requires a considerable amount of energy, produced by mitochondrial respiration and glycolysis, as well as by photosynthesis in green plastids of pericarp. Here, we explain some specifics of energy metabolism of the grain that can be viewed as adaptation to grain structure and endogenous O_2 deprivation.

7.1 Main energy metabolism components and pathway

Photosynthesis uses light energy (and CO_2) to drive carbohydrate synthesis, and releases O_2 in the chloroplasts. Respiration then uses these carbohydrates to produce reducing equivalents and ATP in the mitochondria, supporting growth and cellular maintenance. This process releases CO_2 and consumes O_2. Respiratory carbon oxidation pathways, electron transport, and ATP turnover are tightly coupled processes. Alternatively, so-called alternative oxidases (AOX) could drive the non-energy conserving route in the mitochondrial electron transport chain. Utilizing the AOX pathway may provide some degree of metabolic homeostasis in carbon and energy metabolism (Vanlerberghe, 2013). Glycolysis, the oxidative pentose phosphate pathway, and the mitochondrial TCA cycle are central pathways of energy metabolism and supply carbon intermediates for biosynthesis, as well as coupling carbon oxidation with the reduction of NAD(P) to NAD(P) H. The reducing equivalents are used to support biosynthetic reactions, and are oxidized by the electron transport chain, localized within the inner mitochondrial membrane (Millar et al., 2011).

The barley caryopsis comprises of various tissues that execute distinct types of metabolism. The pericarp includes a number of chlorenchyma layers, whose cells contain chloroplasts. This tissue is highly photosynthetic and is autotrophic, whereas other tissues (e.g. endosperm and embryo) are dependent on carbohydrate import, and are heterotrophic. The activity ratio of photo- and heterotrophic modes changes during development, and leads to metabolic shifts during tissue differentiation. In order to describe the mixed character of whole grain metabolism, the term 'photoheterotrophic metabolism' was established. This describes the grain's demand for organic compounds (delivery of carbohydrates and AA from the maternal plant) to satisfy the main carbon/nitrogen requirements, combined with the capability to utilize light for energy metabolism.

7.2 Barley grain photosynthesis and respiration

Early-stage barley caryopses are green and photosynthetically active. The photosynthetic cell layer is found only in the pericarp of grain and almost completely surrounds the endosperm (Fig. 6a). In barley, photosynthesis is initiated 4–8 DAF, corresponding to an intermediate growth phase. Changes in chlorophyll content do not affect its distribution pattern during development,

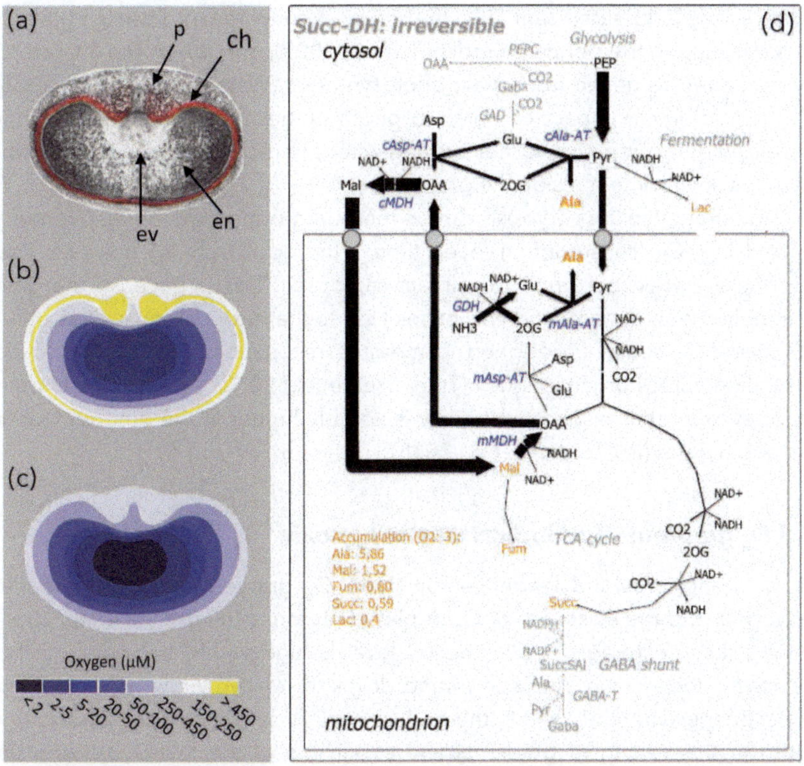

Figure 6 Spatial distribution of oxygen and key fluxes within the primary metabolism of the central (hypoxic) endosperm. (a) cross-section shows structure of caryopsis at grain-filling stage (14 DAP); (b,c) modeled oxygen gradients in measured under lit (b) and non-lit conditions (c); color indicates oxygen concentration; (d) flux distribution map of the central (hypoxic) endosperm region; thickness of each arrow corresponds to the flux value; for details see Rolletschek et al. (2011). Abbreviations: ch-chlorenchyma, en-endosperm, ev-endospermal vacuole, p-pericarp.

but are reflected in changing photosynthetic activity. Chlorenchyma cell plastids are similar to those of leaves and exhibit high-electron transport rates (Tschiersch et al., 2011). Similar to leaves, photosynthetic rates (O_2 release inside the grain) depend on light availability and quality. During the storage phase, when photosynthetic activity is at its peak, the O_2 level within the chlorophyll strands of the pericarp can increase to over twofold of atmospheric levels (Fig. 6b), reflecting a high local O_2 release rate (Rolletschek et al., 2004). This finding is consistent with the known phenomenon of O_2 release from illuminated intact caryopses, reported over 40 years ago (Nutbeam and Duffus, 1978). O_2 produced in the pericarp diffuses not only toward the seed surface, but also inward toward the endosperm. Despite high local photosynthetic rates, suggesting significant CO_2 fixation and concomitant (transient) starch accumulation in the pericarp (Radchuk et al., 2009), only 2% of the final starch was reported to derive from grain photosynthesis (Watson and Duffus, 1988). The physiological impact of pericarp photosynthesis certainly varies during development, but it should not be neglected, particularly the effects on the gaseous environment inside the grain.

Respiration in the caryopsis can be measured using classical respirometry, quantifying O_2 consumption over time. This approach is, however, less informative when performed on intact caryopses. This is because the grain is a multi-organ system, and comprises tissues that produce as well as utilize O_2. Planar O_2 sensors, designed to measure respiration, can provide spatial resolution across organs/tissues (Tschiersch et al., 2012). An O_2 consumption map revealed that respiratory activity is notably higher in the embryo than in the endosperm (Rolletschek et al., 2015).

7.3 O_2 gradients and indications of hypoxia

O_2 mapping provided evidence for steep O_2 gradients inside the barley caryopsis, and the existence of O_2-depleted regions (Rolletschek et al., 2004). During the prestorage phase, the O_2 level is high within the pericarp, but decreases toward the endosperm and drops to relatively low levels (~1% of atmospheric saturation) within the endosperm. Onset of storage is associated with the sharpening of gradients and expansion of the low-O_2 zone in the central region of the endosperm (Fig. 6b and c). In later development, the lowest O_2 concentrations are also in the central endosperm (<0.3 μM), and O_2 shortage is most severe.

Although the embryo exhibits higher respiratory activity than the endosperm, it is less affected by O_2 shortage, possibly due to different gas exchange features. Certain structural features of the seed influence the gas exchange capability, either preventing (lipid layers) or facilitating (pores, intercellular air spaces) it. The outermost seed layers (dense epidermal or compressed cells of pericarp and integuments) effectively form a sealed space separated from the external environment (Cochrane and Duffus, 1979), which is regarded as a barrier to gas exchange. Distinct lipid-containing or suberized cell layers are semipermeable to particular gases, with CO_2 diffusing more readily than O_2. The biochemical composition of 'barriers' is more relevant than tissue thickness for determining gas exchange. Integuments attached to the endosperm allow O_2 diffusion, as derived from O_2 concentration gradients, but accumulation of lipids in aleurone could negatively affect this diffusion. Therefore, 'barriers' cause sharp gradients and contribute to an increase in O_2 depletion in the central part of the caryopsis, where main storage processes occur. The limitation of respiration due to lack of O_2 availability is called hypoxia, which defines a physiological local or generalized condition of tissues. The limitation of respiratory activity is apparent at low µM concentrations found typically in the central endosperm; thus, metabolism of endosperm underlies hypoxic constraints.

7.4 Mechanisms of O_2 sensing and balancing in seeds

Despite low O_2 levels in the grain interior, the levels of fermentation products such as lactate or ethanol remain low. Fermentation and anoxia are avoided, and seeds are able to adjust their metabolism/O_2 demand to the O_2 available. This requires mechanisms for O_2 sensing and balancing. In recent years, important findings have been reported regarding the sensing and management of hypoxia and anoxia (complete O_2 lack). O_2-sensing mechanisms, similar to those in mammalian cells, have been identified (van Dongen and Licausi, 2015; Schmidt et al., 2018a). The low-O_2 signal is transmitted via specific transduction pathways, resulting in characteristic transcriptional responses (Mustroph et al., 2010). Upon hypoxic stress, plants initiate a general set of adaptive responses to circumvent an energy crisis (Schmidt et al., 2018b). Among others, the small gaseous molecule nitric oxide (NO) has been proposed to participate in the O_2 balancing process (Borisjuk et al., 2007). NO is O_2-sensitive and acts to inhibit respiration by binding reversibly to cytochrome oxidase (Yamasaki et al., 2001; Zottini et al., 2002). An increase in endogenous O_2 led to a decline in endogenous NO levels, whereas a decrease notably enhanced the NO levels. We consider NO as an important regulator of O_2 homeostasis in plants (as in mammals; Schumacker, 2003), mediating rapid and reversible O_2 balancing in hypoxic tissues such as the developing barley grain. The dynamics and rapid nature of this balancing is of particular significance, as it allows fast adaptation without the need for changes in the gene expression programs.

Hemoglobins (also known as phytoglobins; Dordas et al., 2003) oxidize NADH, a process that helps maintain the energy and redox homeostasis under hypoxia. The cycling of NO back to nitrate via non-symbiotic class-1 hemoglobins has been proposed to represent an alternative respiratory pathway, possibly explaining the beneficial effects of these hemoglobins under hypoxia (Igamberdiev et al., 2005). Barley hemoglobin Hb1 was the first non-symbiotic plant hemoglobin isolated and characterized from a monocotyledonous plant. It is expressed in grain, but also in root tissues under hypoxia (Taylor et al., 1994). Hemoglobin could act in O_2 balancing by metabolizing (scavenging) NO, and via its effects on reactive oxygen species and hormonal balance. The latter two aspects are currently being investigated, particularly the link between hypoxia, hemoglobins, NO, and PCD (Igamberdiev et al., 2014; Kapoor et al., 2018).

7.5 Management of hypoxia at structural and metabolic levels

Understanding how hypoxia is managed *in vivo* required the consideration of structural tissue features and local O_2 availability. For example, the biosynthesis of storage proteins and lipids is an energetically demanding process compared with starch synthesis (Schwender, 2008). In the barley caryopsis, accumulation of lipid and protein preferentially occurs in the aleurone and embryo (Fig. 2e). Both tissues are peripheral and O_2 is not depleted or absent. *In vivo* analysis of sucrose delivery using MRI (Fig. 4) revealed that the main sugar transport occurs via the well-oxygenated crease vein and nucellar projection into the central endosperm (Melkus et al., 2011), where hypoxia exists. At the same site, enzymatic activity of alanine amino transferase (Ala-AT) is increased, producing alanine (Fig. 6d). It thereby avoids the accumulation of pyruvate and fermentation products, and provides a source of ATP (Rolletschek et al., 2011). The cells in the endosperm are interconnected by multiple plasmodesmata, allowing for an energy-efficient movement of alanine along a concentration gradient toward the more peripheral cells. Since the peripheral cells are not in a hypoxic state, Ala-AT acts to convert alanine back to pyruvate. The reutilization of the ammonia (released by the deteriorating adjacent nucellar tissue; Radchuk et al., 2006) could positively contribute to the efficiency of nitrogen use. Although possibly coincidental, the long-known promoting effect of hypoxia on cell expansion/elongation (Armstrong et al., 1994; Bailey-Serres and Voesenek, 2008) is reflected in the characteristic pattern of the endosperm cell size, where the central hypoxic part of the endosperm is composed of large cells, whereas those at the well-oxygenated periphery are small. This is also observed in developing maize kernels. Region-specific metabolism in the cereal endosperm has been further characterized using a combination of metabolic modeling and biochemical studies (Rolletschek et al., 2011). The data suggest that in the central endosperm, the TCA cycle shifts to a non-cyclic mode (Fig. 6d and further below), elevating glycolytic flux without inducing the γ-aminobutyric acid (GABA) shunt. Transcriptome and proteome analyses of maize endosperm demonstrated that glycolytic enzymes are generally present at markedly higher levels than TCA cycle enzymes (Prioul et al., 2008; Walley et al., 2013). A transcriptome study of the central and peripheral maize endosperm found that genes involved in the glycolysis and TCA cycle are upregulated in the periphery, whereas genes involved in the alanine and fermentative metabolism are upregulated in the center (Gayral et al., 2017). This altogether indicates that the maize endosperm, as in barley, exhibits topical adaptations.

The starchy endosperm exhibits certain metabolic compartmentalization, likely providing a mechanism to ensure metabolic flexibility, and eventually contributes to the high-carbon conversion efficiency shown by the cereal caryopsis (Rolletschek et al., 2015). The compartmentalization can be regarded as an adjustment to the hypoxic environment inside the central endosperm; the suggested mechanisms are as follows: first, glycolysis is upregulated in the central region of the barley endosperm, where the supply of sugar is sufficient (Fig. 4), whereas that of O_2 is limited (Fig. 6b and c). In particular, succinate dehydrogenase, which is involved in the TCA cycle and the respiratory chain, is inactive under hypoxia (Fig. 6d). In turn, the flux through the TCA cycle is decreased. The possible mechanisms behind such modification are explained elsewhere (Sweetlove et al., 2010). Secondly, the barley caryopsis neither accumulates GABA nor invokes the GABA shunt, which catalyzes the irreversible decarboxylation of glutamate to produce GABA (Bailey-Serres and Voesenek, 2008; Miyashita and Good, 2008). Instead, the level of the reaction product (succinic semialdehyde) increases in the hypoxic region of the endosperm. Therefore, it is unlikely that the GABA shunt (prominent pathway under hypoxia in vegetative tissues) can carry any significant flux in the endosperm. This might represent an endosperm-specific adaptation to long-lasting, rather than transient, O_2 deficiency (Rolletschek et al., 2004). Thirdly, in the interior (hypoxic) region of the endosperm, the assimilation of ammonia may operate via glutamate dehydrogenase (GDH), which is upregulated under hypoxia (Branco-Price et al., 2008; Bailey-Serres and Voesenek, 2008), rather than by the glutamate synthase/glutamine synthetase (GS/GOGAT) system. The GDH product delivers a more energy-efficient nitrogen assimilation than GS/GOGAT (Qiu et al., 2009). Fourthly, by removing pyruvate, the activation of ethanol/lactate fermentation is inhibited, and the stimulation of AOX activity can be avoided (Rolletschek et al., 2011). Alanine synthesis under O_2 limitation is regarded as a means to maintain glycolysis and energy provision. The caryopsis can maintain its energy homeostasis and thereby enhance productivity and performance. The endosperm-specific Ala-AT does not respond to O_2 supply, unlike its isoforms in vegetative tissues (Miyashita et al., 2007). A plausible interpretation is that, in the endosperm, posttranslational regulation of Ala-AT is of particular significance. This represents an efficient strategy given the high-energy costs of shifting protein abundance via transcription/translation (Piques et al., 2009).

To summarize, as a result of the architecture of the seed and the non-uniform availability of O_2 and sucrose (the main carbohydrate supplied by the mother plant), metabolic constraints are not uniformly experienced by different grain tissues. Therefore, local metabolic adjustments resolve imbalances *in vivo*. Notably, hypoxic acclimations of the developing endosperm could be comparable to the classical anoxic tissue, waterlogged plant roots. However, the seed tissue is permanently hypoxic, whereas roots experience only transient hypoxia during the flooding event. This could mark an important difference regarding the O_2-signaling components involved, as well as the resulting adjustments in growth and metabolism. Endosperm metabolism should be viewed as an organ-specific system permanently acclimated to low-O_2 concentration, rather than a process that responds rapidly to induced hypoxia.

8 Functional orchestration of the barley grain

The integration of physiological and developmental processes relies on the transmission of hormonal, metabolic, and other signals through the vascular system (van Bel et al., 2013). In the barley grain, the vascular network does not extend beyond the pericarp. Hence, the communication between the pericarp, embryo, and endosperm occurs via the apoplast. Each organ follows its own genetic program and drives distinct metabolic routes, but together they form an interactive system. It is of great relevance to assess this system in action, to evaluate its metabolic capacity and the efficiency of each individual organ, and to discover how it responds to a fluctuating environment that includes conditions such as drought or diurnal changes. The supply of assimilate to the developing grain is assumed to underlie significant diurnal variations (Grafahrend-Belau et al., 2013), although experimental validation of this remains difficult. Whereas diurnal variation in transcript abundance in barley grains has been reported (Mangelsen et al., 2010), it remains unclear whether these changes are mirrored by equivalent changes in the metabolic activity. State-of-the-art experimental methods cannot currently provide solutions for investigating the 'system grain' in all its complexity. However, it is possible to provide new insights into the tripartite interaction inside living grains by combining noninvasive and modeling approaches, as exemplified below.

The growing barley spike and individual grains were investigated *in vivo* using noninvasive NMR technology (Rolletschek et al., 2015). Temporal variability in solute flow toward the entire spike was monitored continuously by measuring solute flow in the peduncle. Diurnal rhythms in assimilate flow were recognized clearly, pointing to the established photoperiod effect. Notably, the positive response of the flux to light was instantly abolished when the spike was kept in the dark (while the rest of the plant remained exposed to the diurnal light/dark cycle). Once the spike was exposed to light again, the normal light response was restored. This experiment implies that the cereal spike can quickly and autonomously regulate flow toward developing grains in response to ambient light conditions. MRI imaging further demonstrated that individual grains responded to illumination by increasing the influx of water/solutes, evident by local changes of tissue parameters along the main transport route.

Cereal grain is considered by some researchers to grow under saturated assimilate supply (Borras et al., 2003), that is, there is a sink limitation. When assimilates are taken up by the grain, they are shared by the embryo and endosperm. The investigation into this competition requires further study, but when *in vivo* analysis is challenging, it may be possible to exploit *in silico* metabolic modeling. This approach has been validated in the form of flux balance analysis (FBA; Baghalian et al., 2014), which represents an established technique of modeling flux distributions in metabolic networks. Such FBA models have been established for barley grain (Grafahrend-Belau et al., 2009) and its endosperm (Rolletschek et al., 2011), and recently also for the metabolism in the endosperm, embryo, and pericarp, combined in a grain model (Rolletschek et al., 2015).

The endosperm receives the majority of assimilate, channeling sucrose mostly into starch. The embryo, which receives under 15% of the assimilate passing through the pericarp, uses it to produce a mixture of starch, proteins, and lipids. The pericarp appears to drive a daily starch turnover. While the embryo and endosperm differ in certain aspects of metabolic activity, their general mode of operation and pathway activity were reported to be similar (Fig. 7). Prominently, the light-induced shift in solute influx resulted in a proportional adjustment of the metabolic flux in both organs. In other words, the developmental stage-specific set of genes encoding transporters and enzymes define the metabolic network structure, but the pathway activity is regulated posttranscriptionally, particularly by the supply rates of sugars, AA, and possibly other constituents.

The next step in the FBA modeling approach was to recognize the separate metabolic contribution made by each of the three grain components (embryo, endosperm, and pericarp), and to explore their response to fluctuating environmental conditions. The key drivers of metabolic behavior and the major bottlenecks were analyzed, as they were viewed as viable targets for intervention. Combining organ-specific models to form a multilayered representation is a great challenge (Grafahrend-Belau et al., 2013; Gomes de Oliveira Dal'Molin et al., 2015). The new multicomponent model of the barley grain (Rolletschek et al., 2015) provided an insider's view on the grain metabolic architecture. The elevated influx of solutes in response to light has little effect on the steady-state levels of metabolites or transcripts, or the distribution of sugar within the grain. Irrespective of the lighting regime, the concentration of sucrose remained high throughout the crease region, and its distribution was unaffected. This suggests the presence of mechanisms able to ensure metabolic homeostasis during short-term environmental fluctuation. What became evident was that the pericarp plays a central role through its sensing of light signals, which then orchestrate the organ-specific metabolic responses. The pericarp behaves as if it were an autotrophic leaf enveloping the heterotrophic storage organs; notably, photosynthetic electron transport rates are as high as those in the flag leaf. The developing grain benefits from

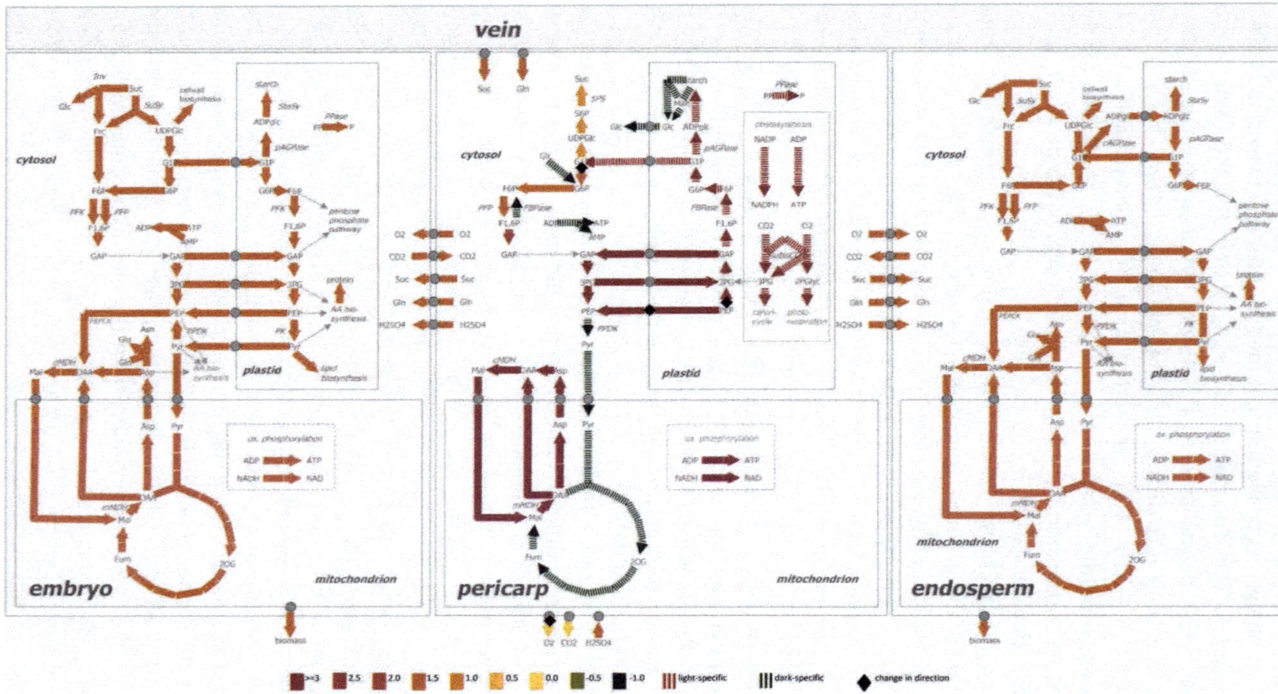

Figure 7 Predicted log-ratio flux maps (day vs. night conditions) of the endosperm, embryo, and pericarp during the grain's main storage phase. Green-dashed arrows indicate metabolic routes only active under lit (day) conditions, and red-dashed ones indicate those exclusive to non-lit (night) conditions. The direction of the metabolic flux indicated by arrowheads refers to flux under lit conditions. Changes in flux direction induced by the switch from lit to non-lit conditions are indicated by black diamonds. Numerical flux values (color-coded) indicate x-fold changes in flux value. Data were derived from flux balance analysis; for details see Rolletschek et al. (2015).

the capacity to anticipate a burst of sugar arriving via the phloem, because this facilitates its rapid adjustment of storage product synthesis. A further conclusion was that, under natural conditions, the grain grows faster during the day than during the night (gain in biomass). Various short-term changes imposed by environmental fluctuations can be evened out via posttranslational metabolic control. Rapid acclimation to changing light supply could be key to achieving high energy and carbon efficiency. The net result is that the developing cereal caryopsis can achieve the extraordinarily high-carbon conversion efficiency of up to 95%. The conclusion is that the architecture of the developing barley grain is an important determinant of its high-carbon conversion efficiency and, therefore, of the crop's yield potential.

9 Conclusion

Barley is cultivated primarily for its grain – a highly structured unit. Its end-use quality is defined in the process of grain development, while its role in sexual reproduction is strictly predetermined and reflected in the sequence of developmental events. Hence, engineering of desired features might interfere with the grain's functionality. Nowadays, easy access to whole-genome information and large transcriptomic datasets allow for an unprecedented view of the molecular machinery that 'makes the grain'. However, to integrate and rank this information, we require a comprehensive understanding of the physiological background. High-resolution imaging and modeling technology help to further explore the functioning of grain as a living system, along with its interactions with changing environments. This is becoming an important issue in biotechnological design and when choosing appropriate breeding strategies. Given the future (predicted) climate conditions and their negative effect on harvestable starch, protein, and other valuable components of the grain, further attention is required for improvements in barley crops. Therefore, we discuss approaches and achievements of various studies on barley grain physiology. We also highlight current technological limitations in the field, and we argue for the necessary application of noninvasive, topographical, and analytical technologies, along with various -omics approaches. The linking of phenomics and genomics becomes most meaningful when considering a thorough understanding of plant physiology, particularly the inner life of the grain–which holds the key to sustain human life.

10 Acknowledgements

We warmly thank Stefan Ortleb, Andre Guendel, Eberhard Munz, Twan Rutten and Steffen Wagner for valuable contributions to barley research.

11 Where to look for further information

Several valuable books and articles describe different aspects of barley grain development:

- Kumlehn and Stein (2014) – Biotechnologically oriented barley research.
- Newman and Newman (2008) – An excellent coverage of chemistry, breeding, utilization of barley and few selected recipes from around the world.
- Shewry and Ullrich (2014) – A comprehensive review about development, structure, composition, or end use properties of the barley grain for cultivation, trade, and utilization.
- Sreenivasulu and Wobus (2013) – A good comparative review on the developmental programs between dicot and monocot seed development.

Conferences and online resources

- International Barley Genetics Symposium (IBSG): symposium on barley genetics held every 4 years.
- The International Society for Seed Science (ISSS) is a professional organization of seed scientists committed to promoting research, education and communication in the scientific understanding of different aspects of seeds. The organization publishes a monthly scientific journal 'Seed Science Research' and organizes three-annual international conferences.
- Examples of project activities and breeding programs are found here:
  - o https://www.cwrdiversity.org/
  - o https://www.cwrdiversity.org/crop/barley/
  - o https://mel.cgiar.org/projects/741

12 References

Ahmed, Z., Tetlow, I. J., Ahmed, R., Morell, M. K. and Emes, M. J. 2015. Protein-protein interactions among enzymes of starch biosynthesis in high-amylose barley genotypes reveal differential roles of heteromeric enzyme complexes in the synthesis of A and B granules. *Plant Sci.* 233, 95–106. doi:10.1016/j.plantsci.2014.12.016.

Ahmed, Z., Tetlow, I. J., Falk, D. E., Liu, Q. and Emes, M. J. 2016. Resistant starch content is related to granule size in barley. *Cereal Chem.* 93(6), 618–30. doi:10.1094/CCHEM-02-16-0025-R.

Armstrong, W., Brändle, R. and Jackson, M. B. 1994. Mechanisms of flood tolerance in plants. *Acta Bot. Neerl.* 43(4), 307–58. doi:10.1111/j.1438-8677.1994.tb00756.x.

Asare, E. K., Jaiswal, S., Maley, J., Båga, M., Sammynaiken, R., Rossnagel, B. G. and Chibbar, R. N. 2011. Barley grain constituents, starch composition, and structure affect starch *in vitro* enzymatic hydrolysis. *J. Agric. Food Chem.* 59(9), 4743–54. doi:10.1021/jf200054e.

Baghalian, K., Hajirezaei, M. R. and Schreiber, F. 2014. Plant metabolic modeling: achieving new insight into metabolism and metabolic engineering. *Plant Cell* 26(10), 3847-66. doi:10.1105/tpc.114.130328.

Bailey-Serres, J. and Voesenek, L. A. 2008. Flooding stress: acclimations and genetic diversity. *Ann. Rev. Plant Biol.* 59, 313-39. doi:10.1146/annurev.arplant.59.032607.092752.

Bønsager, B. C., Finnie, C., Roepstorff, P. and Svensson, B. 2007. Spatio-temporal changes in germination and radical elongation of barley seeds tracked by proteome analysis of dissected embryo, aleurone layer, and endosperm tissues. *Proteomics* 7(24), 4528–40. doi:10.1002/pmic.200700766.

Borén, M., Höglund, A. S., Bozhkov, P. and Jansson, C. 2006. Developmental regulation of a VEIDase caspase-like proteolytic activity in barley caryopsis. *J. Exp. Bot.* 57(14), 3747–53. doi:10.1093/jxb/erl136.

Borén, M., Glaring, M. A., Ghebremedhin, H., Olsson, H., Blennow, A. and Jansson, C. 2008. Molecular and physicochemical characterization of the high-amylose barley mutant *Amo1*. *J. Cereal Sci.* 47(1), 79–89. doi:10.1016/j.jcs.2007.01.009.

Borisjuk, L., Macherel, D., Benamar, A., Wobus, U. and Rolletschek, H. 2007. Low oxygen sensing and balancing in plant seeds - a role for nitric oxide. *New Phytol.* 176(4), 813–23. doi:10.1111/j.1469-8137.2007.02226.x.

Borisjuk, L., Rolletschek, H. and Neuberger, T. 2012. Surveying the plant's world by magnetic resonance imaging. *Plant J.* 70(1), 129–46. doi:10.1111/j.1365-313X.2012.04927.x.

Borrás, L., Westgate, M. E. and Otegui, M. E. 2003. Control of kernel weight and kernel water relations by post-flowering source-sink ratio in maize. *Ann. Bot.* 91(7), 857–67. doi:10.1093/aob/mcg090.

Boudet, J., Merlino, M., Plessis, A., Gaudin, J. C., Dardevet, M., Perrochon, S., Alvarez, D., Risacher, T., Martre, P. and Ravel, C. 2019. The bZIP transcription factor SPA heterodimerizing protein represses glutenin synthesis in *Triticum aestivum*. *Plant J.* 97(5), 858–71. doi:10.1111/tpj.14163.

Branco-Price, C., Kaiser, K. A., Jang, C. J., Larive, C. K. and Bailey-Serres, J. 2008. Selective mRNA translation coordinates energetic and metabolic adjustments to cellular oxygen deprivation and reoxygenation in *Arabidopsis thaliana*. *Plant J.* 56(5), 743–55. doi:10.1111/j.1365-313X.2008.03642.x.

Bull, H., Casao, M. C., Zwirek, M., Flavell, A. J., Thomas, W. T. B., Guo, W., Zhang, R., Rapazote-Flores, P., Kyriakidis, S., Russell, J., Druka, A., McKim, S. M. and Waugh, R. 2017. Barley SIX-ROWED SPIKE3 encodes a putative Jumonji C-type H3K9me2/me3 demethylase that represses lateral spikelet fertility. *Nat. Commun.* 8(1), 936. doi:10.1038/s41467-017-00940-7.

Burton, R. A., Jenner, H., Carrangis, L., Fahy, B., Fincher, G. B., Hylton, C., Laurie, D. A., Parker, M., Waite, D., van Wegen, S., Verhoeven, T. and Denyer, K. 2002. Starch granule initiation and growth are altered in barley mutants that lack isoamylase activity. *Plant J.* 31(1), 97–112. doi:10.1046/j.1365-313X.2002.01339.x.

Carciofi, M., Shaik, S. S., Jensen, S. L., Blennow, A., Svensson, J. T., Vincze, É and Hebelstrup, K. H. 2011. Hyperphosphorylation of cereal starch. *J. Cereal Sci.* 54(3), 339–46. doi:10.1016/j.jcs.2011.06.013.

Carciofi, M., Blennow, A., Jensen, S. L., Shaik, S. S., Henriksen, A., Buléon, A., Holm, P. B. and Hebelstrup, K. H. 2012. Concerted suppression of all starch branching enzyme genes in barley produces amylose-only starch granules. *BMC Plant Biol.* 12, 223. doi:10.1186/1471-2229-12-223.

Chen, L. Q., Cheung, L. S., Feng, L., Tanner, W. and Frommer, W. B. 2015. Transport of sugars. *Annu. Rev. Biochem.* 84, 865–94. doi:10.1146/annurev-biochem-060614-033904.

Chichkova, N. V., Shaw, J., Galiullina, R. A., Drury, G. E., Tuzhikov, A. I., Kim, S. H., Kalkum, M., Hong, T. B., Gorshkova, E. N., Torrance, L., Vartapetian, A. B. and Taliansky, M. 2010. Phytaspase, a relocalisable cell death promoting plant protease with caspase specificity. *EMBO J.* 29(6), 1149–61. doi:10.1038/emboj.2010.1.

Cho, S. Y. and Rhee, C. 2009. Functional and film-forming properties of fractionated barley proteins. *Food Sci. Biotechnol.* 18, 889– 94.

Chope, G. A., Wan, Y., Penson, S. P., Bhandari, D. G., Powers, S. J., Shewry, P. R. and Hawkesford, M. J. 2014. Effects of genotype, season, and nitrogen nutrition on gene expression and protein accumulation in wheat grain. *J. Agric. Food Chem.* 62(19), 4399–407. doi:10.1021/jf500625c.

Cochrane, M. P. and Duffus, C. M. 1979. Morphology and ultrastructure of immature cereal grains in relation to transport. *Ann. Bot.* 44(1), 67–72. doi:10.1093/oxfordjournals.aob.a085707.

Coffeen, W. C. and Wolpert, T. J. 2004. Purification and characterization of serine proteases that exhibit caspase-like activity and are associated with programmed cell death in *Avena sativa*. *Plant Cell* 16(4), 857–73. doi:10.1105/tpc.017947.

Cooper, M., Woodruff, D. R., Phillips, I. G., Basford, K. E. and Gilmour, A. R. 2001. Genotype-by-management interactions for grain yield and grain protein concentration of wheat. *Field Crops Res.* 69(1), 47–67. doi:10.1016/S0378-4290(00)00131-3.

Cuesta-Seijo, J. A., Nielsen, M. M., Marri, L., Tanaka, H., Beeren, S. R. and Palcic, M. M. 2013. Structure of starch synthase I from barley: insight into regulatory mechanisms of starch synthase activity. *Acta Crystallogr. D Biol. Crystallogr.* 69(6), 1013–25. doi:10.1107/S090744491300440X.

Dai, H. X., Zhang, X. Q., Harasymow, S., Roumeliotis, S., Broughton, S., Eglinton, J., Wu, F. B. and Li, C. D. 2014. MALDI-TOF mass spectrometry provides an efficient approach to monitoring protein modification in the malting process. *Int. J. Mass Spectrom.* 371, 8–16.

Daneva, A., Gao, Z., Van Durme, M. and Nowack, M. K. 2016. Functions and regulation of programmed cell death in plant development. *Annu. Rev. Cell Dev. Biol.* 32, 441–68. doi:10.1146/annurev-cellbio-111315-124915.

Degterev, A., Boyce, M. and Yuan, J. 2003. A decade of caspases. *Oncogene* 22(53), 8543–67. doi:10.1038/sj.onc.1207107.

Domínguez, F., Moreno, J. and Cejudo, F. J. 2001. The nucellus degenerates by a process of programmed cell death during the early stages of wheat grain development. *Planta* 213(3), 352–60. doi:10.1007/s004250000517.

Dordas, C., Rivoa, J. and Hill, R. D. 2003. Plant haemoglobins, nitric oxide and hypoxic stress. *Ann. Bot.* 91, 173–8. doi:10.1093/aob/mcf115.

Emes, M. J., Bowsher, C. G., Hedley, C., Burrell, M. M., Scrase-Field, E. S. and Tetlow, I. J. 2003. Starch synthesis and carbon partitioning in developing endosperm. *J. Exp. Bot.* 54(382), 569–75. doi:10.1093/jxb/erg089.

Engell, K. 1989. Embryology of barley: time course and analysis of controlled fertilization and early embryo formation based on serial sections. *Nord. J. Bot.* 9(3), 265–80. doi:10.1111/j.1756-1051.1989.tb01002.x.

Evans, D. E., Li, C. and Eglinton, J. K. 2009. The properties and genetics of barley malt starch degrading enzymes. In: Zhang, G. and Li, C. (Eds), *Genetics and Improvement of Barley Malt Quality. Advanced Topics in Science and Technology in China*. Springer, Berlin, Heidelberg, pp. 143–89.

Evers, A. D., O'Brien, L. and Blakeney, A. B. 1999. Cereal structure and composition. *Aust. J. Agric. Res.* 50(5), 629–50. doi:10.1071/AR98158.

Fahy, B., Siddiqui, H., David, L. C., Powers, S. J., Borrill, P., Uauy, C. and Smith, A. M. 2018. Final grain weight is not limited by the activity of key starch-synthesising enzymes during grain filling in wheat. *J. Exp. Bot.* 69(22), 5461–75. doi:10.1093/jxb/ery314.

Finnie, C. and Svensson, B. 2014. Barley grain proteins. In: Shewry, P. R. and Ulrich, S. E. (Eds), *Barley Chemistry and Technology* (2nd edn.). AACC International, St. Paul, MN, pp. 123–68.

Finnie, C., Melchior, S., Roepstorff, P. and Svensson, B. 2002. Proteome analysis of grain filling and seed maturation in barley. *Plant Physiol.* 129(3), 1308– 19. doi:10.1104/pp.003681.

Fluitt, A., Pienaar, E. and Viljoen, H. 2007. Ribosome kinetics and aa-tRNA competition determine rate and fidelity of peptide synthesis. *Comput. Biol. Chem.* 31(5–6), 335–46. doi:10.1016/j.compbiolchem.2007.07.003.

Fox, G. P. 2010. Chemical composition in barley grains and malt quality. In: Zhang, G. and Li, C. (Eds), *Genetics and Improvement of Barley Malt Quality*. Springer-Verlag, Berlin and Heidelberg, pp. 63–98.

Franckowiack, J. and Konishi, T. 1997. Naked caryopsis. *Barley Genet. Newsl.* 26, 51–2.

Gayral, M., Elmorjani, K., Dalgalarrondo, M., Balzergue, S. M., Pateyron, S., Morel, M. H., Brunet, S., Linossier, L., Delluc, C., Bakan, B. and Marion, D. 2017. Responses to hypoxia and endoplasmic reticulum stress discriminate the development of vitreous and floury endosperms of conventional maize (*Zea mays*) inbred lines. *Front. Plant Sci.* 8, 557. doi:10.3389/fpls.2017.00557.

Ge, Y., Cai, Y. M., Bonneau, L., Rotari, V., Danon, A., McKenzie, E. A., McLellan, H., Mach, L. and Gallois, P. 2016. Inhibition of cathepsin B by caspase-3 inhibitors blocks programmed cell death in Arabidopsis. *Cell Death Differ.* 23(9), 1493–501. doi:10.1038/cdd.2016.34.

Gingold, H. and Pilpel, Y. 2011. Determinants of translation efficiency and accuracy. *Mol. Syst. Biol.* 7, 481. doi:10.1038/msb.2011.14.

Goldstein, A., Annor, G., Putaux, J. L., Hebelstrup, K. H., Blennow, A. and Bertoft, E. 2016. Impact of full range of amylose contents on the architecture of starch granules. *Int. J. Biol. Macromol.* 89, 305–18. doi:10.1016/j.ijbiomac.2016.04.053.

Gomes de Oliveira Dal'Molin, C., Quek, L. E., Saa, P. A. and Nielsen, L. K. 2015. A multi-tissue genome-scale metabolic modeling framework for the analysis of whole plant systems. *Front Plant Sci.* 6, 4.

Gous, P. W., Warren, F., Mo, O. W., Gilbert, R. G. and Fox, G. P. 2015. The effects of variable nitrogen application on barley starch structure under drought stress. *J. Inst. Brew.* 121(4), 502–9. doi:10.1002/jib.260.

Grafahrend-Belau, E., Schreiber, F., Koschützki, D. and Junker, B. H. 2009. Flux balance analysis of barley seeds: a computational approach to study systemic properties of central metabolism. *Plant Physiol.* 149(1), 585–98. doi:10.1104/pp.108.129635.

Grafahrend-Belau, E., Junker, A., Eschenröder, A., Müller, J., Schreiber, F. and Junker, B. H. 2013. Multiscale metabolic modeling: dynamic flux balance analysis on a whole-plant scale. *Plant Physiol.* 163(2), 637–47. doi:10.1104/pp.113.224006.

Gubatz, S. and Shewry, P. R. 2011. The development, structure, and composition of the barley grain. In: Ullrich, S. E. (Ed.), *Barley: Production, Improvement, and Uses*. Wiley-Blackwell Publishing Ltd., Chichester, UK, pp. 391-477.

Gubatz, S. and Weschke, W. 2014. Barley grain: development and structure. In: Shewry, P. R. and Ullrich, S. (Eds), *Barley: Chemistry and Technology* (2nd edn.). AACC International Inc., St. Paul, MN, pp. 11-54.

Guendel, A., Rolletschek, H., Wagner, S., Muszynska, A. and Borisjuk, L. 2018. Micro imaging displays the sucrose landscape within and along its allocation pathways. *Plant Physiol.* 178(4), 1448-60. doi:10.1104/pp.18.00947.

Han, J. J., Lin, W., Oda, Y., Cui, K. M., Fukuda, H. and He, X. Q. 2012. The proteasome is responsible for caspase-3-like activity during xylem development. *Plant J.* 72(1), 129-41. doi:10.1111/j.1365-313X.2012.05070.x.

Hansen, M., Lange, M., Friis, C., Dionisio, G., Holm, P. B. and Vincze, E. 2007. Antisense-mediated suppression of C-hordein biosynthesis in the barley grain results in correlated changes in the transcriptome, protein profile, and amino acid composition. *J. Exp. Bot.* 58(14), 3987-95. doi:10.1093/jxb/erm254.

Harwood, W. A. (Ed.). 2019. *Barley: Methods and Protocols* (Series: Methods in molecular biology, vol. 1900). Humana Press, New York, NY, pp. 195-215. doi:10.1007/978-1-4939-8944-7_13, 978-1-4939-8944-7.

Hatsugai, N., Yamada, K., Goto-Yamada, S. and Hara-Nishimura, I. 2015. Vacuolar processing enzyme in plant programmed cell death. *Front. Plant Sci.* 6, 234. doi:10.3389/fpls.2015.00234.

Hu, M., Zhao, X., Liu, Q., Hong, X., Zhang, W., Zhang, Y., Sun, L., Li ,H. and Tong, Y. 2018. Transgenic expression of plastidic glutamine synthetase increases nitrogen uptake and yield in wheat. *Plant Biotechnol. J.* 16(11), 1858-67. doi:10.1111/pbi.12921.

Hughes, N., Askew, K., Scotson, C. P., Williams, K., Sauze, C., Corke, F., Doonan, J. H. and Nibau, C. 2017. Non-destructive, high-content analysis of wheat grain traits using X-ray micro computed tomography. *Plant Methods* 13, 76. doi:10.1186/s13007-017-0229-8.

Ibba, M. and Söll, D. 2000. Aminoacyl-tRNA synthesis'. *Annu. Rev. Biochem.* 69, 617-50. doi:10.1146/annurev.biochem.69.1.617.

Igamberdiev, A. U., Baron, K., Manac'h-Little, N., Stoimenova, M. and Hill, R. D. 2005. The haemoglobin/nitric oxide cycle: involvement in flooding stress and effects on hormone signalling. *Ann. Bot.* 96(4), 557-64. doi:10.1093/aob/mci210.

Igamberdiev, A. U., Stasolla, C. and Hill, R. D. 2014. Low oxygen stress, nonsymbiotic hemoglobins, NO, and programmed cell death. In: van Dongen, J. T. and Licausi, F. (Eds), *Low Oxygen Stress in Plants*. Plant Cell Monographs 21. Springer, Wien, pp. 41-58.

Jaiswal, S., Båga, M., Ahuja, G., Rossnagel, B. G. and Chibbar, R. N. 2014. Development of barley (*Hordeum vulgare* L.) lines with altered starch granule size distribution. *J. Agric. Food Chem.* 62(10), 2289-96. doi:10.1021/jf405424x.

Johnson, P. E., Patron, N. J., Bottrill, A. R., Dinges, J. R., Fahy, B. F., Parker, M. L., Waite, D. N. and Denyer, K. 2003. A low-starch barley mutant, *Riso16*, lacking the cytosolic small subunit of ADP-glucose pyrophosphorylase, reveals the importance of the cytosolic isoform and the identity of the plastidial small subunit. *Plant Physiol.* 131(2), 684-96. doi:10.1104/pp.013094.

Julian, I., Gandullo, J., Santos-Silva, L. K., Diaz, I. and Martinez, M. 2013. Phylogenetically distant barley legumains have a role in both seed and vegetative tissues. *J. Exp. Bot.* 64(10), 2929-41. doi:10.1093/jxb/ert132.

Källman, A., Bertoft, E., Koch, K., Sun, C., Åman, P. and Andersson, R. 2015. Starch structure in developing barley endosperm. *Int. J. Biol. Macromol.* 81, 730–5. doi:10.1016/j.ijbiomac.2015.09.013.

Kanno, Y., Oikawa, T., Chiba, Y., Ishimaru, Y., Shimizu, T., Sano, N., Koshiba, T., Kamiya, Y., Ueda, M. and Seo, M. 2016. AtSWEET13 and AtSWEET14 regulate gibberellin-mediated physiological processes. *Nat. Commun.* 7, 13245. doi:10.1038/ncomms13245.

Kapoor, K., Mira, M. M., Ayele, B. T., Nguyen, T. N., Hill, R. D. and Stasolla, C. 2018. Phytoglobins regulate nitric oxide-dependent abscisic acid synthesis and ethylene-induced program cell death in developing maize somatic embryos. *Planta* 247(6), 1277–91. doi:10.1007/s00425-018-2862-5.

Korthout, H. A., Berecki, G., Bruin, W., van Duijn, B. and Wang, M. 2000. The presence and subcellular localization of caspase 3-like proteinases in plant cells. *FEBS Lett.* 475(2), 139–44. doi:10.1016/s0014-5793(00)01643-4.

Kumlehn, J. and Stein, N. (Eds). 2014. *Biotechnological Approaches to Barley Improvement*. Series: Biotechnology in Agriculture and Forestry (vol. 69). Springer, Berlin, pp. 139–59. doi:10.1007/978-3-662-44406-1_8 ISBN 978-3-662-44405-4.

Lange, M., Vincze, E., Wieser, H., Schjoerring, J. K. and Holm, P. B. 2007. Suppression of C-hordein synthesis in barley by antisense constructs results in a more balanced amino acid composition. *J. Agric. Food Chem.* 55(15), 6074–81. doi:10.1021/jf0709505.

Langridge, P. 2018. Economic and academic importance of barley. In: Stein, N. and Muehbauer, G. J. (Eds), *The Barley Genome*. Springer International Publishing AG, pp. 1–10.

Li, Z., Li, D., Du, X., Wang, H., Larroque, O., Jenkins, C. L., Jobling, S. A. and Morell, M. K. 2011. The barley *amo1* locus is tightly linked to the starch synthase IIIa gene and negatively regulates expression of granule-bound starch synthetic genes. *J. Exp. Bot.* 62(14), 5217–31. doi:10.1093/jxb/err239.

Linnestad, C., Doan, D. N., Brown, R. C., Lemmon, B. E., Meyer, D. J., Jung, R. and Olsen, O. A. 1998. Nucellain, a barley homolog of the dicot vacuolar-processing protease, is localized in nucellar cell walls. *Plant Physiol.* 118(4), 1169–80. doi:10.1104/pp.118.4.1169.

Lundqvist, U., Franckowiak, J. D. and Konishi, T. 1997. New and revised descriptions of barley genes. *Barley Genet. Newsl.* 26, 22–516.

Luo, H., Harasymow, S., Paynter, B., MacLeod, A., Izydorczyk, M. S., O'Donovan, J. T. and Li, C. 2019. Genetic and environmental impact on protein profiles in barley and malt. *J. Inst. Brew.* 125(1), 28–38. doi:10.1002/jib.532.

Mahalingam, R. 2017. Shotgun proteomics of the barley seed proteome. *BMC Genomics* 18(1), 44. doi:10.1186/s12864-016-3408-5.

Mangelsen, E., Wanke, D., Kilian, J., Sundberg, E., Harter, K. and Jansson, C. 2010. Significance of light, sugar, and amino acid supply for diurnal gene regulation in developing barley caryopses. *Plant Physiol.* 153(1), 14–33. doi:10.1104/pp.110.154856.

Mascher, M., Gundlach, H., Himmelbach, A., Beier, S., Twardziok, S. O., Wicker, T., Radchuk, V., Dockter, C., Hedley, P. E., Russell, J., Bayer, M., Ramsay, L., Liu, H., Haberer, G., Zhang, X. Q., Zhang, Q., Barrero, R. A., Li, L., Taudien, S., Groth, M., Felder, M., Hastie, A., Šimková, H., Staňková, H., Vrána, J., Chan, S., Muñoz-Amatriaín, M., Ounit, R., Wanamaker, S., Bolser, D., Colmsee, C., Schmutzer, T., Aliyeva-Schnorr, L., Grasso, S., Tanskanen, J., Chailyan, A., Sampath, D., Heavens, D., Clissold, L., Cao,

S., Chapman, B., Dai, F., Han, Y., Li, H., Li, X., Lin, C., McCooke, J. K., Tan, C., Wang, P., Wang, S., Yin, S., Zhou, G., Poland, J. A., Bellgard, M. I., Borisjuk, L., Houben, A., Doležel, J., Ayling, S., Lonardi, S., Kersey, P., Langridge, P., Muehlbauer, G. J., Clark, M. D., Caccamo, M., Schulman, A. H., Mayer, K. F. X., Platzer, M., Close, T. J., Scholz, U., Hansson, M., Zhang, G., Braumann, I., Spannagl, M., Li, C., Waugh, R. and Stein, N. 2017. A chromosome conformation capture ordered sequence of the barley genome. *Nature* 544(7651), 427–33. doi:10.1038/nature22043.

Melkus, G., Rolletschek, H., Fuchs, J., Radchuk, V., Grafahrend-Belau, E., Sreenivasulu, N., Rutten, T., Weier, D., Heinzel, N., Schreiber, F., Altmann, T., Jakob, P. M. and Borisjuk, L. 2011. Dynamic 13C/1H NMR imaging uncovers sugar allocation in the living seed. *Plant Biotechnol. J.* 9(9), 1022–37. doi:10.1111/j.1467-7652.2011.00618.x.

Millar, A. H., Whelan, J., Soole, K. L. and Day, D. A. 2011. Organization and regulation of mitochondrial respiration in plants. *Annu. Rev. Plant Biol.* 62, 79–104. doi:10.1146/annurev-arplant-042110-103857.

Miyashita, Y. and Good, A. G. 2008. Glutamate deamination by glutamate dehydrogenase plays a central role in amino acid catabolism in plants. *Plant Signal. Behav.* 3(10), 842–3. doi:10.4161/psb.3.10.5936.

Miyashita, Y., Dolferus, R., Ismond, K. P. and Good, A. G. 2007. Alanine aminotransferase catalyses the breakdown of alanine after hypoxia in *Arabidopsis thaliana*. *Plant J.* 49(6), 1108–21. doi:10.1111/j.1365-313X.2006.03023.x.

Mock, H.-P., Finnie, C., Witzel, K. and Svensson, B. 2018. Barley proteomics. In: Stein, N. and Muehlbauer, G. J. (Eds), *The Barley Genome* (1st edn.) (Series: Kole, C (Ed.): Compendium of Plant Genomes). Springer, Cham, pp. 155–69.

Morell, M. K., Kosar-Hashemi, B., Cmiel, M., Samuel, M. S., Chandler, P., Rahman, S., Buleon, A., Batey, I. L. and Li, Z. 2003. Barley *sex6* mutants lack starch synthase IIa activity and contain a starch with novel properties. *Plant J.* 34(2), 173–85. doi:10.1046/j.1365-313X.2003.01712.x.

Mustroph, A., Lee, S. C., Oosumi, T., Zanetti, M. E., Yang, H., Ma, K., Yaghoubi-Masihi, A., Fukao, T. and Bailey-Serres, J. 2010. Cross-kingdom comparison of transcriptomic adjustments to low-oxygen stress highlights conserved and plant-specific responses. *Plant Physiol.* 152(3), 1484–500. doi:10.1104/pp.109.151845.

Nakamura, Y. 2015. Biosynthesis of reserve starch. In: Nakamura, Y. (Ed.), *Starch Metabolism and Structure*. Springer, Tokyo, pp. 161–209.

Newman, R. K. and Newman, C. W. 2008. *Barley for Food and Health: Science, Technology, and Products*. Wiley, Hoboken, NJ. ISBN: 978-0-470-10249-7.

Newton, A. C., Flavell, A. J., George, T. S., Leat, P., Mullholland, B., Ramsay, L., Revoredo-Giha, C., Russell, J., Steffenson, B. J., Swanston, J. S., Thomas, W. T. B., Waugh, R., White, P. J. and Bingham, I. J. 2011. Crops that feed the world 4. Barley: a resilient crop? Strengths and weaknesses in the context of food security. *Food Sec.* 3(2), 141–78. doi:10.1007/s12571-011-0126-3.

Nutbeam, A. R. and Duffus, C. M. 1978. Oxygen exchange in the pericarp green layer of immature cereal grains. *Plant Physiol.* 62(3), 360–2. doi:10.1104/pp.62.3.360.

Offler, C. E., McCurdy, D. W., Patrick, J. W. and Talbot, M. J. 2003. Transfer cells: cells specialized for a special purpose. *Annu. Rev. Plant Biol.* 54(1), 431–54. doi:10.1146/annurev.arplant.54.031902.134812.

Olsen, O. A. and Weschke, W. 2014. Genetic and molecular aspects of barley grain development. In: Shewry, P. R. and Ullrich, S. (Eds), *Barley: Chemistry and Technology* (2nd edn.). AACC International Inc., St. Paul, MN, pp. 55–70.

Oparka, K. J. and Gates, P. 1984. Sink anatomy in relation to solute movement in rice (*Oryza sativa* L.): a summary of findings. *Plant Growth Regul.* 2(4), 297–307. doi:10.1007/BF00027289.

Østergaard, O., Melchior, S., Roepstorff, P. and Svensson, B. 2002. Initial proteome analysis of mature barley seeds and malt. *Proteomics* 2(6), 733–9. doi:10.1002/1615-9861(200206)2:6<733::AID-PROT733>3.0.CO;2-E.

Patron, N. J., Smith, A. M., Fahy, B. F., Hylton, C. M., Naldrett, M. J., Rossnagel, B. G. and Denyer, K. 2002. The altered pattern of amylose accumulation in the endosperm of low-amylose barley cultivars is attributable to a single mutant allele of granule-bound starch synthase I with a deletion in the 5'-non-coding region. *Plant Physiol.* 130(1), 190–8. doi:10.1104/pp.005454.

Patron, N. J., Greber, B., Fahy, B. F., Laurie, D. A., Parker, M. L. and Denyer, K. 2004. The *lys5* mutations of barley reveal the nature and importance of plastidial ADP-Glc transporters for starch synthesis in cereal endosperm. *Plant Physiol.* 135(4), 2088–97. doi:10.1104/pp.104.045203.

Peng, X. and Sun, M. X. 2018. The suspensor as a model system to study the mechanism of cell fate specification during early embryogenesis. *Plant Reprod.* 31(1), 59–65. doi:10.1007/s00497-018-0326-5.

Peukert, M., Thiel, J., Peshev, D., Weschke, W., Van den Ende, W., Mock, H. P. and Matros, A. 2014. Spatio-temporal dynamics of fructan metabolism in developing barley grains. *Plant Cell* 26(9), 3728–44. doi:10.1105/tpc.114.130211.

Pierucci, V. R., Tilley, M., Graybosch, R. A., Blechl, A. E., Bean, S. R. and Tilley, K. A. 2009. Effects of overexpression of high molecular weight glutenin subunit 1Dy10 on wheat tortilla properties. *J. Agric. Food Chem.* 57(14), 6318–26. doi:10.1021/jf900629s.

Piques, M., Schulze, W. X., Höhne, M., Usadel, B., Gibon, Y., Rohwer, J. and Stitt, M. 2009. Ribosome and transcript copy numbers, polysome occupancy and enzyme dynamics in *Arabidopsis*. *Mol. Syst. Biol.* 5, 314. doi:10.1038/msb.2009.68.

Prioul, J. L., Mechin, V., Lessard, P., Thevenot, C., Grimmer, M., Chateau-Joubert, S., Coates, S., Hartings, H., Kloiber-Maitz, M., Murigneux, A., Sarda, X., Damerval, C. and Edwards, K. J. 2008. A joint transcriptomic, proteomic and metabolic analysis of maize endosperm development and starch filling. *Plant Biotechnol. J.* 6(9), 855–69. doi:10.1111/j.1467-7652.2008.00368.x.

Qiu, X., Xie, W., Lian, X. and Zhang, Q. 2009. Molecular analyses of the rice glutamate dehydrogenase gene family and their response to nitrogen and phosphorous deprivation. *Plant Cell Rep.* 28(7), 1115–26. doi:10.1007/s00299-009-0709-z.

Radchuk, V. and Borisjuk, L. 2014. Physical, metabolic and developmental functions of the seed coat. *Front. Plant Sci.* 5, 510. doi:10.3389/fpls.2014.00510.

Radchuk, V., Borisjuk, L., Radchuk, R., Steinbiss, H. H., Rolletschek, H., Broeders, S. and Wobus, U. 2006. Jekyll encodes a novel protein involved in the sexual reproduction of barley. *Plant Cell* 18(7), 1652–66. doi:10.1105/tpc.106.041335.

Radchuk, V. V., Borisjuk, L., Sreenivasulu, N., Merx, K., Mock, H. P., Rolletschek, H., Wobus, U. and Weschke, W. 2009. Spatiotemporal profiling of starch biosynthesis and degradation in the developing barley grain. *Plant Physiol.* 150(1), 190–204. doi:10.1104/pp.108.133520.

Radchuk, V., Weier, D., Radchuk, R., Weschke, W. and Weber, H. 2011. Development of maternal seed tissue in barley is mediated by regulated cell expansion and cell disintegration and coordinated with endosperm growth. *J. Exp. Bot.* 62(3), 1217–27. doi:10.1093/jxb/erq348.

Radchuk, V., Riewe, D., Peukert, M., Matros, A., Strickert, M., Radchuk, R., Weier, D., Steinbiß, H. H., Sreenivasulu, N., Weschke, W. and Weber, H. 2017. Down-regulation of the sucrose transporters *HvSUT1* and *HvSUT2* affects sucrose homeostasis along its delivery path in barley grains. *J. Exp. Bot.* 68(16), 4595–612. doi:10.1093/jxb/erx266.

Radchuk, V., Tran, V., Radchuk, R., Diaz-Mendoza, M., Weier, D., Fuchs, J., Riewe, D., Hensel, G., Kumlehn, J., Munz, E., Heinzel, N., Rolletschek, H., Martinez, M. and Borisjuk, L. 2018. Vacuolar processing enzyme 4 contributes to maternal control of grain size in barley by executing programmed cell death in pericarp. *New Phytol.* 218(3), 1127–42. doi:10.1111/nph.14729.

Radchuk, V., Sharma, R., Potokina, E., Radchuk, R., Weier, D., Munz, E., Schreiber, M., Mascher, M., Stein, N., Wicker, T., Kilian, B. and Borisjuk, L. 2019. The highly divergent *Jekyll* genes, required for sexual reproduction, are lineage specific for the related grass tribes Triticeae and Bromeae. *Plant J.* 98(6), 961–74. doi:10.1111/tpj.14363.

Ramage, R. T. and Crandall, C. R. 1981. Shrunken endosperm mutant *seg8*. *Barley Genet. Newslett.* 11, 34.

Regina, A., Kosar-Hashemi, B., Ling, S., Li, Z., Rahman, S. and Morell, M. 2010. Control of starch branching in barley defined through differential RNAi suppression of starch branching enzyme IIa and IIb. *J. Exp. Bot.* 61(5), 1469–82. doi:10.1093/jxb/erq011.

Rolletschek, H., Weschke, W., Weber, H., Wobus, U. and Borisjuk, L. 2004. Energy state and its control on seed development: starch accumulation is associated with high ATP and steep oxygen gradients within barley grains. *J. Exp. Bot.* 55(401), 1351–9. doi:10.1093/jxb/erh130.

Rolletschek, H., Melkus, G., Grafahrend-Belau, E., Fuchs, J., Heinzel, N., Schreiber, F., Jakob, P. M. and Borisjuk, L. 2011. Combined noninvasive imaging and modeling approaches reveal metabolic compartmentation in the barley endosperm. *Plant Cell* 23(8), 3041–54. doi:10.1105/tpc.111.087015.

Rolletschek, H., Grafahrend-Belau, E., Munz, E., Radchuk, V., Kartäusch, R., Tschiersch, H., Melkus, G., Schreiber, F., Jakob, P. M. and Borisjuk, L. 2015. Metabolic architecture of the cereal grain and its relevance to maximize carbon use efficiency. *Plant Physiol.* 169(3), 1698–713. doi:10.1104/pp.15.00981.

Rösti, S., Rudi, H., Rudi, K., Opsahl-Sorteberg, H. G., Fahy, B. and Denyer, K. 2006. The gene encoding the cytosolic small subunit of ADP-glucose pyrophosphorylase in barley endosperm also encodes the major plastidial small subunit in the leaves. *J. Exp. Bot.* 57(14), 3619–26. doi:10.1093/jxb/erl110.

Ruan, Y. L. 2014. Sucrose metabolism: gateway to diverse carbon use and sugar signaling. *Annu. Rev. Plant Biol.* 65, 33–67. doi:10.1146/annurev-arplant-050213-040251.

Rubio-Somoza, I., Martinez, M., Diaz, I. and Carbonero, P. 2006a. HvMCB1, a R1MYB transcription factor from barley with antagonistic regulatory functions during seed development and germination. *Plant J.* 45(1), 17–30. doi:10.1111/j.1365-313X.2005.02596.x.

Rubio-Somoza, I., Martinez, M., Abraham, Z., Diaz, I. and Carbonero, P. 2006b. Ternary complex formation between HvMYBS3 and other factors involved in transcriptional control in barley seeds'. *Plant J.* 47(2), 269–81. doi:10.1111/j.1365-313X.2006.02777.x.

Sabelli, P. A., Liu, Y., Dante, R. A., Lizarraga, L. E., Nguyen, H. N., Brown, S. W., Klingler, J. P., Yu, J., LaBrant, E., Layton, T. M., Feldman, M. and Larkins, B. A. 2013. Control of cell proliferation, endoreduplication, cell size, and cell death by the retinoblastoma-related pathway in maize endosperm. *Proc. Natl. Acad. Sci. U. S. A.* 110(19), E1827–36. doi:10.1073/pnas.1304903110.

Sakuma, S., Lundqvist, U., Kakei, Y., Thirulogachandar, V., Suzuki, T., Hori, K., Wu, J., Tagiri, A., Rutten, T., Koppolu, R., Shimada, Y., Houston, K., Thomas, W. T. B., Waugh, R., Schnurbusch, T. and Komatsuda, T. 2017. Extreme suppression of lateral floret development by a single amino acid change in the *VRS1* transcription factor. *Plant Physiol.* 175(4), 1720–31. doi:10.1104/pp.17.01149.

Schmidt, R. R., Fulda, M., Paul, M. V., Anders, M., Plum, F., Weits, D. A., Kosmacz, M., Larson, T. R., Graham, I. A., Beemster, G. T. S., Licausi, F., Geigenberger, P., Schippers, J. H. and van Dongen, J. T. 2018a. Low-oxygen response is triggered by an ATP-dependent shift in oleoyl-CoA in *Arabidopsis*. *Proc. Natl. Acad. Sci. U. S. A.* 115(51), E12101–10. doi:10.1073/pnas.1809429115.

Schmidt, R. R., Weits, D. A., Feulner, C. F. J. and van Dongen, J. T. 2018b. Oxygen sensing and integrative stress signaling in plants. *Plant Physiol.* 176(2), 1131–42. doi:10.1104/pp.17.01394.

Schnurrbusch, T. 2019. Wheat and barley biology: towards new frontiers. *J. Integr. Plant Biol.* 61(3), 198–203. doi:10.1111/jipb.12782.

Schumacker, P. T. 2003. Current paradigms in cellular oxygen sensing. *Adv. Exp. Med. Biol.* 543, 57–71. doi:10.1007/978-1-4419-8997-0_5.

Schwender, J. 2008. Metabolic flux analysis as a tool in metabolic engineering of plants. *Curr. Opin. Biotechnol.* 19(2), 131–7. doi:10.1016/j.copbio.2008.02.006.

Scobie, M. N. and Jones, M. G. K. 2009. Identification of barley varieties by protein profiling. In: Zhang, G. and Li, C. (Eds), *Genetics and Improvement of Barley Malt Quality*. Springer, Berlin, Heidelberg, pp. 99–112. doi:10.1007/978-3-642-01279-2_4.

Scossa, F., Laudencia-Chingcuanco, D., Anderson, O. D., Vensel, W. H., Lafiandra, D., D'Ovidio, R. and Masci, S. 2008. Comparative proteomic and transcriptional profiling of a bread wheat cultivar and its derived transgenic line overexpressing a low molecular weight glutenin subunit gene in the endosperm. *Proteomics* 8(14), 2949–66.

Shaik, S. S., Obata, T., Hebelstrup, K. H., Schwahn, K., Fernie, A. R., Mateiu, R. V. and Blennow, A. 2016. Starch granule re-structuring by starch branching enzyme and glucan water dikinase modulation affects caryopsis physiology and metabolism. *PLoS ONE* 11(2), e0149613. doi:10.1371/journal.pone.0149613.

Shan, J., Zhang, F., Sharkey, J., Tang, T. A., Örd, T. and Kilberg, M. S. 2016. The C/ebp-Atf response element (CARE) location reveals two distinct Atf4-dependent, elongation-mediated mechanisms for transcriptional induction of aminoacyl-tRNA synthetase genes in response to amino acid limitation. *Nucleic Acids Res.* 44(20), 9719–32. doi:10.1093/nar/gkw667.

Shewry, P. R. and Halford, N. G. 2002. Cereal seed storage proteins: structures, properties and role in grain utilization. *J. Exp. Bot.* 53(370), 947–58. doi:10.1093/jexbot/53.370.947.

Shewry, P. R. and Ullrich, S. (Eds). 2014. *Barley: Chemistry and Technology* (2nd edn.). AACC International Inc, St. Paul, MN, pp. 11–54.

Slee, E. A., Harte, M. T., Kluck, R. M., Wolf, B. B., Casiano, C. A., Newmeyer, D. D., Wang, H. G., Reed, J. C., Nicholson, D. W., Alnemri, E. S., Green, D. R. and Martin, S. J. 1999. Ordering the cytochrome c-initiated caspase cascade: hierarchical activation of caspases-2, -3, -6, -7, -8, and -10 in a caspase-9-dependent manner. *J. Cell Biol.* 144(2), 281–92. doi:10.1083/jcb.144.2.281.

Sosso, D., Luo, D., Li, Q. B., Sasse, J., Yang, J., Gendrot, G., Suzuki, M., Koch, K. E., McCarty, D. R., Chourey, P. S., Rogowsky, P. M., Ross-Ibarra, J., Yang, B. and Frommer, W. B. 2015. Seed filling in domesticated maize and rice depends on SWEET-mediated hexose transport. *Nat. Genet.* 47(12), 1489–93. doi:10.1038/ng.3422.

Sparla, F., Falini, G., Botticella, E., Pirone, C., Talamé, V., Bovina, R., Salvi, S., Tuberosa, R., Sestili, F. and Trost, P. 2014. New starch phenotypes produced by tilling in barley. *PLoS ONE* 9(10), e107779. doi:10.1371/journal.pone.0107779.

Sreenivasulu, N. and Wobus, U. 2013. Seed-development programs: a systems biology-based comparison between dicots and monocots. *Annu. Rev. Plant Biol.* 64, 189–217.

Sreenivasulu, N., Radchuk, V., Strickert, M., Miersch, O., Weschke, W. and Wobus, U. 2006. Gene expression patterns reveal tissue-specific signaling networks controlling programmed cell death and ABA-regulated maturation in developing barley seeds. *Plant J.* 47(2), 310–27. doi:10.1111/j.1365-313X.2006.02789.x.

Sreenivasulu, N., Radchuk, V., Alawady, A., Borisjuk, L., Weier, D., Staroske, N., Fuchs, J., Miersch, O., Strickert, M., Usadel, B., Wobus, U., Grimm, B., Weber, H. and Weschke, W. 2010. Deregulation of abscisic acid contents causes abnormal endosperm development in the barley mutant *seg8*. *Plant J.* 64(4), 589–603. doi:10.1111/j.1365-313X.2010.04350.x.

Srichuwong, S. and Jane, J. L. 2007. Physicochemical properties of starch affected by molecular composition and structures: a review. *Food Sci. Biotechnol.* 16, 663– 74.

Stanca, A. M., Gianinetti, A., Rizza, F. and Terzi, V. 2016. Barley: an overview of a versatile cereal grain with many food and feed uses. In: Wrigley, C. W., Corke, H., Seetharaman, K. and Faubion, J. (Eds), *Encyclopedia of Food Grains* (2nd edn.). Elsevier, Oxford, pp. 147-52.

Sterna, V., Zute, S., Jansone, I. and Kantane, I. 2017. Chemical composition of covered and naked spring barley varieties and their potential for food production. *Polish J. Food Nutr. Sci.* 67(2), 151–8. doi:10.1515/pjfns-2016-0019.

Stevnebø, A., Sahlström, S. and Svihus, B. 2006. Starch structure and degree of starch hydrolysis of small and large starch granules from barley varieties with varying amylose content. *Anim. Feed Sci. Technol.* 130(1–2), 23–38. doi:10.1016/j.anifeedsci.2006.01.015.

Sweetlove, L. J., Beard, K. F. M., Nunes-Nesi, A., Fernie, A. R. and Ratcliffe, R. G. 2010. Not just a circle: flux modes in the plant TCA cycle. *Trends Plant Sci.* 15(8), 462–70. doi:10.1016/j.tplants.2010.05.006.

Taketa, S., Amano, S., Tsujino, Y., Sato, T., Saisho, D., Kakeda, K., Nomura, M., Suzuki, T., Matsumoto, T., Sato, K., Kanamori, H., Kawasaki, S. and Takeda, K. 2008. Barley grain with adhering hulls is controlled by an ERF family transcription factor gene regulating a lipid biosynthesis pathway. *Proc. Natl. Acad. Sci. USA* 105(10), 4062–7. doi:10.1073/pnas.0711034105.

Taylor, E. R., Nie, X. Z., MacGregor, A. W. and Hill, R. D. 1994. A cereal haemoglobin gene is expressed in seed and root tissues under anaerobic conditions. *Plant Mol. Biol.* 24(6), 853–62. doi:10.1007/BF00014440.

Tetlow, I. J., Liu, F. and Emes, M. J. 2015. Protein-protein interactions during starch biosynthesis. In: Nakamura, Y. (Ed.), *Starch Metabolism and Structure*. Springer, Tokyo, pp. 291–313.

Thiel, J., Weier, D., Sreenivasulu, N., Strickert, M., Weichert, N., Melzer, M., Czauderna, T., Wobus, U., Weber, H. and Weschke, W. 2008. Different hormonal regulation of cellular differentiation and function in nucellar projection and endosperm transfer cells - a microdissection-based transcriptome study of young barley grains. *Plant Physiol.* 148(3), 1436–52. doi:10.1104/pp.108.127001.

Thiel, J., Müller, M., Weschke, W. and Weber, H. 2009. Amino acid metabolism at the maternal-filial boundary of young barley seeds: a microdissection-based study. *Planta* 230(1), 205–13. doi:10.1007/s00425-009-0935-1.

Thiel, J., Riewe, D., Rutten, T., Melzer, M., Friedel, S., Bollenbeck, F., Weschke, W. and Weber, H. 2012. Differentiation of endosperm transfer cells of barley: a comprehensive analysis at the micro-scale. *Plant J.* 71(4), 639–55. doi:10.1111/j.1365-313X.2012.05018.x.

Thorbjørnsen, T., Villand, P., Kleczkowski, L. A. and Olsen, O. A. 1996. A single gene encodes two different transcripts for the ADP-glucose pyrophosphorylase small subunit from barley (Hordeum vulgare). *Biochem. J.* 313(1), 149–54. doi:10.1042/bj3130149.

Tomlinson, K. and Denyer, K. 2003. Starch synthesis in cereal grains. *Adv. Bot. Res.* 40, 1–61. doi:10.1016/S0065-2296(05)40001-4.

Topping, D. and Morell, M. 2014. Barley foods and public heals. In: Shewry, P. and Ullrich, S. E. (Eds), *Barley Chemistry and Technology*. AACC International, Inc., Eagan, MN, pp.223–33.

Tran, V., Weier, D., Radchuk, R., Thiel, J. and Radchuk, V. 2014. Caspase-like activities accompany programmed cell death events in developing barley grains. *PLoS ONE* 9(10), e109426. doi:10.1371/journal.pone.0109426.

Tschiersch, H., Borisjuk, L., Rutten, T. and Rolletschek, H. 2011. Gradients of seed photosynthesis and its role for oxygen balancing. *Biosystems* 103(2), 302–8. doi:10.1016/j.biosystems.2010.08.007.

Tschiersch, H., Liebsch, G., Borisjuk, L., Stangelmayer, A. and Rolletschek, H. 2012. An imaging method for oxygen distribution, respiration and photosynthesis at a microscopic level of resolution. *New Phytol.* 196(3), 926–36. doi:10.1111/j.1469-8137.2012.04295.x.

van Bel, A. J., Helariutta, Y., Thompson, G. A., Ton, J., Dinant, S., Ding, B. and Patrick, J. W. 2013. Phloem: the integrative avenue for resource distribution, signaling, and defense. *Front. Plant Sci.* 4, 471. doi:10.3389/fpls.2013.00471.

Van Dongen, J. T. and Licausi, F. 2015. Oxygen sensing and signaling. *Annu. Rev. Plant Biol.* 66, 345–67. doi:10.1146/annurev-arplant-043014-114813.

van Doorn, W. G., Beers, E. P., Dangl, J. L., Franklin-Tong, V. E., Gallois, P., Hara-Nishimura, I., Jones, A. M., Kawai-Yamada, M., Lam, E., Mundy, J., Mur, L. A., Petersen, M., Smertenko, A., Taliansky, M., Van Breusegem, F., Wolpert, T., Woltering, E., Zhivotovsky, B. and Bozhkov, P. V. 2011. Morphological classification of plant cell deaths. *Cell Death Differ.* 18(8), 1241–6. doi:10.1038/cdd.2011.36.

Vanlerberghe, G. C. 2013. Alternative oxidase: a mitochondrial respiratory pathway to maintain metabolic and signaling homeostasis during abiotic and biotic stress in plants. *Int. J. Mol. Sci.* 14(4), 6805–47. doi:10.3390/ijms14046805.

Vercammen, D., Declercq, W., Vandenabeele, P. and Van Breusegem, F. 2007. Are metacaspases caspases? *J. Cell Biol.* 179(3), 375–80. doi:10.1083/jcb.200705193.

Waduge, R. N., Hoover, R., Vasanthan, T., Gao, J. and Li, J. 2006. Effect of annealing on the structure and physicochemical properties of barley starches of varying amylose content. *Food Res. Intl.* 39(1), 59–77. doi:10.1016/j.foodres.2005.05.008.

Walley, J. W., Shen, Z., Sartor, R., Wu, K. J., Osborn, J., Smith, L. G. and Briggs, S. P. 2013. Reconstruction of protein networks from an atlas of maize seed proteotypes. *Proc. Natl. Acad. Sci. USA* 110(49), E4808–17. doi:10.1073/pnas.1319113110.

Wang, H. H., Wang, Z., Wang, F., Gu, Y. J. and Liu, Z. 2012. Development of basal endosperm transfer cells in *Sorghum bicolor* (L.) Moench and its relationship with caryopsis growth. *Protoplasma* 249(2), 309–21. doi:10.1007/s00709-011-0281-6.

Wang, J., Hu, P., Chen, Z., Liu, Q. and Wie, C. 2017. Progress in high-amylose cereal crops through inactivation of starch branching enzymes. *Front. Plant Sci.* 8, 469. doi:10.3389/fpls.2017.00469.

Watson, P. A. and Duffus, C. M. 1988. Carbon dioxide fixation by detached cereal caryopses. *Plant Physiol.* 87(2), 504–9. doi:10.1104/pp.87.2.504.

Weschke, W., Panitz, R., Sauer, N., Wang, Q., Neubohn, B., Weber, H. and Wobus, U. 2000. Sucrose transport into barley seeds: molecular characterization of two transporters and implications for seed development and starch accumulation. *Plant J.* 21(5), 455–67. doi:10.1046/j.1365-313x.2000.00695.x.

Wilkinson, L., Tucker, M. and Bird, D. 2018. Exploring the role of the ovule in cereal grain development and reproductive stress tolerance. *Annual. Plant Rev.* 1, 1–35.

Witzel, K., Surabhi, G. K., Jyothsnakumari, G., Sudhakar, C., Matros, A. and Mock, H. P. 2007. Quantitative proteome analysis of barley seeds using ruthenium(II)-tris-(batho phenanthroline-disulphonate) staining. *J. Proteome Res.* 6(4), 1325–33. doi:10.1021/pr060528o.

Witzel, K., Weidner, A., Surabhi, G. K., Varshney, R. K., Kunze, G., Buck Sorlin, G. H., Börner, A. and Mock, H. P. 2010. Comparative analysis of the grain proteome fraction in barley genotypes with contrasting salinity tolerance during germination. *Plant Cell Environ.* 33(2), 211–22. doi:10.1111/j.1365-3040.2009.02071.x.

Xi, D. M. and Zheng, C. C. 2011. Transcriptional regulation of seed storage protein genes in Arabidopsis and cereals. *Seed Sci. Res.* 21(4), 247–54. doi:10.1017/S0960258511000237.

Yamamoto, M. P., Onodera, Y., Touno, S. M. and Takaiwa, F. 2006. Synergism between RPBF Dof and RISBZ1 bZIP activators in the regulation of rice seed expression genes. *Plant Physiol.* 141(4), 1694–707. doi:10.1104/pp.106.082826.

Yamasaki, H., Shimoji, H., Ohshiro, Y. and Sakihama, Y. 2001. Inhibitory effects of nitric oxide on oxidative phosphorylation in plant mitochondria. *Nitric Oxide* 5(3), 261–70. doi:10.1006/niox.2001.0353.

Yangcheng, H., Gong, L., Zhang, Y. and Jane, J. L. 2016. Physicochemical properties of Tibetan hull-less barley starch. *Carbohydr. Polym.* 137, 525–31. doi:10.1016/j.carbpol.2015.10.061.

Yu, X., Chen, X., Wang, L., Yang, Y., Zhu, X., Shao, S., Cui, W. and Xiong, F. 2017. Novel insights into the effect of nitrogen on storage protein biosynthesis and protein body development in wheat caryopsis. *J. Exp. Bot.* 68(9), 2259–74. doi:10.1093/jxb/erx108.

Zhu, F. 2017. Barley starch: composition, structure, properties, and modifications. *Compr. Rev. Food Sci. Food Saf.* 16(4), 558–79. doi:10.1111/1541-4337.12265.

Zottini, M., Formentin, E., Scattolin, M., Carimi, F., Lo Schiavo, F. and Terzi, M. 2002. Nitric oxide affects plant mitochondrial functionality *in vivo*. *FEBS Lett.* 515(1–3), 75–8. doi:10.1016/s0014-5793(02)02438-9.

Zwirek, M., Waugh, R. and McKim, S. M. 2019. Interaction between row-type genes in barley controls meristem determinacy and reveals novel routes to improved grain. *New Phytol.* 221(4), 1950–65. doi:10.1111/nph.15548.

Chapter 4

Exploring barley germplasm for yield improvement under sulphur-limiting environments

Tefera Tolera Angessa, Murdoch University, Australia; Kefei Chen, Curtin University, Australia; David Farleigh, Jenifer Bussanich and Lee-Anne McFawn, Department of Primary Industries and Regional Development-Western Australia, Australia; Kevin Whitfield, CSBP Limited, Australia; Brendon Weir, Mullewa, Australia; Steve Cosh, Department of Primary Industries and Regional Development-Western Australia, Australia; Achalu Chimdi, Gudeta Nepir Gurmu and Tadesse Kenea Amentae, Ambo University, Ethiopia; and Chengdao Li, Murdoch University, Australia

1 Introduction
2 The origins of barley
3 Genetic diversity in barley
4 Using genetic diversity in breeding
5 The role of sulphur in barley growth
6 Assessing the effects of sulphur nutrition on barley and wheat grain yield
7 The effects of sulphur on yield, quality and response to stress
8 Farming systems and sulphur nutrition
9 Genotypic differences in sulphur use
10 Conclusion
11 Acknowledgement
12 References

1 Introduction

Barley (*Hordeum vulgare* L.) is an important cereal crop worldwide. It has established its importance beyond its confirmed centre of origin, the Fertile Crescent, and potential additional centre of origins in the Central Asia and the Tibetan Plateau-China Region. Secondary centres of origin, also called centres of diversity, have been established in countries such as Ethiopia, where diverse barley landraces are grown on ranges of high-altitude environments where

http://dx.doi.org/10.19103/AS.2019.0060.04

frost incidence is very frequent during the vegetative and reproductive growth period of barley. Barley breeders have successfully developed improved varieties that are high in grain yield and quality traits. Success in developing improved varieties is partly attributed to results from researches conducted on diverse barley germplasm with focus on traits important to agriculture. These include grain yield and quality parameters, adaptation to different growing conditions such as tolerance to soil acidity and aluminium toxicity, soil alkalinity, boron toxicity, resistance to foliar and root diseases and so forth. Improvement for nutrient-use efficiency, such as uptake of nitrogen, phosphorus, potassium and other nutrients, has received substantial degree of attention. Genotypes with superior characteristics have been identified for several important traits but their use in breeding programmes is yet to be fully realized due to the linkage drag associated with many genotypes from the landraces and wild types.

Despite the importance of sulphur in plant growth and development, yield and quality parameters, research attention paid to sulphur nutrition in plants and identification of germplasm associated with better adaptation to various levels of sulphur nutrition is very minimal. Sulphur deficiency results in greater losses of nitrogen to the environment. As sulphur has been associated with environmental pollution, its application in crop production has been reduced substantially leading to its deficiency in arable lands and thus effects on crop yield and quality traits (Blake-Kalff et al., 2001; Howarth et al., 2009; O'Connor et al., 1974). This chapter presents result from a preliminary study conducted with 15 barley varieties and five wheat varieties in the Midwest grain-growing region of Australia. Based on reports from elsewhere, this chapter highlights the importance of sulphur deficiency on crop yield and quality parameters, and reports genes associated with sulphur acquisition and use efficiency.

2 The origins of barley

Barley (*Hordeum vulgare* L.) is one of the most important cereal crops with diverse uses beyond its centre of origin across the globe. There are several ways of categorizing barley genetic resources into different groups. Barley genetic resources are generally categorized into three major groups. These are:

(1) wild germplasm (*Hordeum spontaneum* C Koch) that are progenitors of modern barley varieties (*Hordeum vulgare* L.),
(2) barley landraces that are commonly cultivated in countries where subsistence agriculture is practiced and in regions considered as secondary centres of origin,
(3) improved varieties or advanced breeding lines with near-to-perfect characteristics expected in modern commercial barley varieties.

The Fertile Crescent region where the ancestor of wild barley (*Hordeum spontaneum* C Koch) is widely distributed has been accepted as the centre of origin of cultivated barley (Badr et al., 2000). Angessa and Li (2017) summarized previous reports including the review by Nevo (2012) and references therein, and suggested the three unique regional distribution areas of wild barley that may coincide with the three different centres of origin of modern barley. These are:

(1) the Fertile Crescent characterized by warm and dry climates and rarely found above 1500-m altitude,
(2) Central Asia,
(3) China, particularly the Tibetan Plateau, characterized by cold and dry climates as high as 4000-m altitude.

Badr et al. (2000) also agreed that wild barley from Israel, Turkey and Iran was genetically variable. However, they argued that barley has a monophyletic origin with domestication in the Fertile Crescent region and then distributed to secondary habitats including Central Asia and the Himalayas.

Domestication of wild barley resulted in loss of adaptive traits that existed in wild barley suitable for survival under natural environments. Wild barley's adaptive traits are lost through selection for traits that are suited for modern agriculture. Such lost traits included non-brittle, non-shattering, easily threshable ears and other agronomically and qualitatively important traits of cultivated barley that have been developed during the last 10 000 years (Nevo, 2012). This resulted in the development of landraces characterized by high diversity and genetic variation between individual plants in the same accession. Such genetic diversity between individuals in the same accession has served as a risk management mechanism against diseases and insect pests that are common in regions where landraces are commonly grown. This risk management mechanism based on high genetic variations among individual plants in an accession is an insurance system for farmers with access to limited agricultural inputs such as fertilizers and fungicides.

High genetic diversity in barley landraces laid foundations for improved barley varieties that have been developed through conventional and modern breeding technologies. The potential to develop improved varieties using molecular genetics has further advanced to genomic breeding with the recent full sequencing of barley genome (Mascher et al., 2017). The barley genome sequence forms an important platform to link important agronomic and quality traits with the whole barley genome.

3 Genetic diversity in barley

The importance of barley genetic resources including barley germplasm has been well understood and efforts have been made to safeguard their availability in future breeding programmes. The germplasm collection at the Svalbard Global Seed Vault in Norway is the best example of investing heavily in germplasm conservation to serve the future generation. Unique germplasm sets are also kept at local or regional genebanks. They include the barley germplasm collection at Australian Grains Genebank (AGG), the collection at the Ethiopian Biodiversity Institute, ICARDA's barley germplasm collection, the collection at IPK Gatersleben in Germany, the USDA's 33 176 accessions categorized as landraces or cultivars/breeding lines obtained from more than 100 countries (Muñoz-Amatriaín et al., 2014) and Japan's Okayama University Barley Germplasm Resources collection. Barley's genetic resources in many of these institutions have been studied to discover economically important traits for use in breeding programmes. However, the full potential of many of these collections is yet to be explored.

Several studies have reported a very high genetic diversity in barley genetic resources that includes both cultivated types and wild barleys (Angessa and Li, 2017). Different methods are used to estimate the level of genetic diversity, including the barley SNP iSelect platform used to develop a core collection (Muñoz-Amatriaín et al., 2014), target enrichment of phenology genes (Hill et al., 2018, 2019), phenotypic traits, allozymes and various molecular markers (Cross, 1994; Badr et al., 2000; Mikel and Kolb, 2008; Nevo, 2012). Genetic diversity across the genome of wild barley is driven, to a substantial yet unknown degree, by natural selection that leads to strongly divergent phenotypes and genotypes even at microsites (Nevo, 2012 and references therein). It has been reported that wild barley from Israel displays remarkable variations in phenotypic traits that include germination, earliness, biomass and yield in accordance with climatic and edaphic variations that have exerted natural selection in the region. Badr et al. (2000) reported that barley lines from Iran, to which the Himalayan accessions are genetically related to, and those from Israel, to which lines from secondary habitats in the Mediterranean region belong to, were two genetically distinct wild barley gene pools.

Unlike wild barley, the genetic diversity of cultivated barley has been modified by selection for high yield and quality traits. Barley landraces still have considerable genetic diversity compared to modern commercial cultivars. An agro-morphological and electrophoresis study conducted on diverse collection of 1118 spring grown barley cultivars identified genotypes from Algeria, Afghanistan, Argentina, Ethiopia and India as genetically distinct (Cross, 1994). This study has revealed different pattern and groups of genetic diversity that reflect countries of germplasm origins. Mikel and Kolb (2008) studied genetic diversity in 363 contemporary North American barley cultivars registered during 1970–2006 using coefficient of parentage, which is the probability that two alleles at a locus are identical by descent. They reported that genetic diversity was greater among feed than malting barley cultivars.

Genetic diversity was much lower in 23 elite barley cultivars that constituted 87% of the US harvested barley varieties from 2001 to 2005. Within these 23 elite barley cultivars, diversity was substantially lower in six-row malting types than two-row malting types.

4 Using genetic diversity in breeding

Selections of landraces (also called primitive varieties) from their wild ancestors laid the foundation for the modern varieties. The selection practices from wild plant ancestors expanded plant species beyond their centre of origin. In modern agriculture, an improvement in grain yield and grain quality characteristics in barley and other grain crops have been the key breeding objectives. Additionally, although agricultural chemicals have been used heavily for controlling various diseases, development of genetically disease-resistant varieties has been identified as a more economical and environmentally healthy mechanism of disease control in barley and other agricultural crops.

Improved adaptation, grain yield and quality traits of barley will remain the major breeding objectives. Despite recent advances in technologies that have enhanced the efficiency of breeding programmes, genetic gain in grain yield appears to have stagnated. The yield plateaus that appeared to have been reached in many agricultural crops are indicative of the potential that a diverse gene pool can play in breeding programmes. The problem of crops such as barley yielding below their genetic potential is often referred to as yield gap. Overall factors causing the yield gap can be grouped into three broad categories, that is (i) regionally distinguishable factors that can be planned for on-broad scale that include factors associated with latitudinal ranges for example photoperiod, (ii) micro-environment limited factors that are mainly associated with soil factors and (iii) farm management and planning factors that include crop rotations, fertilizer inputs and disease management.

Developed countries such as Australia consider all potential factors causing yield gaps to maximize returns from agricultural production including barley. This is in contrast to the practice in developing countries such as Ethiopia where barley production still relies on landraces that have limited yield potential but are more resilient in response to biotic and abiotic stresses. These landraces are good sources of germplasm for stress factors such as frost damage (Reinheimer et al., 2004), drought stress, disease resistance and boron toxicity tolerance (Hayes et al., 2015). As an example, Negassa (1985) reported that the Ethiopian barley landraces were sources of disease resistance that included the most durable ml-o genes sourced exclusively barleys from the Ethiopian highlands. Reinheimer et al. (2004) indicated that much of the barley-growing area in Ethiopia has periods of extreme cold which has resulted in adaptation to low temperatures. The long season Australian barley variety Franklin derived from the crossing of Australian barley Shannon (Proctor*4/CI3208-1) and the German variety Triumph (Diamant/St 1402964/6) has a potentially frost tolerant CI3208-1 gene which was originated from Ethiopia.

Although breeding programmes have started to focus more on abiotic stress and development of tolerant/resistant varieties, this is still not a major focus in many cases. One of the best examples of this is the role of sulphur nutrition in barley growth and development. Traits related with sulphur, many of which have been the research focus for decades but with limited progress, are important in terms of both quantity and quality of crops (Koprivova and Kopriva, 2016). Although sulphur deficiency is typically observed as pale green particularly on young plant leaves, levels of sulphate deficiency sufficient to have a negative effect on crop yield may not produce visible symptoms (Howarth et al., 2009).

5 The role of sulphur in barley growth

Sulphur is the sixteenth most abundant element in the Earth's crust and is present in many compounds (Ceccotti et al., 1998). It is increasingly recognized as the fourth major plant nutrient after nitrogen, phosphorus and potassium (Jamal et al., 2010). However, sulphur production and consumption declined steadily during the early 1990s following increasing environmental pressures particularly in industrialized countries (Blake-Kalff et al., 2001; Ceccotti et al., 1998; Howarth et al., 2009). Also, incidental sulphur contents in single superphosphate (12% S) and ammonium sulphate (24% S) with dominant phosphorus and nitrogen contents masked soil depletions. However, the increased popularity of fertilizers that contained virtually no sulphur, such as urea, triple superphosphate and ammoniated phosphates, resulted in a drastic reduction in sulphur application as a percentage of the total fertilizer applied in world agriculture (Ceccotti, 1996).

The most important source of sulphur for plants is sulphate which can be taken up by roots in contrast to the phytotoxic-reduced sulphur in the rhizosphere (Ernst, 1998; Ceccotti et al., 1998). Ernst (1998) reported sulphate uptake kinetics (Vmax) estimated to 0.07–0.20 mol per gram fresh weight/hr in barley roots was lower than that estimated for phosphate (1.17–3.68), manganese (1.30–2.44), potassium (2.65–14.20), chloride (4.10–12.20) and sodium (5.48–9.00). Sulphur concentration in plants varies between 0.1% and 1.5% of dry weight (Ahmad et al., 2005).

Photosynthetic plants reduce essential inorganic nutrient sulphur available in the soil as sulphate and produce a large number of primary and secondary metabolites (Calderwood et al., 2014; Takahashi et al., 2011). Once absorbed and integrated into the photosynthates, old and mature leaves become an important source of sulphur for seed filling. However, sulphur fluxes are highly dependent on sulphur availability and on the timing of the deficiency, with late deficiency having less impact on seeds sulphate content (Koprivova and Kopriva, 2016).

Sulphur uptake through plant leaves has been described as a flux of gaseous sulphur compounds into plant leaves. The process has been ranked in the order of SO_2-sulphur dioxide > H_2S-hydrogen sulphide > COS-carbonyl sulphide > CH_3SH-methylmercaptan > CS_2-carbondisulphide (Ernst, 1998; Ceccotti et al., 1998 and references therein). Sulphur in its gaseous form particularly the SO_2 form has often been linked with atmospheric pollution. O'Connor et al. (1974) highlighted the impact of sulphur on human health and environmental pollution in the heavily populated and industrialized urban areas of the world. Due to a very strong association between sulphur compounds and atmospheric pollution, the use of sulphur fertilizer has declined in EU countries such as the UK that are committed to the reduction of SO_2 emissions (Howarth et al., 2009). SO_2 emissions have decreased in Europe due to most EU nations commitment to reduce emission levels (Wilson and Murray, 1990). For instance, the sulphur deposition at the Woburn Farm in the UK in 2018 was <5 kg per ha per year compared with 85 kg in 1970 (Holland et al., 2019). The 1996 protocol of the Economic Commission for Europe set to reduce SO_2 emissions by 80% compared to the 1980 base level from 2.5 million tons in 2000 to 0.98 million tons in 2010 (Blake-Kalff et al., 2001).

Despite sulphur being an essential nutrient in plant growth with similar requirement to phosphorus, it received little attention from researchers for many years (Jamal et al., 2010). Environmental mitigation strategies that targeted the reduction in emission of sulphur compounds, and the misconception that fertilizers and atmospheric inputs supplied the soil with adequate amounts, has resulted in a widespread sulphur deficiency throughout the world (Jamal et al., 2010). Over decades, as the composition of fertilizers changed, crop production intensified, and sulphur dioxide emissions decreased, sulphur deficiencies have become a serious problem (Ceccotti et al., 1998). Sulphur deficiency in agricultural crops has been predicted to increase significantly in Western Europe, partly because of (1) reduced sulphur dioxide (SO_2) emissions and subsequent reduction in sulphur depositions and (2) partly because of changes in fertilizers application practices. Wilson and Murray (1990) recommended that the process of assessment of crop responses to SO_2 must consider important factors such as cultivar and climate on a micro scale (local and regional), and that broad generalizations to a macro scale, involving state or national averaging, must be avoided. To demonstrate the effect of sulphur nutrition on grain yield, developmental stage and harvest time plant height, a preliminary study was conducted in the Midwest grain-growing region of Western Australia (WA) during the 2018 growing season.

6 Assessing the effects of sulphur nutrition on barley and wheat grain yield

6.1 Materials and methods

A preliminary study was conducted during the 2018 growing season with 15 barley and five wheat commercial varieties (Table 1) in the Midwest region of WA. The nearby Bureau of Meteorology's weather station (BOM, 2019) data indicated that the site received more or less optimal rainfall distributed across the growing season. Temperature recorded with Tinytag TGP-4500 at the site showed that temperature during vegetative growth period ranged between

Table 1 Commercial Barley and Wheat varieties evaluated for response to sulphur fertilizer rates

No	Variety	Crop	Year of release	Pedigree	Owning/Promoting company
1	Bass	Barley	2012	B28719/Alexis	InterGrain Pty Ltd
2	Baudin	Barley	2002	Stirling/Franklin	InterGrain Pty Ltd
3	Commander	Barley	2008	Keel/Sloop//Galaxy	University of Adelaide
4	Compass	Barley	2012	County/WI3416// Commander	University of Adelaide
5	Flinders	Barley	2015	Baudin/Cooper	InterGrain Pty Ltd
6	Granger	Barley	2013	Braemar/Adonis	Limagrain
7	Hamelin	Barley	2002	Stirling/Harrington	DPIRD/Ex DAFWA
8	La Trobe	Barley	2013	Dash/VB9409	InterGrain Pty Ltd
9	Mundah	Barley	1995	O'Connor/Yagan	DPIRD/Ex DAFWA
10	RGT Planet	Barley	2017	Tamtam/Concerto	Seed Force Australia
11	Rosalind	Barley		Lockyer/Dash	InterGrain Pty Ltd
12	Scope CL	Barley	2013	Franklin/VB9104// VB9104	Agri. VIC Services Pty Ltd
13	Sloop	Barley	1997	RL1577/84/Schooner	University of Adelaide
14	Spartacus CL	Barley		Scope/4*Hindmarsh// HMVB0325-106	InterGrain Pty Ltd
15	Vlamingh	Barley	2006	WABAR0570/TR118	DPIRD/Ex DAFWA
16	Bindawarra	Wheat	NA	Mexico 120/Koda// Raven	
17	Cutlass	Wheat	NA	RAC1316/2*RAC1400	AGT
18	Grenade	Wheat	NA		
19	Mace	Wheat	NA	CO6127;WAWHT2212/ RAC892;/WAWHT2212	AGT
20	Scepter	Wheat	NA	RAC1480/2*RAC1372	AGT

2 and 26°C with some exceptional days with the temperature dropping to below 0°C or increasing to higher than 26°C. During the flowering period and grain filling stages, the temperature at the site did not drop to below 0°C but increased to higher than 30°C on few occasions.

A Split-plot design with three sulphur levels as the main plot and 20 barley and wheat varieties as a subplot in three replicates was used in the study (Fig. 1). The trial was sown on 29 May 2018 in plots with 12 metres length and 1.1 metres width. The plots length was reduced to 10 metres with the pathway spray 2 months after seeding, and all the grains were harvested on 13 November 2018. Grain yield on per hectare basis was estimated from grains harvested from individual plots.

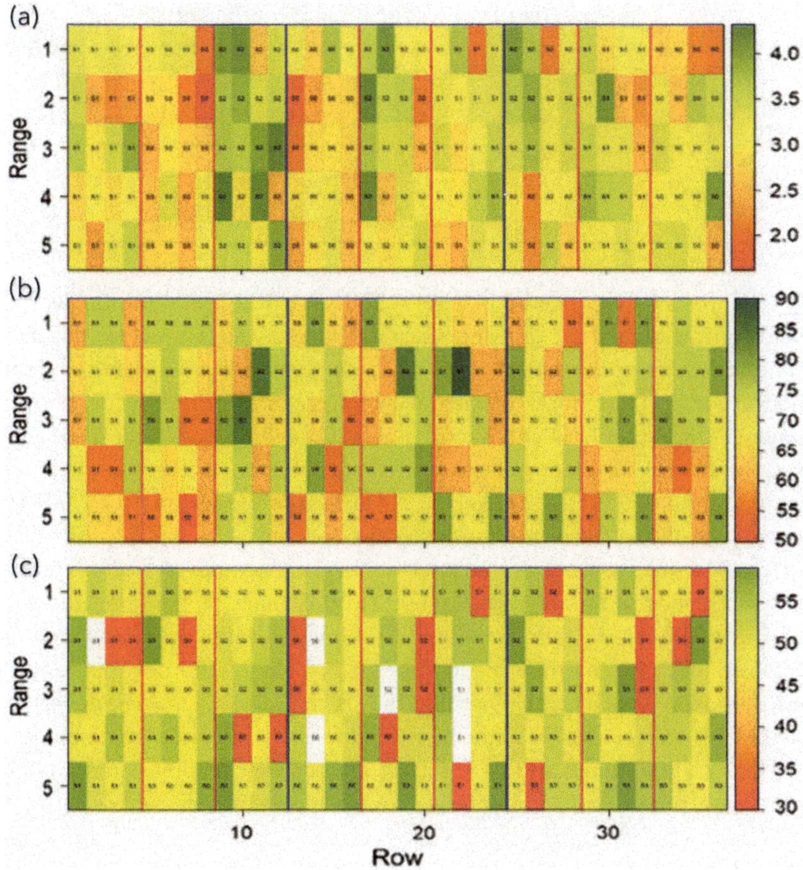

Figure 1 Heatmap (a) GrYld - grain yield in t/ha, (b) PlHt - plant height at maturity measured in cm, and (C) ZGS - Zadoks Growth Stage measured on the 86th day after planting. Each coloured rectangular unit represented subplots, while the red highlighted blocks represented main plot sulphur treatments, while the blue highlighted blocks represented the three replications.

Table 2 Sulphur fertilizer rates treatments applied at ZS02-04 (Zadoks growth stage 12 to 14)

Treatment	Time of application	Product	Rate (kg/ha)	Nitrogen (kg/ha)	Phosphorus (kg/ha)	Sulphur (kg/ha)
S0	At seeding	MAP	45	4.95	10.26	0.54
	At seeding	Urea	12	5.52		
	ZS12-14	Urea	64	29.44		
				39.91	*10.26*	*0.54*
S1	At seeding	MAP	45	4.95	10.26	0.54
	At seeding	Urea	12	5.52		
	At seeding	Agstar	73	10.44	10.22	7.01
	ZS12-14	Urea	64	29.44		
				50.35	*20.48*	*7.55*
S2	At seeding	MAP	45	4.95	10.26	0.54
	At seeding	Urea	12	5.52		
	At seeding	Agstar	73	10.44	10.22	7.01
	ZS12-14	Gran NS	33	6.3		7.2
	ZS12-14	Urea	50	23		
				50.21	*20.48*	*14.75*

The trial site was located in WA's AGZONE 1/AGZONE 4 region, which is characterized by a short growing season. The site was chosen for this study due to low sulphur content of 1 mg/kg content in the top 0–10 cm soil depth. The top soil 0–10 cm depth of the trial site was characterized by low nitrogen content (1 mg/kg of NH4 and 3 mg/kg of NO_3), sufficient phosphorus (15 mg/kg-Colwell) and sufficient potassium (75 mg/kg-Colwell). The soil pH ($CaCl_2$) was in a range of 6.1 (0–10 cm depth), 5.4 (10–20 cm depth) and 6.1 (20–30 cm depth) with the corresponding electrical conductivity of 0.0021, 0.011 and 0.015 dS/m, respectively. A rate of 7 kg/ha sulphur and 43 kg/ha nitrogen were recommended to mitigate the nutrient deficiency on the farm where the trial site was located. The recommended sulphur nutrition level was the basis of determining treatment levels in this study (Table 2). The only nutrient that was purposely applied at varying levels was sulphur which was applied at a rate of 0.54 kg/ha included in the base fertilizer MAP at the S0 treatment level, 7.55 kg/ha S1 treatment and 14.75 kg/ha S2 treatment. All other nutrients such as nitrogen, phosphorus and potassium were applied at sufficient or high rate.

6.2 Data analysis

Linear mixed models (LMM) were fitted with ASReml-R (Butler et al., 2018), where the variance parameters in the mixed model were estimated using the residual maximum likelihood (REML) procedure of Patterson and Thompson

(Patterson and Thompson, 1971). Residual diagnostics were performed to examine the validity of the model assumption (normality and homogeneity of variance). In the LMM, the fixed treatment structure is Crop/Variety + Sulphur + Crop:Variety:Sulphur, where Crop was a two-level factor describing the crop for barley and wheat; Variety was a 20-level factor describing 15 barley and five wheat commercial varieties; and Sulphur was a factor with three levels at negligible amounts of 0.54 kg/ha in the base fertilizer, 7.55 kg/ha and 14.75 kg/ha. The random block structure is the Block/Main plot, where the Block/Main plot described the three replicate blocks for each sulphur treatment levels and the complete set of 20 barley and wheat varieties included in the current study.

6.3 Results

Among the three traits (i.e. grain yield (Fig. 2a), plant height at maturity (Fig. 2b) and developmental stages scored on the 86th day after planting (Fig. 2c)), the current study demonstrated that sulphur treatment had a significant effect on grain yield as demonstrated by an associated P-value of 0.028. The distribution of the overall yield as reflected by the violin plot (Fig. 2a) that displayed in the same way as a boxplot was approximately normal with a skewness of −0.26 and kurtosis of −0.19. A highly significant varietal grain yield response to sulphur (S) treatments was observed in wheat and barley varieties included in this preliminary study. The interaction effect between sulphur treatment and crop type was on the borderline of significance with associated P-value of 0.066. This suggests that the sulphur effect potentially differs between barley and wheat crops.

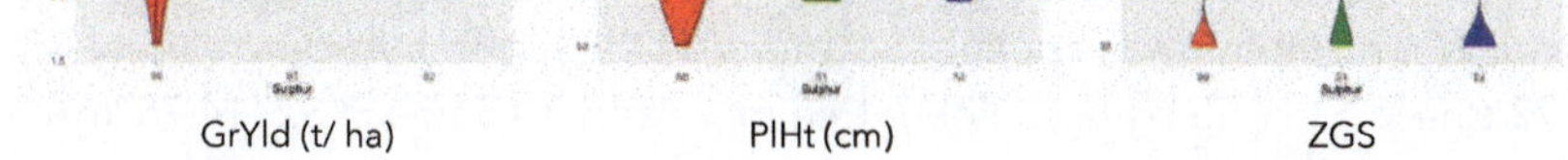

Figure 2 Violin plot (a) GrYld - grain yield, (b) PlHt - plant height, and (c) ZGS - Zadoks Growth Stage measured on the 86th day after planting as a mirrored density plot displayed in the same way as boxplots of observed traits by three levels of sulphur treatments. [Note: The median is represented by the line subdividing the box. The mean is represented by the dot in the box. The length of the box represents the interquartile range (IQR). One whisker extends to include all data points within 1.5 IQR of the upper quartile and stops at the largest value excluding outliers, and the other whisker extends to all data within 1.5 IQR of the lower quartile and stops at the smallest value excluding outliers.]

Overall there was a 0.30 t/ha increase in grain yield with S1 treatment compared to the S0 treatment. Compared to the S1 treatment, the overall increase in grain yield with S2 treatment was 0.25 t/ha. Despite differences in the number of varieties, barley varieties (n = 15) yielded much higher than wheat varieties (n = 5) in all three sulphur treatment levels. This was observed from the predicted mean grain yield of 3.05, 3.36, and 3.65 t/ha of all barley varieties compared with 2.24, 2.53 and 2.66 t/ha wheat yield recorded, respectively, with sulphur treatment of S0 (0.54 kg/ha), S1 (7.55 kg/ha) and S2 (14.75 kg/ha) levels. Compared with the negligible S0 sulphur level, the grain yield increased by 10.02% in barley and by 12.87% in wheat with the application of 7.55 kg/ha sulphur (S1 treatment). Increasing the sulphur level to 14.75 kg/ha (S2 treatment) has substantially increased the grain yield by 19.52% in barley and 18.68% in wheat compared to those in sulphur level S0.

All the wheat and barley varieties responded positively to an increase in sulphur level with an increase in grain yield. This included barley and wheat varieties that are normally recommended for the environments akin to the current trial site included in this study. Barley varieties that have been developed for such environments with respective yield levels in S0, S1 and S2 sulphur levels, Rosalind (3.40, 3.59, 4.03 t/ha), Scope CL (2.95, 3.25, 3.57 t/ha) and Compass (2.90, 3.37, 3.50 t/ha), responded substantially to sulphur application. An enormous increase of 1.0 t/ha grain yield was observed in Spartacus CL that recorded grain yield levels of 2.73, 3.13, 3.73 t/ha, respectively, in S0, S1 and S2 treatments. Among the barley varieties included in the current study, Mundah with 3.67 t/ha under negligible S0 level and 4.03 kg/ha under S2 level was the highest yielding variety (Fig. 3). In the S1 treatment with 7.55 kg/ha sulphur level, RGT Planet with 3.90 t/ha was the highest yielding variety.

Australian Premium White (APW) wheat variety Cutlass with mid-long maturity, and two Australian Hard (AH) varieties Mace and Scepter with short-mid maturity (DPIRD, 2019) were included in this study. Overall, the wheat varieties were the lowest yielding cereal entries in this preliminary trial. Mace with 2.61 t/ha was the highest yielding wheat variety in contrast to Bindawarra that yielded 1.85 t/ha grain yield under negligible S0 (0.54 kg/ha) sulphur level. When sulphur was applied, Scepter was the highest yielding variety with 2.92 t/ha in S1 (7.55 kg/ha) sulphur treatment and 3.13 t/ha in S2 (14.75 kg/ha) sulphur treatment.

In summary, this preliminary study demonstrated the importance of sulphur nutrition on cereal grain yield in the Midwest WA's grain-growing region. Overall, barley varieties yielded much higher than wheat varieties in an environment predominated with wheat and canola production, the later with very high demand for sulphur nutrition. Result from this study is indicative of the role that barley can play in increase agricultural productivity in the

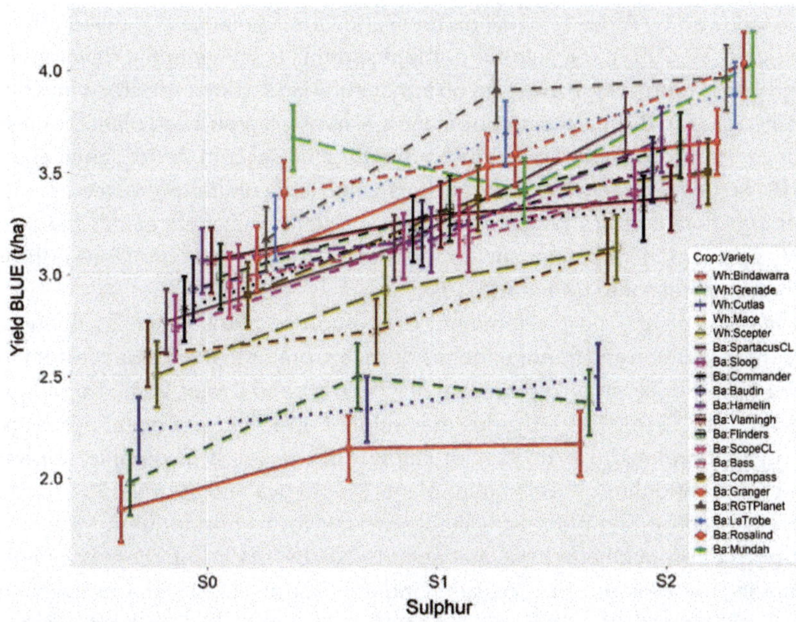

Figure 3 Line plot of predicted grain yield in wheat (5) and barley (15) varieties with three levels of sulphur treatments, S0 (0.54 kg/ha), S1 (7.55 kg/ha) and S3 (14.75 kg/ha). Error bars represented 1 × SED (standard error of difference).

region characterized by short growing season. Further investigation with more diverse barley and wheat germplasm on the importance of sulphur nutrition in the region will ascertain the current preliminary finding and paves a way for the efficient identification of genotypes with genes better adapted to soils and production systems with various sulphur levels. Identification of genes associated with sulphur nutrition and incorporation into adapted varieties will contribute to the improvement of barley yield plateau. Development of such varieties will improve agricultural productivity in the farming systems where sulphur deficiency is a contributing factor of low yield.

7 The effects of sulphur on yield, quality and response to stress

Sulphur is susceptible to leaching in agricultural soils and its deficiency occurs in all agricultural production systems across the world. In the soils of humid climatic zones, sulphate does not accumulate due to leaching with the annual loss estimated between 20 and 120 kg sulphur per ha/year in contrast to the arid conditions where sulphate can accumulate in the surface soil and finally precipitate as gypsum (Schachtschabel et al., 1992 in Ernst, 1998). The regional plant sulphur nutrient deficiency was estimated to be severe in Asia (6 million tons) followed by Africa (1.6 million tons) and North America (1.4 million tons; Messick et al., 2005). The sulphur nutrient shortfall is severe in the developing world where there is high pressure on agricultural land to increase crop yield per unit area to feed the increasing population. Harvested grains and plant biomass remove greater quantities of sulphur from the soil (Ceccotti, 1996; Jamal et al., 2010). Removal of sulphur through harvesting agricultural produces and its leaching characteristics beyond the reach of plants necessitates the need to apply sulphur fertilizers annually depending on specific soil conditions and/or agro-ecological zones (Menna, 2019).

In most plants, the predominant proportion of >90% organic sulphur is present as cysteine and methionine residues in proteins that are essential for the entire biological kingdom (Zhao et al., 1999; Wirtz and Droux, 2005; Koprivova and Kopriva, 2016). This indicates that sulphur and nitrogen metabolism are strongly interrelated. Up to 95% of sulphur in the soil is in organic sulphur compounds and thus unavailable to plants (Koprivova and Kopriva, 2016). The process of converting plant-unavailable organic forms of sulphur to inorganic oxidized anion sulphates that are absorbable by the plant roots is mainly controlled by biological processes (Lindeman et al., 1991). The conversion process is dependent on the size and activity of the microbial biomass, sulphur particle size, soil temperature, soil moisture and fertilizer formulation.

Once absorbed by the plant roots, sulphur is assimilated into a variety of cellular metabolites and amino acids cysteine and methionine, co-enzymes and prosthetic groups, sulpholipids, sulphate peptides and diverse secondary metabolites (Ernst, 1998; De Kok et al., 2011; Koprivova and Kopriva, 2016). Sulphur is important in vitamin synthesis and is essential for the synthesis of coenzyme A, the latter being required for fatty acid biosynthesis and oxidation, oxidation of intermediates of the citric acid cycle and ferredoxin that plays vital roles in photosynthesis and biological nitrogen fixation (Havlin et al., 2005). Sulphur- and nitrogen-containing thiol compounds, also called secondary sulphur compounds, play important roles in the protection of plants against stress and pests (De Kok et al., 2011). Studies conducted with sulphur in various environments indicate that sulphur deficiency limits crop yield and quality traits due to its role in many cellular metabolites and synthesis of essential amino acids.

7.1 Yield

Sulphur determines yield and yield-associated traits. Sulphur deficiency leads to leaf chlorosis affecting photosynthesis and yield loss. A study conducted to evaluate the effect of sources of nitrogen on sulphur use revealed that barley plants grown on sulphur-deficient soil for 7 weeks and supplemented with nitrogen increased biomass and sulphur concentrations in shoots of nitrate- and urea-supplied plants to the same extent (De Bona et al., 2011). However, nitrate-supplied plants under sulphur deficiency accumulated more nitrogen in the form of nitrate and asparagine than urea-supplied plants. Young, developing leaves from 14-day-old barley plants grown hydroponically with mild sulphate deficiency exhibited no symptom of sulphate deficiency, while plants with severe sulphate deficiency exhibited visible chlorosis with 40% lower SPAD units, has shown fresh weight reduction of 30% in leaves and 10% in roots (Astolfi et al., 2006). Sulphur-deficient plants are characteristically small and spindly with the younger leaves appearing light green to yellowish in colour (Ceccotti, 1996). Sulphur promoted tillering appearance rate under high nitrogen conditions (Alzueta et al., 2012) and increased the number of tillers per emerged leaf by 24% (Salvagiotti and Miralles, 2007). Salvagiotti and Miralles (2007) reported that the highest 11.6 tillers per plant was achieved in wheat when sulphur and nitrogen were not deficient, while nitrogen and sulphur deficiencies reduced tiller production, respectively, by 37% and 15%. They reported that under nitrogen deficiency, tiller production stopped when the apex reached the double-ridge stage in contrast to under higher nitrogen supply where tillering stopped at terminal spikelet. Lack of difference in phenology due to different levels of nitrogen and sulphur applied at sowing time was reported in wheat and barley (Alzueta et al., 2012). Unlike under field conditions where nitrogen and sulphur did not modify the duration of different phenological stages in wheat, a high sulphur rate under optimum nitrogen condition delayed the period from emergence to anthesis by 65° days in a controlled environment (Salvagiotti and Miralles, 2007).

7.2 Quality

Sulphur plays important roles in barley malt quality. It has been accepted that the malt extract is negatively related with the grain protein content and positively with the grain size, the negative effect of proteins being reported due to the main storage protein hordeins in barley grains enveloping starch granules and thus inhibiting their hydrolysis during maceration (Prystupa et al., 2019 and references therein). A field-based study revealed that sulphur fertilization at a rate of 10 kg/ha under different levels of nitrogen availability affected hordein composition, increased malt extract and apparent attenuation limit and decreased malt hardness (Prystupa et al., 2019). Hordeins are classified into B (rich in sulphur), C (poor in sulphur), D (intermediate sulphur content) and γ (rich in sulphur) fractions with 70–80% B fraction and 10–20% C fraction accounting for the total hordein content, while the D and γ groups are quantitatively minor components (Prystupa et al., 2019 and references therein).

Sulphur plays key roles in amino acid synthesis and thus the protein quality. In wheat, the reduction in sulphur availability was very problematic from the beginning of the twenty-first century as reflected by the frequent acrylamide detection in processed foods. The review of earlier studies by Koprivova and Kopriva (2016) reported that the Maillard reaction between free amino acids, mainly asparagine, and the reducing sugars during high-temperature processing produce acrylamide due to the content of free asparagine in wheat flour. In barley, urea supply under limited sulphur availability prevented excess accumulation of asparagine but did not improve growth and nitrogen-use efficiency (De Bona et al., 2011). Adequate amount of sulphur is required to synthesize sulphur-containing amino acids including cysteine, which is required for the formation of disulphide bonds (Shewry, 2011). Sulphur deficiency in *B. napus* reduced nitrogen-use efficiency and the incorporation of nitrogen and sulphur into proteins by 29% and 62%, respectively (Lee et al., 2016).

7.3 Adaptation to abiotic and biotic stress factors

Sulphur plays important roles in resistance to diseases, tolerance to extreme temperature levels and salinity stress. Many stress defence reactions are dependent on sulphur-containing metabolites that include glutathione due to its role in the Halliwell-Asada cycle for detoxification of reactive oxygen species. Glutathione is involved in chilling tolerance of maize and response to salinity in rice (Koprivova and Kopriva, 2016). Environmental factors such as drought affect sulphur-containing amino acids and proteins. Compared with control plants, polyethylene glycol-induced drought stressed plants had their sulphur-containing amino acids decreased significantly 72 h after treatment by 48% (leaves) and 39% (roots) in Mosa, and 37% (leaves) and 36% (roots) in Saturnin (Lee et al., 2016). Drought stress decreased sulphur-containing proteins by 42% (leaves) and 38% (roots) in Mosa compared with 27% (leaves) and 21% (roots) in Saturnin 72 h after treatment compared to control plants (Lee et al., 2016). A review by Ihsan et al. (2019) summarized the role of sulphur biomolecules in response to heat stress that included detoxification of reactive oxygen species and redox control of proteins.

Sulphur modulates the stress response and removes reactive oxygen species through thiol-containing compounds, especially reduced glutathione, which is sensitive to oxidized environments (Lee et al., 2016). Sulphur is an essential nutrient for the synthesis of methionine which in turn is required in the biosynthesis pathway of phytosiderophores, the metabolic modifications necessary to cope with iron deficiency (Astolfi et al., 2006; Lee et al., 2016). Astolfi et al. (2006) observed in barley variety Europa experimented with the contrasting supply of iron and sulphur and reported that sulphur availability can influence either the release of phytosiderophores or the ability to take up iron from an external solution. Plant roots secretion of phytosiderophores to the rhizosphere that include mugineic acid (MA), deoxymugineic acid (DMA) and epi-hydroxymugineic acid (epi-HMA) is an important step in the acquisition of soluble soil iron in graminaceous species under iron deficiency (Astolfi et al., 2006). Sulphur has been reported to play roles in combating arsenic toxicity. Smith et al. (1997) reported that all arsenic toxicity-combating compounds contain thiol groups. It was reported that the high tissue concentrations of arsenic observed in many plants necessitate a higher uptake of sulphur for thiol synthesis to which AsIII can be conjugated to.

8 Farming systems and sulphur nutrition

Sulphur requirements in plants vary substantially among species. For instance, the amount of sulphur required to produce 1 ton of seed is about 3–4 kg in cereals (range 1–6), 8 kg in legume crops (range 5–13) and 12 kg in oil crops (range 5–20) (Jamal et al., 2010). Oil seed rape and Brassica species have characteristically high sulphur demand during vegetative growth for the synthesis of proteins (Blake-Kalff et al., 2001; Calderwood et al., 2014; Koprivova and Kopriva, 2016; Zhao et al., 1999). In addition to nitrogen, oil crops require about the same amount of sulphur or more phosphorus to achieve high yield and quality levels (Jamal et al., 2010).

Barley is perceived to perform well on poor and low-fertility soils. Even though barley yields are comparable to those of wheat on fertile soils, fertilizer rates to barley crops have been low (GRDC, 2018). It has also been reported that including sulphur, nitrogen, phosphorus, potassium, zinc, copper, manganese and molybdenum are some of the limiting nutrients in some barley-growing soils of Australia. Sulphur deficiency in barley limits the synthesis of metabolites that play important roles in plants' defensive mechanisms against biotic and abiotic stress factors (Calderwood et al., 2014; Koprivova and Kopriva, 2016).

Farming systems in different regions have their peculiar characteristics in terms of the agricultural inputs used, objectives of production, and most importantly they differ substantially by the production environments. The production environments are very complex due to the varying degrees of interacting abiotic and biotic stress factors determining yield and quality levels. These interacting biotic and abiotic stress factors play significant roles on the efficiency and use of plant nutrients that include sulphur.

In an intercropping farming system, growing a mixture of two or more companion crops increases yield and improves nutrient availability to the plants. For instance, an organically cultivated intercropping of barley and pea produced maximum relative seed and shoot yields; accumulated higher nitrogen and sulphur in the shoot than pure stands of pea and spring barley (Pötzsch et al., 2019). In the monocropping system, various crop species are grown in temporal sequence with the aim to maximize yield, improve nutrients availability and break various diseases cycle. Results from six sites from the

Northern Australian grain zone reported that chickpeas grown following wheat crop provided about 42 kg nitrogen/ha which was equivalent to 1.0 t/ha grain yield benefit. However, the rotational benefit of pulses on wheat yields tends to last for one season only, where the residual benefits of chickpeas for a second wheat crop were small and inconsistent (GRDC, 2011).

Canola is one of the most important oil crops with very high demand for sulphur. Canola production in Australia has increased enormously from 50 000 ha (78 000 tons grain) in 1989 to 3 171 000 ha (3 893 000 tons grain) in 2017 (ABARES, 2019). However, the national average grain yield of 1.8 t/ha recorded in 1989 has not improved over the 30-year period. Though improvement for grain yield is largely driven by the development of improved varieties by the breeding programmes, agronomic management practices that include use of appropriate rate of fertilizers, disease control and proper temporal crop rotation system contribute to yield increase. A typical crop rotation of wheat-wheat-canola-barley is a predominant system used for disease and weeds burden break in Australia. The crop rotation farming system practices consisting of 4 years of cereal and 1 year of canola, or 3 years of cereal and 2 years of canola, consistently provided the greatest gross margins and thus was the best performing rotation across all agricultural regions in WA (DPIRD, 2015, https://www.agric.wa.gov.au/news/media-releases/five-years-data-gives-insight-wa-crop-rotations).

The preliminary study on 15 barley varieties and five wheat varieties demonstrated genotypic differences in response to different sulphur levels. This observation has been reported from elsewhere for crops that included barley, wheat and canola. Sulphur being one of the most important plant nutrients that limits yield and quality, its most dynamic nature of leaching out of the reach of plant roots necessitates the effort to develop sulphur-efficient varieties. Sulphur-use efficiency (SUE) is rather a complex trait that involves the uptake and utilization efficiencies, may reduce the dependence on sulphur fertilizer and reduce associated environmental risks (Koprivova and Kopriva, 2016).

9 Genotypic differences in sulphur use

9.1 Varietal differences

Variation in agricultural crops response to growth environments are the basis of improved varieties development. Environments determine availability of nutrients such as sulphur and nitrogen and thus affect yield and quality traits. For instance, under polyethylene glycol-induced drought stress environment, significant varietal differences were observed in total sulphur content in *B. napus* varieties Saturnin (50%) and Mosa (80%) leaves, and in Mosa (41%) roots (Lee et al., 2016). Estimated SUE based on sulphur uptake and sulphur assimilation showed that SUE was much higher in Saturnin than in Mosa, and Saturnin with higher SUE was more tolerant to PEG-induced drought stress (Lee et al., 2016). It was reported that the total amount of sulphur incorporated into amino acids and proteins was generally higher in Saturnin (664 mg sulphur per plant) than in Mosa (337 mg sulphur per plant). Environmental factors such as water shortage strongly affect SUE. The inhibition of photosynthetic activity including Rubisco protein degradation caused by drought stress was much lower in Saturnin with higher SUE than Mosa (Lee et al., 2016). Drought stress decreased the amount of sulphur assimilated into amino acids and proteins with the total amount of newly absorbed sulphur varying considerably between varieties Mosa (−64%) and Saturnin (−41%; Lee et al., 2016).

Dari et al. (2019) conducted a field study on high pH calcareous soil during the 2015 and 2016 growing seasons in Idaho, USA, with two barley cultivars ABI-Voyager and Moravian 69 with three rates of sulphur (0, 112, 224 kg/ha) and five rates of phosphorus (0, 37, 73, 110 and 147 kg/ha) treatments. They reported that sulphur had no significant effect on biomass yield, grain yield and harvest index, while phosphorus application has improved the crop performance.

Unlike reports on shortage of sulphur in the soil, toxicity due to excess amount of atmospheric sulphur has been reported to limit crops growth and development. A study conducted with five spring barley varieties revealed an order of tolerance in the varieties with Midas > Tyra > Golden Promise > Patty > Koru in response to SO_2 toxicity. SO_2 at 0.08 and 0.12 μL per litre significantly reduced the growth of every part in all the varieties except for the number of tillers in Patty in 0.08 μL per litre SO_2 (Pande, 1985). In a low concentration of 0.4 μL per litre SO_2, no significant effect was observed on tillering number of any of the five spring barley varieties, while the number of leaves was significantly reduced in variety Koru and the leaf dry weights were not much affected in Midas, Patty and Tyra (Pande, 1985). The root weight rather decreased by more than 26% in Koru and 25% in Golden Promise but slightly increased in Midas (4%) and Tyra (7%).

In wheat, a study conducted with variety Banks exposed to five SO_2 concentrations that ranged from 0.004–0.517 μL per litre for 79 days for 4 h per day revealed that SO_2 level of 0.042 μL per litre or more significantly reduced plant height, shoot weight, developmental stage, tiller number, ear weight per plant, average ear weight and total ear number (Wilson and Murray, 1990). A summary of earlier works by Wilson and Murray (1990) indicated that SO_2 concentration level of 0.10 μL per litre SO_2 reduced shoot dry weight in wheat variety Halberd; reduced ears, grain and shoots dry weight by ~25% in variety Eradu; reduced yield in variety Yecora Rojo (0.030 μL per litre) and in variety Vona (0.039, 0.166 or 0.363 μL per litre). When plants other than agricultural crops were considered, O'Connor et al. (1974) reported that casuarina species were most resistant while Eucalyptus were the most sensitive tree species to SO_2 toxicity. Acacia exhibited an acute injury in response to over 3 h exposure to 1 ppm of SO_2. Wilson and Murray (1990) indicated that SO_2 severity on winter cereals depended on the growth stages and other environmental factors such as low temperature exacerbating the level of damage.

In general, improvement for plant sulphate uptake can be achieved in three different ways. These are (1) through genetic mechanisms with sulphate transporters that have affinity to low sulphur concentration, (2) through agronomic management by increasing sulphur availability in the soil and (3) by improving plant sulphur nutrition by biofertilization with appropriate microorganisms of certain Pseudomonas and mycorrhiza fungi (Koprivova and Kopriva, 2016; Takahashi et al., 2011). However, SUE is a complex trait for which a step improvement is inconceivable mainly because associated key genes are unknown currently (Koprivova and Kopriva, 2016).

9.2 Genes associated with sulphur use

Sulphate is taken up from soil and distributed around the plant by a family of membrane-bound sulphate transporters. An assay of hydroponically grown barley plants for their nitrate and nitrite reductase activities in response to sulphur supply revealed that nitrate reductase activity sharply decreased under limiting sulphur supply, while nitrite reductase activity did not respond to sulphur supply, indicating that nitrate reduction rather than nitrite reduction represents the sulphur-limited assimilatory process (De Bona et al., 2011; Fazili et al., 2008). Sulphate transporters and the two enzymes of sulphate assimilation, ATP sulfurylase and APS reductase, are key genes associated with SUE pathways (Koprivova and Kopriva, 2016). A cDNA encoding a high-affinity sulphate transporter isolated from barley, designated HVST1, that encodes a polypeptide of 660 amino acids has been reported to be strong pH-dependent proton/sulphate co-transporter, expressed specifically in root tissues with sulphur starvation increasing abundance of mRNA corresponding to HVST1 and the capacity of roots to take up sulphate (Smith et al., 1997). They reported that adequate sulphur supply reduced the abundance of mRNA corresponding to HVST1 and the capacity of root sulphur uptake. The identity between HVST1 and the plant H+/sulphate co-transporters SHST1, SHST2, SHST3 and AST56 was 69.2%, 68.6%, 51.9% and 47.2%, respectively, where the highest identities were with the high-affinity plant H+/sulphate co-transporters, SHST1 and SHST2 (Smith et al., 1997). The effect of sulphate, cysteine and glutathione possibly be candidates for negative metabolic regulators of sulphate transporter gene expression may have been overridden by the addition of the cysteine precursor O-acetylserine that lead to increases in sulphate transporter mRNA, sulphate uptake rates and tissue contents of glutathione and cysteine (Smith et al., 1997).

In wheat variety Hereward grown under nutrient-deficient field environment, a gene induced by sulphate deficiency, *sdi1*, has been reported as a sensitive indicator of sulphur nutritional status (Howarth et al., 2009). The gene *sdi1* was induced in leaf and root tissues in response to sulphate deficiency, but was not induced by nitrogen, phosphorus, potassium or magnesium deficiency. It was also reported to increase in plant tissues as the external sulphate concentration in hydroponically grown plants was reduced from 1.0 to 0.0 mM. The *Arabidopsis thaliana* gene most highly homologous to *sdi1* was reported to be At5g48850, and the Atsdi1 T-DNA 'knockout' mutants were shown to maintain higher tissue sulphate concentrations than wild-type plants under sulphur-limiting conditions.

10 Conclusion

Sulphur is one of the most important plant nutrients with substantial effect on growth and development of agricultural crops that include barley, wheat and canola. Efforts that targeted sulphur compounds to mitigate environmental pollution have resulted in sulphur deficiency in agriculture. The economic cost of widespread sulphur deficiency is yet to be estimated, but research findings indicate that sulphur deficiency causes substantial yield loss and reduces quality. A preliminary study with 15 barley varieties and five wheat varieties conducted in WA demonstrated a grain yield increase of 10.02% in barley and 12.87% in wheat by the application of 7.55 kg/ha sulphur. The grain yield increase was tremendous when the sulphur level was increased to 14.75 kg/ha which was predicted to be 19.52% in barley and 18.68% in wheat when compared with the yield level with S0 (0.54 kg/ha sulphur) treatment. This result is indicative of genotypic differences in both barley and wheat in response to sulphur nutrition. Most importantly, wheat being the most dominant crop grown in the Midwest grain-growing region of WA, a substantial yield advantage of barley over wheat under different sulphur nutrition implies the potential that barley can play to increase agricultural productivity in the region. This observation should be further expanded to include more diverse barley and wheat genotypes and environments to identify genes associated with sulphur acquisition and transport in cereal crops.

11 Acknowledgement

This study is part of the Grains Research and Development Corporation (GRDC)-funded barley project UMU00049 – Maintenance of grain plumpness and transfer of heat tolerance into Australian barley germplasm.

12 References

ABARES - Australian Bureau of Agricultural and Resource Economics and Sciences (2019), Australian Crop Report, Canberra, June. CC BY 4.0, https://doi.org/10.25814/5cf8e645b377e.

Ahmad, A., Khan, I., Anjum, N. A., Abrol, Y. P. and Iqbal, M. (2005), Role of sulphate transporter systems in sulphur efficiency of mustard genotypes. *Plant Science* 169: 842–6.

Alzueta, I., Abeledo, L. G., Mignone, C. M. and Miralles, D. J. (2012), Differences between wheat and barley in leaf and tillering coordination under contrasting nitrogen and sulfur conditions. *European Journal of Agronomy* 41: 92–102, http://dx.doi.org/10.1016/j.eja.2012.04.002.

Angessa, T. T. and Li, C. (2017), Exploration and utilization of genetic diversity exotic Germplasm for barley improvement. In G. Zhang and C. Li (Eds), *Exploration, Identification and Utilization of Barley Germplasm*, Elsevier Inc., Zhejiang University Press Co., Ltd., China, ISBN: 978-0-12-802922-0, pp. 223–40.

Astolfi, S., Cesco, S., Zuchi, S., Neumann, G. and Roemheld, V. (2006), Sulfur starvation reduces phytosiderophores release by iron-deficient barley plants. *Soil Science and Plant Nutrition* 52: 43–48, doi:10.1111/j.1747-0765.2006.00010.x.

Badr, A., Mueller, K., Schaefer-Pregl, R., El Rabey, H., Effgen, S., et al. (2000), On the origin and domestication history of barley (*Hordeum vulgare*). *Molecular Biology and Evolution* 17: 499–510.

Blake-Kalff, M. M. A., Zhao, F.-J., Hawkesford, M. J. and McGrath, S. P. (2001), Using plant analysis to predict yield losses caused by sulphur deficiency. *Annals of Applied Biology* 138: 123–7.

BOM - Common Wealth Government Bureau of Meteorology (2019), http://www.bom.gov.au/jsp/ncc/cdio/weatherData/av?p_display_type=dataSGraph&p_stn_num=008292&p_nccObsCode=136&p_month=13&p_startYear=2018.

Butler, D. G., Cullis, B. R., Gilmour, A. R. and Gogel, B. J. (2018). ASReml-R 4 reference manual: mixed models for S language environments: Queensland Department of Primary Industries and Fisheries.

Calderwood, A., Morris, R. J. and Kopriva, S. (2014), Predictive sulfur metabolism – a field in flux. *Frontiers in Plant Science* 5: 646, doi:10.3389/fpls.2014.00646.

Ceccotti, S. P. (1996), Plant nutrient sulphur – A review of nutrient balance, environmental impact and fertilizers. *Fertilizer Research* 43: 117–25.

Ceccotti, S. P., Morris, R. J. and Messick, D. L. (1998), A global overview of the sulphur situation: Industry's background, market trends, and commercial aspects of sulphur fertilizers. In E. Schnug (Eds), *Sulphur in Agroecosystems. Nutrients in Ecosystems*, vol. 2. Springer, Dordrecht, https://doi.org/10.1007/978-94-011-5100-9_6.

Cross, R. J. (1994), Geographical trends within a diverse spring barley collection as identified by agro/morphological and electrophoretic data. *Theoretical and Applied Genetics* 88: 597–603.

Dari, B., Rogers, C. W. and Liang, X. (2019), Plant, grain, and soil response of irrigated malt barley as affected by cultivar, phosphorus, and sulfur applications on an alkaline soil. *Journal of Plant Nutrition*, https://doi.org/10.1080/01904167.2019.1589504.

De Bona, F. D., Fedoseyenko, D., von Wirén, N. and Monteiro, F. A. (2011), Nitrogen utilization by sulfur-deficient barley plants depends on the nitrogen form. *Environmental and Experimental Botany* 74: 237–44.

De Kok, L. J., Stulen, I. and Hawkesford, M. J. (2011), Sulfur nutrition in crop plants. In M. J. Hawkesford and P. Barraclough (Eds), *The Molecular and Physiological Basis of Nutrient Use Efficiency in Crops*, 1 st edn, John Wiley & Sons, Inc., UK, pp. 295–310.

DPIRD – Department of Primary Industries and Regional Development (2015), https://www.agric.wa.gov.au/news/media-releases/five-years-data-gives-insight-wa-crop-rotations.

DPIRD – Department of Primary Industries and Regional Development (2019), Wheat varieties sowing guide for Western Australia. *Bulletin 4894*. ISSN No. 1833 7236, https://www.agric.wa.gov.au/sites/gateway/files/2019%20Wheat%20Variety%20Guide-web.pdf.

Ernst, W. H. O. (1998), Sulfur metabolism in higher plants: Potential for phytoremediation. *Biodegradation* 9: 311–18.

Fazili, I. S., Jamal, A., Ahmad, S., Masoodi, M., Khan, J. S. and Abdin, M. Z. (2008), Interactive effect of sulfur and nitrogen on nitrogen accumulation and harvest in oilseed crops differing in nitrogen assimilation potential. *Journal of Plant Nutrition* 31(7): 1203–20, doi:10.1080/01904160802134905.

GRDC – Grains Research and Development Corporation (2011), *Choosing Rotation Crops Fact Sheet*, https://grdc.com.au/data/assets/pdf_file/0024/223683/grdcfsbreakcropsnorthpdf.pdf.

GRDC – Grains Research and Development Corporation (2018), *Grownotes: Section 5: Barley – Nutrition and Fertiliser*, https://grdc.com.au/__data/assets/pdf_file/0026/369233/GrowNote-Barley-South-5-Nutrition-and-fertiliser.pdf, accessed 31 July 2019.

Havlin, J. L., Beaton, J. D., Tisdale, S. L. and Nelson, W. L. (2005), *Soil Fertility and Fertilizers: An Introduction to Nutrient Management*, 7th edn, Pearson Prentice Hall, New Jersey, 528pp.

Hayes, J. E., Pallotta, M., Garcia, M., Öz, M. T., Rongala, J. and Sutton, T. (2015), Diversity in boron toxicity tolerance of Australian barley (*Hordeum vulgare* L.) genotypes. *BMC Plant Biology* 15: 231, doi:10.1186/s12870-015-0607-1.

Hill, C., Angessa, T. T., McFawn, L., Wong, D., Tibbits, J., et al. (2018), Hybridisation-based target enrichment of phenology genes to dissect the genetic basis of yield and adaptation in barley. *Plant Biotechnology Journal*, https://doi.org/10.1111/pbi.13029.

Hill, C. B., Wong, D., Josquin, T., Kerrie, F., Matthew, H., et al. (2019), Targeted enrichment by solution-based hybrid capture to identify genetic sequence variants in barley. *Nature Scientific Data* 6: 12, https://doi.org/10.1038/s41597-019-0011-z.

Holland, J. E., White, P. J., Glendining, M. J., Goulding, K. W. T. and McGrath, S. P. (2019), Yield responses of arable crops to liming – An evaluation of relationships between yields and soil pH from a long-term liming experiment. *European Journal of Agronomy*, https://doi.org/10.1016/j.eja.2019.02.016.

Howarth, J. R., Parmar, S., Barraclough, P. B. and Hawkesford, M. J. (2009), A sulphur deficiency-induced gene, sdi1, involved in the utilization of stored sulphate pools under sulphur-limiting conditions has potential as a diagnostic indicator of sulphur nutritional status. *Plant Biotechnology Journal* 7: 200–9, doi: 10.1111/j.1467-7652.2008.00391.x.

Ihsan, M. Z., Daur, I., Alghabari, F., Alzamanan, S., Rizwan, S., et al. (2019), Heat stress and plant development: Role of sulphur metabolites and management strategies. *Acta Agriculturae Scandinavica, Section B – Soil & Plant Science* 69(4): 332–42, https://doi.org/10.1080/09064710.2019.1569715.

Jamal, A., Moon, Y.-S. and Abdin, M. Z. (2010), Sulfur - a general overview and interaction with nitrogen. *Australian Journal of Crop Science* 4(7): 523-9.

Koprivova, A. and Kopriva, S. (2016), Sulfur metabolism and its manipulation in crops. *Journal of Genetics and Genomics* 43(2016): 623-9.

Lee, B.-R., Zaman, R., Avice, J.-C., Ourry, A. and Kim, T.-H. (2016), Sulfur use efficiency is a significant determinant of drought stress tolerance in relation to photosynthetic activity in Brassica Napus cultivars. *Frontiers in Plant Science* 7: 459, doi: 10.3389/fpls.2016.00459.

Lindeman, W. C., Aburto, J. J., Haffner, W. M. and Bona, A. A. (1991), Effect of sulfur source on sulfur oxidation. *Soil Science Society of American Journals* 55: 85-90.

Mascher, M., Gundlach, H., Himmelbach, A., Beier, S., Twardziok, S. O., et al. (2017), A chromosome conformation capture ordered sequence of the barley genome. *Nature* 544: 427-33, doi:10.1038/nature22043.

Menna, A. (2019), Assessment of sulfur deficiency in soils through plant analysis in three representative areas of the central highlands of Ethiopia-IV. *Journal of Agriculture and Ecology Research International* 12(2): 1-13, doi:10.9734/JAERI/2017/34287.

Messick, D. L., Fan, M. X. and de Brey, C. (2005), Global sulphur requirement and sulphur fertilizers. *Landbauforschung Völkenrode*, Special Issue 283.

Mikel, M. A. and Kolb, F. L. (2008), Genetic diversity of contemporary North American Barley. *Crop Science* 48: 1399-407.

Muñoz-Amatriaín, M., Cuesta-Marcos, A., Endelman, J. B., Comadran, J., Bonman, J. M., et. al. (2014), The USDA Barley core collection: Genetic diversity, population structure, and potential for genome-wide association studies. *PLoS ONE* 9(4): e94688, doi:10.1371/journal.pone.0094688.

Negassa, M. (1985), Genetics of resistance to powdery mildew in some Ethiopian barleys. *Hereditas* 102: 123-38.

Nevo, E. (2012), Evolution of wild barley and barley improvement. In *Advances in Barley Sciences: Proceedings of 11th International Barley Genetics Symposium*, Zhejiang University Press, China, pp. 1-16.

O'Connor, J. A., Parbery, D. G. and Strauss, W. (1974), The effects of phytotoxic gases on native Australian plant species: Part I. Acute effects of sulphur dioxide. *Environmental Pollution* 7(1): 7-23.

Pande, P. C. (1985), An examination of the sensitivity of five barley cultivars to S02. *Pollution Environmental Pollution* (Series A) 37: 27-41.

Patterson, H. and Thompson, R. (1971), Recovery of inter-block information when block sizes are unequal. *Biometrika* 58: 545-54.

Pötzsch, F., Lux, G., Lewandowska, S., Bellingrath-Kimura, S. D. and Schmidtke, K. (2019), Optimizing relative seed frequency of intercropped pea and spring barley. *European Journal of Agronomy* 105: 32-40, https://doi.org/10.1016/j.eja.2019.02.009.

Prystupa, P., Peton, A., Pagano, E. and Gutierrez-Boem, F. H. G. (2019), Sulphur fertilization of barley crops improves malt extract and fermentability. *Journal of Cereal Science* 85: 228-35, https://doi.org/10.1016/j.jcs.2018.12.014.

Reinheimer, J. L., Barr, A. R. and Eglinton, J. K. (2004), QTL mapping of chromosomal regions conferring reproductive frost tolerance in barley (*Hordeum vulgare* L.). *Theoretical and Applied Genetics* 109: 1267-74, doi 10.1007/s00122-004-1736-3.

Salvagiotti, F. and Miralles, D. J. (2007), Wheat development as affected by nitrogen and sulfur nutrition. *Australian Journal of Agricultural Research* 58: 39-45.

Shewry, P. R. (2011), Effects of nitrogen and sulfur nutrition on grain composition and properties of wheat and related cereals. In M. J. Hawkesford and P. Barraclough (Eds), *The Molecular and Physiological Basis of Nutrient Use Efficiency in Crops*, 1st edn, John Wiley & Sons, Inc., UK, pp. 103-20.

Smith, F. W., Hawkesford, M. J., Ealing, P. M., Clarkson, D. T., Berg, P. J. V., et al. (1997), Regulation of expression of a cDNA from barley roots encoding a high affinity sulphate transporter. *The Plant Journal* 12(4): 875–84.

Takahashi, H., Kopriva, S., Giordano, M., Saito, K. and Hel, R. (2011), Sulfur assimilation in photosynthetic organisms: Molecular functions and regulations of transporters and assimilatory enzymes. *Annual Review of Plant Biology* 62: 157–84, doi: 10.1146/annurev-arplant-042110-103921.

Wilson, S. A. and Murray, F. (1990), SO_2-induced growth reductions and sulphur accumulation in wheat. *Environmental Pollution* 66: 179–91.

Wirtz, M. and Droux, M. (2005), Review – synthesis of the sulfur amino acids: Cysteine and methionine. *Photosynthesis Research* 86: 345–62, doi: 10.1007/s11120-005-8810-9.

Zhao, F. J., Hawkesford, M. J. and McGrath, S. P. (1999), Sulphur assimilation and effects on yield and quality of wheat. *Journal of Cereal Science* 1: 1–17, doi: 10.1006/jcrs.1998.0241.

Chapter 5

Mapping and exploiting the barley genome: techniques for mapping genes and relating them to desirable traits

Hélène Pidon and Nils Stein, Leibniz Institute of Plant Genetics and Crop Plant Research (IPK), Germany

1 Introduction
2 New possibilities for genetic mapping in the genomics era
3 Classical mapping strategies and their improvement in the genomics era
4 The association mapping boom
5 Multiparental populations: the perfect balance?
6 From an interval to the causal gene: from high-resolution mapping to gene cloning
7 Emerging mapping strategies: fast NGS-enabled technologies
8 Conservation of barley germplasm
9 Genetic and genomic resources of barley
10 Case study: from *rym4* to *rym11*, illustration of paradigm shift in disease resistance mapping and cloning
11 Conclusion and future trends
12 Acknowledgement
13 Where to look for further information
14 References

1 Introduction

Barley has been proposed as a model species for Triticeae (Linde-Laursen et al., 1997) given its high economic importance, being the fourth most cultivated cereal (FAOSTAT, 2018) and at the same time having a less complex genome than the tetra- and hexaploid wheats. It is diploid and highly inbreeding, and a large number of barley genetic stocks are available. Barley can adapt to various biotic and abiotic stresses and has various uses, from malting to animal feeding, thus breeders target a large number of traits. The molecular mapping of genes

http://dx.doi.org/10.19103/AS.2019.0060.05

and traits started more than three decades ago and, since then, genetic and genomic resources for barley were systematically developed and improved. Marker-assisted selection was implemented to accelerate breeding; however, low marker density and only loose linkage of markers to functional genes were serious limitations in early attempts to isolate functional genes underlying traits of interest. The complexity of the 5.1 Gbp barley genome, packed full with repetitive elements (International Barley Genome Sequencing Consortium, 2012; Mascher et al., 2017), rendered this step slower than in smaller genome plant species such as *Arabidopsis* or rice. The first gene in barley isolated by positional cloning was *Mlo*, a gene conferring resistance to powdery mildew (Büschges et al., 1997). As of today at least 65 cloned barley genes have been documented, including 28 for architecture traits alone (Hansson et al., 2018). Forty-two of them were cloned after 2010, demonstrating the importance of improved genomic resources in barley gene isolation. The introduction of next-generation sequencing (NGS) technologies constituted a paradigm shift, bringing barley genetics into the genomic era where marker discovery and genotyping is easy and fast. Recently, an annotated reference sequence was published (Mascher et al., 2017), enabling a range of new methods to rapidly and precisely map genes at moderate cost, and to provide breeders with tools for genomics-assisted breeding. Species diversity and germplasm collections are key for finding new diversity for traits desirable in breeding but, despite the careful survey of the species that started centuries ago, genebanks' descriptions remain limited. NGS also opened the door to their complete genetic characterization, providing better-informed access and use of genetic resources in breeding programmes.

In this chapter, we review the methods currently available for mapping genes and quantitative trait loci (QTLs) in barley, from high-throughput genotyping to gene identification, with a focus on how the application of NGS technologies in barley studies can accelerate gene mapping. We conclude by considering prospects for the future.

2 New possibilities for genetic mapping in the genomics era

When an interesting phenotype has been identified, the first steps are to determine whether it is mono- or polygenic, and its dominance. This is achieved by studying the segregation of the trait by phenotyping a segregating population, typically derived from crossing homozygous parental genotypes. Segregating progeny are then genotyped to map the gene. Genetic markers used in gene mapping have undergone constant progress for almost 50 years. From the use of metabolites, allozymes and isozymes, DNA markers emerged almost 40 years ago, allowing higher marker density and flexibility.

The first DNA markers were restriction fragment length polymorphisms (RFLPs) based on fragment length differentiation of DNA digested by restriction endonuclease (Botstein et al., 1980; Graner et al., 1991; Tanksley et al., 1989). For visualization, a Southern blot of the digested fragments was hybridized to a radioactively labelled DNA probe and then exposed to X-ray films, resulting in differential banding profiles of the autoradiogram. Widely used in the 1980s, hybridization-based markers entailed laborious procedures and labelling of probes with radioisotopes. The development of PCR technology in 1985 (Saiki et al., 1985) prompted the replacement of those markers by PCR-based markers (Fig. 1). Initially, RFLP DNA fragments were sequenced (Michalek et al., 1999) and converted into PCR-based markers like sequence-tagged site (STS) markers (Olson et al., 1989). Since that time, the development of alternative PCR-based assays such as random amplified polymorphic DNA (RAPD) (Kleinhofs et al., 1993; Williams et al., 1990) or amplified fragment length polymorphisms (AFLP) (Becker et al., 1995; Vos et al., 1995), both of which are more frequent in the genome, diminished the importance of RFLPs and their derived STS markers for genetic analysis. However, those new markers were also labour intensive and poorly reproducible in the case of RAPD (Table 1). The development of microsatellites (or simple sequence repeats, SSR) (Becker and Heun, 1995; Litt and Luty, 1989; Saghai Maroof et al., 1994) was another PCR-facilitated breakthrough for genotyping. Based on the length of a tandemly repeated nucleotide motif of one to six bases, SSR markers have been the markers of choice for trait mapping for years and are still used in numerous current projects because of their advantages, i.e. they are: abundant, co-dominant, polyallelic universal, robust, easy to automate and transferable between genotypes (Gupta and Varshney, 2000).

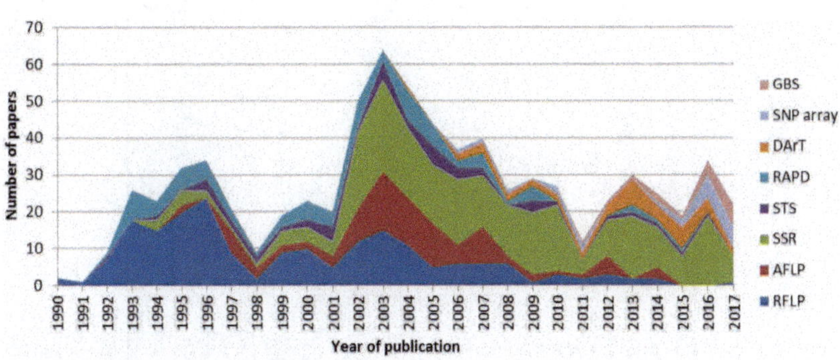

Figure 1 Cumulative number of papers per year per type of marker. Based on the number of papers in PubMed using the word 'barley' as well as the name of the different markers in their title or abstract.

Table 1 Characteristics of the most-used markers in barley, currently or in the past

Marker	RFLP	AFLP	RAPD	SSR	DArT	SNP array	GBS
Co-dominant/dominant	Co-dominant	Dominant	Dominant	Co-dominant	Dominant	Co-dominant	Co-dominant
Genome abundance	High	Very high	Very high	Medium	Very high	Very high	Very high
Reproducibility	High	High	Low	High	High	High	High
Number of markers scored per assay	Low	Moderate	Low	Low	Moderate	High	High
Technical requirement	High	Moderate	Low	Moderate	High	Low	Moderate
Cost per data point	High	Moderate	Low	Low	Very low	Low	Very low
Automation	Difficult	Moderate	Difficult	Moderate	Easy	Easy	Easy
Prior sequence knowledge	No	No	No	Yes	No	Yes	Preferable
Marker development	Fast	Moderate	Long	Long	Long	Long	Moderate
Ease of analysis	Difficult	Difficult	Difficult	Easy	Easy	Easy	Easy
Quantity of DNA required (μg)	10 000	500–1000	20	50	50–100	200	20
Equipment cost	Moderate	Moderate	Low	Low	High	High	High
Throughput	Low	Moderate	Moderate	Moderate	High	High	High

Genotyping entered the high-throughput era with diversity array technology (DArT) (Jaccoud et al., 2001), developed in 2004 for barley (Wenzl et al., 2004). This hybridization-based method generates whole-genome fingerprints by scoring the presence/absence of DNA fragments and provides genotype info at several thousand loci in parallel. DArT markers are reliable and were extensively used in several species for establishing high-density consensus maps (Wenzl et al., 2006). But, with the development of automated, multiplex-sequencing technologies, and the inherent reduction of costs, higher-throughput genotyping technologies emerged. Through Sanger sequencing technology, expansive expressed sequence tag (EST) (Zhang et al., 2004) datasets were produced from different genotypes. They were gathered in databases like HarvEST (http://www.harvest-web.org/) and provided the first generic genome-wide datasets for single nucleotide polymorphism (SNP) mining and marker development, eventually leading to the development of the first SNP arrays in barley. SNPs represent the most abundant source of variation in the genome, making them very good polymorphisms for gene mapping. SNP arrays allow the genotyping of a large number of polymorphisms by array capture of total DNA, but require prior SNP discovery through DNA or RNA sequencing. Successive EST assemblies provided such a basis for the development of Illumina platforms BOPA1 and BOPA2, later merged into the Illumina GoldenGate assay containing almost 3000 SNPs (Close et al., 2009; Muñoz-Amatriaín et al., 2011). The barley SNP array resources became precursors to higher-density formats like the Illumina 9k iSelect chip (Comadran et al., 2012) and, more recently, the 50k iSelect SNP array (Bayer et al., 2017) that allows for parallel genotyping of almost 45 000 SNPs.

The SNP arrays allow for high-throughput genotyping, but their production requires a lot of resources and includes a bias towards alleles that were present in the population used for polymorphism detection. Further improvement of NGS technologies introduced an ongoing decline in costs. This provided the basis for establishing the first barley genome sequences (International Barley Genome Sequencing Consortium, 2012; Mascher et al., 2017), which enable sequence-based genotyping at the genome scale. This has removed the bottleneck of the marker discovery process and allows for direct SNP identification in the population of interest. Deep sequencing of the whole barley genome entails a heavy sequencing load, and its resolution would be higher than that needed for the purpose of SNP calling and genetic mapping. Here, several complexity reduction methods helped to introduce more cost-efficient solutions. One is to sequence only the coding fraction of the genome, either through RNA-Seq or through exome-capture sequencing (Mascher et al., 2013). Exome capture is based on the use of oligonucleotide baits specific to a set of exons predicted from the sequence of cv. Morex (International Barley Genome Sequencing Consortium, 2012). Those baits are then hybridized to genomic DNA to capture the respective gene space for sequencing. The region targeted for sequencing is thus greatly reduced, and sequencing load similarly so, allowing for higher-coverage sequencing and, ultimately, highly accurate genotype calling. However, even if this is an efficient method for some applications (Mascher et al., 2013), it is labour intensive. Another approach involves randomly selecting regions for sequencing based on the presence of restriction sites, greatly lowering the cost and labour intensity. Three such methods are used in barley: RAD-seq (Chutimanitsakun et al., 2011), genotyping by sequencing (GBS) (Poland et al., 2012) and DArTseq (DArT sequencing: www.diversityarrays.com). RAD-seq consists in cutting the DNA with restriction enzymes like *Sbf*I or *Eco*RI (Baird et al., 2008). The fragments are then barcoded through ligation to adapters, and the samples from several individuals are pooled. Subsequent steps are performed on the pooled samples, decreasing the labour and cost. The pooled fragments are sheared and size selected to achieve a length suitable to the sequencing platform, followed by a selective PCR used to select only the fragment harbouring the restriction site. GBS follows a similar protocol, but with fewer steps. It utilizes two enzymes: a rare-cutter (*PstI*) and a frequent-cutter (*MspI*). Fragments are barcoded as follows: forward adaptors are designed with the *PstI* adaptor, whereas reverse adaptors are matching *MspI*. Unlike RAD-seq, no shearing or size-selection step is required after pooling the fragments. The short *PstI-MspI* fragments are instead selected by a selective PCR amplification step with a short extension time, and sequenced on an NGS platform. DArTseq has a similar protocol to GBS, but uses different enzymes in pairs with *PstI* (Wenzl et al., 2004). Both GBS and DaRTseq complexity reduction are slightly deviated to genic regions because of the use of enzymes more efficient in under-methylated regions.

The prevalence of these different methods over time is illustrated in Fig. 1, and their distinctive characteristics are listed in Table 1. The polymorphisms detected by these sequencing-based methods can be used directly for mapping, or to design markers to genotype a larger number of plants with a lower-marker density. Those markers are classically used in the process of high-resolution mapping (see Section 6). The most popular ones are *cleaved amplified polymorphic sequences* (Konieczny and Ausubel, 1993), but newer methods like detection of SNP-specific hybridization by fluorophores is growing. Of those systems, the two most frequent are Taqman (Holland et al., 1991) and KASPar (LGC genomics, Semagn et al., 2014) markers. The genotype at a locus is determined in a PCR with pairs of fluorescently labelled primers or probes, with one of each pair being specific to an allele. After a single PCR, the reading of the fluorescence is performed in a qPCR thermal cycler or a fluorescent plate reader. The relative fluorescence levels from each member of a marker pair allows clear distinction of heterozygotes and homozygotes. These systems thus allow a rapid, reliable and efficient genotyping of a large number of samples at a limited number of loci.

3 Classical mapping strategies and their improvement in the genomics era

Access to high-throughput genotyping conferred considerable benefits for linkage mapping. It accelerated data generation for mapping by generating high numbers of markers in a single experiment, and allowed an increase in population size owing to reduced labour intensity in comparison to previous genotyping methods. Linkage mapping of a given trait is usually performed with the help of designed biparental mapping populations. The first step of mapping is to assess whether the trait is controlled primarily by a single major gene, or if it is a quantitative trait, likely influenced by many genes. This is determined by studying the segregation and distribution of the phenotype in developed populations. If the trait is a controlled single gene, the segregation pattern is also the way to determine the status of dominance of the gene. A trait governed by a single gene can be mapped in a simple population like F_2 or back-cross. But in case of a quantitative trait, the unmonitored segregation in the numerous heterozygous regions limits the replication of phenotyping and can impact its reproducibility, ultimately reducing the power of QTL detections. Recombinant inbred lines (RILs) (Burr et al., 1988), obtained by repeatedly selfing F_2 lines, were created to increase the total number of recombination events while obtaining mostly homozygous lines. Due to this last characteristic, phenotyping can then be repeated several times with a high level of reproducibility. However, the generation of this kind of population is tedious as it requires six to eight generations and has encouraged the development of faster ways to obtain homozygous lines. One of these is the doubled haploid (DH) lines (Clapham et al., 1973; Kao et al., 1991), which are produced by inducing haploid lines from F_1 plants and converting them into diploids. DH lines are based on a single gamete, therefore they only have one meiosis to accumulate recombinations. The total number of recombination is therefore lower than in RILs, but it is the fastest way to obtain homozygous lines and is now widely used.

Linkage mapping is used for both single genes and QTLs and can be performed with different algorithms. The single-marker analysis identifies the gene interval or QTLs though calculation of the effect of each marker of the trait by *t*-test or ANOVA (Beckmann and Soller, 1988; Edwards et al., 1987; Thoday, 1961), but it lacks precision. Interval mapping (Haley and Knott, 1992; Lander and Botstein, 1989) is based on the simultaneous analysis of two markers at a time, where the two marker positions define an interval within which the likelihood of a gene or QTL being present is estimated. The resulting position and effect of each detected locus is thus more precise. The composite interval mapping (CIM) (Zeng, 1993) approach takes into account that a trait can be influenced by several QTLs, thus reducing the error. To improve the power of linkage mapping, several statistical models, such as multiple interval mapping (Kao et al., 1999), were derived from CIM, including more factors such as additive, dominance or epistasis effects.

The availability of high-throughput genotyping and access to a reference genome sequence of barley now allows better linkage mapping by increasing the number of markers included in such studies, while making reference-guided mapping possible, using the information of the reference sequence to define the order of markers. But, even if increasing the number of markers can improve the mapping resolution, this improvement is limited by the number of recombination events that occurred at the locus, and the genetic resolution depends mostly on the size of the population and the frequency of recombination.

4 The association mapping boom

The genetic resolution achieved by linkage mapping is limited. It is directly linked to the recombination frequency in the population, which remains low because of the low number of generations that can be achieved. To alleviate this problem a natural population can be used to perform genome-wide association studies (GWAS). First used in humans where experimental populations cannot be developed (Spielman et al., 1993), it also has advantages in plant traits mapping: the recombination events that can be identified in natural populations are the historical recombinations that occurred during the evolutionary history of the species, representing a much higher number of generations than the few collected in experimental populations (Zhu et al., 2008). High-throughput genotyping allows to easily retrieve a large number of polymorphisms in a population and the genomic era became the one of GWAS. This mapping strategy is detailed elsewhere this volume and will only be generally described here.

GWAS can be applied for major genes but the kind of population used make it better suited to quantitative traits. It is taking advantage of the linkage disequilibrium (LD) that exists between the causative locus and genotyped loci nearby to identify small haplotype blocks associated with the phenotype of interest in the population. The resolution depends on the extent of LD in the genomic region of interest, and on the density of the genotyping. The optimal marker density is then linked to LD in the population: the higher the LD extent, the lower the optimal number of markers. One big advantage of GWAS is the use of a population that does not need to be created by crossing, saving time. However, the use of natural populations can also be a drawback as LD can be affected by the genetic structure of the population. In each subpopulation, allele frequencies evolve independently, creating false haplotype associations caused by the coincidence over-representation of the phenotype in a subpopulation that also bears similar genotypes at many loci - and not to any true association between the marker and the phenotype. This bias can be controlled by adding parameters like genetic structure or relatedness as cofactors to the model (mixed linear model, MLM) (Yu et al., 2006). These corrections may also increase the risk of false negative associations, as traits really linked to the population structure will not be detected. Other models like CMLM (Zhang et al., 2010), ECMLM (Li et al., 2014) or MLMM (Segura et al., 2012) were developed to try to further compensate for population structure.

In linkage mapping, the size of the population is crucial, as it counterbalances population structure. Compared with linkage mapping, GWAS usually allows a better resolution than linkage mapping but has a reduced capacity to detect effects linked to rare variants, which for certain traits can represent the largest source of phenotypic variation. If the allele frequency at the locus is too low in the population, either the quality filter applied before the analysis may delete the informative markers (the minimum allele frequency is often set at 0.05) or the statistical power of the analysis itself will not allow detection. GWAS are thus better suited for either large-effect-low-frequency or low-effect-large-frequency loci.

5 Multiparental populations: the perfect balance?

To combine the advantages of biparental and natural-diversity populations, multiparental populations were developed. The allele frequency and the QTL effects are higher than in natural-diversity panels while the effects of population structure are alleviated, so it is easier to detect rare variants or QTLs with smaller effect compared with classical GWAS. Compared to classical QTL mapping, having more than two parents also adds some historical recombination events to the population, thus increasing the resolution of the QTLs. It also widens the represented diversity and can allow access to more causal loci than those segregating in a biparental population. However, some drawbacks of biparental and natural populations still remain. As the number of population founders is limited, they do not represent the same proportion of the species' diversity than a classical GWAS panel. Moreover, as for linkage mapping population, the creation of a multiparental population requires several generations of crosses.

The two major types of multiparental populations were developed in barley: Multi-parent advanced generation inter-cross (MAGIC) (Sannemann et al., 2015) and two nested association mapping (NAM) (Maurer et al., 2015; Nice et al., 2016) populations. NAM populations are made by crossing multiple inbred lines with a single one and subsequently deriving RILs or DH lines from the progeny (Yu et al., 2008). The common parent genotype normalizes the genetic background, and common-parent-specific markers are used to genotype the population. So even as diversity is introduced by the diverse parental lines, the use of a common parent reduces the creation of new haplotypes, since no new haplotypes are recreated between the unique parents. MAGIC populations (reviewed by Huang et al., 2015) imply a greater number of crosses than NAM populations and are thus more labour intensive to make, but include more diversity. These populations are made by intercrossing several founders (typically eight) and selfing the lines, then using the progeny of those crosses to obtain RILs. Allele frequency is more balanced than in NAM populations and haplotype diversity is higher, but these populations are more time-consuming to develop and cannot be extended to additional founders as NAM populations can.

6 From an interval to the causal gene: from high-resolution mapping to gene cloning

Whether linkage mapping or GWAS is used, the resolution achieved is rarely sufficient to directly identify the causal gene or polymorphism. High-resolution mapping of the locus of interest is most often needed to reduce the region and obtain an interval containing a limited number of candidate genes. However, in the case of a quantitative trait, not all QTLs are suitable for high-resolution mapping. QTLs with small effects or low statistical support should be avoided, as the chance of success in high-resolution mapping is low. High-resolution mapping relies on increasing recombination in a gene interval saturated with markers. Single-gene traits can be mapped at a high resolution in F_2 populations. To obtain higher resolution, RILs can be derived specifically from plants recombining in the gene interval (segmental recombinants to obtain so-called segmental RILs). However, for quantitative traits, the locus should first be 'mendelized', meaning that a high-resolution mapping population must be developed where the inheritance of the phenotypic variation follows Mendel's law: two distinct phenotypic classes, easily distinguishable, must be obtained. To reach this goal, the number of segregating loci affecting the trait must be reduced by inserting the two alleles at a locus in a common genetic background. Near isogenic lines (NILs) are thus produced. NILs are preferably homozygous lines, often containing a single locus from a donor parent (RILs, DHs, F_2/F_3 or original accessions containing the trait of interest) in the genetic background of a phenotypically distinct recurrent parent. NILs are produced by crossing those two parents, then performing backcrossing and/or selfing. A population is then created by crossing the obtained NIL with the recurrent parent, and high-resolution mapping is performed in this new population.

With the tools of today, saturating a locus interval with markers is usually unproblematic. Most of the time, high-resolution mapping will result in an interval containing several candidate genes. The next step is typically a candidate-gene approach. With the availability of an annotated reference genome, retrieving lists of candidate genes is straightforward. If one (or a small number) of those genes stands out for its predicted function, it can be prioritized for further analysis. However, this simple case may not occur. It is possible that the causal gene has no obvious predicted function. Another possible reason is that phenotypic traits can be due to presence/absence variations (PAV) (Feuk et al., 2006) and the sequencing of the interval in parents may still be needed to find the right candidate gene. With the decreasing cost of *de novo* assemblies, we will, in the very near future, see even several high-quality genome sequences published per species. Already initiated in maize (Lu et al., 2015; Hirsch et al., 2016; Yang et al., 2017a; Sun et al., 2018; Springer et al., 2018), this development empowers us with possibilities to compare the structure and genes present in mapping intervals in a greater proportion of the diversity, and thus decreasing the risk of missing the causal gene that can be absent in one genome (Monat et al., 2018; Worley et al., 2017). To identify the polymorphisms that can explain the phenotypic differences between parents, a preferred candidate gene (or the complete set of genes in the interval) is usually resequenced. Since the trait can also be due to differential gene expression, it is possible to investigate gene expression differences by using microarray (Close et al., 2004), qPCR or RNA-Seq (Costa-Silva et al., 2017).

Such approaches allow a researcher to identify a reduced number of candidate causal genes for further validation. A common gene validation method in barley is mutant analysis to identify the effects of mutations that impair or alter a gene's function. This approach typically relies on methods like RNA interference, chemical or physical mutagenesis or T-DNA insertion. In the case of chemical and physical mutagenesis, the mutation is not targeted and screening of the mutants can be performed by targeting induced local lesions in genomes (TILLING, later detailed). Another approach is the use of genetic transformation to either overexpress the candidate gene in a plant without the trait of interest or supplement a mutant. Contemporary genome-editing technologies like TALENS (Gurushidze et al., 2014; Wendt et al., 2013) and CRISPR-Cas9 (Lawrenson et al., 2015) are becoming a method of choice to generate mutants in barley at specific loci (Lawrenson et al., 2015; Holme et al., 2017; Kapusi et al., 2017; Kumar et al., 2018). These methods should allow us to achieve a larger spectrum of genome editing in the near future, from complete gene insertion or knockout to single-base replacement (Bortesi and Fischer, 2015), which would allow to unravel the role of each candidate polymorphism in the trait by single-nucleotide edition.

7 Emerging mapping strategies: fast NGS-enabled technologies

While the standard methods described so far are very efficient, even faster methods were introduced in the genomics era. The classic quantitative genetic method to identify rapidly a gene or QTL interval is by bulk segregant analysis (BSA) (Giovannoni et al., 1991; Michelmore et al., 1991). It is based on the genotyping of individuals selected from the two tails of the phenotypic distribution of a mapping population or of a diversity set (Fig. 2). Plants from both ends of the distribution can be genotyped individually, but further

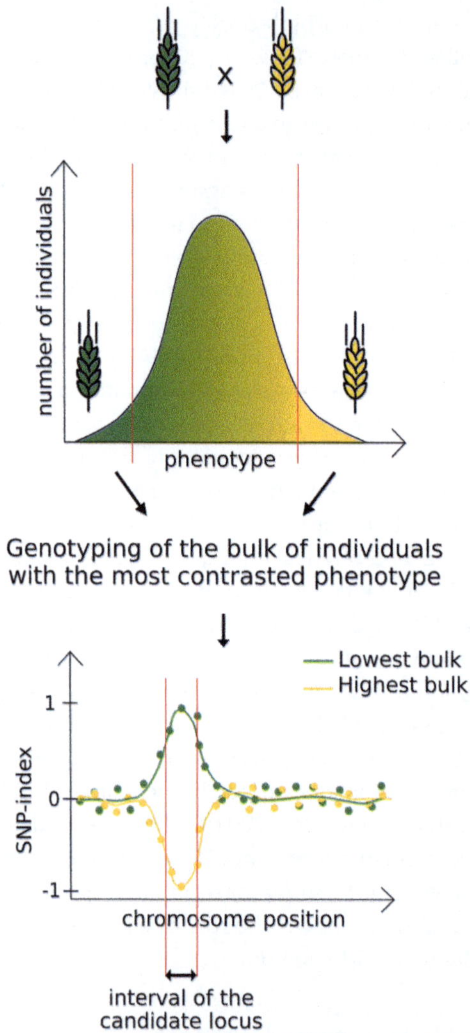

Figure 2 Schematic representation of bulk segregant analysis (BSA). A population is generated by crossing two parents with distinct phenotypes. The offspring are phenotyped and two bulks, comprising the individuals presenting the most extreme phenotypes of each class, are created. The two bulks are sequenced and their genotypes are graphed along the chromosomes. Candidate intervals are defined as loci where the two bulks show a divergent genotype.

economic savings can be made by genotyping pooled DNA from groups of individuals with similar phenotypes. The frequency or the strength of signal at markers near the causal locus shows a significant deviation between the two pools, allowing identification of the genes or QTL loci, with a very limited genotyping effort. BSA was used for the first time in the early 1990s in barley (Barua et al., 1993) but, while early applications were confounded by insufficient marker density resulting in a high rate of false positives, state-of-the-art high-throughput genotyping methods have alleviated this problem. The accurate phenotyping of the entire original population and the real cost and labour are the only remaining limitations (Gallais et al., 2007; Sun et al., 2010).

Another fruitful method is the use of mutants exhibiting a trait of interest, or the loss of a trait of interest, to find the mutation responsible for the trait in a reverse genetics approach. TILLING (McCallum et al., 2000) combines chemical mutagenesis with high-throughput genome-wide screening for point mutations in genes of interest. The original TILLING method used denaturing HPLC for mutation discovery. For higher throughput, a protocol using Li-Cor DNA Analyzer was soon developed (Till et al., 2003). Until recently, barley TILLING populations were mostly characterized either by dHPLC (Caldwell et al., 2004) or LI-COR (Gottwald et al., 2009; Kurowska et al., 2012; Lababidi et al., 2009; Szarejko et al., 2017; Talamè et al., 2008). However, both still involve inconvenient steps, such as acrylamide gel preparation and amplicon purification, and require labelled primers that can impact mutation detection by reducing PCR efficiency. For these reasons, protocol using Fragment Analyzer™ (Advanced Analytical Technologies) were developed (Kang et al., 2018; Szurman-Zubrzycka et al., 2018). A further future direction of TILLING in barley is NGS sequencing-based approaches. Such a protocol was developed by Tsai et al. (2011), consisting in sequencing different PCR target amplicons in DNA pools. This method is still expensive and so far restricted to species where the mutation density is high, such as wheat (Krasileva et al., 2017). With decreasing costs of sequencing, it is probable that TILLING will move to detection of mutations by sequencing in the near future.

Combining sequencing and mutant screening is the goal of mapping-by-sequencing, following SHOREmap (Schneeberger et al., 2009) or similar analysis like MutMap (Abe et al., 2012). The idea is to combine EMS mutagenesis with whole-genome sequencing (WGS) in an approach similar to BSA. A mutant is generated, most often by EMS. As EMS mutation causes several mutations in the genome, the simple sequencing of one mutant is not sufficient to identify the causal mutation; the mutant must be crossed with the non-mutated parent and a BSA is performed. Complete sequencing of the genome of the mutants allows to retrieve all polymorphisms (Abe et al., 2012; Schneeberger et al., 2009). Since the barley genome is large and complex, less-expensive methods using genome complexity reduction were developed.

Mapping-by-sequencing was developed and performed successfully in barley (Jost et al., 2016; Mascher et al., 2014; Pankin et al., 2014). After phenotyping, plants with the mutant phenotypes and those with wild-type phenotypes were separated into two pools. DNAs from both pools were subjected to exome capture (Mascher et al., 2013) as a complexity reduction method and subsequently sequenced on an Illumina platform. The obtained reads were mapped against the reference genome and SNPs were detected. Allele frequency in the two pools was then mapped along the genome and a clear imbalance of allele frequency between the two pools could be observed. This method allows precise mapping of a gene with one-sixth of the sequencing load needed for similar results using WGS. However, the barley exome capture assay published in Mascher et al. (2013) was estimated to capture around 86% of the high-confidence exons and no exome-capture assay can capture 100% of the genes, thus there is always an increased risk of missing the target gene by this approach.

For the specific case of pathogen resistance, Steuernagel et al. (2016) described a method combining sequencing of captured targets in the genome, and mutagenesis to clone genes. Successfully applied in bread wheat, this method, called MutRenSeq, aimed at cloning dominant pathogen resistance genes. Those genes are most frequently represented by NLR genes, whose structure is highly conserved in the genome and can be enriched in sequencing libraries and specifically sequenced in a capture assay called RenSeq (Jupe et al., 2013). MutRenSeq consists in producing loss of resistance mutants by EMS mutagenesis. Independent susceptible mutant lines are then sequenced using the RenSeq protocol (Fig. 3). The NLR genes are assembled and aligned to those of the resistant parent. If a gene harbours polymorphisms in all susceptible mutant lines, it is identified as the candidate gene for resistance. The application of RenSeq reduces the amount of data to be sequenced to less than 0.1% of the genome. However, even in the case of dominant resistance, the causal gene may belong to a different gene family. Thus, there is also in this case an elevated risk of missing the target gene.

To alleviate the risk of missing the target gene, either because it is missing from the capture but also in case it is due to PAV or large structural variation that cannot be detected by conventional mapping to a reference genome, methods using local *de novo* assembly were designed. One of these methods is MutMap-Gap that combines the sequencing reads mapped in the gene interval with the unmapped reads to perform a small-scale *de novo* assembly (Takagi et al., 2013). But while this method is efficient in a small genome such as rice, the sequencing load to achieve a sufficient coverage in barley would be very high and a complexity reduction method is required. Taking advantage of chromosome flow-sorting technology and efficient *de novo* assembly methods (ChromSeq) (The International Wheat Genome Sequencing Consortium, 2014),

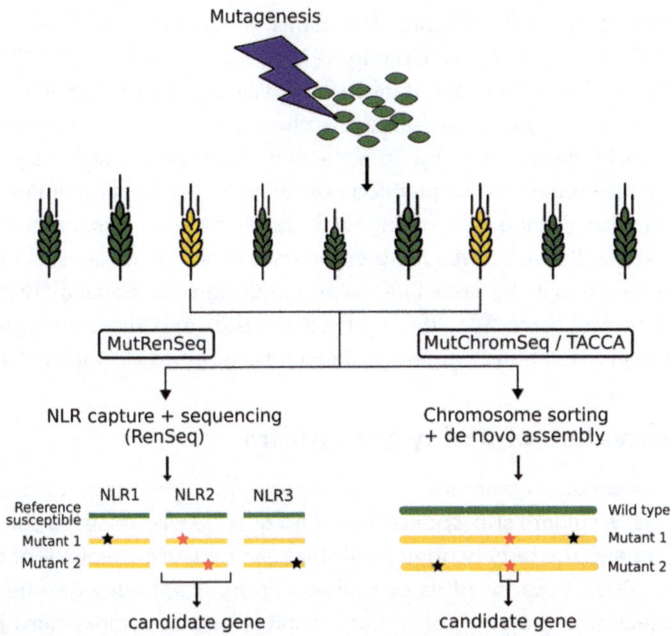

Figure 3 Schematic representation of two fast-enabled gene-mapping methods. Seeds from the accession presenting the trait of interest are mutagenized, usually by EMS, and the mutants are phenotyped. A few mutants presenting a loss of function for this trait are selected. MutRenSeq consists in capturing the NLR genes in these mutants and sequencing them in a method called RenSeq. NLR sequences are then *de novo* assembled and aligned to the susceptible reference. The candidate genes are the NLRs presenting a mutation in all the loss-of-function mutants sequenced. MutChromSeq and TACCA consist in performing chromosome sorting on the loss-of-function mutants, as well as the wild-type parent, to select the chromosome harbouring the locus coding for the trait of interest. These sorted chromosomes are sequenced, *de novo* assembled and aligned. The candidate genes are the ones showing a mutation in all loss-of-function mutants.

Sánchez-Martín et al. (2016) proposed a method called MutChromSeq. It consists in generating EMS mutants and screening those for loss of the desirable trait, then flow sorting the chromosomes of the interesting mutants to select the chromosome that contains the locus of interest (Fig. 3). Then, this single chromosome is sequenced and assembled *de novo*. The mutation overlap in independent mutants allows rapid and precise identification of the gene underlying the trait. While chromosome flow sorting is generally complex, it is well optimized for barley (Doležel et al., 2012), reduces the cost of sequencing and achieves a greater depth of sequencing to perform a better assembly. These methods are very efficient for genes producing strong phenotypes, but more quantitative traits still elude such approaches. Moreover, they depend on the identification of loss-of-function mutants, which can be difficult to obtain. Targeted chromosome-based cloning via long-range assembly (TACCA) (Thind et al., 2017) is designed to avoid these flaws. Like MutChromSeq, it is based on sequencing of the flow-sorted candidate chromosome and its assembly, which needs to reach high contiguity. To achieve it, Dovetail Chicago™ sequencing libraries are prepared and sequenced in addition to the classic Illumina libraries to perform long-range scaffolding with state-of-the-art assembly pipelines (Putnam et al., 2016). The resulting chromosome sequence is used to resolve structural variation in the gene interval and to design additional markers. While slower than MutChromSeq, this method is faster and more precise than a classical high-resolution mapping and can be applied to all kinds of traits.

8 Conservation of barley germplasm

The genus *Hordeum* comprises cultivated barley *H. vulgare* spp. *vulgare*, its wild subspecies *H. vulgare* spp. *spontaneum* and over 30 wild relatives constituting the second and the tertiary gene pools that can be a useful source of diversity (Blattner, 2009). Because of its economic importance, barley germplasm has been collected and stored for about a century and now ranks third in terms of the number of accessions kept in *ex situ* genebanks after wheat and rice (Commission on Genetic Resources for Food and Agriculture; FAO, 2010). Knüpffer (2009) estimated that over 450 000 *Hordeum* accessions are kept in *ex situ* collections distributed over 204 genebanks, including almost 300 000 accessions from *H. vulgare* spp. *vulgare* and around 32 000 of *H. vulgare* spp. *spontaneum*. Among these genebanks, seven currently hold more than 20 000 accessions: the PGRC in Canada (http://www.agr.gc.ca/pgrc-rpc), the NSGC in the United States (http://www.ars-grin.gov/npgs), the EMBRAPA CENARGEN in Brazil (www.embrapa.br/en/recursos-geneticos-e-biotecnologia), the ICARDA in Syria (www.icarda.org/research-sub/biodiversity-and-its-utilization), the John Innes Center in the United Kingdom (www.jic.ac.uk/research/germplasm-resources-unit/), the IPK Gatersleben in Germany (www.ipk-gatersleben.de/en/gbisipk-gaterslebendegbis-i/) and the VIR in Russia (www.vir.nw.ru/). These resources are available to any researcher, and allow the mining of new traits, genes and alleles.

From those genebanks, panels of germplasm and core collections are identified to represent the largest diversity possible from a size-limited subsample of accessions. Core collections allow the screening of a species' diversity with minimal effort. The International Barley Core Collection is constituted of about 1500 accessions, including some *H. vulgare* spp. *spontaneum* and some wild relatives (Knüpffer and van Hintum, 2003). It was characterized genetically and phenotypically for some traits. Other core collections include the NSGC Barley Core (Muñoz-Amatriaín et al., 2014) with 1860 accessions or the Landrace Collection LRC1485 that consists of 1485 spring barley landraces selected specifically for the study of adaptation to climate (Pasam et al., 2014). Some of these panels are also constituted to represent the diversity in the wild relatives like the Wild Barley Diversity Collection that consist in 318 *H. vulgare* spp. *spontaneum* accessions (Steffenson et al., 2007) or the Barley1K collection that comprises 1020 wild barleys sampled in 75 sites in Israel (Hübner et al., 2009). All these collections are genetically characterized, for example by the 9k iSelect array in the case of the NSGC Barley Core (Comadran et al., 2012), as well as phenotypically for some traits, and have been used for GWAS or for germplasm mining.

But, as carefully as a core collection is constructed, it cannot contain all the diversity contained in a species, therefore limiting its use for discovery of very rare alleles. IPK Gatersleben recently completed GBS of its entire barley collection (Milner et al., 2019). The related webtool (http://bridge.ipk-gatersleben.de) permits to mine the germplasm of the barley collection and design subsets of it based on criteria such as genetic or geographic diversity, but also to deliver direct access to the SNPs discovered via an intuitive visual interface. This initiative aims at unlocking the collection for allele mining and diversity studies and is already applied to other genebanks, expanding the genomic information of genetic resources and allowing association mapping to be performed on a subset of a genebank without additional genotyping.

9 Genetic and genomic resources of barley

Characterization of barley germplasm has been ongoing for a long time and resources are stored in online databases. GrainGenes (https://wheat.pw.usda.gov/GG3/barley_blvd) (Carollo et al., 2005) is a database harbouring molecular and phenotypic information of Triticeae and *Avena*, including genetic maps, markers and germplasm information. Some databases like PLEXdb (http://www.plexdb.org/) (Wise et al., 2007) or GENEVESTIGATOR (https://genevestigator.com/gv/doc/content.jsp) provide access to the pattern of expression of the genes represented on the Affymetrix microarray. RNA-Seq data from Morex cultivar is also available at morexGenes (https://ics.hutton.ac.uk/morexGenes) as well as a barley epigenome browser (https://ics.hutton.ac.uk/barley-epigenome).

The systematic sequencing of the barley genome (International Barley Genome Sequencing Consortium, 2012; Mascher et al., 2017; Mayer et al., 2011) ran in parallel with the development of new tools to access the data. BARLEX (http://barlex.barleysequence.org) (Colmsee et al., 2015) includes the different types of data produced for the barley sequence assembly (Mascher et al., 2017). Centred on the minimum tilling path of the sequenced BACs, it gives access to information such as the number of contigs and sequences of each individual BAC assembly, tables of published annotated genes, mapping of the 9k iSelect markers on the genomes and a link to a BLAST server (http://webblast.ipk-gatersleben.de/barley_ibsc/) allowing for homology searches against the different published assemblies as well as other databases like full-length cDNA, ESTs or regions sequenced by exome capture.

A second way to access the genome assembly data is through the genome browser displayed by the common tool of Gramene (http://www.gramene.org) (Tello-Ruiz et al., 2016) and Ensembl Plants (http://plants.ensembl.org) (Kersey et al., 2016), where annotation of genes, as well as full cDNA and transcriptomic data, can be visualized along the genome. These web tools also allow comparative analysis between genomes to be performed. A third useful database for exploring barley genomic data is PGSB PlantsDB (http://pgsb.helmholtz-muenchen.de/plant/barley/index.jsp) (Spannagl et al., 2016). It contains visualization of the different versions of sequence assembly, information on annotated genes and synteny comparison with the genomes of *Brachypodium distachyon* and *Oryza sativa*.

10 Case study: from *rym4* to *rym11*, illustration of paradigm shift in disease resistance mapping and cloning

To illustrate recent advances in positional cloning, we compared the mapping of two resistance genes, performed in the same labs. The first is the *rym4/rym5* locus that was cloned in 2005 (Stein et al., 2005). The second is *rym11*, cloned almost 10 years later, in 2014 (Yang et al., 2014b). Both genes confer complete recessive resistance to soil-borne bymoviruses BaMMV and BaYMV and encode for susceptibility factors.

The *rym4* gene was introgressed from landrace 'Ragusa' into commercial German barley varieties as early as the 1980s but its genetic basis was unknown. This recessive resistance gene was proven to be linked or allelic to *Rym1* by allelic test in a cross with *Rym1* variety Mokusekko 3 (Friedt et al., 1987). Crosses between the *rym4* variety Franka and barley accessions exhibiting a strong dominant phenotype mapped to a chromosome (genetic marker stocks) proved that *rym4* was neither on chromosomes 4H, 1H nor 5H. (Kaiser and Friedt, 1992, 1989), but placed *rym4* distal on the long arm of chromosome 3H by segregation analysis in crosses between the *rym4* 'Ogra' and 'Sonate' varieties and trisomic lines (lines trisomic for specific chromosomes) and telotrisomic lines. From these early attempts, successive innovations in molecular markers allowed the precise characterization of the genetic interval. In the 1990s, several studies used RFLP, isozymes and RAPD markers and mapped *rym4* and *rym5* in the same interval at the distal end of chromosome 3HL (Graner et al., 1995, 1999; Graner and Bauer, 1993; Konishi et al., 1997; Ordon et al., 1995). Despite the increase in numbers of marker types and of population sizes, the initial interval of 2.4 cM (Graner and Bauer, 1993) was not significantly reduced during this decade. The first high-resolution mapping of the locus *rym4/rym5* (Pellio et al., 2005) took advantage of advances in molecular markers to screen large populations: 3884 F_2 plants for *rym5* and 1040 F_2 individuals for *rym4* mapping. DNA bulks from individuals of these DH populations were used for AFLP, RAPD and RFLP marker saturation of the regions and RILs were derived from recombinants between the common flanking markers of *rym4* and *rym5*. The new markers were converted into STSs and used to genotype the RILs, reducing *rym4* and *rym5* intervals to 0.05 and 0.013 cM, eventually leading to the cloning of the gene (Stein et al., 2005). The Morex BAC library (Yu et al., 2000) was screened with the markers of the *rym4/rym5* region, and a 650-kb physical map was constructed. The six BAC clones cosegregating with resistance were sequenced to full length and annotated (Wicker et al., 2005), and two open reading frames (ORFs) were found. One of them, *Hv-eIF4E*, was homologous to genes involved in recessive *Potyvirus* resistance in dicotyledonous species and was sequenced in 56 barley accessions known to carry either *rym4*, *rym5*, an unknown source of resistance or no resistance. This sequencing revealed several non-synonymous polymorphisms in *rym4* and *rym5* accessions and this gene was confirmed by complementation via *Agrobacterium*-mediated transformation of a *rym4* NIL with either the full-length cDNA or the genomic DNA of a susceptible accession, thus inducing susceptibility. Kanyuka et al. (2005) performed sequencing of RT-PCR products of different variants and evidenced that *rym6* is another natural allele of *Hv-eIF4E*. They also carried out TILLING on the gene and retrieved one mutant allele, with which resistance segregates. Additional allele mining was later performed, identifying several other alleles of resistance at this gene (Yang et al., 2014a, 2017b).

In contrast, *rym11* mapping was comparatively fast. The preliminary mapping of *rym11* was performed on two populations of 48 F_2 and 101 BC_1F_2, respectively (Bauer et al., 1997). The authors performed BSA with RFLPs, RAPDs and one SSR marker, which localized the gene to within a large interval of 16.4 cM, close to the centromeric region of chromosome 4HL. The first fine-mapping study (Nissan-Azzouz et al., 2005) analysed three DH populations of 57, 191 (IPK1) and 161 (IPK2) DH lines, respectively. The authors successively performed an initial mapping with RFLP markers, two BSA with RAPD and AFLP markers respectively and a more complete mapping with SSR, RAPD and AFLP markers. Based on the results, *rym11* was mapped in a 3.7 cM interval in IPK1 and a 10.7 cM interval in IPK2 population. Lüpken et al. (2013) used a genomics-informed approach to resolve the interval in a population comprising 5102 F_2 plants. To carry out the mapping, they used publically available SSR and SNP markers and designed additional SNP and STS markers with the information available in EST databases, in the barley genome zipper (Mayer et al., 2011), as well as in the WGS assembly of Morex (International Barley Genome

Sequencing Consortium, 2012). Polymorphisms were revealed either on gels or by pyrosequencing, leading to a final interval of 0.074 cM. The gene *rym11* was eventually isolated by chromosome walking (Yang et al., 2014b). Barley BAC clones (Ariyadasa et al., 2014; Schulte et al., 2011), sequenced in conjunction with the barley genome-sequencing project (Mascher et al., 2017), were identified by sequence comparison to the flanking markers. Two overlapping BAC contigs were identified, resulting in a physical contig covering the complete genetic interval. Based on the available BAC sequence data, new markers were designed to reduce the interval to 0.0196 cM, representing 1.25 Mbp, using the same 5102 F_2 population previously explored by Lüpken et al. (2013). Annotation of this interval revealed four ORFs. These ORFs were resequenced, and only one gene exhibited polymorphism between susceptible and resistant genotypes. This gene encoded a putative protein disulphide isomerase like 5-1 (HvPDIL5-1), and the resistant plants' allele contained a 1.3 kb deletion in its promoter and first exons. It was validated as the resistance gene by TILLING in an EMS-induced mutant population and by complementation via *Agrobacterium*-mediated transformation of the resistant genotype with the cDNA of the gene from a susceptible accession. Moreover, the gene *HvPDIL5-1* was sequenced in 1732 diverse barley accessions, both domesticated (*H. vulgare* ssp. *vulgare*) and wild (*H. vulgare* ssp. *spontaneum*), and identified three more natural haplotypes conferring resistance, all harbouring a premature stop codon in *HvPDI5-1*. This allele mining was later completed in a larger collection (Yang et al., 2017b).

These studies illustrate how developments in molecular marker technology have revolutionized gene-mapping studies, allowing the genotyping of large populations more easily and, by increasing the number of useful markers available, allowing the discovery of intervals smaller than 1 cM. Increased availability of sequence data represented a second game changer. It enabled for instance the design of numerous markers in the interval of *rym11*, sufficient to achieve a mapping resolution below 0.1 cM. It has also accelerated the acquisition of physical maps. The tedious chromosome walking carried out for *rym4* included the screening of a BAC library and the sequencing of the identified BACs, whereas for *rym11* the sequences were already available and BAC clones screening was performed *in silico*. Moreover, TILLING and allele mining for both loci provided a fast way of validation as well as alternative alleles that could be of use for breeding. The *rym11* mapping is now already 5 years old and new methods of sequencing have rendered gene-mapping studies even faster and more precise (Jost et al., 2016; Mascher et al., 2014).

11 Conclusion and future trends

Advances in sequencing technology and the availability of the barley reference genome have underpinned genomics-informed barley gene mapping and cloning. Genotyping is progressively faster, easier and relatively more cost-efficient. New NGS-based gene-mapping methods are in place and more are appearing. A paradigm changer for gene mapping in the coming years will be the spread of *de novo* assemblies of complete barley genomes or flow-sorted chromosomes. More reference-quality *de novo* assemblies of different barley haplotypes are likely to come as a result of pan-genome analysis (Hurgobin and Edwards, 2017). Already initiated (https://www.pflanzenforschung.de/de/plant-2030/fachinformationen/projektdatenbank/strukturelle-genomvariation-haplotypendiversitaumlt-und-419), which will continue to unlock the diversity of barley, including structural changes, so far unreachable with a single reference, and increasing the efficiency of gene cloning.

In addition to classical gene and QTL mapping, the future will probably see an increase in expression QTL (eQTL) (Damerval et al., 1994) analysis. Barley gene expression is affected by complex regulation. More than half of the genes have been found to be differently regulated between samples, and more than two-thirds have been shown to exhibit alternative splicing (International Barley Genome Sequencing Consortium, 2012). Moreover, quantitative traits are often regulated by the differential expression of genes. In eQTL analysis, gene expression is quantified and related to the phenotype. Whole-genome eQTL analysis has been performed on barley (Potokina et al., 2008) using RNA binding on an Affymetrix chip. However, it is now possible to perform it by RNA-Seq, achieving a better throughput. Moreover, the methylome can also be sequenced with NGS. Called whole-genome bisulphite-sequencing (BS-seq), this method is based on sodium-bisulphite treatment that converts unmethylated cytosine of the genomic DNA to uracil, followed by WGS. With this method, a single-base methylome map of *Arabidopsis thaliana* was generated (Lister et al., 2008). It was also performed recently on barley Morex accession (Wicker et al., 2017), and could thus be applied to trait mapping in the near future.

However, despite the huge progress made in 30 years, efficient gene mapping is still facing several bottlenecks. Time of generation during population construction, or to obtain progenies for phenotyping, is slowing down a lot of projects. Watson et al. (2018) developed 'speed-breeding' to shorten the breeding cycle. This is achieved by increasing daily light exposure and harvesting the plants at early growth stages. By this process, the number of barley generations in a year was doubled, from three to six. A low-cost version of 'speed-breeding' makes this protocol available to the whole community and will probably be intensively used in the upcoming years (Ghosh et al., 2018). Another challenge faced by trait mapping is computing power. The amount of sequencing data available grows as the price of sequencing decreases. Previous data analysis systems are no longer sufficient to process the data generated (for review see Yin et al., 2017), and storage quickly becomes a limiting factor in a lot of labs. To keep up with modern analysis and data requirements, labs must invest in appropriate bioinformatics platforms and trained staff.

As genotyping methods develop and the cost of sequencing decreases, genotyping ceases to be a primary bottleneck in trait genetics. However, whether it is for GWAS or for mutant screening, many of the plants need to be phenotyped and this task can be arduous. Therefore, phenotyping has become a more primary concern, particularly in the case of quantitative yield and stress-response traits that necessitate field phenotyping, and are strongly influenced by GxE effects. Progress in imaging using sensors and drones (for review, see Tardieu et al., 2017) is beginning to alleviate this problem, but development of so-called phenomics is still in its infancy and remains very costly. However, even if limited in the traits to which it can be applied, phenomics will become a major part of trait mapping in the future, allowing mapping of some complex traits currently far out of reach.

12 Acknowledgement

We warmly thank Mark Timothy Rabanus-Wallace for language editing.

13 Where to look for further information

13.1 Further reading

- Bazakos et al. (2017) - Annual review of plant biology: a good review on quantitative genetics methods from phenotyping, to populations and QTL analysis.
- Grover and Sharma (2016) - A good review on the different type of markers and their compared qualities.
- Jiao and Schneeberger (2017) - A review summarizing the third generation sequencing that renders assembly of barley possible.
- Rifkin (2012) - A very complete book on QTL mapping, from population construction to marker selection and statistical analysis of the data.
- Stein and Muehlbauer (2018) - Summarizes the state-of-the-art in barley genome analysis.

13.2 Key journals/conferences

- PAG (Plant and Animal Genome): the biggest conference on genetics and genomics, held in January every year in San Diego, United States
- International Barley Genetics symposium (IBSG): symposium on barley genetics held every 4 years
- *Theoretical and Applied Genetics* (*TAG*)

- *Molecular Breeding*
- *Plant Breeding*
- *Molecular Plants*
- *Plant Journal*
- *Plant Physiology*
- *PNAS*
- *Plant Cell*
- *Nature Genetics*

13.3 Major international research projects

International Barley Sequencing Consortium (IBSC)

14 References

Abe, A., Kosugi, S., Yoshida, K., Natsume, S., Takagi, H., Kanzaki, H., Matsumura, H., Yoshida, K., Mitsuoka, C., Tamiru, M., et al. 2012. Genome sequencing reveals agronomically important loci in rice using MutMap. *Nat. Biotechnol.* 30(2), 174–8. doi:10.1038/nbt.2095.

Ariyadasa, R., Mascher, M., Nussbaumer, T., Schulte, D., Frenkel, Z., Poursarebani, N., Zhou, R., Steuernagel, B., Gundlach, H., Taudien, S., et al. 2014. A sequence-ready physical map of barley anchored genetically by two million single-nucleotide polymorphisms. *Plant Physiol.* 164(1), 412–23. doi:10.1104/pp.113.228213.

Baird, N. A., Etter, P. D., Atwood, T. S., Currey, M. C., Shiver, A. L., Lewis, Z. A., Selker, E. U., Cresko, W. A. and Johnson, E. A. 2008. Rapid SNP discovery and genetic mapping using sequenced RAD markers. *PLoS ONE* 3(10), e3376. doi:10.1371/journal.pone.0003376.

Barua, U. M., Chalmers, K. J., Hackett, C. A., Thomas, W. T. B., Powell, W. and Waugh, R. 1993. Identification of RAPD markers linked to a *Rhynchosporium secalis* resistance locus in barley using near-isogenic lines and bulked segregant analysis. *Heredity (Edinb)* 71(2), 177–84. doi:10.1038/hdy.1993.122.

Bauer, E., Weyen, J., Schiemann, A., Graner, A. and Ordon, F. 1997. Molecular mapping of novel resistance genes against Barley Mild Mosaic Virus (BaMMV). *Theor. Appl. Genet.* 95(8), 1263-9. doi:10.1007/s001220050691.

Bayer, M. M., Rapazote-Flores, P., Ganal, M., Hedley, P. E., Macaulay, M., Plieske, J., Ramsay, L., Russell, J., Shaw, P. D., Thomas, W., et al. 2017. Development and evaluation of a barley 50k iSelect SNP array. *Front. Plant Sci.* 8, 1792. doi:10.3389/fpls.2017.01792.

Bazakos, C., Hanemian, M., Trontin, C., Jiménez-Gómez, J. M. and Loudet, O. 2017. New strategies and tools in quantitative genetics: how to go from the phenotype to the genotype. *Annu. Rev. Plant Biol.* 68, 435–55. doi:10.1146/annurev-arplant-042916-040820.

Becker, J. and Heun, M. 1995. Barley microsatellites: allele variation and mapping. *Plant Mol. Biol.* 27(4), 835–45. doi:10.1007/BF00020238.

Becker, J., Vos, P., Kuiper, M., Salamini, F. and Heun, M. 1995. Combined mapping of AFLP and RFLP markers in barley. *Mol. Gen. Genet.* 249(1), 65–73. doi:10.1007/BF00290237.

Beckmann, J. S. and Soller, M. 1988. Detection of linkage between marker loci and loci affecting quantitative traits in crosses between segregating populations. *Theor. Appl. Genet.* 76(2), 228–36. doi:10.1007/BF00257850.

Blattner, F. R. 2009. Progress in phylogenetic analysis and a new infrageneric classification of the barley genus *Hordeum* (Poaceae: Triticeae). *Breed. Sci.* 59(5), 471–80. doi:10.1270/jsbbs.59.471.

Bortesi, L. and Fischer, R. 2015. The CRISPR/Cas9 system for plant genome editing and beyond. *Biotechnol. Adv.* 33(1), 41–52. doi:10.1016/j.biotechadv.2014.12.006.

Botstein, D., White, R. L., Skolnick, M. and Davis, R. W. 1980. Construction of a genetic linkage map in man using restriction fragment length polymorphisms. *Am. J. Hum. Genet.* 32(3), 314–31.

Burr, B., Burr, F. A., Thompson, K. H., Albertson, M. C. and Stuber, C. W. 1988. Gene mapping with recombinant inbreds in maize. *Genetics* 118(3), 519–26. doi:10.1063/1.3634023.

Büschges, R., Hollricher, K., Panstruga, R., Simons, G., Wolter, M., Frijters, A., van Daelen, R., van der Lee, T., Diergaarde, P., Groenendijk, J., et al. 1997. The barley *Mlo* gene: a novel control element of plant pathogen resistance. *Cell* 88(5), 695–705. doi:10.1016/S0092-8674(00)81912-1.

Caldwell, D. G., McCallum, N., Shaw, P., Muehlbauer, G. J., Marshall, D. F. and Waugh, R. 2004. A structured mutant population for forward and reverse genetics in barley (*Hordeum vulgare* L.). *Plant J.* 40(1), 143–50. doi:10.1111/j.1365-313X.2004.02190.x.

Carollo, V., Matthews, D. E., Lazo, G. R., Blake, T. K., Hummel, D. D., Lui, N., Hane, D. L. and Anderson, O. D. 2005. GrainGenes 2.0. An improved resource for the small-grains community. *Plant Physiol.* 139(2), 643–51. doi:10.1104/pp.105.064485.

Chutimanitsakun, Y., Nipper, R. W., Cuesta-Marcos, A., Cistué, L., Corey, A., Filichkina, T., Johnson, E. A. and Hayes, P. M. 2011. Construction and application for QTL analysis of a Restriction Site Associated DNA (RAD) linkage map in barley. *BMC Genomics* 12, 4. doi:10.1186/1471-2164-12-4.

Clapham, D., Edwards, J. and Portlock, P. 1973. Haploid Hordeum plants from anthers *in vitro*. *Z. Pflanzenzucht* 69, 142-55.

Close, T. J., Wanamaker, S. I., Caldo, R. A., Turner, S. M., Ashlock, D. A., Dickerson, J. A., Wing, R. A., Muehlbauer, G. J., Kleinhofs, A. and Wise, R. P. 2004. A new resource for cereal genomics: 22K barley GeneChip comes of age. *Plant Physiol.* 134(3), 960–8. doi:10.1104/pp.103.034462.

Close, T. J., Bhat, P. R., Lonardi, S., Wu, Y., Rostoks, N., Ramsay, L., Druka, A., Stein, N., Svensson, J. T., Wanamaker, S., et al. 2009. Development and implementation of high-throughput SNP genotyping in barley. *BMC Genomics* 10, 582. doi:10.1186/1471-2164-10-582.

Colmsee, C., Beier, S., Himmelbach, A., Schmutzer, T., Stein, N., Scholz, U. and Mascher, M. 2015. BARLEX – the barley draft genome explorer. *Mol. Plant* 8(6), 964–6. doi:10.1016/j.molp.2015.03.009.

Comadran, J., Kilian, B., Russell, J., Ramsay, L., Stein, N., Ganal, M., Shaw, P., Bayer, M., Thomas, W., Marshall, D., et al. 2012. Natural variation in a homolog of Antirrhinum CENTRORADIALIS contributed to spring growth habit and environmental adaptation in cultivated barley. *Nat. Genet.* 44(12), 1388–92. doi:10.1038/ng.2447.

Commission on Genetic Resources for Food and Agriculture; FAO. 2010. The state of *ex situ* conservation. In: *The Second Report on the State of the World's Plant Genetic Resources for Food and Agriculture*. FAO, Rome, pp. 54–90.

Costa-Silva, J., Domingues, D. and Lopes, F. M. 2017. RNA-Seq differential expression analysis: an extended review and a software tool. *PLoS ONE* 12(12), e0190152. doi:10.1371/journal.pone.0190152.

Damerval, C., Maurice, A., Josse, J. M. and de Vienne, D. 1994. Quantitative trait loci underlying gene product variation: a novel perspective for analyzing regulation of genome expression. *Genetics* 137(1), 289–301.

Doležel, J., Vrána, J., Šafář, J., Bartoš, J., Kubaláková, M. and Šimková, H. 2012. Chromosomes in the flow to simplify genome analysis. *Funct. Integr. Genomics* 12(3), 397–416. doi:10.1007/s10142-012-0293-0.

Edwards, M. D., Stuber, C. W. and Wendel, J. F. 1987. Molecular-marker-facilitated investigations of quantitative-trait loci in maize. I. Numbers, genomic distribution and types of gene action. *Genetics* 116(1), 113–25.

FAOSTAT. 2018. *Statistical Databases*. Food and Agricultural Organization of the United Nations.

Feuk, L., Carson, A. R. and Scherer, S. W. 2006. Structural variation in the human genome. *Nat. Rev. Genet.* 7(2), 85–97. doi:10.1038/nrg1767.

Friedt, W., Kaiser, R., Götz, R., Umbach, H. and Foroughi-Wehr, B. 1987. Genetic basis of breeding for resistance to barley yellow mosaic virus (BaYMV). In: Cooper, J. I. and Asher, M. J. C. (Eds), *Proceedings of a Conference at the University of St Andrews*. Association of Applied Biologists, Warwick, UK.

Gallais, A., Moreau, L. and Charcosset, A. 2007. Detection of marker-QTL associations by studying change in marker frequencies with selection. *Theor. Appl. Genet.* 114(4), 669–81. doi:10.1007/s00122-006-0467-z.

Giovannoni, J. J., Wing, R. A., Ganal, M. W. and Tanksley, S. D. 1991. Isolation of molecular markers from specific chromosomal intervals using DNA pools from existing mapping populations. *Nucleic Acids Res.* 19(23), 6553–8. doi:10.1093/nar/19.23.6553.

Ghosh, S., Watson, A., Gonzalez-Navarro, O. E., Ramirez-Gonzalez, R. H., Yanes, L., Mendoza-Suárez, M., Simmonds, J., Wells, R., Rayner, T., Green, P., et al. 2018. Speed breeding in growth chambers and glasshouses for crop breeding and model plant research. *Nat. Protoc.* 13(12), 2944–63. doi:10.1038/s41596-018-0072-z.

Gottwald, S., Bauer, P., Komatsuda, T., Lundqvist, U. and Stein, N. 2009. TILLING in the two-rowed barley cultivar "Barke" reveals preferred sites of functional diversity in the gene *HvHox1*. *BMC Res. Notes* 2, 258. doi:10.1186/1756-0500-2-258.

Graner, A. and Bauer, E. 1993. RFLP mapping of the ym4 virus resistance gene in barley. *Theor. Appl. Genet.* 86(6), 689–93. doi:10.1007/BF00222657.

Graner, A., Jahoor, A., Schondelmaier, J., Siedler, H., Pillen, K., Fischbeck, G., Wenzel, G. and Herrmann, R. G. 1991. Construction of an RFLP map of barley. *Theor. Appl. Genet.* 83(2), 250–6. doi:10.1007/BF00226259.

Graner, A., Bauer, E., Kellermann, A., Proeseler, Wenzel, G. and Ordon, F. 1995. RFLP analysis of resistance to the barley yellow mosaic virus complex. *Agronomie* 15(7–8), 475–9. doi:10.1051/agro:19950716.

Graner, A., Streng, S., Kellermann, A., Schiemann, A., Bauer, E., Waugh, R., Pellio, B. and Ordon, F. 1999. Molecular mapping and genetic fine-structure of the *rym5* locus encoding resistance to different strains of the Barley Yellow Mosaic Virus Complex. *Theor. Appl. Genet.* 98(2), 285–90. doi:10.1007/s001220051070.

Grover, A. and Sharma, P. C. 2016. Development and use of molecular markers: past and present. *Crit. Rev. Biotechnol.* 36(2), 290–302. doi:10.3109/07388551.2014.959891.

Gupta, P. K. and Varshney, R. K. 2000. The development and use of microsatellite markers for genetic analysis and plant breeding with emphasis on bread wheat. *Euphytica* 113(3), 163–85. doi:10.1023/A:1003910819967.

Gurushidze, M., Hensel, G., Hiekel, S., Schedel, S., Valkov, V. and Kumlehn, J. 2014. True-breeding targeted gene knock-out in barley using designer TALE-nuclease in haploid cells. *PLoS ONE* 9(3), e92046. doi:10.1371/journal.pone.0092046.

Haley, C. S. and Knott, S. A. 1992. A simple regression method for mapping quantitative trait loci in line crosses using flanking markers. *Heredity* 69(4), 315–24 [doi:10.1038/hdy.1992.131].

Hansson, M., Komatsuda, T., Stein, N. and Muehlbauer, G. J. 2018. Molecular mapping and cloning of genes and QTLs. In: Stein, N. and Muehlbauer, G. (Eds), *The Barley Genome. Compendium of Plant Genomes*. Springer, Cham, pp. 139–54. doi:10.1007/978-3-319-92528-8_10.

Hirsch, C. N., Hirsch, C. D., Brohammer, A. B., Bowman, M. J., Soifer, I., Barad, O., Shem-Tov, D., Baruch, K., Lu, F., Hernandez, A. G., et al. 2016. Draft assembly of elite inbred line PH207 provides insights into genomic and transcriptome diversity in maize. *Plant Cell* 28(11), 2700–14 [doi:10.1105/tpc.16.00353].

Holland, P. M., Abramson, R. D., Watson, R. and Gelfand, D. H. 1991. Detection of specific polymerase chain reaction product by utilizing the 5′----3′ exonuclease activity of *Thermus aquaticus* DNA polymerase. *Proc. Natl. Acad. Sci. U.S.A.* 88(16), 7276–80. doi:10.1073/pnas.88.16.7276.

Holme, I. B., Wendt, T., Gil-Humanes, J., Deleuran, L. C., Starker, C. G., Voytas, D. F. and Brinch-Pedersen, H. 2017. Evaluation of the mature grain phytase candidate *HvPAPhy_a* gene in barley (*Hordeum vulgare* L.) using CRISPR/Cas9 and TALENs. *Plant Mol. Biol.* 95(1–2), 111–21. doi:10.1007/s11103-017-0640-6.

Huang, B. E., Verbyla, K. L., Verbyla, A. P., Raghavan, C., Singh, V. K., Gaur, P., Leung, H., Varshney, R. K. and Cavanagh, C. R. 2015. MAGIC populations in crops: current status and future prospects. *Theor. Appl. Genet.* 128(6), 999–1017. doi:10.1007/s00122-015-2506-0.

Hübner, S., Höffken, M., Oren, E., Haseneyer, G., Stein, N., Graner, A., Schmid, K. and Fridman, E. 2009. Strong correlation of wild barley (*Hordeum spontaneum*) population structure with temperature and precipitation variation. *Mol. Ecol.* 18(7), 1523–36. doi:10.1111/j.1365-294X.2009.04106.x.

Hurgobin, B. and Edwards, D. 2017. SNP discovery using a pangenome: has the single reference approach become obsolete? *Biology (Basel)* 6(1), 21. doi:10.3390/biology6010021.

International Barley Genome Sequencing Consortium. 2012. A physical, genetic and functional sequence assembly of the barley genome. *Nature* 491(7426), 711–6. doi:10.1038/nature11543.

Jaccoud, D., Peng, K., Feinstein, D. and Kilian, A. 2001. Diversity arrays: a solid state technology for sequence information independent genotyping. *Nucleic Acids Res.* 29(4), E25. doi:10.1371/journal.pone.0019379.

Jiao, W. B. and Schneeberger, K. 2017. The impact of third generation genomic technologies on plant genome assembly. *Curr. Opin. Plant Biol.* 36, 64–70. doi:10.1016/j.pbi.2017.02.002.

Jost, M., Taketa, S., Mascher, M., Himmelbach, A., Yuo, T., Shahinnia, F., Rutten, T., Druka, A., Schmutzer, T., Steuernagel, B., et al. 2016. A homolog of Blade-On-Petiole 1 and 2 (BOP1/2) controls internode length and homeotic changes of the barley inflorescence. *Plant Physiol.* 171(2), 1113–27. doi:10.1104/pp.16.00124.

Jupe, F., Witek, K., Verweij, W., Śliwka, J., Pritchard, L., Etherington, G. J., Maclean, D., Cock, P. J., Leggett, R. M., Bryan, G. J., et al. 2013. Resistance gene enrichment sequencing (RenSeq) enables reannotation of the NB-LRR gene family from sequenced plant genomes and rapid mapping of resistance loci in segregating populations. *Plant J.* 76(3), 530–44. doi:10.1111/tpj.12307.

Kaiser, R. and Friedt, W. 1989. Chromosomal location of resistance to barley yellow mosaic virus in German winter-barley identified by trisomic analysis. *Theor. Appl. Genet.* 77(2), 241–5. doi:10.1007/BF00266193.

Kaiser, R. and Friedt, W. 1992. Gene for resistance to barley mild mosaic virus in German winter barley located on chromosome 3L. *Plant Breed.* 108(2), 169–72. doi:10.1111/j.1439-0523.1992.tb00116.x.

Kang, H. S., Kim, S. H., Lee, S. W., Kim, S. W., Ryu, J., Kim, J.-B., Yeom, S.-I., Kang, S.-Y. and Jo, Y. D. 2018. Optimization of TILLING system based on capillary electrophoresis for targeted selection of pepper gene mutants. *Hortic. Environ. Biotechnol.* 59(3), 447–60. doi:10.1007/s13580-018-0049-4.

Kanyuka, K., Druka, A., Caldwell, D. G., Tymon, A., McCallum, N., Waugh, R. and Adams, M. J. 2005. Evidence that the recessive bymovirus resistance locus *rym4* in barley corresponds to the eukaryotic translation initiation factor 4E gene. *Mol. Plant Pathol.* 6(4), 449–58. doi:10.1111/j.1364-3703.2005.00294.x.

Kao, K. N., Saleem, M., Abrams, S., Pedras, M., Horn, D. and Mallard, C. 1991. Culture conditions for induction of green plants from barley microspores by anther culture methods. *Plant Cell Rep.* 9(11), 595–601. doi:10.1007/BF00231796.

Kao, C. H., Zeng, Z. B. and Teasdale, R. D. 1999. Multiple interval mapping for quantitative trait loci. *Genetics* 152(3), 1203–16. doi:10.1016/0046-8177(92)90086-I.

Kapusi, E., Corcuera-Gómez, M., Melnik, S. and Stoger, E. 2017. Heritable genomic fragment deletions and small indels in the putative ENGase gene induced by CRISPR/Cas9 in barley. *Front. Plant Sci.* 8, 540. doi:10.3389/fpls.2017.00540.

Kersey, P. J., Allen, J. E., Armean, I., Boddu, S., Bolt, B. J., Carvalho-Silva, D., Christensen, M., Davis, P., Falin, L. J., Grabmueller, C., et al. 2016. Ensembl Genomes 2016: more genomes, more complexity. *Nucleic Acids Res.* 44(D1), D574–80. doi:10.1093/nar/gkv1209.

Kleinhofs, A., Kilian, A., Saghai Maroof, M. A., Biyashev, R. M., Hayes, P., Chen, F. Q., Lapitan, N., Fenwick, A., Blake, T. K., Kanazin, V., et al. 1993. A molecular, isozyme and morphological map of the barley (*Hordeum vulgare*) genome. *Theor. Appl. Genet.* 86(6), 705–12. doi:10.1007/BF00222660.

Knüpffer, H. 2009. *Triticeae* genetic resources in *ex situ* Genebank Collections. In: Muehlbauer, G. J. and Feuillet, C. (Eds), *Genetics and Genomics of the Triticeae*. Springer US, New York, NY, pp. 31–79. doi:10.1007/978-0-387-77489-3_2.

Knüpffer, H. and van Hintum, T. J. L. 2003. Summarised diversity – the barley core collection. In: von Bothmer, R., van Hintum, T., Knüpffer, H. and Sato, K. (Eds), *Diversity in Barley (Hordeum vulgare)*. Elsevier Science, Amsterdam, pp. 259–67.

Konieczny, A. and Ausubel, F. M. 1993. A procedure for mapping *Arabidopsis* mutations using co-dominant ecotype-specific PCR-based markers. *Plant J.* 4(2), 403–10. doi:10.1046/j.1365-313X.1993.04020403.x.

Konishi, T., Ban, T., Iida, Y. and Yoshimi, R. 1997. Genetic analysis of disease resistance to all strains of BaYMV in a Chinese barley landrace, Mokusekko 3. *Theor. Appl. Genet.* 94(6–7), 871–7. doi:10.1007/s001220050489.

Krasileva, K. V., Vasquez-Gross, H. A., Howell, T., Bailey, P., Paraiso, F., Clissold, L., Simmonds, J., Ramirez-Gonzalez, R. H., Wang, X., Borrill, P., et al. 2017. Uncovering hidden variation in polyploid wheat. *Proc. Natl. Acad. Sci. U.S.A.* 114(6), E913–21. doi:10.1073/pnas.1619268114.

Kumar, N., Galli, M., Ordon, J., Stuttmann, J., Kogel, K. H. and Imani, J. 2018. Further analysis of barley MORC1 using a highly efficient RNA-guided Cas9 gene-editing system. *Plant Biotechnol. J.* 16(11), 1892-903. doi:10.1111/pbi.12924.

Kurowska, M., Labocha-Pawłowska, A., Gnizda, D., Maluszynski, M. and Szarejko, I. 2012. Molecular analysis of point mutations in a barley genome exposed to MNU and gamma rays. *Mutat. Res. Mol. Mech. Mutagen.* 738–739, 52–70. doi:10.1016/j.mrfmmm.2012.08.008.

Lababidi, S., Mejlhede, N., Rasmussen, S. K., Backes, G., Al-Said, W., Baum, M. and Jahoor, A. 2009. Identification of barley mutants in the cultivar 'Lux' at the *Dhn* loci through TILLING. *Plant Breed.* 128(4), 332–6. doi:10.1111/j.1439-0523.2009.01640.x.

Lander, E. S. and Botstein, D. 1989. Mapping mendelian factors underlying quantitative traits using RFLP linkage maps. *Genetics* 121(1), 185–99. doi:10.1038/hdy.2014.4.

Lawrenson, T., Shorinola, O., Stacey, N., Li, C., Østergaard, L., Patron, N., Uauy, C. and Harwood, W. 2015. Induction of targeted, heritable mutations in barley and *Brassica oleracea* using RNA-guided Cas9 nuclease. *Genome Biol.* 16, 258. doi:10.1186/s13059-015-0826-7.

Li, M., Liu, X., Bradbury, P., Yu, J., Zhang, Y. M., Todhunter, R. J., Buckler, E. S. and Zhang, Z. 2014. Enrichment of statistical power for genome-wide association studies. *BMC Biol.* 12, 73. doi:10.1186/s12915-014-0073-5.

Linde-Laursen, I., Heslop-Harrison, J. S., Shepherd, K. W. and Taketa, S. 1997. The barley genome and its relationship with the wheat genomes. A survey with an internationally agreed recommendation for barley chromosome nomenclature. *Hereditas* 126(1), 1–16. doi:10.1111/j.1601-5223.1997.00001.x.

Lister, R., O'Malley, R. C., Tonti-Filippini, J., Gregory, B. D., Berry, C. C., Millar, A. H. and Ecker, J. R. 2008. Highly integrated single-base resolution maps of the epigenome in *Arabidopsis*. *Cell* 133(3), 523–36. doi:10.1016/j.cell.2008.03.029.

Litt, M. and Luty, J. A. 1989. A hypervariable microsatellite revealed by in vitro amplification of a dinucleotide repeat within the cardiac muscle actin gene. *Am. J. Hum. Genet.* 44(3), 397–401.

Lu, F., Romay, M. C., Glaubitz, J. C., Bradbury, P. J., Elshire, R. J., Wang, T., Li, Y., Li, Y., Semagn, K., Zhang, X., et al. 2015. High-resolution genetic mapping of maize pan-genome sequence anchors. *Nat. Commun.* 6(1), 6914. doi:10.1038/ncomms7914.

Lüpken, T., Stein, N., Perovic, D., Habekuß, A., Krämer, I., Hähnel, U., Steuernagel, B., Scholz, U., Zhou, R., Ariyadasa, R., et al. 2013. Genomics-based high-resolution mapping of the BaMMV/BaYMV resistance gene *rym11* in barley (*Hordeum vulgare* L.). *Theor. Appl. Genet.* 126(5), 1201–12. doi:10.1007/s00122-013-2047-3.

Mascher, M., Richmond, T. A., Gerhardt, D. J., Himmelbach, A., Clissold, L., Sampath, D., Ayling, S., Steuernagel, B., Pfeifer, M., D'Ascenzo, M., et al. 2013. Barley whole exome capture: a tool for genomic research in the genus *Hordeum* and beyond. *Plant J.* 76(3), 494–505. doi:10.1111/tpj.12294.

Mascher, M., Jost, M., Kuon, J. E., Himmelbach, A., Aßfalg, A., Beier, S., Scholz, U., Graner, A. and Stein, N. 2014. Mapping-by-sequencing accelerates forward genetics in barley. *Genome Biol.* 15(6), R78. doi:10.1186/gb-2014-15-6-r78.

Mascher, M., Gundlach, H., Himmelbach, A., Beier, S., Twardziok, S. O., Wicker, T., Radchuk, V., Dockter, C., Hedley, P. E., Russell, J., et al. 2017. A chromosome conformation capture ordered sequence of the barley genome. *Nature* 544(7651), 427–33. doi:10.1038/nature22043.

Maurer, A., Draba, V., Jiang, Y., Schnaithmann, F., Sharma, R., Schumann, E., Kilian, B., Reif, J. C. and Pillen, K. 2015. Modelling the genetic architecture of flowering time control in barley through nested association mapping. *BMC Genomics* 16, 290. doi:10.1186/s12864-015-1459-7.

Mayer, K. F. X., Martis, M., Hedley, P. E., Šimková, H., Liu, H., Morris, J. A., Steuernagel, B., Taudien, S., Roessner, S., Gundlach, H., et al. 2011. Unlocking the barley genome by chromosomal and comparative genomics. *Plant Cell* 23(4), 1249–63. doi:10.1105/tpc.110.082537.

McCallum, C. M., Comai, L., Greene, E. A. and Henikoff, S. 2000. Targeting Induced Local Lesions IN Genomes (TILLING) for plant functional genomics. *Plant Physiol.* 123(2), 439–42. doi:10.1104/pp.123.2.439.

Michalek, W., Künzel, G. and Graner, A. 1999. Sequence analysis and gene identification in a set of mapped RFLP markers in barley (*Hordeum vulgare*). *Genome* 42(5), 849–53.

Michelmore, R. W., Paran, I. and Kesseli, R. V. 1991. Identification of markers linked to disease-resistance genes by bulked segregant analysis: a rapid method to detect markers in specific genomic regions by using segregating populations. *Proc. Natl. Acad. Sci. U.S.A.* 88(21), 9828–32. doi:10.1073/pnas.88.21.9828.

Milner, S. G., Jost, M., Taketa, S., Mazón, E. R., Himmelbach, A., Oppermann, M., Weise, S., Knüpffer, H., Basterrechea, M., König, P., et al. 2019. Genebank genomics highlights the diversity of a global barley collection. *Nat. Genet.* 51(2), 319–26. doi:10.1038/s41588-018-0266-x.

Monat, C., Schreiber, M., Stein, N. and Mascher, M. 2018. Prospects of pan-genomics in barley. *Theor. Appl. Genet.*, 1–12. doi:10.1007/s00122-018-3234-z.

Muñoz-Amatriaín, M., Moscou, M. J., Bhat, P. R., Svensson, J. T., Bartoš, J., Suchánková, P., Šimková, H., Endo, T. R., Fenton, R. D., Lonardi, S., et al. 2011. An improved consensus linkage map of barley based on flow-sorted chromosomes and single nucleotide polymorphism markers. *Plant genome J.* 4, 238. doi:10.3835/plantgenome2011.08.0023.

Muñoz-Amatriaín, M., Cuesta-Marcos, A., Endelman, J. B., Comadran, J., Bonman, J. M., Bockelman, H. E., Chao, S., Russell, J., Waugh, R., Hayes, P. M., et al. 2014. The USDA barley core collection: genetic diversity, population structure, and potential for genome-wide Association Studies. *PLoS ONE* 9(4), e94688. doi:10.1371/journal.pone.0094688.

Nice, L. M., Steffenson, B. J., Brown-Guedira, G. L., Akhunov, E. D., Liu, C., Kono, T. J. Y., Morrell, P. L., Blake, T. K., Horsley, R. D., Smith, K. P., et al. 2016. Development and genetic characterization of an Advanced Backcross-Nested Association Mapping (AB-NAM) population of wild x cultivated barley. *Genetics* 203(3), 1453–67. doi:10.1534/genetics.116.190736.

Nissan-Azzouz, F., Graner, A., Friedt, W. and Ordon, F. 2005. Fine-mapping of the BaMMV, BaYMV-1 and BaYMV-2 resistance of barley (*Hordeum vulgare*) accession PI1963. *Theor. Appl. Genet.* 110(2), 212–8. doi:10.1007/s00122-004-1802-x.

Olson, M., Hood, L., Cantor, C. and Botstein, D. 1989. A common language for physical mapping of the human genome. *Science* 245(4925), 1434–5. doi:10.1126/science.2781285.

Ordon, F., Bauer, E., Friedt and Graner, A. 1995. Marker-based selection for the ym4 BaMMV-resistance gene in barley using RAPDs. *Agronomie* 15(7–8), 481–5. doi:10.1051/agro:19950717.

Pankin, A., Campoli, C., Dong, X., Kilian, B., Sharma, R., Himmelbach, A., Saini, R., Davis, S. J., Stein, N., Schneeberger, K., et al. 2014. Mapping-by-sequencing identifies *HvPHYTOCHROME C* as a candidate gene for the *early maturity 5* locus modulating the circadian clock and photoperiodic flowering in barley. *Genetics* 198(1), 383–96. doi:10.1534/genetics.114.165613.

Pasam, R. K., Sharma, R., Walther, A., Özkan, H., Graner, A. and Kilian, B. 2014. Genetic diversity and population structure in a legacy collection of spring barley landraces adapted to a wide range of climates. *PLoS ONE* 9(12), e116164. doi:10.1371/journal.pone.0116164.

Pellio, B., Streng, S., Bauer, E., Stein, N., Perovic, D., Schiemann, A., Friedt, W., Ordon, F. and Graner, A. 2005. High-resolution mapping of the *Rym4/Rym5* locus conferring resistance to the barley yellow mosaic virus complex (BaMMV, BaYMV, BaYMV-2) in barley (*Hordeum vulgare* ssp. *vulgare* L.). *Theor. Appl. Genet.* 110(2), 283–93. doi:10.1007/s00122-004-1832-4.

Poland, J. A., Brown, P. J., Sorrells, M. E. and Jannink, J. L. 2012. Development of high-density genetic maps for barley and wheat using a novel two-enzyme genotyping-by-sequencing approach. *PLoS ONE* 7(2), e32253. doi:10.1371/journal.pone.0032253.

Potokina, E., Druka, A., Luo, Z., Wise, R., Waugh, R. and Kearsey, M. 2008. Gene expression quantitative trait locus analysis of 16 000 barley genes reveals a complex pattern of genome-wide transcriptional regulation. *Plant J.* 53(1), 90–101. doi:10.1111/j.1365-313X.2007.03315.x.

Putnam, N. H., O'Connell, B. L., Stites, J. C., Rice, B. J., Blanchette, M., Calef, R., Troll, C. J., Fields, A., Hartley, P. D., Sugnet, C. W., et al. 2016. Chromosome-scale shotgun assembly using an in vitro method for long-range linkage. *Genome Res.* 26(3), 342–50. doi:10.1101/gr.193474.115.

Rifkin, S. A. 2012. *Quantitative Trait Loci (QTL): Methods and Protocols*. Humana Press, Totowa, NJ. doi:10.1007/978-1-61779-785-9.

Saghai Maroof, M. A., Biyashev, R. M., Yang, G. P., Zhang, Q. and Allard, R. W. 1994. Extraordinarily polymorphic microsatellite DNA in barley: species diversity, chromosomal locations, and population dynamics. *Proc. Natl. Acad. Sci. U.S.A.* 91(12), 5466–70. doi:10.1073/pnas.91.12.5466.

Saiki, R. K., Scharf, S., Faloona, F., Mullis, K. B., Horn, G. T., Erlich, H. A. and Arnheim, N. 1985. Enzymatic amplification of beta-globin genomic sequences and restriction site analysis for diagnosis of sickle cell anemia. *Science* 230(4732), 1350–4. doi:10.1126/science.2999980.

Sánchez-Martín, J., Steuernagel, B., Ghosh, S., Herren, G., Hurni, S., Adamski, N., Vrána, J., Kubaláková, M., Krattinger, S. G., Wicker, T., et al. 2016. Rapid gene isolation in barley and wheat by mutant chromosome sequencing. *Genome Biol.* 17(1), 221. doi:10.1186/s13059-016-1082-1.

Sannemann, W., Huang, B. E., Mathew, B. and Léon, J. 2015. Multi-parent advanced generation inter-cross in barley: high-resolution quantitative trait locus mapping for flowering time as a proof of concept. *Mol. Breed.* 35(3), 86. doi:10.1007/s11032-015-0284-7.

Schneeberger, K., Ossowski, S., Lanz, C., Juul, T., Petersen, A. H., Nielsen, K. L., Jørgensen, J. E., Weigel, D. and Andersen, S. U. 2009. SHOREmap: simultaneous mapping and mutation identification by deep sequencing. *Nat. Methods* 6(8), 550–1. doi:10.1038/nmeth0809-550.

Schulte, D., Ariyadasa, R., Shi, B., Fleury, D., Saski, C., Atkins, M., DeJong, P., Wu, C. C., Graner, A., Langridge, P., et al. 2011. BAC library resources for map-based cloning and physical map construction in barley (*Hordeum vulgare* L.). *BMC Genomics* 12, 247. doi:10.1186/1471-2164-12-247.

Segura, V., Vilhjálmsson, B. J., Platt, A., Korte, A., Seren, Ü., Long, Q. and Nordborg, M. 2012. An efficient multi-locus mixed-model approach for genome-wide association studies in structured populations. *Nat. Genet.* 44(7), 825–30. doi:10.1038/ng.2314.

Semagn, K., Babu, R., Hearne, S. and Olsen, M. 2014. Single nucleotide polymorphism genotyping using Kompetitive Allele Specific PCR (KASP): overview of the technology and its application in crop improvement. *Mol. Breed.* 33(1), 1–14. doi:10.1007/s11032-013-9917-x.

Spannagl, M., Nussbaumer, T., Bader, K. C., Martis, M. M., Seidel, M., Kugler, K. G., Gundlach, H. and Mayer, K. F. X. 2016. PGSB PlantsDB: updates to the database framework for comparative plant genome research. *Nucleic Acids Res.* 44(D1), D1141–7. doi:10.1093/nar/gkv1130.

Spielman, R. S., McGinnis, R. E. and Ewens, W. J. 1993. Transmission test for linkage disequilibrium: the insulin gene region and insulin-dependent diabetes mellitus (IDDM). *Am. J. Hum. Genet.* 52(3), 506–16.

Springer, N. M., Anderson, S. N., Andorf, C. M., Ahern, K. R., Bai, F., Barad, O., Barbazuk, W. B.,Bass, H. W., Baruch, K., Ben-Zvi, G., et al. 2018. The maize W22 genome provides a foundation for functional genomics and transposon biology. *Nat. Genet.* 50(9), 1282–8. doi:10.1038/s41588-018-0158-0.

Steffenson, B. J., Olivera, P., Roy, J. K., Jin, Y., Smith, K. P. and Muehlbauer, G. J. 2007. A walk on the wild side: mining wild wheat and barley collections for rust resistance genes. *Aust. J. Agric. Res.* 58(6), 532. doi:10.1071/AR07123.

Stein, N. and Muehlbauer, G. J. 2018. *The Barley Genome, Compendium of Plant Genomes*. Springer International Publishing, Cham. doi:10.1007/978-3-319-92528-8.

Stein, N., Perovic, D., Kumlehn, J., Pellio, B., Stracke, S., Streng, S., Ordon, F. and Graner, A. 2005. The eukaryotic translation initiation factor 4E confers multiallelic recessive Bymovirus resistance in *Hordeum vulgare* (L.). *Plant J.* 42(6), 912–22. doi:10.1111/j.1365-313X.2005.02424.x.

Steuernagel, B., Periyannan, S. K., Hernández-Pinzón, I., Witek, K., Rouse, M. N., Yu, G., Hatta, A., Ayliffe, M., Bariana, H., Jones, J. D. G., et al. 2016. Rapid cloning of disease-resistance genes in plants using mutagenesis and sequence capture. *Nat. Biotechnol.* 34(6), 652–5. doi:10.1038/nbt.3543.

Sun, Y., Wang, J., Crouch, J. H. and Xu, Y. 2010. Efficiency of selective genotyping for genetic analysis of complex traits and potential applications in crop improvement. *Mol. Breed.* 26(3), 493–511. doi:10.1007/s11032-010-9390-8.

Sun, S., Zhou, Y., Chen, J., Shi, J., Zhao, H., Zhao, H., Song, W., Zhang, M., Cui, Y., Dong, X., et al. 2018. Extensive intraspecific gene order and gene structural variations between Mo17 and other maize genomes. *Nat. Genet.* 50(9), 1289–95. doi:10.1038/s41588-018-0182-0.

Szarejko, I., Szurman-Zubrzycka, M., Nawrot, M., Marzec, M., Gruszka, D., Kurowska, M., Chmielewska, B., Zbieszczyk, J., Jelonek, J. and Maluszynski, M. 2017. Creation of a TILLING population in barley After chemical mutagenesis with sodium azide and MNU. In: Jankowicz-Cieslak, J., Tai, T. H., Kumlehn, J. and Till, B. J. (Eds), *Biotechnologies for Plant Mutation Breeding*. Springer International Publishing, Cham, pp. 91–111. doi:10.1007/978-3-319-45021-6_6.

Szurman-Zubrzycka, M. E., Zbieszczyk, J., Marzec, M., Jelonek, J., Chmielewska, B., Kurowska, M. M., Krok, M., Daszkowska-Golec, A., Guzy-Wrobelska, J., Gruszka, D., et al. 2018. HorTILLUS–A rich and renewable source of induced mutations for forward/reverse genetics and pre-breeding programs in barley (*Hordeum vulgare* L.). *Front. Plant Sci.* 9, 216. doi:10.3389/fpls.2018.00216.

Takagi, H., Uemura, A., Yaegashi, H., Tamiru, M., Abe, A., Mitsuoka, C., Utsushi, H., Natsume, S., Kanzaki, H., Matsumura, H., et al. 2013. MutMap-Gap: whole-genome resequencing of mutant F2 progeny bulk combined with *de novo* assembly of gap regions identifies the rice blast resistance gene *Pii*. *New Phytol.* 200(1), 276–83. doi:10.1111/nph.12369.

Talamè, V., Bovina, R., Sanguineti, M. C., Tuberosa, R., Lundqvist, U. and Salvi, S. 2008. TILLMore, a resource for the discovery of chemically induced mutants in barley. *Plant Biotechnol. J.* 6(5), 477–85. doi:10.1111/j.1467-7652.2008.00341.x.

Tanksley, S. D., Young, N. D., Paterson, A. H. and Bonierbale, M. W. 1989. RFLP mapping in plant breeding: new tools for an old science. *Nat. Biotechnol.* 7(3), 257–64. doi:10.1038/nbt0389-257.

Tardieu, F., Cabrera-Bosquet, L., Pridmore, T. and Bennett, M. 2017. Plant phenomics, From sensors to knowledge. *Curr. Biol.* 27(15), R770–83. doi:10.1016/j.cub.2017.05.055.

Tello-Ruiz, M. K., Stein, J., Wei, S., Preece, J., Olson, A., Naithani, S., Amarasinghe, V., Dharmawardhana, P., Jiao, Y., Mulvaney, J., et al. 2016. Gramene 2016: comparative plant genomics and pathway resources. *Nucleic Acids Res.* 44(D1), D1133–40. doi:10.1093/nar/gkv1179.

The International Wheat Genome Sequencing Consortium. 2014. Ancient hybridizations among the ancestral genomes of bread wheat. *Science* 345(6194), 1250092. doi:10.1126/science.1250092.

Thind, A. K., Wicker, T., Šimková, H., Fossati, D., Moullet, O., Brabant, C., Vrána, J., Doležel, J. and Krattinger, S. G. 2017. Rapid cloning of genes in hexaploid wheat using cultivar-specific long-range chromosome assembly. *Nat. Biotechnol.* 35(8), 793–6. doi:10.1038/nbt.3877.

Thoday, J. M. 1961. Location of polygenes. *Nature* 191(4786), 368–70. doi:10.1038/191368a0.

Till, B. J., Reynolds, S. H., Greene, E. A., Codomo, C. A., Enns, L. C., Johnson, J. E., Burtner, C., Odden, A. R., Young, K., Taylor, N. E., et al. 2003. Large-scale discovery of induced point mutations with high-throughput TILLING. *Genome Res.* 13(3), 524–30. doi:10.1101/gr.977903.

Tsai, H., Howell, T., Nitcher, R., Missirian, V., Watson, B., Ngo, K. J., Lieberman, M., Fass, J., Uauy, C., Tran, R. K., et al. 2011. Discovery of rare mutations in populations: TILLING by sequencing. *Plant Physiol.* 156(3), 1257–68. doi:10.1104/pp.110.169748.

Vos, P., Hogers, R., Bleeker, M., Reijans, M., Van De Lee, T., Hornes, M., Friters, A., Pot, J., Paleman, J., Kuiper, M., et al. 1995. AFLP: a new technique for DNA fingerprinting. *Nucleic Acids Res.* 23(21), 4407–14. doi:10.1093/nar/23.21.4407.

Watson, A., Ghosh, S., Williams, M. J., Cuddy, W. S., Simmonds, J., Rey, M.-D., Asyraf Md Hatta, M., Hinchliffe, A., Steed, A., Reynolds, D., et al. 2018. Speed breeding is a powerful tool to accelerate crop research and breeding. *Nat. Plants* 4, 23-29. doi:10.1038/s41477-017-0083-8.

Wendt, T., Holm, P. B., Starker, C. G., Christian, M., Voytas, D. F., Brinch-Pedersen, H. and Holme, I. B. 2013. TAL effector nucleases induce mutations at a pre-selected location in the genome of primary barley transformants. *Plant Mol. Biol.* 83(3), 279-85. doi:10.1007/s11103-013-0078-4.

Wenzl, P., Carling, J., Kudrna, D., Jaccoud, D., Huttner, E., Kleinhofs, A. and Kilian, A. 2004. Diversity Arrays Technology (DArT) for whole-genome profiling of barley. *Proc. Natl. Acad. Sci. U.S.A.* 101(26), 9915–20. doi:10.1073/pnas.0401076101.

Wenzl, P., Li, H., Carling, J., Zhou, M., Raman, H., Paul, E., Hearnden, P., Maier, C., Xia, L., Caig, V., et al. 2006. A high-density consensus map of barley linking DArT markers to SSR, RFLP and STS loci and agricultural traits. *BMC Genomics* 7, 206. doi:10.1186/1471-2164-7-206.

Wicker, T., Zimmermann, W., Perovic, D., Paterson, A. H., Ganal, M., Graner, A. and Stein, N. 2005. A detailed look at 7 million years of genome evolution in a 439 kb contiguous sequence at the barley Hv-eIF4E locus: recombination, rearrangements and repeats. *Plant J.* 41(2), 184–94. doi:10.1111/j.1365-313X.2004.02285.x.

Wicker, T., Schulman, A. H., Tanskanen, J., Spannagl, M., Twardziok, S., Mascher, M., Springer, N. M., Li, Q., Waugh, R., Li, C., et al. 2017. The repetitive landscape of the 5100 Mbp barley genome. *Mob. DNA* 8, 22. doi:10.1186/s13100-017-0102-3.

Williams, J. G. K., Kubelik, A. R., Livak, K. J., Rafalski, J. A. and Tingey, S. V. 1990. DNA polymorphisms amplified by arbitrary primers are useful as genetic markers. *Nucleic Acids Res.* 18(22), 6531-5. doi:10.1093/nar/18.22.6531.

Wise, R. P., Caldo, R. A., Hong, L., Shen, L., Cannon, E. and Dickerson, J. A. 2007. BarleyBase/PLEXdb. In: Edwards, D. (Ed.), *Plant Bioinformatics*. Humana Press, Totowa, NJ, pp. 347–63. doi:10.1007/978-1-59745-535-0_17.

Worley, K. C., Richards, S. and Rogers, J. 2017. The value of new genome references. *Exp. Cell Res.* 358(2), 433–8. doi:10.1016/j.yexcr.2016.12.014.

Yang, P., Habekuß, A., Ordon, F. and Stein, N. 2014a. Analysis of bymovirus resistance genes on proximal barley chromosome 4HL provides the basis for precision breeding for BaMMV/BaYMV resistance. *Theor. Appl. Genet.* 127(7), 1625–34. doi:10.1007/s00122-014-2324-9.

Yang, P., Lüpken, T., Habekuss, A., Hensel, G., Steuernagel, B., Kilian, B., Ariyadasa, R., Himmelbach, A., Kumlehn, J., Scholz, U., et al. 2014b. PROTEIN DISULFIDE ISOMERASE LIKE 5-1 is a susceptibility factor to plant viruses. *Proc. Natl. Acad. Sci. U. S. A.* 111(6), 2104-9. doi:10.1073/pnas.1320362111.

Yang, N., Xu, X. W., Wang, R. R., Peng, W. L., Cai, L., Song, J. M., Li, W., Luo, X., Niu, L., Wang, Y., et al. 2017a. Contributions of *Zea mays* subspecies mexicana haplotypes to modern maize. *Nat. Commun.* 8(1), 1874. doi:10.1038/s41467-017-02063-5.

Yang, P., Habekuß, A., Hofinger, B. J., Kanyuka, K., Kilian, B., Graner, A., Ordon, F. and Stein, N. 2017b. Sequence diversification in recessive alleles of two host factor genes suggests adaptive selection for bymovirus resistance in cultivated barley from East Asia. *Theor. Appl. Genet.* 130(2), 331–44. doi:10.1007/s00122-016-2814-z.

Yin, Z., Lan, H., Tan, G., Lu, M., Vasilakos, A. V. and Liu, W. 2017. Computing platforms for big biological data analytics: perspectives and challenges. *Comput. Struct. Biotechnol. J.* 15, 403–11. doi:10.1016/j.csbj.2017.07.004.

Yu, Y., Tomkins, J. P., Waugh, R., Frisch, D. A., Kudrna, D., Kleinhofs, A., Brueggeman, R. S., Muehlbauer, G. J., Wise, R. P. and Wing, R. A. 2000. A bacterial artificial chromosome library for barley (*Hordeum vulgare* L.) and the identification of clones containing putative resistance genes. *Theor. Appl. Genet.* 101(7), 1093–9. doi:10.1007/s001220051584.

Yu, J., Pressoir, G., Briggs, W. H., Vroh Bi, I., Yamasaki, M., Doebley, J. F., McMullen, M. D., Gaut, B. S., Nielsen, D. M., Holland, J. B., et al. 2006. A unified mixed-model method for association mapping that accounts for multiple levels of relatedness. *Nat. Genet.* 38(2), 203–8. doi:10.1038/ng1702.

Yu, J., Holland, J. B., McMullen, M. D. and Buckler, E. S. 2008. Genetic design and statistical power of nested association mapping in maize. *Genetics* 178(1), 539–51. doi:10.1534/genetics.107.074245.

Zeng, Z. B. 1993. Theoretical basis for separation of multiple linked gene effects in mapping quantitative trait loci. *Proc. Natl. Acad. Sci. U.S.A.* 90(23), 10972–6. doi:10.1073/pnas.90.23.10972.

Zhang, H., Sreenivasulu, N., Weschke, W., Stein, N., Rudd, S., Radchuk, V., Potokina, E., Scholz, U., Schweizer, P., Zierold, U., et al. 2004. Large-scale analysis of the barley transcriptome based on expressed sequence tags. *Plant J.* 40(2), 276–90. doi:10.1111/j.1365-313X.2004.02209.x.

Zhang, Z., Ersoz, E. S., Lai, C. Q., Todhunter, R. J., Tiwari, H. K., Gore, M. A., Bradbury, P. J., Yu, J., Arnett, D. K., Ordovas, J. M., et al. 2010. Mixed linear model approach adapted for genome-wide association studies. *Nat. Genet.* 42(4), 355–60. doi:10.1038/ng.546.

Zhu, C., Gore, M., Buckler, E. S. and Yu, J. 2008. Status and prospects of association mapping in plants. *Plant Genome J.* 1, 5. doi:10.3835/plantgenome2008.02.0089.

Part 2

Advances in breeding

Chapter 6

Advanced designs for barley breeding experiments

Alison Kelly, Queensland Department of Agriculture and Fisheries and Queensland Alliance for Agriculture and Food Innovation (QAAFI), The University of Queensland, Australia; and Clayton Forknall, Queensland Department of Agriculture and Fisheries, Australia

1 Introduction
2 Background to experimental design of field trials
3 Designs for late-generation field trials
4 Designs for early-generation field trials
5 Incorporating a genetic relationship matrix
6 Multi-phase design for laboratory experiments
7 Conclusions
8 References

1 Introduction

Cereal breeding programmes typically test genotypes in a series of field trials across multiple locations and years (known as multi-environment trials or METs). These plant breeding programmes aim to release new genotypes that are both high yielding and have superior grain quality. Selection for grain yield and grain size has improved through the use of statistical methods for spatial design and analysis of field trials (Cullis and Gleeson, 1989; Gilmour et al., 1997; Fox et al., 2006) and through advanced models for genotype-by-environment interaction (Smith et al., 2001a). These methods are both intuitively appealing and statistically efficient as they consider the nature of the underlying variation within a field trial and genetic response patterns across METs. Genetic gains have been realised in traits associated with field trials due to the accuracy of phenotyping and the ability to partition genetic and non-genetic variation. The statistical advances for experimental design addressed in this chapter relate to breeding for annual crops in general, but will be presented with a focus on barley breeding applications.

Plant breeders aim to release superior genotypes with improved genetic potential, advancing through selection from a broad range of genetics in early

http://dx.doi.org/10.19103/AS.2019.0060.06

generation populations, and then refining to later-stage selection based on an assessment of broad adaptation to the target production regions. The trialling structure of the plant breeding programme resembles a breeding funnel (Fig. 1), where early generation trials are based on a large number of genotypes in a small number of trials with limited replication, and later-stage testing is based on a smaller number of highly selected genotypes across many trials with increased levels of replication. Experimental design changes across the stages of the breeding programme, to balance the trade-off in resource allocation of numbers of genotypes against level of replication, while maintaining accuracy of estimates of genotypic performance.

In plant breeding trials, genotype comparisons are of prime importance, and statistical models are used to predict the genotypic performance. These predictions must be unbiased and efficient (accurate and precise). A model that captures the 'true' genetic potential must accurately partition the three effects of genotype, systematic trend and experimental error. Systematic trend includes specific conditions pertaining to the trial as well as other management and micro-environmental factors. In a field trial, for example, systematic trend may arise from underlying soil fertility and moisture gradients, together with management practices associated with trial conduct. Similarly in the laboratory, day and batch effects for chemical procedures may systematically influence the experimental results. Experimental design mitigates this potential bias, through the processes of replication and randomisation. Advanced methods of statistical analysis increase the accuracy and precision of estimates of genotypic performance through more plausible models for systematic trend in the experimental material, be it field trend or variation in the laboratory process.

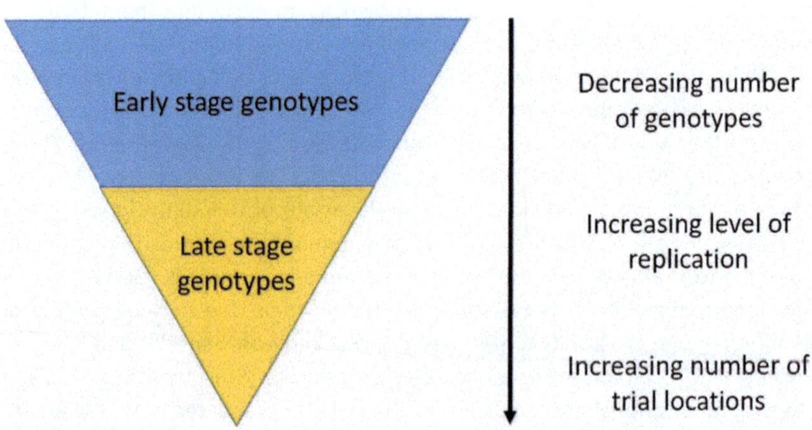

Figure 1 Plant breeding funnel showing the change in the number of breeding population genotypes and trialling resources for replicates and locations, from the early to later stages of testing.

Field trials form the basis of selection for agronomic traits with yield being the key economic driver. In subsistence production systems, biomass may be of equal importance for animal feed, whereas in developed economies, maximum yield and machine harvestability may be of prime importance. Fundamentally, all grain is produced for food or feed, beverage or fuel, so the quality of grain for the intended purpose is vitally important. Assessment of physical grain traits (such as grain size and grain weight) is commonplace in a breeding programme. Further assessment of cereal chemistry traits and the quality of end products from the grain is also important, but is often undertaken with less awareness of the potential to induce systematic bias during the experimental process. There is potentially large genetic gain to be made by studying the nature of variation in the quality assessment process, and using this to design efficient laboratory experiments. However, genetic gain in selection for genotypes with superior grain quality has been limited by low heritability associated with some of the grain-quality traits (Cullis et al., 2003).

This chapter provides a history of the developments in experimental design, including a collection of key references. A brief history to set the scene is presented in Section 1, and an excellent background to this earlier work is given in Kempton and Fox (1997). Sections 2–4 cover the background of experimental design for field trials. The discussion highlights the key principles that are still fundamental for modern comparative experiments, even in the world of data-rich science. Sections 5 introduce model-based design as a new technology for dealing with complex experiments where the data is correlated, or where the design is unbalanced. This section also explores the quantification of genetic relationships through either pedigree or molecular marker information. It demonstrates how these relationships can be incorporated into modern experimental design to improve the randomisation of genetic material across expected experimental trend. Section 6 presents the principles of multi-phase experiments for testing material both in the field and in the laboratory. Multi-phase experimental design allows for the systematic variation in the laboratory procedure to be estimated, and then partitioned from genetic information to ensure reproducible results. In addition, three case studies are presented at the end of specific sections. These case studies highlight non-standard experimental designs that should be in the toolkit of every agricultural scientist, and are essential for modern plant breeding programmes.

2 Background to experimental design of field trials

Experimental design forms the foundation of sound experimental process, and ensures robust and rigorous research methodology producing accurate results and valid scientific inference. The principles of design, and indeed the science of biometrics, had its beginnings with the work of R. A. Fisher at Rothamsted Experimental Station in the early 1920s. Fisher was a quantitative geneticist who was grappling with variability surrounding experimental conditions, and hence pioneered the science of biometrics in agricultural research. He devised experimental design and associated analysis methods based on variance partitioning to provide some certainty around cause and effect in his experimental results.

The key principles of experimental design for comparative studies are well summarised in many textbooks (e.g. Mead, 1990; Cobb, 2008) with a special treatment of experimental designs for agricultural field experiments presented in the books by Pearce (1983) and Petersen (1994). Common throughout these references and underpinning any rigorous experimentation are the processes of replication and randomisation. The theory of randomisation is implemented to eliminate bias in experimental results. It has formed the basis of experimental designs (Bailey, 1981) and analysis (Nelder, 1965a,b) over the past century. Replication ensures an underlying estimate of experimental error for assessing the significance of treatment effects. Both are essential components of any experimental method, particularly in agricultural and biological science, where the experimental units to which treatments are applied may be highly variable.

The presence of field heterogeneity at experimental sites was reported in the literature over 100 years ago (Mercer and Hall, 1911). For this reason, the experimental area was divided in small units of land representative of the trial site, termed plots, and these plots were introduced as the experimental unit, to which treatments were applied (Pearce, 1983). An example of the potential heterogeneity measured between experimental units in the field is demonstrated for a barley uniformity trial in Fig. 2. In this 'experiment', one barley genotype was grown in a field at Hermitage Research Station, Warwick, Qld (28.22°S, 152.03°E), and then harvested in small plots of 5 metres x 2 metres. The yield measured on a small-plot basis in this experiment varies from approximately 2.0 to 3.5 t/ha. Conducting comparative experiments in this background of field variation is challenging, and requires careful management practices as well as sound experimental design techniques.

Experimental design and analysis was founded by Fisher in 1925, when he formalised the techniques of replication and randomisation. Fisher's aim was to provide accurate estimates of the treatment effects, and at the same time to provide a measure of the precision of these estimates (Pearce, 1983). However, it was the need to control for variation in experimental units that drove many advances in both the design and analysis of agricultural field experiments (Mead, 1990). Blocking was devised to handle potential bias from systematic variation. Fisher (1925) introduced the randomised complete block (RCB) design, where each block contained a complete set of treatments. Later,

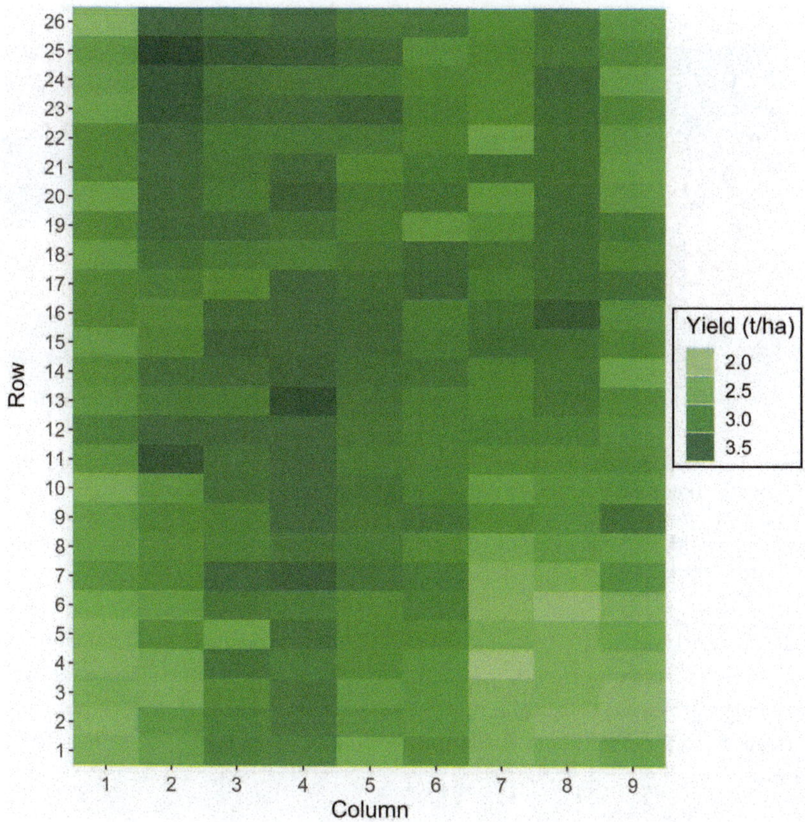

Figure 2 Uniformity trial showing grain yield of one barley genotype across 9 columns × 26 rows, harvested in small plots that were 5 metres long × 2 metres wide.

Yates (1933) introduced a double blocking system in the form of latin square designs to account for variation in two dimensions. Figure 3 presents layouts for a hypothetical experiment where treatments were randomised according to an RCB design and then following the more restricted randomisation of a latin square design. As experiments grew in size, Yates (1936a) realised the difficulty of selecting uniform complete blocks, and introduced incomplete block (IB) designs, where the smaller blocks each contained a subset of the treatments. If the incomplete blocks could be combined to form a complete block, the design was known as resolvable.

Modern designs in plant breeding applications are more closely aligned with the spatial arrangement of experimental units, or plots, in a field setting. These designs are based on a contiguous arrangement of plots in a two-dimensional array in the field, indexed by rows and columns. These modern developments recognise that many trial management operations are aligned

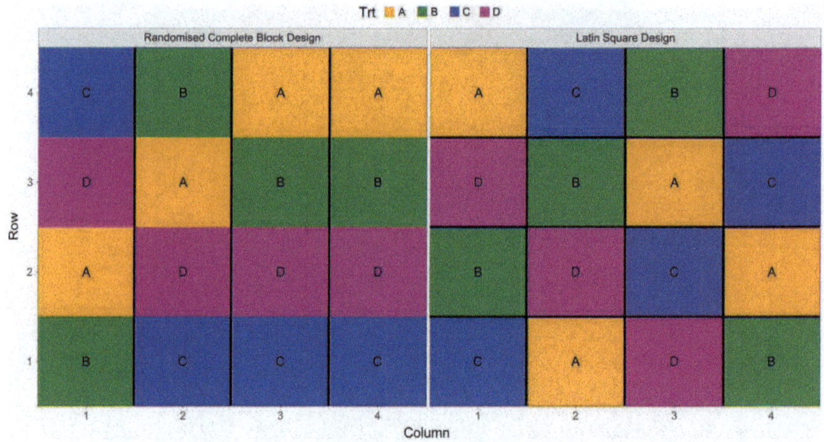

Figure 3 Layouts of a hypothetical experiment with treatments randomised according to a randomised complete block design or a latin square design. The experiment consists of four replicates of four treatments, laid out as a two-dimensional array of 4 columns × 4 rows in the field, forming a total of 16 plots (experimental units). Thick bold lines denote replicate block boundaries.

with rows and columns in the field, such as planting, irrigating and harvesting, as well as measurement procedures such as manual rating of plots. More recently, the imaging of experimental units is also aligned with these trial dimensions, perpetuating the usefulness and importance of these design solutions into the future.

3 Designs for late-generation field trials

The major grouping of IB designs applicable to larger-scale field trials was presented by Yates (1937) as pseudo-factorial designs, later known as lattices, and then grouped in two dimensions, as lattice square designs (Yates, 1940). The restriction of these designs was the requirement that the number of treatments, v, had to be a perfect square, due to the partial balance properties required between the incomplete blocks. An extension to this class of design was provided through rectangular lattices, where the number of treatments is the product of two consecutive integers, $v = k(k+1)$ (Harshbarger, 1949). The most general extension to IB designs was through alpha-lattice designs (Patterson and Williams, 1976), and these are still widely used in replicated plant breeding experiments. The development of alpha designs ensured the wide applicability of IB designs, by increasing the range of values for v to any non-prime number of treatments. These developments arose from the shortage of suitable IB designs available for large-scale cereal trials, such as the statutory field trials conducted in the United Kingdom (Patterson and Hunter, 1976).

Row-column alpha designs were introduced by John and Eccleston (1986), and then extended by Williams (1986a) as latinised row-column designs. These latinised designs relied on the principle of each treatment only occurring once in each row and column of the design, forming a binary design across these trial dimensions. Partial balance across two dimensions of the field was a desirable property to guard against any systematic trend induced through trial management aligned with rows and columns.

In the same decade, research into spatial analysis of field trials was becoming popular. The move toward spatial analysis was driven by a realisation that experimental units could not be considered independent, and evidence mounted for association between plots that were close together in a field trial (Wilkinson et al., 1983; Besag and Kempton, 1986; Williams, 1986b; Gleeson and Cullis, 1987). Methods for spatial design which accounted for an underlying correlation model in the design process began to emerge (Martin, 1986), and these methods form the basis of current model-based designs (Martin and Eccleston, 1993; Chan and Eccleston, 2003; Butler et al., 2008). Model-based designs, by definition, rely on an underlying variance model for estimating the set of treatment effects, together with an optimality criterion. The objective function is typically chosen to be the A-optimality criterion, which is based on an underlying linear mixed model. This criterion is calculated as the average pairwise variance of genotype comparisons (Martin et al., 2006).

The spatial model, like the row-column design, is based on a contiguous arrangement of plots in a two-dimensional array in the field, denoted by rows and columns. In spatial design, the residual variance is modelled as an underlying correlated process. One model that has been widely adopted to describe the relationship between neighbouring plots is an autoregressive process, based on the distance between neighbouring plots in each dimension of the field. This robust model arises from the spatial analysis model of Gilmour et al. (1997) with the associated implementation to field trial design proposed by Butler et al. (2014). The plot dimensions in field trials are typically rectangular, and this requires a separable model for the covariance structure in each dimension (Martin, 1986). In this model, different spatial correlation parameters across the row and column dimensions model decreasing association as plots become further apart. The correlations between rows and columns are then combined through a separable variance model for spatial designs, and this aligns with the separable model for spatial analysis presented by Gilmour et al. (1997) and implemented in the optimal design, *od*, package (Butler, 2018) for the R statistical computing environment (R Core Team, 2018).

To complete this section on designs for late-generation field trials, we consider a case study using a non-standard blocking technique.

3.1 Case study 1: A 2-year trial blocking on a measured trait

Due to the nature of field heterogeneity, it is commonly accepted that neighbouring plots (experimental units) in the field will be more alike (homogeneous) than plots located further apart. As such, the blocking techniques referred to previously rely on grouping sets of neighbouring or closely situated plots, to partition potential systematic sources of variation from experimental error and enable more precise treatment comparisons. However, blocking is often enforced with no pre-existing information of the potential sources of field heterogeneity. In a case where some level of prior information is available regarding potential sources of field heterogeneity, can a more informed method of defining blocks of plots be used to increase the precision of treatment comparisons?

The motivating data for this example arose during the conduct of a 2-year field experiment. The experiment consisted of a two-way factorial treatment structure, whereby four treatments were applied to plots in the first year of experimentation (labelled Trt A to D), while plots were over-sown with six additional treatments in the second year (labelled Trt 1 to 6). The treatments were initially randomised to plots according to a randomised block design, where each of the resultant 24 unique treatment combinations were replicated four times. At the conclusion of the first year of experimentation and prior to sowing the second-year treatments, soil samples taken from each plot revealed that the treatments applied in the first year had differentially influenced the populations of a particular species of root lesion nematode, known to cause significant yield losses in Australia. However, rather than the nematode populations being confounded with the first-year treatments (i.e. first-year treatments resulting in distinct 'levels' of nematode populations), the populations resulting from the treatments tended to overlap and formed a continuous range (Fig. 4).

Given the nematode populations were not completely confounded with the first-year treatments, there was potential that through simple randomisation of the second-year treatments according to the originally proposed randomised block design, some treatments may have been biased (advantaged or disadvantaged) by their assignment to plots. To mitigate this potential biasing, a blocking strategy based upon the measured nematode populations (observed quantity) was imposed. To begin, plots were ordered from smallest to largest, according to the nematode populations measured. Once ordered, blocks were defined to consist of groups of six plots, where the size of the block (six plots) corresponded to the number of treatments to be tested in the second year of experimentation. In this way, plots which displayed homogeneous nematode populations were grouped together in the same block, while plots exhibiting heterogeneous nematode populations were in

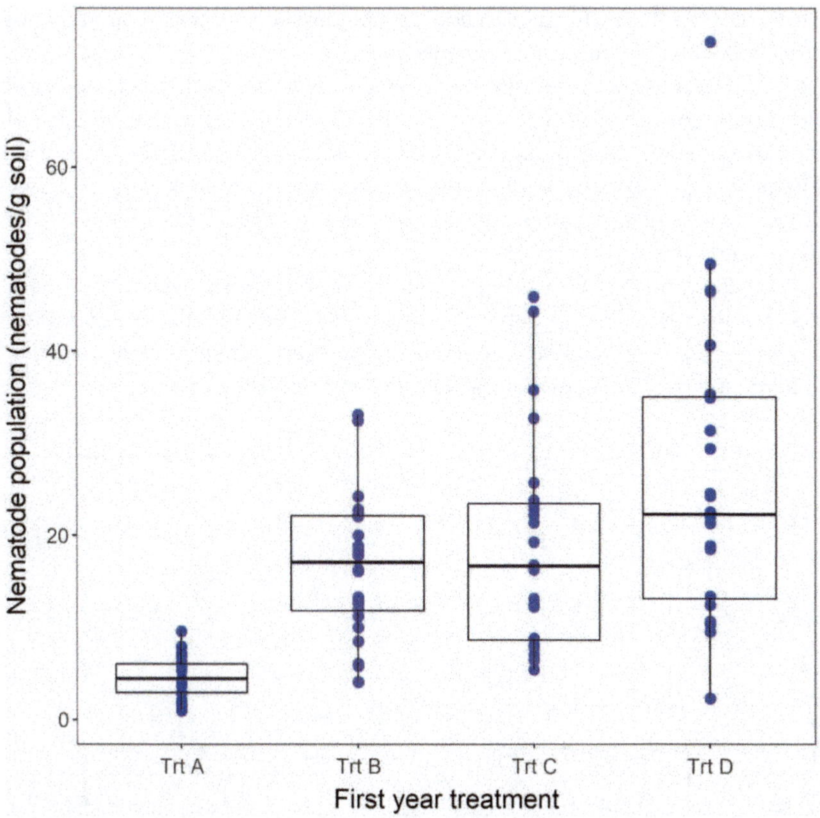

Figure 4 Distribution of the nematode populations measured in each plot resulting from the first-year treatments of the example field experiment in case study 1. Nematode populations are expressed as the number of nematodes per gram of soil.

separate blocks. Although blocking according to an observed quantity is not commonplace in agricultural field experiments, the technique is widely used in experiments involving animals, where the observed quantity is often the live weight of animals (Bailey, 2008).

In addition to the known differences in nematode populations resulting from the first-year treatments, there was also potential that these treatments had influenced unmeasured characteristics of the experimental material (e.g. soil moisture and soil nutrient levels). As such, as a secondary constraint in the design process, it was desired that the second-year treatments also be balanced across their allocation to first-year treatments. Finally, the replicate blocks imposed through the original randomisation of first-year treatments according to a randomised block design were also respected and retained in the final design, both to ensure resolvability of the second-year treatments and also to provide the ability to partition any systematic sources of variation not associated with the first-year treatments.

The experimental design for the second-year treatments was constructed using the model-based design package *od* (Butler, 2018) in the R statistical computing environment (R Core Team, 2018). Figure 5 presents the distribution of nematode populations to which the second-year treatments were exposed, demonstrating relative consistency between the distributions in terms of the median and inter-quartile range.

With the growth of precision agriculture and the advancement of on-board data capture devices on the harvesting equipment, the potential to incorporate observed quantities expected to induce significant heterogeneity between plots into the design of agricultural field experiments is increasing.

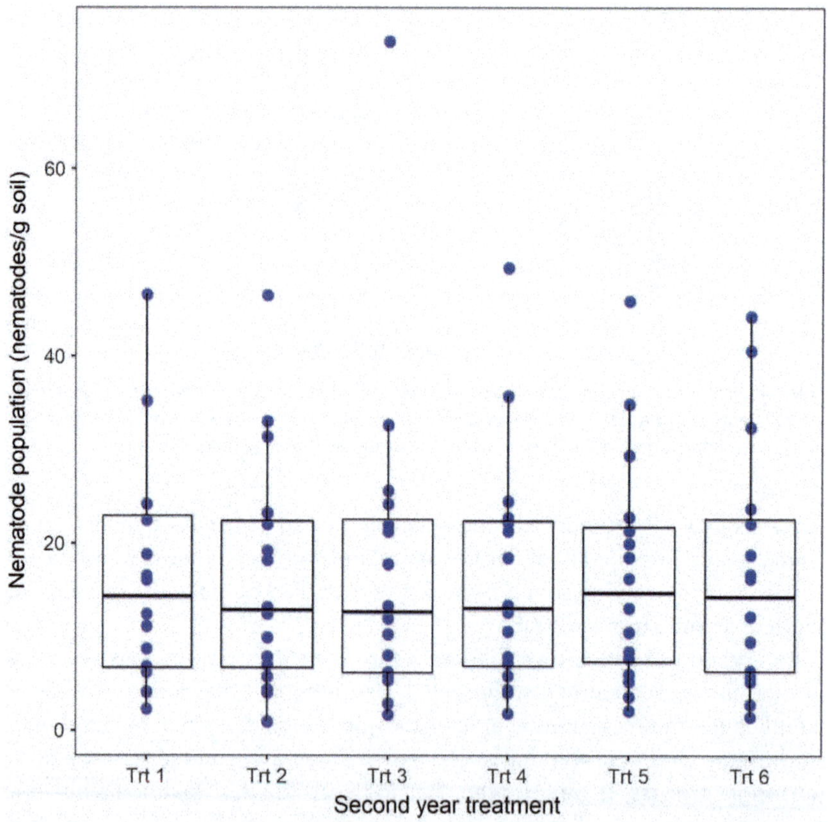

Figure 5 Distribution of the nematode populations to which each of the second-year treatments were exposed following application of a blocking strategy based on observed quantities in case study 1. Nematode populations are expressed as the number of nematodes per gram of soil.

4 Designs for early-generation field trials

Designs comprised of fully replicated sets of genotypes are widely used in the later stages of plant breeding programmes, but do not generally suit the requirements of early generation trials in which limited seed supply of test genotypes restricts complete replication. More importantly, the aim of the breeding programme is to maximise genetic gain and this requires assessment of a large number of genotypes to allow for selection of the 'best' genotypes to enter the later stages of testing (Kempton and Fox, 1997). In a field trial, where trend is known to exist but can be accounted for in a statistical model, it is more beneficial to test a large number of genotypes with limited accuracy, than to obtain highly accurate estimates of only a few test genotypes through replication (Cullis et al., 2006).

Experimental design layouts that were first implemented for early generation trials differentiated between control genotypes and test genotypes. Yates (1936b) proposed the use of replicated plots that are treated with a control genotype and placed systematically throughout the trial area. The unreplicated test genotypes are then randomised to single plots between the replicated plots of the control genotype. Subsequently, a 'fertility index' is derived from the replicated plots of the control genotype and used as a covariate in the analysis of plot yields (Yates, 1936b). An example of a trial with systematic placement of two control genotypes, extending the approach of Yates (1936b), is shown in Fig. 6. One limitation of this design approach is that a large number of plots are allocated to one or more control genotypes. Alternatively, if these plots were used to replicate the test genotypes, there would be increased precision for comparison of the test genotypes. Furthermore, the local control of trend is only effective if the control genotype/s react to the underlying field trend in the same way as the test genotypes. If this is not the case, the response of one, or a small number of control genotypes, could bias the estimate of trend, due to micro- genotype-by-environment interaction effects (Besag and Kempton, 1986).

In contrast to the systematic placement of a control genotype, Federer (1956) presented augmented designs for the randomisation of genotypes in field trials where a large number of the genotypes are unreplicated. In these designs, a standard block design is formed for the replicated control genotypes, and then each block is augmented with unreplicated test genotypes. The test genotype effects are then adjusted for block differences using the control genotype value(s) as the block mean. As an extension, Federer and Raghavarao (1975) advocated the use of classical square designs, allowing for two-directional adjustment in the field. Also, Lin and Poushinsky (1983) laid control plots out in a standard latin square design with systematic placement of the control plots at the centre of the block (Fig. 7). While the

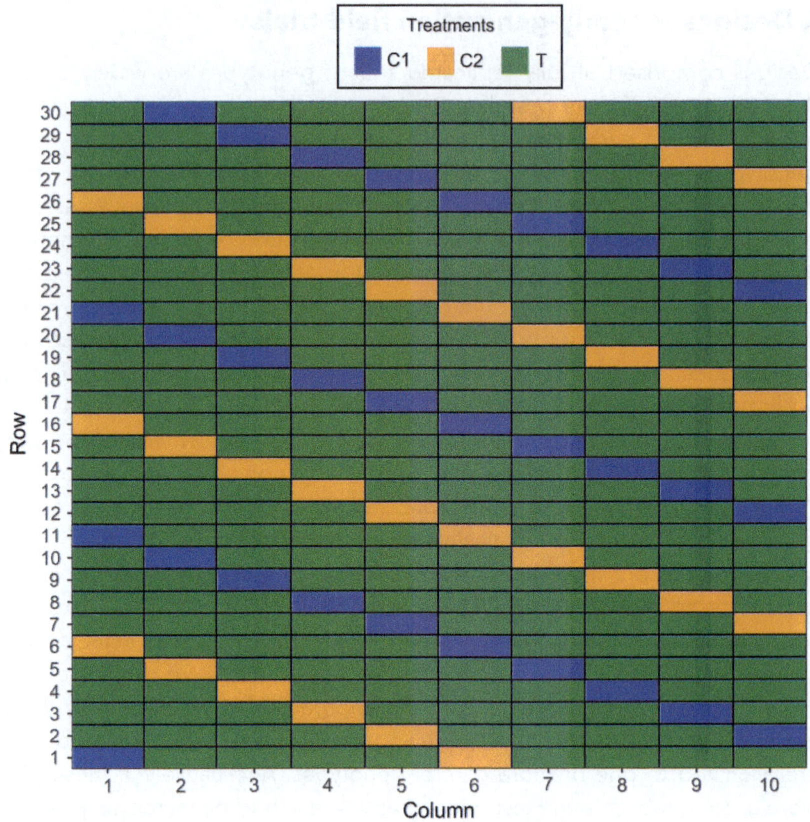

Figure 6 A systematic design for 10 columns × 30 rows with two repeated control treatments (C1 and C2), laid out in a systematic pattern across the field. The unreplicated treatments are labelled as T.

technology is ageing, these designs are still currently in use in many breeding programmes.

Partially replicated (*p-rep*) designs provide the most recent development to address the specific requirements of early generation trials, where the aim is to maximise genetic gain (Cullis et al., 2006). These designs replace a large number of repeated control genotypes by minimal replication of a subset of the test genotypes undergoing selection. In this way, genetic variance and, hence, genetic diversity can be maintained. Additionally, the adjustment for underlying field trend should be improved through replication of a large number of test genotypes rather than a small number of control genotypes, resulting in greater response to selection (Cullis et al., 2006).

The strength of partially replicated designs lies in their application to a MET scenario. For example, consider an early generation trial with 600 genotypes

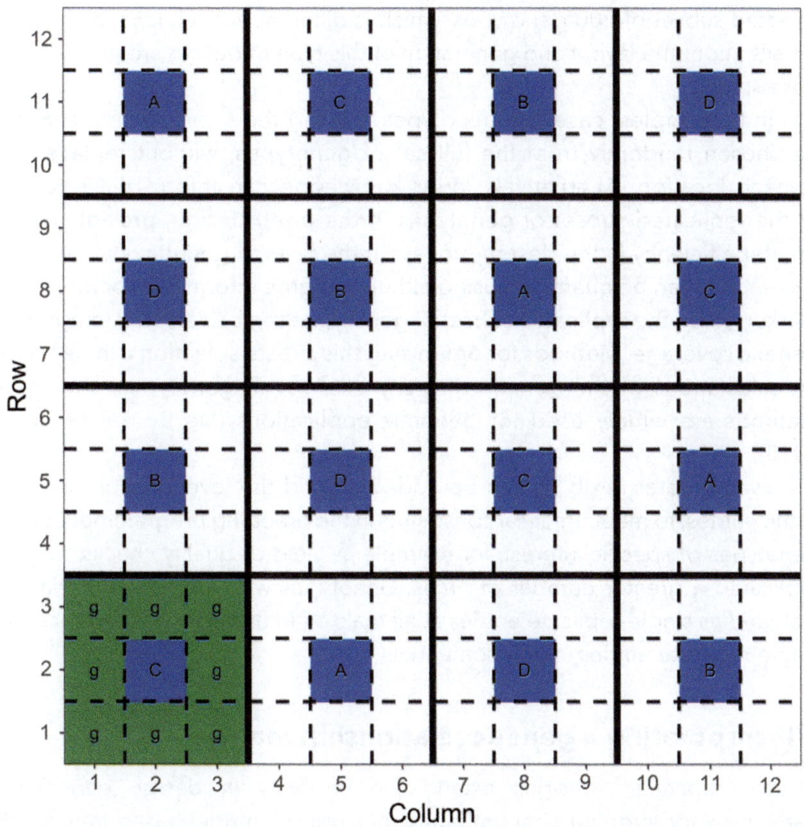

Figure 7 An augmented design based on a latin square of four replicates × four control treatments (A,B,C,D), with the unreplicated treatments (g) allocated as plots (dotted lines) within blocks (solid lines) around the replicated control treatments.

to be grown at three locations. Given that seed is not limiting, the set of 600 genotypes is (randomly) divided into three subsets, of 200 genotypes each (Table 1). Each trial can then be planned with 800 plots, where 400 genotypes are grown as single-replicate plots, and two replicate plots are planted for the

Table 1 An example of genotype subsetting showing the number of replicate plots involved in the formation of a partially replicated design across trials at three locations

Genotype subsets	Location 1	Location 2	Location 3
1-200	2	1	1
201-400	1	2	1
401-600	1	1	2
Total plots	800	800	800

selected subset of 200 genotypes, which is different at each location. Further details about the layout and generation of this type of design are presented in Case study 2.

In the simplest case, the genotypes forming the replicated subsets can be chosen randomly from the full set of genotypes, without replacement at each location. Alternatively, prior knowledge can inform the selection of the replicated subset of genotypes in the *p-rep* design, providing even greater efficiency in the design process. If the level of genetic relationship is known, and can be quantified using either pedigree information or molecular markers, then the replicated subset of genotypes can be selected to optimise genetic coverage. Methods for optimising this subset selection can be based on calculations of the genetic diversity of a set of genotypes, and these methods are widely used for genomic applications (De Beukelaer et al., 2018).

Even greater flexibility can be added around the level of replication for some entries, to meet practical constraints in the breeding programme. Control genotypes of specific interest, for example as yield or quality checks, may be replicated a greater number of times. Genotypes with limited seed may be included as single-replicate entries at all trials or, in extreme circumstances, as single-replicate entries in only some trials.

5 Incorporating a genetic relationship matrix

Partially replicated designs extend the flexibility in design generation, particularly for large numbers of genotypes using a model-based framework. Trials with large numbers of genotypes are commonly found in crop breeding programmes, where the genotypes come from a known crossing programme with varying levels of genetic relationship. Again, advances in statistical models which enable the incorporation of this genetic structure into the analysis of such trials has also resulted in a corresponding improvement in the design of these trials, and is made possible through the use of a model-based design framework (Butler et al., 2014).

It has been long established that selection decisions in a plant or animal breeding programme can be improved by incorporating information on genetic relatedness. The calculation of genetic relatedness can be based on using pedigree information to derive the average proportion of genetic material in common between each pair of genotypes, that is, identity by descent (Henderson, 1976). More recently, genetic relationship can be derived using molecular marker information, as a similarity index based on markers, (see e.g. Van Raden, 2008), and this derives genetic relatedness based on an identity-by-state measure.

The information on genetic relationship is naturally incorporated into the linear mixed model by partitioning the total genetic effect into additive and non-additive effects. One of the most plausible genetic models appearing in the plant breeding literature is that presented by Oakey et al. (2006). These authors present a joint model for additive and non-additive genotype effects, with application to a replicated field trial for an inbred population. Additive effects can be estimated if pedigree information is available and, if genotypes are replicated, as they commonly are in plant breeding trials, non-additive effects can also be partitioned from residual error. The attraction of this genetic model is that it provides predictions of breeding values through the additive term, and predictions of individual merit through the combined additive and non-additive effects. Furthermore, the genetic model presented by Oakey et al. (2007) partitions the non-additive genetic effects into dominance effects and higher-order interaction effects (Falconer and Mackay, 1996). While the dominance effects are zero for inbred crops, the form of the dominance matrix is important for hybrid crops.

Incorporation of the relationship between genetic effects into the model-based design framework requires the introduction of random effects into the design. While this topic is beyond the scope of this chapter, related references for design with random effects based on blocking principles are Bueno Filho and Gilmour (2003), Bueno Filho and Gilmour (2007) and Piepho and Williams (2006), while Butler et al. (2014) considers designs with genetic structure and correlated residual errors. A case study is now presented to demonstrate the principles of partial replication and genetic relatedness introduced in the previous two sections.

5.1 Case study 2: A partially replicated design incorporating genetic relatedness

In most experimental design packages, an assumption of independence is made between the levels of the treatment. More recently, packages have implemented model-based design to allow for correlation between treatment levels (Butler et al., 2014). In this case study, we present a design approach for when the treatments consist of genotypes with a known genetic relationship structure. The design model includes a genetic relationship matrix, $\boldsymbol{K}$, that is formed from the similarity between genotypes derived from molecular markers. This application in plant breeding offers an opportunity for improved design efficiency in the background of expected field variation.

The motivating design for this case study arises from a series of field trials at two locations aiming to test 198 doubled haploid (DH) barley genotypes derived from the cross ND24260 x Flagship, together with the two parent genotypes (Robinson et al., 2016). All 200 genotypes were previously genotyped with 1300 Diversity Arrays Technology (DArT) polymorphic markers, and the DArT marker data was reduced to 605 polymorphic (DArT) markers for the DH genotypes, following the linkage map for the ND24260 x Flagship DH population. A kinship matrix between genotypes was calculated from the marker data using the *rrBLUP* package (Endelman, 2011) in the R statistical computing environment (R Core Team, 2018). The relationships within this matrix are depicted through a dendrogram in Fig. 8. This hierarchical grouping highlights the entries more similar to each of the parent genotypes, ND24260 and Flagship, and also those genotypes that are intermediate in genetic composition between the parent genotypes.

In order to generate a partially replicated design, a subset of genotypes must be selected for replication. In this example, 100 genotypes are chosen from the 200 genotypes to form the replicated set at one trial location. The remaining 100 genotypes then form the replicated set at the second trial location.

The kinship matrix is used in two ways in the experimental design process. First, clusters within the kinship matrix are formed to describe groups of like genotypes. These clusters are then proportionally sampled, without replacement for each trial location, to obtain a subset of genetically diverse entries for replication in each trial. Secondly, the kinship matrix can be incorporated in the model-based design process, to account for correlation between the levels of genotypes when determining the allocation of genotypes to plots in the randomisation process.

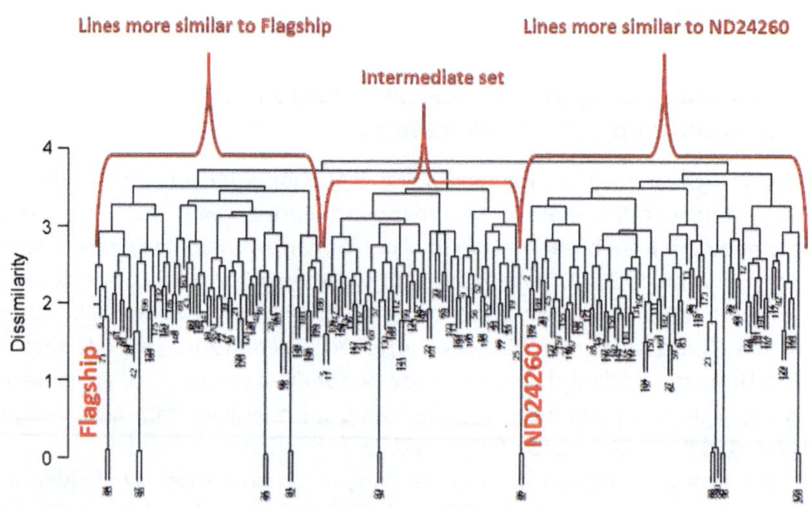

Figure 8 A dendrogram showing the hierarchical classification formed from the genetic relationship matrix between the 198 doubled haploid genotypes and the two parent genotypes, considered in case study 2.

The optimal design package, *od* (Butler, 2018), in the R statistical computing environment (R Core Team, 2018), allows for specification of the relationship matrix in the design process, where the design model is formulated in a linear mixed model framework. This package allows for specification of either additive effects or total genetic effects as the treatment effects for optimisation. The package also allows for resolvability into replicate blocks for the replicated genotypes and latinisation across rows and columns in the design through specification of these factors as random effects in the linear mixed model framework. The output of the randomisation from *od* showing the replicated and unreplicated genotypes in the DH barley field trial design at one trial location is given in Fig. 9. Note that the *od* package is available at: https://mmade.org/.

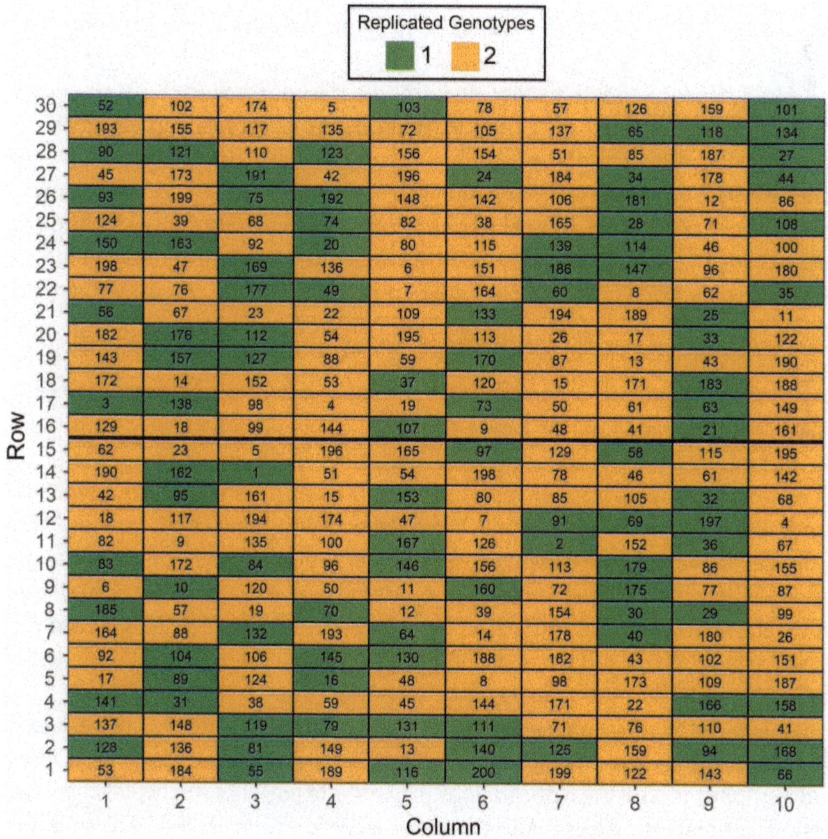

Figure 9 Field randomisation of the partially replicated design for 200 genotypes at one trial location from Case study 2. The trial consists of 10 columns × 30 rows, forming a two-dimensional array of 300 field plots. Different colours correspond to those genotypes that are replicated once or twice in the trial.

6 Multi-phase design for laboratory experiments

Multi-phase experiments are defined by Brien and Bailey (2006) as ones with a valid experiment in phase one, followed by the randomisation of experimental units from phase one to units in the second phase of the experiment. The simplest case of a two-phase experiment was first introduced by McIntyre (1955), using the example of comparing sugar content of sugar cane genotypes. He proposed that the two requirements for valid genotype comparisons were ensuring a valid field trial with replicated genotypes in phase one of the experiment, and then the processing of produce from each of the field plots in phase two of the experiment. Phase two was undertaken in the laboratory and involved the chemical analysis of the produce from each plot, which could be undertaken on duplicate or triplicate samples. McIntyre (1955) proposed that replication in the second phase was not essential, but was highly desirable if there was a large source of variance associated with the laboratory phase of the experiment.

Multi-phase experimental design forms the basis for testing cereal chemistry parameters and end product traits in all crops, both cereals and legumes. To formalise this experimental procedure for grain-quality testing, we note that data are usually obtained from a multi-phase process. Genotypes are grown in replicated plots in a field trial in phase one and grain from these field plots is processed in the laboratory in phase two. In this second phase, duplicate grain samples from the field plots (experimental units) of phase one should be formed and allocated as replicates in a second experimental design. The grain samples taken from the experimental units of phase one are randomly allocated to laboratory order, the experimental units of phase two. To form a valid replicated experiment in phase two, grain samples from the field plots must be split into replicate samples, as depicted in Fig. 10. If complete replicates of all experimental units from phase one are formed and tested in phase two, the experiment rapidly grows beyond a cost-effective and practical solution. The proposal of *p-q* replicate designs by Smith et al. (2006) offers an efficient design framework for the construction of multi-phase experiments. In such a multi-phase study, an experimental design is used to replicate and randomise samples in each phase of the process and the resulting data can be analysed using an underlying statistical model based on the design parameters (Brien and Bailey, 2006).

While these multi-phase experiments are formed using the same basic design principles as for single-phase experiments, implementing replication and randomisation at each phase does cause some additional complexity to the design generation process. A useful framework for defining the experimental process around these more complex experiments was proposed by Brien (1983) through the definition of tiers, where tiers were defined as the set of factors involved in one phase of the randomisation process. Building on this

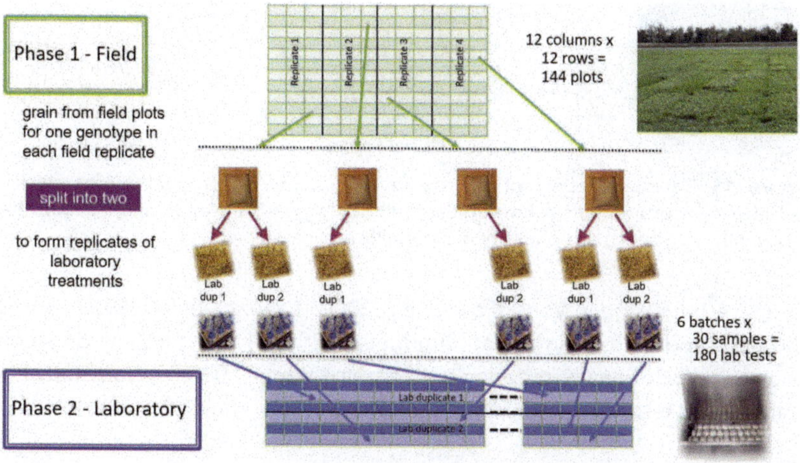

Figure 10 A graphical depiction of a two-phase design for a laboratory experiment. Phase one is a field trial of 4 replicates × 36 genotypes. Phase two is a partially replicated design of 144 field plots by partial replicates processed as 6 batches of 30 laboratory samples, forming a total of 180 laboratory tests.

definition, Brien and Bailey (2006) proposed a graphical method of linking factors across tiers through the randomisation process. In this way, the design anatomy of the multi-phase experiment, and hence the terms in the analysis model, could be developed following some simple rules. We will demonstrate this process in the multi-phase example presented in Case study 3.

The mixed model approach for multi-phase experiments allows for simultaneous modelling of field and laboratory variation together with genetic effects. A statistical design adopting the standard principles of replication and randomisation at the laboratory phase is essential for accurate estimation of genetic effects. Design ensures first that data errors and outliers can be detected, and for the complex processes involved in measuring some grain-quality traits, the error rate is often higher than desirable. Furthermore, the standard procedure of using laboratory controls as checks has limited power in detecting errors in the test samples (Cullis et al., 2003). Design also ensures that consistent variation due to batches or order of processing can be detected and accounted for as a non-genetic effect. Finally, design at each phase of the process (field and laboratory) allows for the adjustment of field trend influences on grain-quality traits. The importance of dealing with variation in grain quality arising from field trend in phase one has been demonstrated in a number of publications (Cullis et al., 2003; Fox et al., 2006).

Very few published studies consider multi-phase experiments for comparing genotypes when assessing grain quality and end product traits measured in the laboratory. Some examples are malting and mashing to

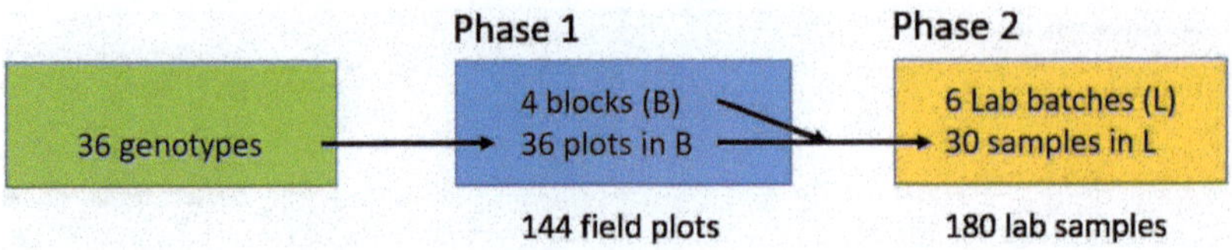

Figure 11 Design anatomy of the two-phase laboratory experiment for chickpea carbohydrate components, following the methodology of Brien and Bailey (2006). Phase one is fully replicated, and phase two is partially replicated.

compare barley genotypes (Fox et al., 2007; Panozzo et al., 2007), milling and end product traits in wheat (Smith et al., 2001b) and germination studies investigating dormancy in barley genotypes (Osama, 2019). A case study from grain-quality testing in a chickpea physiology study is now presented.

6.1 Case Study 3: Multi-phase design in the laboratory

An example of a multi-phase design is presented for a chickpea experiment, where the aim is to compare genotypes for different carbohydrate components in the harvested peas.

In phase one of the experiment, a field trial was undertaken testing 36 genotypes. The field trial was designed as a randomised block design, laid out in the field as a two-dimensional array of 12 columns × 12 rows, forming a total of 144 field plots. Chickpeas were harvested from the plants in the field plots and retained for subsequent testing in the laboratory. Phase two, by definition, takes the experimental units from phase one, and assigns these as treatments in phase two of the experiment.

The second phase of the experiment was designed to test material from phase one when taken to the laboratory. The sampling of field plots and duplication of these samples in the plots is similar to that depicted in Fig. 10, and the design anatomy as defined by Brien and Bailey (2006) is given in Fig. 11.

7 Conclusions

Experimental design has a firm basis in randomisation theory, with an intrinsic link between design and analysis. This foundation results in valid inference with appropriate error estimates when comparing treatments (Nelder, 1965a,b). More recently, model-based designs offer greater flexibility for experiments where the assumption of independent errors, or indeed independent treatment effects, is violated. Additionally, these model-based approaches can easily generate designs without complete replication, either in a field or a multi-phase laboratory experiment. However, for model-based designs, care needs to be exercised in specifying an appropriate design generation model, and these models should be validated to be robust against alternative dependence assumptions.

8 References

Bailey, R. A. (1981), A unified approach to design of experiments, *Journal of the Royal Statistical Society, Series A*, 144, 214–23.

Bailey, R. A. (2008), *Design of Comparative Experiments*, Cambridge University Press, Cambridge.

Besag, J. and Kempton, R. A. (1986), Statistical analysis of field experiments using neighbouring plots, *Biometrics*, 42, 231–51.

Brien, C. J. (1983), Analysis of variance tables based on experimental structure, *Biometrics*, 39, 53–9.

Brien, C. J. and Bailey, R. A. (2006), Multiple randomizations, *Journal of the Royal Statistical Society, Series B*, 68, 571–609.

Bueno Filho, J. S. S. and Gilmour, S. G. (2003), Planning incomplete block experiments when treatments are genetically related, *Biometrics*, 59, 375–81.

Bueno Filho, J. S. S. and Gilmour, S. G. (2007), Block designs for random treatment effects, *Journal of Statistical Planning and Inference*, 137, 1446–51.

Butler, D. G. (2018), OD: Generate optimal experimental designs, R package version 2.0.0, Working Paper 04-18, National Institute for Applied Statistics Research Australia, University of Wollongong.

Butler, D. G., Eccleston, J. A. and Cullis, B. R. (2008), On an approximate optimality criterion for the design of field experiments under spatial correlation, *Australian and New Zealand Journal of Statistics*, 50 (4), 295–307.

Butler, D. G., Smith, A. B. and Cullis, B. R. (2014), On the design of field experiments with correlated treatment effects, *Journal of Agricultural, Biological and Environmental Statistics*, 19(4), 541–57.

Chan, B. S. P. and Eccleston, J. A. (2003), On the construction of nearest-neighbour balanced row-column designs, *Australian and New Zealand Journal of Statistics*, 45, 97–106.

Cobb, G. W. (2008), *Introduction to Design and Analysis of Experiments*, Wiley, New York.

Cullis, B. R. and Gleeson, A. C. (1989), Efficiency of neighbour analysis for replicated field trials in Australia, *Journal of Agricultural Science*, 113, 233–9.

Cullis, B. R., Smith, A. B., Panozzo, J. F. and Lim, P. (2003), Barley malting quality: Are we selecting the best?, *Australian Journal of Agricultural Research*, 54, 1261–75.

Cullis, B. R., Smith, A. B. and Coombes, N. E. (2006), On the design of early generation genotype trials with correlated data, *Journal of Agricultural, Biological, and Environmental Statistics*, 11, 381–93.

De Beukelaer, H., Davenport, G. F. and Fack, V. (2018), Core Hunter 3: flexible core subset selection, *BMC Bioinformatics*, 19, 203. doi: 10.1186/s12859-018-2209-z

Endelman, J. B. (2011), Ridge regression and other kernels for genomic selection with R package rrBLUP, *Plant Genome*, 4, 250–5. doi: 10.3835/plantgenome2011.08.0024

Falconer, D. S. and Mackay, T. F. C. (1996), *Introduction to Quantitative Genetics*, 4th edn. Longman Group Ltd.

Federer, W. T. (1956), Augmented (or Hoonviaku) designs, *Hawaiian Planters' Record*, 55, 191-208.
Federer, W. T. and Raghavarao, D. (1975), On augmented designs, *Biometrics*, 31, 29-35.
Fisher, R. A. (1925), *Statistical Methods for Research Workers*, 1st edn. Oliver and Boyd, Edinburgh.
Fox, G. P., Kelly, A. M., Poulsen, D. M. E., Inkermann, P. A. and Henry, R. J. (2006), Selecting for increased barley grain size, *Journal of Cereal Science*, 43, 198-208.
Fox, G. P., Osborne, B. G., Bowman, J. G. P., Kelly, A. M., Cakir, M., Poulsen, D. M. E., Inkerman, P. A. and Henry, R. J. (2007), Measurement of genetic and environmental variation in barley grain hardness, *Journal of Cereal Science*, 48, 82-92.
Gilmour, A. R., Cullis, B. R. and Verbyla, A. P. (1997), Accounting for natural and extraneous variation in the analysis of field experiments, *Journal of Agricultural, Biological, and Environmental Statistics*, 3, 269-93.
Gleeson, A. C. and Cullis, B. R. (1987), Residual maximum likelihood estimation of a neighbour model for field experiments, *Biometrics*, 43, 277-88.
Harshbarger, B. (1949), Triple rectangular lattices, *Biometrics*, 5, 1-13.
Henderson, C. R. (1976), A simple method for computing the inverse of a numerator relationship matrix used in prediction of breeding values, *Biometrics*, 32(1), 69-83.
John, J. A. and Eccleston, J. A. (1986), Row-column α-designs, *Biometrika*, 73, 301-6.
Kempton, R. A. and Fox, P. N. (1997), *Statistical Methods for Plant Variety Evaluation*, Chapman and Hall, London.
Lin, C. S. and Poushinsky, G. (1983), A modified augmented design for an early stage of plant selection involving a large number of test lines without replication, *Biometrics*, 39, 553-61.
Martin, R. J. (1986), On the design of experiments under spatial correlation, *Biometrika*, 73, 247-77.
Martin, R. J. and Eccleston, J. A. (1993), Incomplete block designs with spatial layouts when observations are dependent, *Journal of Statistical Planning and Inference*, 35, 77-91.
Martin, R. J., Eccleston, J. A., Chauhan, N. and Chan, B. S. P. (2006), Some results on the design of field experiments for comparing unreplicated treatments, *Journal of Agricultural, Biological and Environmental Statistics*, 11(4), 394-410.
McIntyre, G. A. (1955), Design and analysis of two-phase experiments, *Biometrics*, 11, 324-34.
Mead, R. (1990), *The Design of Experiments*, Cambridge University Press, Cambridge.
Mercer, W. B. and Hall, A. D. (1911), The experimental error of field trials, *Journal of Agricultural Science*, 4, 107-32.
Nelder, J. A. (1965a), The analysis of randomised experiments with orthogonal block structure. I. Block structure and the null analysis of variance, *Proceedings of the Royal Society of London, Series A*, 283, 147-62.
Nelder, J. A. (1965b), The analysis of randomised experiments with orthogonal block structure. II. Treatment structure and the general analysis of variance, *Proceedings of the Royal Society of London, Series A*, 283, 163-78.
Oakey, H., Verbyla, A., Pitchford, W., Cullis, B. and Kuchel, H. (2006), Joint modelling of additive and non-additive genetic line effects in single field trials, *Theoretical and Applied Genetics*, 113, 809-19.
Oakey, H., Verbyla, A., Cullis, B., Wei, X. and Pitchford, W. (2007), Joint modelling of additive and non-additive (genetic line) effects in multi-environment trials, *Theoretical and Applied Genetics*, 114, 1319-32.

Osama, S. (2019), Seed dormancy in malting barley, PhD Thesis, The University of Queensland. In preparation.

Panozzo, J. F., Eckermann, P. J., Mather, D. E., Moody, D. B., Black, C. K., Collins, H. M., Barr, A. R., Lim, P. and Cullis, B. R. (2007), QTL analysis of malting quality traits in two barley populations. *Australian Journal of Agricultural Research*, 58, 858–66.

Patterson, H. D. and Hunter, E. A. (1976), The efficiency of incomplete block designs in national list and recommended list cereal variety trials, *Journal of Agricultural Science*, 101, 427–33.

Patterson, H. D. and Williams, E. R. (1976), A new class of resolvable incomplete block designs, *Biometrika*, 63, 83–92.

Pearce, S. C. (1983), *The Agricultural Field Experiment*, John Wiley and Sons Inc., New York.

Petersen, R. G. (1994), *Agricultural Field Experiments: Design and Analysis*, John Wiley and Sons Inc., New York.

Piepho, H-P. and Williams, E. R. (2006), A comparison of experimental designs for selection in breeding trials with nested treatment structure, *Theoretical and Applied Genetics*, 113: 1505–13. doi: 10.1007/s00122-006-0398-8

R Core Team (2018), R: A Language and Environment for Statistical Computing, R Foundation for Statistical Computing, Vienna, Austria.

Robinson, H. M., Hickey, L. T., Richard, C. A., Mace, E. S., Kelly, A. M., Borrell, A. K., Franckowiak, J. and Fox, G. P. (2016), Genomic regions influencing seminal root traits in barley, *The Plant Genome*, 9. doi: 10.3835/plantgenome2015.03.0012

Smith, A. B., Cullis, B. R. and Thompson R. (2001a), Analysing variety by environment data using multiplicative mixed models and adjustments for spatial field trend, *Biometrics*, 57, 1138–47.

Smith, A. B., Cullis, B. R., Appels, R., Campbell, A. W., Cornish, G. B., Martin, D. and Allen, H. M. (2001b), The statistical analysis of quality traits in plant improvement programs with application to the mapping of milling yield in wheat, *Australian Journal of Agricultural Research*, 52, 1207–19.

Smith, A. B., Lim, P. and Cullis, B. R. (2006), On the design of multi-phase experiments for quality trait data, *Journal of Agricultural Science*, 144, 393–409.

Van Raden, P. M. (2008), Efficient methods to compute genomic predictions, *Journal of Dairy Science*, 91(11), 4414–23.

Wilkinson, G. N., Eckert, S. R., Hancock, T. W. and Mayo, O. (1983), Nearest neighbour (NN) analysis of field experiments (with discussion), *Journal of the Royal Statistical Society, Series B*, 45, 151–211.

Williams, E. R. (1986a), Row and column designs with contiguous replicates, *Australian Journal of Statistics*, 28, 154–63.

Williams, E. R. (1986b), A neighbour model for field experiments, *Biometrika*, 73, 279–87.

Yates, F. (1933), The formation of latin squares for use in field experiments, *Empire Journal of Experimental Agriculture*, 1, 235–44.

Yates, F. (1936a), Incomplete randomised blocks. *Annals of Eugenics*, 7, 121–40.

Yates, F. (1936b), A new method of arranging variety trials involving a large number of varieties. *Journal of Agricultural Science*, 26, 424–55.

Yates, F. (1937), A new further note on the arrangement of variety trials involving a large number of genotypes. *Annals of Eugenics*, 7, 319–32.

Yates, F. (1940), Lattice designs. *Journal of Agricultural Science*, 30, 672–87.

Chapter 7

Advances in molecular breeding techniques for barley: genome-wide association studies (GWAS)

W. T. B. Thomas, James Hutton Institute, UK

1 Introduction
2 Progress in barley breeding
3 Mapping of malting quality and yield traits
4 Genome-wide association studies (GWAS) mapping in barley
5 Application of results from genome-wide association studies (GWAS) in barley improvement
6 Conclusion and future trends
7 Acknowledgements
8 References

1 Introduction

Barley breeders choose parental combinations that complement each other for desirable characteristics for their target environments and then rely on recombination to provide the right re-assortment in breeding lines that could potentially be developed into new varieties. This is essentially a filtration process that used to rely on effective phenotypic screens applied to segregating generations as lines became more homozygous. Barley breeders initially used the pedigree inbreeding scheme or variations upon it. Techniques such as Single Seed Descent (Knott and Kumar, 1975) and the production of 'instant inbred lines' through doubled haploidy (Thomas et al., 2003) have been used to speed up the advance to homozygosity. Despite the reduction in the breeding cycle that these methods can offer, the central problem remains that phenotyping and hence selection for the key characters such as yield and malting quality cannot be conducted until sufficient seed is produced for each line. Breeders therefore seek to identify suitable surrogate tests that can be applied early in the breeding cycle to select for a component character of for example malting quality. Ideally, identifying a suitable marker to 'tag'

http://dx.doi.org/10.19103/AS.2019.0060.08

key aspects of such a character could be used to begin selection early in a programme and so much research has been conducted to identify markers, beginning with morphological ones running through isozymes to molecular markers (Thomas, 2003).

2 Progress in barley breeding

Before considering the advances that molecular techniques have brought to barley breeding it is instructive to look at the genetic progress that has been achieved so far. In the United Kingdom, variety testing of potential new cultivars is carried out on a nationwide basis. Breeders enter potential new varieties into National List trials, which are conducted over two years at an average of 11 (spring) and 8 (winter) sites each year. If the lines have agronomic merit and also pass Distinctness, Uniformity and Stability (DUS) tests, they are placed on the National List and awarded Plant Breeders Rights. The best of these lines are then entered as candidates into Recommended List Trials (RLT), which are grown at an average of 23 (spring) and 20 (winter) sites each year. All trials are grown with a prophylactic fungicide management regime and a typical nitrogen fertiliser management for the area in which the trial is located. If a candidate outperforms control varieties for key phenotypic characters in its first year of RLT, it will receive a provisional recommendation and progress to a full recommendation after another two years of RLT. Since 1993, RLT has been co-ordinated by the Agricultural and Horticultural Development Board's (AHDB) Cereals and Oilseeds sector. This system provides a comprehensive body of phenotypic data that can be used to monitor breeding progress by using mixed models to derive Best Linear Unbiased Predictions for each line that has been entered into RLT and then regressing it against the year when a variety was first recommended.

Considering the lines that were first recommended in the United Kingdom after 1990, we can see a clear improvement in fungicide treated yield for both winter and spring barley (Fig. 1). The annual rate of improvement in yield since 1990 is 0.03 and 0.04 t/ha for spring and winter barley respectively, but has not been accompanied by any significant changes over time in crop height (Fig. 2). The improvement in yield has also been accompanied by a slight increase in the days from sowing to maturity in the spring crop at a rate of +0.01 days per year (P=0.04), which cannot explain the yield improvement, especially as there was no significant trend in the winter crop (Fig. 3).

In 1990, marker technology for cereals was still largely based upon Restriction Fragment Length Polymorphisms and it wasn't until the advent of Amplified Fragment Length Polymorphisms and Simple Sequence Repeats that markers began to be deployed in barley breeding programmes and even then it was just for specific major gene characters that were difficult to phenotype.

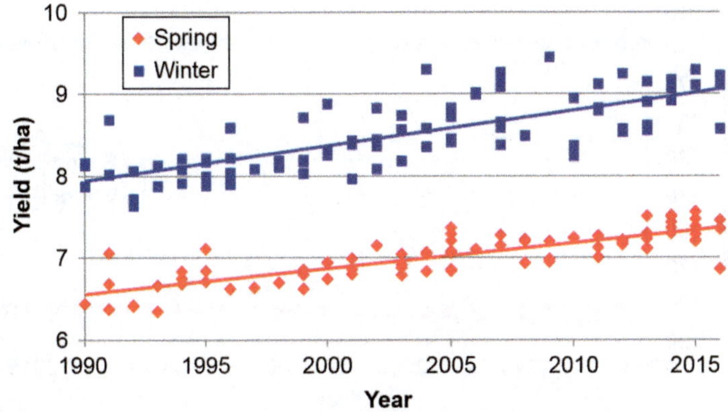

Figure 1 Best linear unbiased prediction of yield in fungicide treated trials regressed against year that a variety was first placed on the AHDB Cereals & Oilseeds Recommended List for UK growers. Raw data from AHDB and BSPB trials.

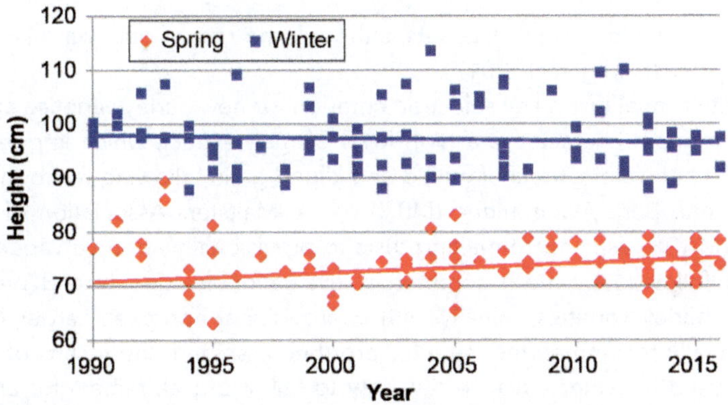

Figure 2 Best linear unbiased prediction of height in fungicide treated trials regressed against year that a variety was first placed on the AHDB Cereals & Oilseeds Recommended List for UK growers. Raw data from AHDB and BSPB trials.

The advent of genome-wide genotyping technologies such as the Barley Oligo Pooled Arrays (Close et al., 2009) enabled greater integration of genotyping with breeding, either through the ability to filter out key markers for tagging specific traits of interest or for monitoring shifts in the elite breeding gene pool. If we consider that this effect would start to impact varieties that have been recommended over the past 10 years, we can still see significant improvements in the yield potential of spring and winter two-row barley. It is worth noting that this period covers the so-called UK 'yield plateau' but RLT data shows that the rate of genetic gain over the past 10 years for fungicide protected yield is the same as it has been over the past 26 years. Analysis of German spring barley official trial and on-farm data over a 32 year period also showed genetic

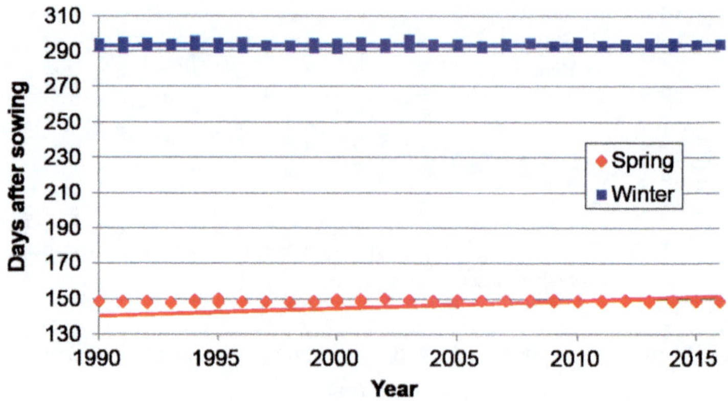

Figure 3 Best linear unbiased prediction of maturity date in fungicide treated trials regressed against year that a variety was first placed on the AHDB Cereals & Oilseeds Recommended List for UK growers. Raw data from AHDB and BSPB trials.

progress for yield in both data sets, although the rate of gain was less in the on-farm data (Laidig et al., 2017).

Malting quality is a key selection criterion for new barley varieties and the most important character is a high level of malt extract, which is measured following hot water extract of milled and kilned grain following micro-malting. The Malting Barley Committee (MBC) of the Maltsters Association of Great Britain (MAGB) uses micro-malting tests to provisionally approve varieties for malting followed by more commercial evaluation to identify and promote malting barley varieties, initially with a provisional approval. Larger lots of provisionally recommended varieties are then assessed, the results of which will be used to either advance a variety to full approval, remain for another year's testing in provisional year 2, or be dropped.

We utilised micro-malting data of all varieties that have been tested by the MBC to determine if significant genetic progress has also been made for malting characters. While small, significant improvements in malt extract have been made in spring ($P<0.05$) and winter ($P<0.05$) barley between 1990 and 2006, no significant improvements have been made for the character in the decade since 2006 (Fig. 4). A significant reduction in grain nitrogen content since 1990 can also be seen in the spring ($P=0.03$) and winter ($P=0.02$) malting barley varieties. Significant improvements in malt extract have also been detected in German spring barley data but some of this improvement appears to be largely due to the number of feed varieties tested prior to 1990 (Laidig et al., 2017).

There is a well-known inverse relationship between hot water extract and grain nitrogen content and it is possible that the trend over time is more a reflection of the reduction in grain nitrogen than a genuine increase in malt

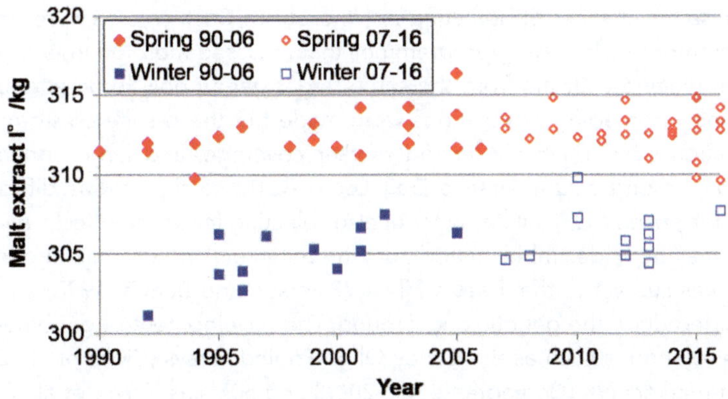

Figure 4 Best linear unbiased prediction of malt extract in fungicide treated trials regressed against year that a variety was first placed on the AHDB Cereals & Oilseeds Recommended List for UK growers. Raw data from AHDB and BSPB trials. Only varieties that were at least considered for acceptance by maltsters included.

extract. Certainly, there are no discernible improvements in malting component characters in the spring crop as malt friability and wort viscosity show no significant relationship with time over the same period. In contrast, the winter crop shows a significant increase in friability and a significant decrease in wort viscosity over the same period, although the fitted means for the winter crop at the end of the survey period are still inferior to those of the spring crop.

Even in cases where we do not detect any significant improvements in characters over time, we do detect significant genetic variation amongst the varieties. In such cases, it can be concluded that phenotypic selection is no longer effective or that the character in question is not essential for the target market in the region surveyed. While it could also be argued that the lack of a significant trend with time indicates that a maximum expression of a phenotype has been reached, the fact that there are still significant differences among lines implies that a better means of identifying the best lines needs to be found to optimise selection of the best in cases where the character is a key for the target market. Genetic mapping of the characters can identify markers that 'tag' barley genomic regions associated with quality improvements, which can then be used to select the best lines and we will examine how this can be done in Sections 3 and 4.

3 Mapping of malting quality and yield traits

Initial attempts to map malting quality characters, reviewed by Mohammadi et al. (2015), were based upon bi-parental random populations usually derived from crosses between good and poor quality lines that were designed more to maximise the chances of detecting polymorphism. Such crosses were atypical of contemporary breeding programmes that focussed more on inter-crossing of elite malting cultivars from the same target gene pool. Some attempts at mapping in breeding populations were made but these suffered from small population sizes, relatively limited marker coverage, and large confidence intervals around a Quantitative Trait Locus (QTL) so that it was difficult to select for positive QTL alleles without also selecting for linked effects, some of which may be undesirable. In addition, the detection and estimation of a QTL effect was subject to the 'Beavis Effect' (Paterson and Beavis, 1998) and also depended upon the genetic background. The development of genome-wide marker platforms such as the Barley Oligo Pooled Arrays (Close et al., 2009), iSelect markers (9K (Comadran et al., 2012) and 50k sets (Bayer et al., 2017)), techniques such as DArTSeq (Cruz et al., 2013) and genotyping by sequencing (Elshire et al., 2011) means that detailed genotyping can be conducted for relatively little cost (Monat et al., 2019).

Bi-parental mapping was quite effective at identifying the major differences in malting quality between different gene pools but this was of little use in providing targets for use in Marker-Assisted Selection (MAS) in elite breeding programmes where the beneficial allele had already been fixed through convention phenotypic selection. One exception where this approach can prove of value is when introgressing alleles from non-native germplasm to improve specific traits, for example malting quality. Several crosses between malting quality parents from Europe, Japan and Canada and local adapted parents were investigated by Australian researchers, who found markers that could potentially be used in introgressing higher malt extract alleles into their gene pool. This approach was then validated in some independent crosses where selection for the marker allele increased malt extract, indicating that they could be used in breeding (Collins et al., 2003). Later, a successful variety from this introgression strategy (Baudin) was crossed to a Canadian malting barley variety (AC Metcalfe) and a doubled population from the cross was used to further investigate malting quality differences between the gene-pools. Just one QTL for malt extract was detected where AC Metcalfe contributed the favourable allele and provides a further example where introgression from outside the local gene pool can be a useful breeding strategy (Zhou et al., 2016).

4 Genome-wide association studies (GWAS) mapping in barley

Researchers had realised the limitations of bi-parental mapping around the same time that high throughput genotyping became more routine and methods for accounting population structure (Pritchard et al., 2000) when surveying germplasm panels of defined lines were developed. This led to many GWAS in barley and these have recently been reviewed by Igartua et al. (2019) so the reader is encouraged to read this reference for a comprehensive survey of GWAS in barley whereas this chapter will concentrate on the aspects of application of GWAS results in breeding, especially with respect to malting barley improvement. Igartua et al. (2019) noted 30 barley GWAS studies that focussed on quality, including malting quality. Mohammadi et al. (2015) described a large scale effort by the Triticeae Co-Ordinated Project (TCAP) to conduct GWAS for malting quality across a number of North American breeding programmes. Despite the diversity that this study covered, it was successful in revealing significant marker trait associations for a range of malting quality parameters, including malt extract (also known as hot water extract), some of which were novel.

The big advance that the Mohammadi et al. (2015) study provided was to identify QTL that were specific to individual breeding programmes and therefore address the challenge of identifying QTL within the elite gene pool. Such QTL were generally of smaller effect and comparatively larger populations were therefore required so that there was enough power to reliably detect them. Assembly of large GWAS panels of a similar genetic background provided the power so the results from Mohammadi et al. (2015) located QTL by GWAS to tightly defined regions. This fine resolution of QTL meant that specific genomic regions could be introgressed in MAS programmes with reduced risk of introducing linked deleterious characters and so provided targets that could be applied immediately in selection for improved malting quality in the programmes under study.

The Mohammadi et al. (2015) study, which focussed on spring germplasm, was later augmented by a relatively smaller study that restricted the TCAP germplasm to facultative or winter six-row barley lines (Belcher et al., 2018). Here again they detected some novel QTL, including one for malt extract, which they attributed to the fact that the germplasm had not been specifically studied in earlier research (Belcher et al., 2018). The study compared germplasm from two programmes, one based in Oregon and the other in Minnesota and while they found that population structure analysis reflected this difference, the differences in phenotypic distributions were much more subtle and they concluded that it was more likely that the two programmes had generated similar phenotypes through the selection for different combinations of favourable alleles at key loci. It is, however, noteworthy that they did not detect any association of diastatic power with variation at the Bmy1 locus that had previously been reported to be a major determinant of the character (Gibson et al., 1995). They attributed this to the likely fixation of variation at the locus in past phenotypic selection programmes resulting from crossing parents that carried the same Bmy1 allele. The fact that they did not detect any effect of the

Vrn2 locus, which is closely linked to Bmy1 (Hackett et al., 1992), on heading date, would justify this conclusion (Belcher et al., 2018).

Interestingly, there was little correlation between population size and marker density and QTLs detected (Igartua et al., 2019). While this does not make statistical sense, there could be a number of reasons for this, including panel composition as a small panel of relatively diverse genotypes may reveal more QTL than a larger panel of relatively similar genotypes. For instance, relatively small populations of mainly six-rowed winter and facultative germplasm were used by Gutierrez et al. (2011) in GWAS to identify QTL for malt extract and other malting quality parameters in the barley Co-ordinated Agricultural Project (CAP). Similarly, Beattie et al. (2010) also used a small population of Canadian two-rowed germplasm to also detect QTL for malt extract. Gutierrez et al. (2011) note that the QTL they detected were mostly co-incident with QTL previously reported from bi-parental studies and were likely to have large effect due to the small population size. To help in judging the value of GWAS studies, a minimum set of reporting standards should be established (Igartua et al., 2019).

Phenotyping malting quality parameters is expensive (Belcher et al., 2018). Specialised equipment is needed to first micro-malt cleaned and graded grain samples under uniform conditions. Then the germination process is halted by blowing air that is heated to closely defined and controlled temperatures through the sample before roots and shoots are removed. The sample is then milled to pass through a sieve to produce the grist, which is then extracted by hot water at a uniform temperature and filtered to provide the wort. It is the wort that is analysed for characters such as malt extract, wort beta-glucan content and soluble nitrogen. This brief description of the characters obtained in micro-malting analysis is provided to give an idea of the equipment, and hence the cost, required to provide malting quality phenotypes that are consistent with process errors minimised.

One of the biggest sources of error in phenotyping malting quality characters is batch size and while there is micro-malting equipment that will handle almost 100 samples in a single run, the size of each sample is relatively small. The testing carried out by maltsters and officially designated laboratories (American Malting Barley Association in the USA, Canadian Malting Barley Technical Centre in Canada, and the Malting and Brewing Industry Barley Technical Committee in Australia) uses fewer samples in each batch but each sample is of a larger size (in the 200–500 g range) and thus is less subject to error. These tests are even more expensive, yet they are routinely conducted as part of the process towards official recognition of malting quality in a country. In the United Kingdom, the malting quality evaluation by the MBC utilises a network of micro-malting laboratories at commercial maltings to test lines in registration trials in batches of 16 entries with common controls to an agreed set protocol (http://www.ukmalt.com/protocols-and-procedures). This system has provided accurate data with least significant differences as low as 1.7 l°/kg (https://projectblue.blob.core.windows.net/media/Default/Imported%20Publication%20Docs/AHDB%20Cereals%20&%20Oilseeds/Varieties/RL2019-20/Table%209.%20Spring%20barley%20Recommended%20List%202019-1.pdf), which enable the most promising lines for malting use to be promoted early in the registration and recommendation process.

The Beattie et al. (2010) study is a good example of the value of utilising a historical data set and the same approach was also exploited by Matthies et al. (2014) to identify QTL for malting quality among malting barley lines that had been entered into official trials in Germany. Despite another relatively small data set, Matthies et al. (2014) were able to identify novel QTL, again highlighting the value of working with a restricted elite gene pool. The detection of significant marker-trait associations in this study was particularly noteworthy as the germplasm consisted of winter and spring barley, which would have been grown in separate trials and while the same phenotypes would have been measured in any one year, they would essentially have been derived from different growing environments. The analyses included a correction for population structure effects (Matthies et al., 2014) that may well have accounted for most of the environmental differences between the two crop types. Nevertheless, it would have been instructive to examine the distribution of favourable and unfavourable alleles at each QTL locus to determine where breeding progress is being made in the different crop types as the allele frequencies given in the supplementary information were not broken down into those for each crop type.

Historical phenotypic data sets can also be used to form pools of extremes for a character which can then be sequenced for marked changes in allele frequencies at SNP loci between the extremes in a similar approach to Bulked Segregant Analysis (Michelmore et al., 1991). This approach, called extreme-phenotype genome-wide association analysis (Yang et al., 2015), was used to produce barley exome capture data for four DNA pools made up from the extreme phenotypes for diastatic power in UK spring and winter barley elite germplasm. The data was then screened to identify SNPs that showed strong contrasts in allele frequencies between the pools within each crop type (Looseley et al., 2017). KASP™ competitive allele-specific polymerase chain reaction (PCR) technology (https://biosearch-cdn.azureedge.net/assetsv6/KASP-brochure.pdf) markers were produced for each SNP and validated against independent test panels of breeding lines provided by several commercial breeders. While the most significant alleles were largely fixed in the spring barley validation set the SNP data for the winter barley lines was able to predict the diastatic power of the winter lines (Russell et al., 2017).

Igartua et al. (2019) also reported 39 association studies of agronomic characters that, from an inspection of the references, include 17 yield studies.

The studies are on quite different germplasm sets ranging from two-rowed germplasm from a single breeding programme (Berger et al., 2013) to Nested Association Mapping populations with wild barley selections crossed to a common cultivar, for example (Merchuk-Ovnat et al., 2018) and, as pointed out by Igartua et al. (2019), variation at genetic loci such as VRS1 are not only drivers of population variation but generally have major effects on yield components, grains per ear in the case of VRS1, and hence on yield itself. This can present problems in detection of significant associations when correcting for population structure and may warrant the use of a naïve model when loci that differentiate germplasm groups also exert a major influence on plant development.

Searching for yield QTL after accounting for any major gene differences is still worthwhile as associating more minor effect yield QTL with yield component QTL (spike number, grains per spike, and grain weight) can be instructive in formulating a strategy for yield improvement in a target environment. The problem is that multi-environment studies usually detect QTL × Environment interactions. For example, Xu et al. (2018a) examined panels of spring and winter barley cultivars grown in trials in Germany, Italy and the United Kingdom over two seasons, detecting 17 QTL but all were interactions with the environment. Most QTL studies, including GWAS, have focussed on additive models of gene action, mainly because the inclusion of epistatic models massively increases the computing power required for analysis. Nevertheless, inclusion of pairwise epistatic effects in a re-analysis of barley CAP data increased the variation accounted for from 31 to 40% (Xu et al., 2018b).

5 Application of results from genome-wide association studies (GWAS) in barley improvement

The interactions of QTL with environment limit the deployment of markers significantly associated with characters such as yield and malt extract in barley breeding. Provided there are no significant cross-over interactions, where a SNP allele is associated with a positive effect in one environment and a negative one in another, then an associated SNP could be used in MAS, especially if the significant effects are in environments that are more typical of the target. It remains unclear, however, whether or not commercial breeders are using MAS for yield associated SNPs but some are definitely using the technique for malting quality parameters.

The Association Genetics of UK Elite Barley project (AGOUEB) used a comprehensive set of historical data from UK official trials in a GWAS approach to identify QTL for malting quality characters (Thomas et al., 2014). QTL were successfully identified for component characters of malting quality in barley such as soluble nitrogen in the malt and wort glucan content (Wang et al., 2012) despite relatively narrow ranges in the phenotypic data. The results also highlighted that the malt extract genetic potential of winter barley was less than that of spring and that improvement in spring barley was notable in three key genomic regions where the allelic patterns (haplotype) were different between the two crops (Thomas, 2018). To test this hypothesis, the IMPROMALT project was set up to introgress spring quality QTL regions into current elite winter barley lines and then compare the malt extracts of the lines with the introgressions with those without. If successful, the strategy would provide a significant increase in the overall malting quality of winter barley that, when combined with the greater yield of the winter crop would result in a big uplift in the production of barrels of beer and whisky per unit of land area used to grow the crop (Thomas, 2018). This project is a good demonstration of the value of using GWAS and population analyses to understand breeding germplasm and develop strategies to exploit diversity from different germplasm groups in an efficient manner in plant breeding.

Producing a stock of a new line that passes VCU tests can be challenging for a barley breeder, especially when the competitive nature of barley breeding means that stocks are being generated from single plants at a generation when the average homozygosity at any one segregating locus is still quite high for example F4 or F5. In such cases, knowledge of the key genomic regions controlling DUS characters is of enormous benefit to barley breeders as they can screen mother plants for homozygosity in such regions to ensure that stocks are Uniform and Stable for the DUS traits as well as using information on other varieties to identify combinations that will be Distinct. In addition, where DUS characters that require microscopic seed examinations have been resolved to candidate genes, stocks can be screened with appropriate high throughput marker systems, such as KASP™, well in advance of harvest to ensure that just the plots with the right genotypes are harvested thus saving time and resources.

Another feature of the AGOUEB project was the use of DUS data in GWAS to identify markers that could be used to genotypically tag 15 out of 32 DUS characters that were surveyed with one being resolved to a validated candidate gene (Cockram et al., 2010). Further work identified KASP™ markers that could be used by breeders to select for these 15 DUS characters (Cockram et al., 2012). Re-analysing the data with a very high density set of almost 140 000 SNPs from barley exome capture data in the INTEGRA project (http://www.barleyhub.org/integra/) identified significant marker-trait associations for another 10 DUS characters with the resolution of all associations greatly improved (Kelly Houston, JHI unpublished data).

While information that are most relevant to barley breeders can be gleaned from the analyses of historical data on elite variety trials, it is unlikely that the most highly associated marker underlying a significant QTL is actually in the causal gene. The fact that the most significant marker is in linkage disequilibrium with a putative causal gene can, however, be used to monitor changes in allele frequency over time and hence the consequences of phenotypic selection for a specific character.

Using UK historical data on scald infection, Looseley et al. (2018) were able to show that the frequency of some markers associated with resistance alleles had increased in the UK elite spring barley gene pool over time, whereas others had either remained fairly constant or decreased. For example, SCRI_RS_221644 associated with the Rrs1 resistance locus was a minor allele (<10%) from the 1970s to approximately 2005, after which it increased in frequency with breeders placing more emphasis in selecting for resistance to scald so that it became the major allele from 2010 to 2014. In contrast, the frequency of SCRI_RS_138723, which was linked with a probable escape mechanism associated with allelic variation at the sdw1 dwarfing locus declined from 0.83 prior to 1980 to 0 by 2000–2005, reflecting continued selection for short straw in commercial spring barley breeding programmes in NW Europe. These statistics represent important information for barley breeders as pools of lines where the frequency of the desirable allele is high means that they have a number of choices of potential parents that can be used in designing crosses to combine, in this example, disease resistance with other economically important characters. Cases where the desirable allele is at a low frequency may mean that it is a relatively recent introduction and therefore might not yet be in an appropriate genetic background without further crossing and selection.

Reporting QTLs from GWAS experiments requires some thought to render the information that are most useful to plant breeders as they want to know what the results mean for the germplasm that they are working with. Noting that a particular base at a locus has a desirable effect upon a target character is useful but doesn't frame the result in a meaningful context for plant breeding programmes. Using the frequency of the desirable associated allele is one way of making the information more relevant but still doesn't directly inform the breeder about which lines carry the desirable alleles without cross-checking the SNP polymorphism with genotypic data files to identify which lines carry the favourable allele or haplotype over the QTL interval. One more meaningful solution is to choose a popular variety amongst a GWAS panel as a reference line. Lines carrying the same allele as the reference line are re-coded to reflect this, for example as a '1', and lines carrying the alternative alleles are coded as for example '0'. This then provides a simple way that the results can be interpreted by the breeder with an appropriate colour coding system. If the desirable allele or haplotype is a '1' then the breeder can combine this information with pedigree data to trace the allele in other germplasm to easily identify suitable parents for use in crossing programmes.

6 Conclusion and future trends

Provided that GWAS experiments are based upon germplasm sets that are representative for a barley breeder's target gene pool, they can make excellent training populations when implementing genomic selection programmes (Heffner et al., 2009). Commercially and publicly funded breeding programmes are using genomic selection in their programmes and a German study of winter six-row barley has shown some quite high levels of prediction accuracy for some characters, although they were not so promising for yield (Thorwarth et al., 2017). What is clear from pedigree analyses is that barley breeders are re-crossing material from within their own programme within two to four years of making the original cross (Fig. 5).

Most programmes are using genomic selection to identify which lines to select as parents for another set of crosses and/or advance into further trials. Looking further back into pedigrees, it can be seen that breeders had been operating such a system to some degree on a few lines prior to the advent of genomic selection but genomic selection has made the reduced cycle far more

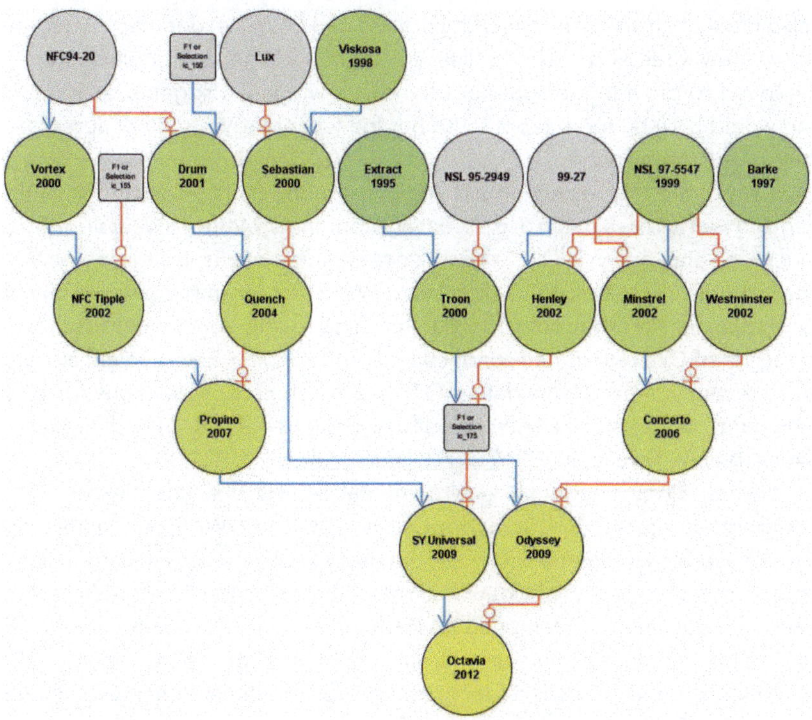

Figure 5 Abbreviated pedigree of the UK spring barley Octavia. Number in each circle is the year a variety was entered into National List Trials in the UK. Colour gets lighter for more recent entries into National List Trials.

widespread and reliable enabling potential parents to be routinely selected for re-crossing at the F4 generation.

The breeding values generated from genomic selection programmes do not provide any information upon the genes that are being manipulated and, from a breeding viewpoint, this can be ignored so long as breeding progress is being made. At some point, however, new variation needs to be brought into a barley breeding programme to maintain genetic variation and/or bring in sources of resistance or tolerance to emerging biotic and abiotic stresses. It is therefore essential that GWAS studies are maintained to provide information about the genetics behind the target characters in barley breeding, especially where genomic regions can be resolved to very small intervals with a few candidate genes. While further studies may be required to resolve a GWAS QTL to a candidate gene, the potential to use sequence information collected on germplasm collections, so-called 'genebank genomics' (Milner et al., 2019), means that suitable variation can be identified in a germplasm collection without phenotyping for the target trait and an introgression programme used to transfer just the specific genomic region to an adapted background. The introgressed line can then be entered into a new training set for genomic selection that has been informed by the use of a GWAS experiment. In regions where new breeding technologies are approved, genome editing can be deployed to provide the superior phenotype within an isogenic background and would be a far more rapid route but the technology is still not acceptable throughout the world.

One of the major problems facing plant breeders is producing lines that perform well across a range of different environments. Methods were developed to model and analyse QTL × Environment interactions using segregating populations (Boer et al., 2007) but most GWAS studies in barley have tended to analyse each environment separately and compare the consistency of detection of QTL rather than conducting a proper QTL × Environment analysis. More recently, methods to detect QTL × Environment interactions for large genotypic data sets have been developed (Millet et al., 2016) and applied to barley (Bustos-Korts et al., 2019).

Methodology to identify environmental factors that are behind QTL x Environment interactions (Romagosa et al., 2009) in GWAS populations can also be used to determine which are the best QTL alleles to select for specific environmental conditions. Evidence suggests that rising temperatures increases the starch gelatinisation temperature (Myllarinen et al., 1998; Rittenauer et al., 2017) and temperature increase is one of the environmental variables that would lead to the predicted 'beer drought' under climate change scenarios (Xie et al., 2018). Using climate variables to understand QTL × Environment interactions could result in a strategy to mitigate against the added potential of increased starch gelatinisation temperatures in brewing and distilling, especially as QTL for pasting temperature have been found to co-locate with malt quality QTL (Wang et al., 2018).

Phenomics platforms have and are being used in GWAS studies and can be used to identify QTL for abiotic stress responses that may help identify germplasm and genetic loci that could be used in breeding for adaptation for climate change (Pham et al., 2019). While these approaches can provide valuable information about plant development in response to specific stresses, breeding for adaptability in the real world requires better and more frequent characterisation of phenotypes obtained upon field grown trials together with integration with detailed characterisation of climatic variables within each environment. Here the use of drones to capture images will provide information to direct future breeding strategies.

7 Acknowledgements

This chapter uses data gathered under the IMPROMALT project, which was funded under the BBSRC LINK scheme (grant BB/K008188/1) with additional financial support from AHDB Cereals & Oilseeds R&D committee (RD-2012-3776) and the Scottish Governments Rural Affairs and the Environment programme. WTBT is funded by the latter organisation.

8 References

Bayer, M. M., Rapazote-Flores, P., Ganal, M., Hedley, P. E., Macaulay, M., Plieske, J., Ramsay, L., Russell, J., Shaw, P. D. and Thomas, W. 2017. Development and evaluation of a barley 50k iSelect SNP array. *Frontiers in Plant Science*, 8, 1792.

Beattie, A. D., Edney, M. J., Scoles, G. J. and Rossnagel, B. C. 2010. Association mapping of malting quality data from Western Canadian two-row Barley cooperative trials. *Crop Science*, 50, 1649–63.

Belcher, A. R., Cuesta-Marcos, A., Smith, K. P. and Hayes, P. M. 2018. TCAP FAC-WIN6 Elite Barley GWAS Panel QTL. II. Malting quality QTL in Elite North American facultative and winter six-rowed barley identified via GWAS. *Crop Science*, 58, 120–32.

Berger, G. L., Liu, S., Hall, M. D., Brooks, W. S., Chao, S., Muehlbauer, G. J., Baik, B., Steffenson, B. and Griffey, C. A. 2013. Marker-trait associations in Virginia Tech winter barley identified using genome-wide mapping. *Theoretical and Applied Genetics*, 126, 693–710.

Boer, M. P., Wright, D., Feng, L., Podlich, D. W., Luo, L., Cooper, M. and Van Eeuwijk, F. A. 2007. A mixed-model quantitative trait loci (QTL) analysis for multiple-environment trial data using environmental covariables for QTL-by-environment interactions, with an example in maize. *Genetics*, 177, 1801–13.

Bustos-Korts, D., Dawson, I. K., Russell, J., Tondelli, A., Guerra, D., Ferrandi, C., Strozzi, F., Nicolazzi, E. L., Molnar-Lang, M., Ozkan, H., Megyeri, M., Miko, P., Cakir, E., Yakisir, E., Trabanco, N., Delbono, S., Kyriakidis, S., Booth, A., Cammarano, D., Mascher, M., Werner, P., Cattivelli, L., Rossini, L., Stein, N., Kilian, B., Waugh, R. and Van Eeuwijk, F.

A. 2019. Exome sequences and multi-environment field trials elucidate the genetic basis of adaptation in barley. *Plant Journal*, 99, 1172-91.

Close, T. J., Bhat, P. R., Lonardi, S., Wu, Y., Rostoks, N., Ramsay, L., Druka, A., Stein, N., Svensson, J. T., Wanamaker, S., Bozdag, S., Roose, M. L., Moscou, M. J., Chao, S., Varshney, R. K., Szűcs, P., Sato, K., Hayes, P. M., Matthews, D. E., Kleinhofs, A., Muehlbauer, G. J., Deyoung, J., Marshall, D. F., Madishetty, K., Fenton, R. D., Condamine, P., Graner, A. and Waugh, R. 2009. Development and implementation of high-throughput SNP genotyping in barley. *BMC Genomics*, 10, 582.

Cockram, J., White, J., Zuluaga, D. L., Smith, D., Comadran, J., Macaulay, M., Luo, Z., Kearsey, M. J., Werner, P., Harrap, D., Tapsell, C., Liu, H., Hedley, P. E., Stein, N., Schulte, D., Steuernagel, B., Marshall, D. F., Thomas, W. T., Ramsay, L., Mackay, I., Balding, D. J., Waugh, R. and O'Sullivan, D. M. 2010. Genome-wide association mapping to candidate polymorphism resolution in the unsequenced barley genome. *Proceedings of the National Academy of Sciences of the United States of America*, 107, 21611-16.

Cockram, J., Jones, H., Norris, C. and O'Sullivan, D. M. 2012. Evaluation of diagnostic molecular markers for DUS phenotypic assessment in the cereal crop, barley (Hordeum vulgare ssp vulgare L.). *Theoretical and Applied Genetics*, 125, 1735–49.

Collins, H. M., Panozzo, J. F., Logue, S. J., Jefferies, S. P. and Barr, A. R. 2003. Mapping and validation of chromosome regions associated with high malt extract in barley (Hordeum vulgare L.). *Australian Journal of Agricultural Research*, 54, 1223–40.

Comadran, J., Kilian, B., Russell, J., Ramsay, L., Stein, N., Ganal, M., Shaw, P., Bayer, M., Thomas, W. and Marshall, D. 2012. Natural variation in a homolog of Antirrhinum Centroradialis contributed to spring growth habit and environmental adaptation in cultivated barley. *Nature Genetics*, 44, 1388.

Cruz, V. M. V., Kilian, A. and Dierig, D. A. 2013. Development of DArT marker platforms and genetic diversity assessment of the us collection of the new oilseed crop lesquerella and related species. *Plos One*, 8.

Elshire, R. J., Glaubitz, J. C., Sun, Q., Poland, J. A., Kawamoto, K., Buckler, E. S. and Mitchell, S. E. 2011. A Robust, Simple Genotyping-by-Sequencing (GBS) approach for high diversity species. *Plos One*, 6.

Gibson, T. S., Solah, V., Holmes, M. R. G. and Taylor, H. R. 1995. diastatic power in malted barley - contributions of malt parameters to its development and the potential of barley-grain beta-amylase to predict malt diastatic power. *Journal of the Institute of Brewing*, 101, 277–80.

Gutierrez, L., Cuesta-Marcos, A., Castro, A. J., Von Zitzewitz, J., Schmitt, M. and Hayes, P. M. 2011. Association mapping of malting quality quantitative trait loci in winter barley: Positive signals from small germplasm arrays. *Plant Genome*, 4, 256–72.

Hackett, C. A., Ellis, R. P., Forster, B. P., McNicol, J. W. and Macaulay, M. 1992. statistical-analysis of a linkage experiment in barley involving quantitative trait loci for height and ear-emergence time and 2 genetic-markers on chromosome-4. *Theoretical and Applied Genetics*, 85, 120–6.

Heffner, E. L., Sorrells, M. E. and Jannink, J. L. 2009. Genomic selection for crop improvement. *Crop Science*, 49, 1-12.

Igartua, E., Cantalapiedra, C. P. and Casas, A. 2019. Genome-wide association studies (GWAS) in barley. In: Ordon, F. and Friedt, W. (Eds), *Advances in Breeding Techniques for Cereal Crops*. Cambridge, UK: Burleigh Dodds Science Publishing Ltd.

Knott, D. R. and Kumar, J. 1975. Comparison of early generation yield testing and a single seed descent procedure in wheat breeding. *Crop Science*, 15, 295–9.

Laidig, F., Piepho, H. P., Rentel, D., Drobek, T. and Meyer, U. 2017. Breeding progress, genotypic and environmental variation and correlation of quality traits in malting barley in German official variety trials between 1983 and 2015. *Theoretical and Applied Genetics*, 130, 2411–29.

Looseley, M. E., Bayer, M., Bull, H., Ramsay, L., Thomas, W., Booth, A., Canto, C. D., Morris, J., Hedley, P. E. and Russell, J. 2017. Association mapping of diastatic power in uk winter and spring barley by exome sequencing of phenotypically contrasting variety sets. *Frontiers in Plant Science*, 8.

Looseley, M. E., Griffe, L. L., Büttner, B., Wright, K. M., Middlefell-Williams, J., Bull, H., Shaw, P. D., Macaulay, M., Booth, A. and Schweizer, G. 2018. Resistance to Rhynchosporium commune in a collection of European spring barley germplasm. *Theoretical and Applied Genetics*, 131, 2513–28.

Matthies, I. E., Malosetti, M., Roder, M. S. and Van Eeuwijk, F. 2014. Genome-Wide association mapping for kernel and malting quality traits using historical european barley records. *Plos One*, 9.

Merchuk-Ovnat, L., Silberman, R., Laiba, E., Maurer, A., Pillen, K., Faigenboim, A. and Fridman, E. 2018. Genome scan identifies flowering-independent effects of barley HsDry2.2 locus on yield traits under water deficit. *Journal of Experimental Botany*, 69, 1765–79.

Michelmore, R. W., Paran, I. and Kesseli, R. V. 1991. Identification of markers linked to disease-resistance genes by bulked segregant analysis - a rapid method to detect markers in specific genomic regions by using segregating populations. *Proceedings of the National Academy of Sciences of the United States of America*, 88, 9828–32.

Millet, E. J., Welcker, C., Kruijer, W., Negro, S., Coupel-Ledru, A., Nicolas, S. D., Laborde, J., Bauland, C., Praud, S., Ranc, N., Presterl, T., Tuberosa, R., Bedo, Z., Draye, X., Usadel, B., Charcosset, A., Van Eeuwijk, F. and Tardieu, F. 2016. Genome-wide analysis of yield in europe: Allelic effects vary with drought and heat scenarios. *Plant Physiology*, 172, 749–64.

Milner, S. G., Jost, M., Taketa, S., Mazon, E. R., Himmelbach, A., Oppermann, M., Weise, S., Knupffer, H., Basterrechea, M., Konig, P., Schuler, D., Sharma, R., Pasam, R. K., Rutten, T., Guo, G. G., Xu, D. D., Zhang, J., Herren, G., Muller, T., Krattinger, S. G., Keller, B., Jiang, Y., Gonzalez, M. Y., Zhao, Y. S., Habekuss, A., Farber, S., Ordon, F., Lange, M., Borner, A., Graner, A., Reif, J. C., Scholz, U., Mascher, M. and Stein, N. 2019. Genebank genomics highlights the diversity of a global barley collection. *Nature Genetics*, 51, 319–+.

Mohammadi, M., Blake, T. K., Budde, A. D., Chao, S., Hayes, P. M., Horsley, R. D., Obert, D. E., Ullrich, S. E. and Smith, K. P. 2015. A genome-wide association study of malting quality across eight US barley breeding programs. *Theoretical and Applied Genetics*, 128, 705–21.

Monat, C., Schreiber, M., Stein, N. and Mascher, M. 2019. Prospects of pan-genomics in barley. *Theoretical and Applied Genetics*, 132, 785–96.

Myllarinen, P., Schulman, A. H., Salovaara, H. and Poutanen, K. 1998. The effect of growth temperature on gelatinization properties of barley starch. *Acta Agriculturae Scandinavica Section B-Soil and Plant Science*, 48, 85–90.

Paterson, A. H. and Beavis, W. D. 1998. QTL analyses: Power, precision, and accuracy. In: Paterson, A. H. (Ed.), *Molecular Dissection of Complex Traits*. New York: CRC Press.

Pham, A. T., Maurer, A., Pillen, K., Brien, C., Dowling, K., Berger, B., Eglinton, J. K. and March, T. J. 2019. Genome-wide association of barley plant growth under drought stress using a nested association mapping population. *Bmc Plant Biology*, 19.

Pritchard, J. K., Stephens, M. and Donnelly, P. 2000. Inference of population structure using multilocus genotype data. *Genetics*, 155, 945–59.

Rittenauer, M., Gladis, S., Gastl, M. and Becker, T. 2017. Critical evaluation of viscometrically determined pasting temperatures in barley malt. *Journal of the Institute of Brewing*, 123, 472–9.

Romagosa, I., Van Eeuwijk, F. A. and Thomas, W. T. 2009. Statistical analyses of genotype by environment data. In: Carina, M. (Ed), *Cereals*. Handbook of plant breeding, 3, 291–331. New York, NY: Springer.

Russell, J., Looseley, M., Bayer, M., Bull, H., Ramsay, L., Thomas, W., Booth, A., Morris, J., Hedley, P., Hoad, S., Hess, L. and Brosnan, J. 2017. PR 583. Project Report No 583: A Genome Wide Analysis of Key Genes Controlling Diastatic Power Activity in UK Barley (DPGENES). AHDB Cereals and Oilseeds. Available: https://cereals.ahdb.org.uk/media/1302880/pr583-final-project-report-dpgenes-v2.pdf.

Thomas, W. 2018. Impromalt: Improving the malting quality of winter barley for sustainability of distilling industry. The Dundee Conference. Crop Production in Northern Britain 2018, Dundee, UK, 27–28 February 2018. The Association for Crop Protection in Northern Britain, 115–20.

Thomas, W., Comadran, J., Ramsay, L., Shaw, P., Marshall, D., Newton, A., O'Sullivan, D., Cockram, J., Mackay, I., Bayles, R., White, J., Kearsey, M., Luo, Z., Wang, M., Tapsell, C., Harrap, D., Werner, P., Klose, S., Bury, P., Wroth, J., Argillier, O., Habgood, R., Glew, M., Bochard, A.-M., Gymer, P., Vequaud, D., Christerson, T., Allvin, B., Davies, N., Broadbent, R., Brosnan, J., Bringhurst, T., Booer, C. and Waugh, R. 2014. *Project Report No. 528: Association Genetics of UK Elite Barley (AGOUEB)*. HGCA. Available: http://cereals.ahdb.org.uk/media/337662/pr528-final-project-report.pdf.

Thomas, W. T. B. 2003. Prospects for molecular breeding of barley. *Annals of Applied Biology*, 142, 1–12.

Thomas, W. T. B., Forster, B. P. and Gertsson, B. 2003. Doubled haploids in breeding. In: Maluszynski, M., Szarejko, I., Kasha, K. J. and Forster, B. P. (Eds), *Doubled Haploid Production in Crop Plants*, 337–50. Vienna: Kluwer Academic Publisher.

Thorwarth, P., Ahlemeyer, J., Bochard, A. M., Krumnacker, K., Blumel, H., Laubach, E., Knochel, N., Cselenyi, L., Ordon, F. and Schmid, K. J. 2017. Genomic prediction ability for yield-related traits in German winter barley elite material. *Theoretical and Applied Genetics*, 130, 1669–83.

Wang, J. M., Yang, J. M., Hua, W., Wu, X. J., Zhu, J. H., Shang, Y. and Zhou, M. X. 2018. QTL mapping reveals the relationship between pasting properties and malt extract in barley. *International Journal of Molecular Sciences*, 19.

Wang, M., Jiang, N., Jia, T., Leach, L., Cockram, J., Waugh, R., Ramsay, L., Thomas, B. and Luo, Z. 2012. Genome-wide association mapping of agronomic and morphologic traits in highly structured populations of barley cultivars. *Theoretical and Applied Genetics*, 124, 233–46.

Xie, W., Xiong, W., Pan, J., Ali, T., Cui, Q., Guan, D. B., Meng, J., Mueller, N. D., Lin, E. and Davis, S. J. 2018. Decreases in global beer supply due to extreme drought and heat. *Nature Plants*, 4, 964–+.

Xu, X., Sharma, R., Tondelli, A., Russell, J., Comadran, J., Schnaithmann, F., Pillen, K., Kilian, B., Cattivelli, L., Thomas, W. T. B. and Flavell, A. J. 2018a. Genome-wide association

analysis of grain yield-associated traits in a pan-european barley cultivar collection. *Plant Genome*, 11.
Xu, Y., Wu, Y. J. and Wu, J. X. 2018b. Capturing pair-wise epistatic effects associated with three agronomic traits in barley. *Genetica*, 146, 161-70.
Yang, J. L., Jiang, H. Y., Yeh, C. T., Yu, J. M., Jeddeloh, J. A., Nettleton, D. and Schnable, P. S. 2015. Extreme-phenotype genome-wide association study (XP-GWAS): A method for identifying trait-associated variants by sequencing pools of individuals selected from a diversity panel. *Plant Journal*, 84, 587-96.
Zhou, G. F., Panozzo, J., Zhang, X. Q., Cakir, M., Harasymow, S. and Li, C. D. 2016. QTL mapping reveals genetic architectures of malting quality between Australian and Canadian malting barley (Hordeum vulgare L.). *Molecular Breeding*, 36.

Chapter 8

Advances in molecular breeding techniques for barley: targeted induced local lesions in genomes (TILLING)

Serena Rosignoli and Silvio Salvi, University of Bologna, Italy

1 Introduction
2 Technical details on artificial mutagenesis and mutation discovery in TILLING
3 TILLING resources in barley
4 Current and future trends of barley TILLING
5 TILLING versus other reverse genetics tools in barley
6 Conclusion
7 Where to look for further information
8 References

1 Introduction

Targeting Induced Local Lesions IN Genomes (TILLING) is a reverse genetics method enabling the identification of individuals carrying chemically induced allelic variants at a gene of interest, within a mutagenized population. TILLING was first described in *Arabidopsis* (McCallum et al., 2000) and *Drosophila* (Bentley et al., 2000), and then extended to several plant species including maize (Till et al., 2004b), wheat (Slade et al., 2005), rice (Suzuki et al., 2008; Till et al., 2007), and others. It can be implemented in any species where a population of mutant individuals can be generated starting from a homogeneous line. Additionally, a version of TILLING has been described to help in discovery of polymorphisms in natural populations or in cultivars collections (EcoTILLING; Comai et al., 2004). Commonly, TILLING a gene implies three consecutive steps, namely (i) production or availability of a mutagenized population, (ii) PCR amplification of the target gene in a pool of individual genomic DNA samples, and (iii) detection of mutant alleles and individuals carrying them by means of a biochemical essay suitable for nucleotide polymorphism (e.g. SNP) identification. Most TILLING resources have been developed using chemical mutagenesis which

http://dx.doi.org/10.19103/AS.2019.0060.10

mostly induces single-point mutations. This provides the investigator with a very informative range of allelic effects, from effective protein knockouts produced by stop-gain mutations to hypomorphic alleles caused by missense (i.e. amino acid substitution) mutations. Such different functional alleles are not usually available within populations generated with other types of mutational agents such as insertional mutagenesis (e.g. T-DNA, transposons) or irradiation (e.g. gamma-ray, fast-neutrons), all of which mostly generate knockouts. One additional positive aspect of TILLING is that it does not require any prior genomic knowledge (except for the target gene sequence, which must be known).

Soon after its introduction, the potential of TILLING for plant genetic improvement was realized. Slade et al. (2005) first applied TILLING to recover novel alleles of *granule-bound starch synthase 1* (*gbss1*), a gene of breeding value as it affects endosperm starch composition. Furthermore, it is not secondary to note that chemically induced TILLING mutants are not regulated as a genetically modified organism (GMO) and can be freely grown both in controlled and field conditions. GMO regulation instead applies for several insertional mutant types (e.g. T-DNA or modified versions of transposons) in certain countries.

Artificial mutagenesis in barley has a very long and successful history in supporting both fundamental plant biology investigations and breeding. However, only TILLING-related aspects of mutagenesis will be discussed here and detailed discussions about barley mutagenesis can be found elsewhere (Druka et al., 2011; Lundqvist, 2014).

2 Technical details on artificial mutagenesis and mutation discovery in TILLING

2.1 Tissue or organ to mutagenize

Tissues or organs undergoing artificial mutagenesis for TILLING ranged from vegetative parts (buds, apical meristems, leaves), to pollen, to seeds, depending of the plant species. For instance, shoot apical meristems were isolated and chemically treated in TILLING efforts in banana (Jankowicz-Cieslak et al., 2012); apple leaf cuts were gamma-irradiated and then underwent tissue culture and regeneration (Salvi et al., 2015); pollen is usually mutagenized in maize (Weil and Monde, 2007). Seed has so far been the most commonly used organ type for mutagenesis for TILLING applications in barley and in related species such as bread and durum wheats.

2.2 Mutagen choice

Choosing the right mutagen agent is an important step and responds to several considerations. First, mutagenesis should produce the desired (and often, as high as possible) mutation density. An ideal mutagenic agent should be characterized by the greatest genotoxicity (i.e. mutagenic effect on cell DNA) and the least cytotoxicity (i.e. negative effects on cell or tissue viability unrelated with DNA damage). Cytotoxicity kills cells irrespective of mutagenesis, thus reducing the efficiency of the mutation treatment and limiting the final mutation density that can be reached. The mutagenic treatment should also guarantee the production of a range of functional alleles, from mild-effect to strong-effect mutations and in this respect chemically induced mutagenesis should be preferred to physical mutagenesis, which is known to induce mostly truncated (knockout) alleles. The treatment protocol should be as simple as possible and operators should be exposed to the least health risks. Finally, it is important to match the type of induced mutations (e.g. single-point mutations or insertion/deletions) with the molecular protocols to be used for mutation identification. Based on all of the considerations mentioned, chemical mutagenesis on seeds has been the choice for TILLING in barley.

In barley, popular chemical mutagens for TILLING applications are Ethyl methanesulfonate (EMS), N-methyl-N-nitroso-urea (MNU), and sodium azide (NaN_3). EMS is the most utilized mutagen in plants. It is an alkylating agent producing chemical modifications on nucleotide bases, which results in base mispairing and base changes. In the great majority (99%) of cases, EMS induces C>T changes resulting in C/G to T/A substitutions (Kim et al., 2006).

Like EMS, MNU is an alkylating agent that causes DNA damage by transferring a methyl (-CH3) group to the oxygen and nitrogen atoms of nucleotide bases. In a detailed analysis of nucleotide substitutions induced by MNU in barley, Kurowska et al. (2012) showed that the majority (63.6%) of the mutations were transitions, with a similar number of G>A and C>T substitutions. However, MNU also generates inversions and translocations at low frequencies (Szarejko and Maluszynski, 1980; Maluszynski and Szarejko, 2003).

NaN_3 is another popular chemical mutagenic agent in barley (Nilan et al., 1973). Its mutagenicity is mediated by a metabolically synthesized organic azide, L-azidoalanine, and its action appears to require an active DNA repair mechanism involved in UV-like damage (Owais and Kleinhofs, 1988). It produces mostly GC>AT transitions (Talamè et al., 2008; Szarejko et al., 2017). NaN_3 complex mechanism of action is likely the reason of its species-specific mutagenic activity. For instance, NaN_3 mutagenicity is high in barley and very low in Arabidopsis (Gichner and Veleminsky, 1977). Likely for the same reason, it has little mutagenic activity in mammals (Slamenová and Gabelová, 1980; Arenaz et al., 1989). The high density of generated mutations (Table 1), the simple and hazardous-free protocol, and the negligible occurrence of chromosome aberrations make NaN_3 mutagenesis commonly used in barley (Lundqvist et al., 1991). Double chemical mutagenic treatment (EMS+MNU) has also

Table 1 Currently described barley TILLING resources

Cultivar	Mutagen	Lines (No.)	Mutation density (kb)	Mutation type (%)[a]	References
Optic	EMS	9216	1/1000	NA	Caldwell et al. (2004)
Morex	NaN_3	4906	1/374	1.8-69.6-28.6	Talamè et al. (2008, 2009)
Barke	EMS	10279	1/500	4.7-45.3-50.0	Gottwald et al. (2009)
Lux	NaN_3	9575	1/2500	NA	Engvild and Rasmussen (2004), Lababidi et al. (2009)
Sebastian	MNU + NaN_3	9600	1/477	1.1-61.0-37.5	Szurman-Zubrzycka et al. (2018)
Golden Promise	EMS	Under completion	1/154	3.3-64.1-32.6	Schreiber et al. (2019)

NA: not available.
[a] Mutation types are provided as percentage in the order nonsense–missense–silent. Only mutations in coding sequence (exons) were considered.

been successfully applied as a strategy to increase mutation density (Szurman-Zubrzycka et al., 2018).

The use of physical mutagenesis (e.g. gamma and/or X-rays, fast neutrons, and/or heavy ions) to generate nucleotide substitutions and small deletions is also possible in plants (Sato et al., 2006; Morita et al., 2009; Yoshihara et al., 2010). However, for TILLING applications in barley, gamma-ray irradiation was shown to be far less efficient than MNU chemical treatment (Kurowska et al., 2012). The reason likely lies in the low frequency of single nucleotide events and the large number of induced chromosome-level mutations which reduces viability and fertility of M_1 and M_2 generations. This notwithstanding, a mutant wheat population was produced using heavy ion irradiation and was shown to provide a useful source of mutant alleles in a TILLING-like reverse genetics context (Fitzgerald et al., 2015).

2.3 Mutation density

The mutation density of a TILLING resource directly affects the number of plants/lines to be generated and screened in order to recover new alleles of a target gene. In diploid species, the average mutation density in TILLING populations is in the order of one mutation per 380 kb (Wang et al., 2012), and this value is approximately observed in barley TILLING populations (Table 1). Densities are much higher in polyploid species reaching 1/32 kb in hexaploids wheat and oat (Wang et al., 2012; discussed in Parag. 5.2). However, it appears possible to reach higher mutation densities in diploid species using some strategies, including selecting M_1 plants with higher than average number of mutations, using phenotypic markers for mutation density (e.g. high level of sterility (Martin et al., 2009)) or by applying double or triple cycles of chemical mutagenesis (e.g. re-treating M1 seeds). Indeed, recently, Schreiber et al. (2019) reached a mutation density of one per 154 kb in barley by using the latter strategy. Based on the expectation that 4.5% randomly induced mutations lead to a truncation (either stop codon or splice junction mutation) and that 20% will be deleterious missense mutations, Wang et al. (2012) estimated that in a diploid species at least 2000 TILLING lines are required to recover a missense deleterious allele at $P=0.8$, while 10000 lines are required to reach the same probability for a truncation. Thus, highly efficient TILLING populations in barley should at least be as large as few thousand plants or lines.

2.4 Searching for mutations

In TILLING, the search and identification of mutations is usually carried out in DNA pools after PCR amplification of the target gene (or portion of it). Approaches include the use of techniques based on endonucleases that cleave single- and/or double-stranded DNA formed by a mutated strand and its wildtype complementary strand (the so-called heteroduplex) at the site of mismatched nucleotides caused by the mutation (Colbert et al., 2001). One widely adopted endonuclease is CEL1, which is extracted from celery (Till et al., 2003), others are CJE, a non-purified celery leaf extract maintaining endonuclease activity (Till et al., 2004a), or ENDO1, an endonuclease originally characterized in *A. thaliana* (Triques et al., 2007). Such cleaved heteroduplexes generate a visible signal that can be detected by polyacrylamide gel electrophoresis or by simple agarose gel electrophoresis (Raghavan et al., 2007). Other heteroduplexes detection techniques that do not rely on enzymatic cleavage are DHPLC–denaturing high-performance liquid chromatography (McCallum et al., 2000), capillary electrophoresis, and HRM–high resolution melting analysis (Gady et al., 2009; Dong et al., 2009).

More recently, the advent of next-generation sequencing (NGS) technologies made DNA sequencing cheaper and faster, thus providing a more efficient way of detecting mutations. TILLING by Sequencing, or TbyS, can be defined as the application of NGS supported by bioinformatics tools helping mutant identification (Kumar et al., 2017). It has been applied for the first time in rice and wheat using an NGS platform on target genes amplified from multidimensionally pooled DNA templates (Tsai et al., 2011). Mutations are identified by comparison with a reference sequence, and sequencing errors can be discriminated from actual point mutations through bioinformatic analysis, if a sufficient sequencing coverage was reached. Screening a mutagenized population for mutation identification by TbyS is more efficient than by heteroduplex/gel-based protocols as it permits multiplexing of gene targets and genomes. A technically unresolved problem in TbyS is that the average TILLING amplicons are longer than NGS (i.e. Illumina) reads. However, ultrasonication for the simultaneous fragmentation of multiplexed PCR amplicons can be applied to streamline this step (Tramontano et al., 2019).

Other protocols potentially interesting for allele identification in TILLING include allele-specific capture (Kohata et al., 2018). In this case, PCR products of mutant alleles are captured by biotin/streptavidin. By repeated captures of each strand and combining both strands, mutant alleles with a concentration of 1/1000 in wild-type alleles resulted detectable by CAPS or dCAPS analysis. Allele-specific detection of rare alleles in large pools can also be obtained by using some CRISPR-Cas9-related technologies as was recently demonstrated (Abudayyeh et al., 2019).

2.5 TILLING in silico

The evolving of high-throughput sequencing platforms (e.g. Illumina, PacBio, Ion Torrent, Oxford Nanopore, and others), and the drop in sequencing costs led to the availability of sequenced reference genomes for most crops. Currently, the genomes of c. 240 angiosperm species have been completely sequenced (Chen et al., 2018) of which 75% are crops or their wild relatives (https://stateoftheworldsplants.org/2017/). Barley too now has a very high-quality reference genome sequence (Mascher et al., 2017) based on the cultivar Morex. This technological revolution also led to an increasing number of whole exome and whole-genome resequencing efforts. As far as TILLING is concerned, the advent of NGS approaches made possible to sequence all genes for all mutant lines of a TILLING population, and store information on all mutations in a searchable database indexed to mutant lines. With this type of resource, the mutation search step is reduced to a simple database search using the target gene model as query. This approach has been named TILLING *in-silico* (Wang et al., 2012). An impressive TILLING *in-silico* resource based on exome sequencing of 1535 and 1200 EMS mutant lines of durum wheat and bread wheat, respectively, has already been made available (Krasileva et al., 2017). This resource provides access to more than 10 million mutations in wheat, corresponding to an average of 23–24 missense and truncation alleles per gene. One truncation or one deleterious missense mutation (at least) is present in 90% of the genes (Krasileva et al., 2017). Similar resources are being completed in *Brachypodium distachyon* (Granier et al., 2015; John Vogel Lab, Berkeley/DOE Joint Genome Institute) and in *Sorghum* (Jiao et al., 2016 and https://www.purdue.edu/sorghumgenomics/). It appears inevitable that TILLING resources in most species, including barley, will soon evolve in TILLING *in-silico*.

3 TILLING resources in barley

In barley, six TILLING resources have been described (Table 1). They were obtained starting from seed chemical mutagenesis and their mutation densities spanned from to 1/154 to 1/2500 mutations/base pair. Five resources are based on two-row cultivars (Barke, Golden Promise, Lux, Optic, Sebastian) and one on the six-row cultivar Morex. Protocols for mutation search were generally based on amplicon design on target gene sequence followed by PCR in DNA pools and heteroduplex processing by endonucleases and gel-electrophoresis or HRM. However, amplicons sequencing on DNA pools was also applied (Schreiber et al., 2019). The predicted percentage of truncation on total mutations in exons ranged from 1.1% to 4.7%, and the percentage of missense mutations ranged from 45% to 70%.

Along with TILLING resources providing access to nuclear genes, an additional barley resource was developed targeting chloroplast mutations (Landau et al., 2016), following an approach called cpTILLING. In this work, the authors took advantage of the barley *chloroplast mutator* (*cpm*) gene, that induces several types of cytoplasmically inherited chlorophyll deficiencies when homozygous. An extensive scanning of 33 plastome genes and intergenic regions provided at least 61 mutational events (Landau et al., 2016).

4 Current and future trends of barley TILLING

At least 22 studies have been published where gene function has been investigated using TILLING in barley (Table 2). In these studies, scientists analyzed genes involved from basic cell biology and/or plant development, such as DNA repair mechanisms (Stolarek et al., 2015a,b), brassinosteroids biosynthesis (Gruszka et al., 2016), or leaf variegation (Li et al., 2019), to genes of more direct breeding interest.

Quite surprisingly, as shown by the examples in the following section, opportunities abound for breeding applications offered by knocking down or out genes. One promising area is the response to abiotic stresses such as drought. Responses to drought include decrease cell, leaf and stem elongation, stomatal dynamics, pollen sterility, selective seed abortion, and others, which provide plants with adaptive mechanisms. However, while some of these responses (e.g., increased stomatal closure, seed abortion, and others) may impact positively on fitness of wild crop relatives in natural environments, the same responses can be maladaptive in crops (i.e. impact negatively on yield) in farming conditions, where stress events are kept shorter or milder by human management. In these conditions, tempering or abolishing genes involved in stress perception and corresponding signal transduction pathway could have a positive effect on yield. For instance, downregulation of ethylene

Table 2 Applications of TILLING in barley, in chronological order

Collection	Trait	Target gene	Reference
Optic	Potyvirus resistance	*eIF4E*	Kanyuka et al. (2005)
Lux	Drought and cold tolerance	*Dhn*	Lababidi et al. (2009)
Morex	New starch phenotypes	Starch synthesis related genes	Sparla et al. (2014)
Ad hoc TILLING (MNU) on DH line	Plant architecture	Brassinosteroid genes	Dockter et al. (2014)
Barke	Leaf initiation	*HvMND*	Mascher et al. (2014)
Barke	Virus resistance	*HvPDIL5-1*	Yang et al. (2014)
Barke	Haploid induction	*CENH3*	Karimi-Ashtiyani et al. (2015)
Barke	Grain Dispersal System	*Btr1, Btr2*	Pourkheirandish et al. (2015)
Morex and Barke	Spike architecture	*COM2*	Poursarebani et al. (2015)
Sebastian	DNA repair	*HvPARP3*	Stolarek et al. (2015a)
Sebastian	DNA repair	*HvKu80*	Stolarek et al. (2015b)
Sebastian	Brassinosteroid synthesis	*HvDWARF*	Gruszka et al. (2016)
Barke	Inflorescence architecture	*HvLax-a*	Jost et al. (2016b)
Barke	Leaf width	*BLF1*	Jöst et al. (`)
Sebastian	Strigolactone signaling	*HvD14*	Marzec et al. (2016)
Sebastian	Waterlogging tolerance	*HvPRT6*	Mendiondo et al. (2016)
Barke	Seed dormancy	*MKK3*	Nakamura et al. (2016)
Barke	Virus resistance	*HvPDIL5, HvEIF4E*	Yang et al. (2017)
Sebastian	β-subunit of farnesyltransferase, ABA-hypersensitivity	*HvERA1*	Daszkowska-Golec et al. (2018)
Barke	Leaf variegation	*HvAST (HvCMF7)*	Li et al. (2019)
Sebastian	Double-strand break	*HvDMC1*	Szurman-Zubrzycka et al. (2019)

This table was compiled after searching Google Scholar (last on July 17, 2019) using 'TILLING and barley' as keywords, and then by manually checking for studies dealing with functional gene analysis.

(a typical stress-related hormone) production or ethylene response by genetic engineering *ZmACS* or *OsERF109* in maize or rice, respectively, showed positive effects on yield in mild water-stress conditions (Habben et al., 2014; Yu et al., 2017). The highly articulate abscisic acid synthesis and responsive pathways could be targeted as well (Vishwakarma et al., 2017). In this direction, Daszkowska-Golec et al. (2018) have already characterized a TILLING (from cv. Sebastian) mutant allele of gene *CBP20* (*Cap-Binding Protein 20*), encoding a protein involved in RNA metabolism and ABA-dependent drought-response mechanisms. The mutant clearly showed a faster response at the onset of water deficit and better fitness-to-stress conditions than the corresponding wild type.

Similarly, defective alleles at key genes were shown to be effective for parasite resistance. A well-known example is provided by the *mildew-resistance locus O (MLO)*. *MLO* appears to be an integral component of an evolutionarily conserved machinery for basal nonhost resistance and likely other unknown cell functions (Acevedo-Garcia et al., 2014); *MLO* knockout alleles in barley or wheat provided resistance to powdery mildew disease, although the resistance was occasionally accompanied by a range of negative pleiotropic effects (Wang et al., 2014; Acevedo-Garcia et al., 2017). In a different strategy, modulators of plant immunity reaction could be targeted. For instance, knocking out the ERF transcription factor gene *OsERF922* strongly enhanced tolerance to rice blast, a destructive rice disease caused by the fungus *Magnaporthe oryzae* (Wang et al., 2016). The many and mostly unknown transcriptional co-regulatory genes involved in plant immunity networks (Tsuda and Somssich, 2015) are potential access points for modifying resistance by TILLING.

Endosperm starch content and composition, mainly in terms of amylose/amylopectin rate, are the most important quality traits in barley. Both low amylose-high amylopectin starch type and its opposite type (high amylose-low amylopectin) are of great industrial interest. High amylose starches have nutritional properties due to their positive correlation with resistant starch and linked health benefits (Sharma et al., 2008), while low-amylose starches find applications in the food or chemical industry (Doblado-Maldonado et al., 2017). A complex network of structural and regulatory gene functions (e.g. phosphorylation enzymes) controls starch traits (Xia et al., 2018). In the first attempt, all genes coding for starch branching enzymes (SBE I, SBE IIa, SBE IIb) in barley have been suppressed by RNAi, resulting in production of amylose-only starch granules in the endosperm (Xia et al., 2018). Sparla et al. (2014) used the Morex-based TILLMore population (Talamè et al., 2008) to till new alleles of known genes involved in starch metabolism in barley. The authors focused on five genes affecting distinctive features of starch metabolism (*beta-amylase* 1–*BMY1*, *granule-bound starch synthase I–GBSSI*, *limit dextrinase 1–LDA 1*, *starch synthase I–SSI* and *starch synthase Iia–SSIIa*), and found 29 new alleles and phenotyped the corresponding mutants for total starch content, amylose content, starch morphology, and starch crystallinity. Seven mutant lines presented starch properties different from wild type and could potentially be exploited in food or industrial applications. While these two studies successfully showed the power of deleting structural (i.e. enzymatic) gene functions, the number of genes in the starch pathway is much higher (e.g. regulatory genes have not been targeted so far). Indeed, the systematic characterization of transcription factors and cis regulatory elements of starch metabolism has been recognized as a promising avenue for rational crop improvement (López-Gonzáles et al. 2019).

5 TILLING versus other reverse genetics tools in barley

5.1 TILLING versus other reverse genetics tools

Of course, TILLING is not the only reverse genetics tool available in barley. Large-scale insertional mutagenesis using transposons was repeatedly tempted, and provided some interesting results. For instance, a set of transposon insertional lines was developed by Zhao et al. (2016) based on the heterologous maize mobile elements system Ac/Ds. More recently, the behavior of Ds elements was characterized in detail, including their frequency and sites of insertion (Brown et al., 2015) and genes involved in malting quality were efficiently tagged using this system (Singh and Singh, 2017). However, when compared with TILLING, this approach suffers major shortcomings. First, the production of saturated mutant collections using heterologous transposons (Ac/Ds or others) appears currently difficult due to inherent low efficiency of several steps including genetic transformation and mobilization of transposons on barley chromosomes; second, DNA insertions mostly cause gene knockout mutations rather than a range of allele effects typically obtained in TILLING; third, barley lines carrying heterologous mobile elements must undergo GMO regulation in several countries.

Different from transposon-based mutagenesis, the bottleneck caused by the arduous barley *in-vitro* cell culturability and regenerability prevented genome-scale application of T-DNA insertional mutagenesis, which was instead implemented in other species such as *Arabidopsis* (Alonso et al., 2003), rice (Sallaud et al., 2004), and *Brachypodium* (Thole et al., 2010). Indeed, barley genetic transformation, although technically possible, is still implemented in a few labs because it is strongly dependent by the genetic background (the cultivar Golden Promise being still one of the few suitable for transformation) and by the availability of well-trained personnel (Kumlehn, 2016).

Gene functional analysis is being revolutionized by the advent of gene-editing protocols based on Clustered Regularly Interspaced Short Palindromic Repeats (CRISPR)-associated (Cas) protein (CRISPR/Cas9 system; Puchta, 2016; Yin et al., 2017). In its biotechnological version, CRISPR/Cas systems include an endonuclease (Cas9 is the most utilized) that cuts double-strand DNA, and a single guide RNA (sgRNA) that drives Cas9 to the target region. Mutations are introduced by DNA repair mechanisms ubiquitous in eukaryotic cells. The advantages of this approach over other methods, including TILLING, are multiple. It enables targeting modifications (mainly simple insertion/deletions) with nucleotide precision and at high efficiency; at the same time, the system is highly flexible in defining the target region given that target specificity is almost completely defined by the sgRNA nucleotide sequence. A related advantage over TILLING is the absence of background mutations in the edited lines (although this is waiting for final empirical confirmation). Such background mutations are instead always present in TILLING lines and their elimination requires some backcross generations. Additionally, and uniquely among reverse genetics methods, CRISPR/Cas9 enables targeting and modifying multiple related sequences contemporarily (i.e. in the same experiment and plant materials). Therefore, gene families and/or homoeologous genes in polyploids can now be more easily analyzed. The first CRISPR/Cas-edited barley plants were described by Lawrenson et al. (2015) and involved targeted mutagenesis of *HvPM19*, a gene coding for an ABA-inducible plasma membrane protein involved in seed dormancy. Since then, several applications have been described (Kapusi et al., 2017; Gasparis et al., 2018). Recently, the Cas endonuclease was modified by fusing with a cytidine deaminase which can perform C-to-T transition without double-strand breaks, improving the CRISPR-Cas system capacity to introduce single-point mutations of the same exact type of those typically induced in TILLING populations (Komor et al., 2016; Zong et al., 2017; Jacob et al., 2018). It is an easy prediction that CRISPR/Cas will play a major role in barley functional genomics in the years to come, as it has already been shown in model species.

5.2 Barley TILLING versus wheat TILLING

Given the narrow phylogenetic relationships of the main crop species (bread wheat, durum wheat, barley, and rye) belonging to Triticeae tribe of the Pooideae subfamily, it is interesting to ask whether functional genomic tools should be developed in most or all species or if tools available in one Triticeae species would significantly help gene function investigations in other species, making other resources useless. Several lines of evidence suggest that while gene function is overall conserved within Triticeae, a gene must be tested in the original species in order to draw final conclusions on its role. First, specific traits often exist or are relevant in one species only (e.g. endosperm-expressed genes which affect malting quality in barley and grain flower characteristics in wheat). Second, horthologuos genes and their functional networks may have sufficiently diverged to blur gene functional prediction across species. For instance, reduction of plant height in wheat commercial breeding is obtained by means of mutated *Reduced height-1* (*Rht1*) genes, which provide a limited response to the growth hormone gibberellin (GA), while the same trait in barley is mostly based on mutant versions of the GA biosynthetic gene *semidwarf1/denso/GA20ox2*, and of *uzu1.a*, a gene involved in brassinosteroid physiology (reviewed in Nadolska-Orczyk et al., 2017).

Another aspect is the difference in ploidy level between barley and wheat. TILLING can be applied to both diploid (i.e. barley) and polyploid (i.e. bread and durum wheat) species, although with important differences in efficiency. On one hand, TILLING populations in polyploid wheats have shown a much higher mutation rate (1/49 kb in tetraploid–durum wheat and 1/32 kb in hexaploids–bread wheat) than diploid species (1/380 kb in barley, on average), likely because of the favorable complementation effect due to the multiple homoeologous copies of the same gene (Tsai et al., 2013). This has the positive effect of reducing the number of plants (or plant pools) to screen. On the other hand, once a polyploid plant is found as carrying a mutation in the target gene, additional molecular and phenotyping work is required to verify gene effects independently from other homoeologous genes. This is not the case in diploid species such as barley, where phenotypes are visible right away. This is one of the reasons why TILLING resources in diploid wheats *Triticum monococcum* (Rawat et al., 2012) and *Aegilops* (*Triticum*) *tauschii* (Rawat et al., 2018), progenitors of A and D polyploid wheat genomes, respectively, were established.

6 Conclusion

TILLING has already shown its great potential for barley reverse genetics. Its popularity among barley geneticists lies in the availability of relatively simple and efficient artificial mutagenesis protocols, and the ease of propagation, description, and storing of barley mutant lines. In addition, because barley is a true diploid species cultivated as an inbred line, the phenotypic analysis of the identified mutants is simple. All this makes producing and managing barley TILLING resources suitable to small- or medium-size research groups.

TILLING is still technically evolving. Mutations detection based on PCR amplicon sequencing of DNA pools is becoming popular, while in the medium term, as sequencing costs will continue to drop, large permanent TILLING resources will likely move to TILLING *in-silico* where all plant genes will be preliminarily sequenced, indexed, and made searchable in online database.

In a future perspective, gene editing is expected to dominate reverse genetics. However, the still relatively high recalcitrance of barley to *in-vitro* tissue culture and genetic transformation, and the current controversial regulation of edited plants will likely keep TILLING among the most important tools for reverse genetics applications.

7 Where to look for further information

Recent high-level scientific review articles covering TILLING applications beyond barley are available (Tadele, 2016; Kumar et al., 2017) including a critical comparison between techniques using random (TILLING) or targeted (Genome editing) mutagenesis (Holme et al., 2019).

TILLING by sequencing resources are not available in barley, however a large resource of this kind, possibly informative for barley geneticists is available for wheat (Krasileva et al., 2017) and can be accessed at: http://www.wheat-tilling.com/and at https://dubcovskylab.ucdavis.edu/wheat-tilling.

A starting point to look for chemically induced barley mutant lines is the International Database for Barley Genes and Barley Genetic Stocks, at https://www.nordgen.org/bgs/.

Preliminarily characterized barley mutant lines in the Morex cv background can also be visualized and accessed at http://www.distagenomics.unibo.it/TILLMore/index.htm.

8 References

Abudayyeh, O. O., Gootenberg, J. S., Kellner, M. J. and Zhang, F. 2019. Nucleic acid detection of plant genes using CRISPR-Cas13. *CRISPR J.* 2, 165–71. doi:10.1089/crispr.2019.0011.

Acevedo-Garcia, J., Kusch, S. and Panstruga, R. 2014. Magical Mystery Tour: MLO proteins in plant immunity and beyond. *New Phytol.* 204(2), 273–81. doi:10.1111/nph.12889.

Acevedo-Garcia, J., Spencer, D., Thieron, H., Reinstädler, A., Hammond-Kosack, K., Phillips, A. L. and Panstruga, R. 2017. mlo-based powdery mildew resistance in hexaploid bread wheat generated by a non-transgenic TILLING approach. *Plant Biotechnol. J.* 15(3), 367–78. doi:10.1111/pbi.12631.

Alonso, J. M., Stepanova, A. N., Leisse, T. J., Kim, C. J., Chen, H., Shinn, P., Stevenson, D. K., Zimmerman, J., Barajas, P., Cheuk, R., Gadrinab, C., Heller, C., Jeske, A., Koesema, E., Meyers, C. C., Parker, H., Prednis, L., Ansari, Y., Choy, N., Deen, H., Geralt, M., Hazari, N., Hom, E., Karnes, M., Mulholland, C., Ndubaku, R., Schmidt, I., Guzman, P., Aguilar-Henonin, L., Schmid, M., Weigel, D., Carter, D. E., Marchand, T., Risseeuw, E., Brogden, D., Zeko, A., Crosby, W. L., Berry, C. C. and Ecker, J. R. 2003. Genome-wide insertional mutagenesis of *Arabidopsis thaliana*. *Science* 301(5633), 653–7. doi:10.1126/science.1086391.

Arenaz, P., Hallberg, L., Mancillas, F., Gutierrez, G. and Garcia, S. 1989. Sodium azide mutagenesis in mammals: inability of mammalian cells to convert azide to a mutagenic intermediate. *Mutat. Res.* 227(1), 63–7. doi:10.1016/0165-7992(89)90070-5.

Bentley, A., Maclennan, B., Calvo, J. and Dearolf, C. R. 2000. Targeted recovery of mutations in *Drosophila*. *Genetics* 156(3), 1169–73.

Brown, R. H., Singh, J., Singh, S., Dahleen, L. S., Lemaux, P. G., Stein, N., Mascher, M. and Bregitzer, P. 2015. Behavior of a modified Dissociation element in barley: a tool for genetic studies and for breeding transgenic barley. *Mol. Breed.* 35(3). doi:10.1007/s11032-015-0193-9.

Caldwell, D. G., McCallum, N., Shaw, P., Muehlbauer, G. J., Marshall, D. F. and Waugh, R. 2004. A structured mutant population for forward and reverse genetics in Barley (*Hordeum vulgare* L.). *Plant J.* 40(1), 143–50. doi:10.1111/j.1365-313X.2004.02190.x.

Chen, F., Dong, W., Zhang, J., Guo, X., Chen, J., Wang, Z., Lin, Z., Tang, H. and Zhang, L. 2018. The sequenced angiosperm genomes and genome databases. *Front. Plant Sci.* 9, 418. doi:10.3389/fpls.2018.00418.

Colbert, T., Till, B. J., Tompa, R., Reynolds, S., Steine, M. N., Yeung, A. T., McCallum, C. M., Comai, L. and Henikoff, S. 2001. High-throughput screening for induced point mutations. *Plant Physiol.* 126(2), 480–4. doi:10.1104/pp.126.2.480.

Comai, L., Young, K., Till, B. J., Reynolds, S. H., Greene, E. A., Codomo, C. A., Enns, L. C., Johnson, J. E., Burtner, C., Odden, A. R. and Henikoff, S. 2004. Efficient discovery of DNA polymorphisms in natural populations by Ecotilling. *Plant J.* 37(5), 778–86. doi:10.1111/j.0960-7412.2003.01999.x.

Daszkowska-Golec, A., Skubacz, A., Sitko, K., Słota, M., Kurowska, M. and Szarejko, I. 2018. Mutation in barley ERA1 (Enhanced Response to ABA1) gene confers better photosynthesis efficiency in response to drought as revealed by transcriptomic and physiological analysis. *Environ. Exp. Bot.* 148, 12–26. doi:10.1016/j.envexpbot.2018.01.003.

Doblado-Maldonado, A. F., Gomand, S. V., Goderis, B. and Delcour, J. A. 2017. Methodologies for producing amylose: a review. *Crit. Rev. Food Sci. Nutr.* 57(2), 407–17. doi:10.1080/10408398.2014.954030.

Dockter, C., Gruszka, D., Braumann, I., Druka, A., Druka, I., Franckowiak, J., Gough, S. P., Janeczko, A., Kurowska, M., Lundqvist, J., Lundqvist, U., Marzec, M., Matyszczak, I., Müller, A. H., Oklestkova, J., Schulz, B., Zakhrabekova, S. and Hansson, M. 2014. Induced variations in brassinosteroid genes define barley height and sturdiness, and expand the green revolution genetic toolkit. *Plant Physiol.* 166(4), 1912–27. doi:10.1104/pp.114.250738.

Dong, C., Vincent, K. and Sharp, P. 2009. Simultaneous mutation detection of three homoeologous genes in wheat by High Resolution Melting analysis and Mutation Surveyor. *BMC Plant Biol.* 9, 143. doi:10.1186/1471-2229-9-143.

Druka, A., Franckowiak, J., Lundqvist, U., Bonar, N., Alexander, J., Houston, K., Radovic, S., Shahinnia, F., Vendramin, V., Morgante, M., Stein, N. and Waugh, R. 2011. Genetic dissection of barley morphology and development. *Plant Physiol.* 155(2), 617–27. doi:10.1104/pp.110.166249.

Engvild, K. C. and Rasmussen, S. K. 2004. Root hair mutants of barley. *Barley Genet. Newsl.* 34, 13-5.

Fitzgerald, T. L., Powell, J. J., Stiller, J., Weese, T. L., Abe, T., Zhao, G., Jia, J., Mcintyre, C. L., Li, Z., Manners, J. M. and Kazan, K. 2015. An assessment of heavy ion irradiation mutagenesis for reverse genetics in wheat (*Triticum aestivum* L.). *PLoS ONE* 10(2), e0117369. doi:10.1371/journal.pone.0117369.

Gady, A. L., Hermans, F. W., Van De Wal, M. H., Van Loo, E. N., Visser, R. G. and Bachem, C. W. 2009. Implementation of two high through-put techniques in a novel application: detecting point mutations in large EMS mutated plant populations. *Plant Methods* 5, 13. doi:10.1186/1746-4811-5-13.

Gasparis, S., Kała, M., Przyborowski, M., Łyżnik, L. A., Orczyk, W. and Nadolska-Orczyk, A. 2018. A simple and efficient CRISPR/Cas9 platform for induction of single and multiple, heritable mutations in barley (*Hordeum vulgare* L.). *Plant Methods* 14, 111. doi:10.1186/s13007-018-0382-8.

Gichner, T. and Veleminsky, J. 1977. The very low mutagenic activity of sodium azide in *Arabidopsis thaliana*. *Biol. Plant.* 19(2), 153–5. doi:10.1007/BF02926758.

Gottwald, S., Bauer, P., Komatsuda, T., Lundqvist, U. and Stein, N. 2009. TILLING in the two-rowed barley cultivar 'Barke' reveals preferred sites of functional diversity in the gene HvHox1. *BMC Res. Notes* 2, 258. doi:10.1186/1756-0500-2-258.

Granier, F., Lemaire, A., Wang, Y., Lebris, P., Antelme, S., Vogel, J., Laudencia-Chingcuanco, D. and Sibout, R. 2015. *Chemical and Radiation Mutagenesis: Induction and Detection by Whole Genome Sequencing*. Springer, Cham.

Gruszka, D., Janeczko, A., Dziurka, M., Pociecha, E., Oklestkova, J. and Szarejko, I. 2016. Barley brassinosteroid mutants provide an insight into phytohormonal homeostasis in plant reaction to drought stress. *Front. Plant Sci.* 7, 1824. doi:10.3389/fpls.2016.01824.

Habben, J. E., Bao, X., Bate, N. J., DeBruin, J. L., Dolan, D., Hasegawa, D., Helentjaris, T. G., Lafitte, R. H., Lovan, N., Mo, H., Reimann, K. and Schussler, J. R. 2014. Transgenic alteration of ethylene biosynthesis increases grain yield in maize under field drought-stress conditions. *Plant Biotechnol. J.* 12(6), 685–93. doi:10.1111/pbi.12172.

Holme, I. B., Gregersen, P. L. and Brinch-Pedersen, H. 2019. Induced genetic variation in crop plants by random or targeted mutagenesis: convergence and differences. *Frontiers in Plant Science* 10, 1468.

Jacob, P., Avni, A. and Bendahmane, A. 2018. Translational research: exploring and creating genetic diversity. *Trends Plant Sci.* 23(1), 42–52. doi:10.1016/j.tplants.2017.10.002.

Jankowicz-Cieslak, J., Huynh, O. A., Brozynska, M., Nakitandwe, J. and Till, B. J. 2012. Induction, rapid fixation and retention of mutations in vegetatively propagated banana. *Plant Biotechnol. J.* 10(9), 1056–66. doi:10.1111/j.1467-7652.2012.00733.x.

Jiao, Y., Burke, J., Chopra, R., Burow, G., Chen, J., Wang, B., Hayes, C., Emendack, Y., Ware, D. and Xin, Z. 2016. A sorghum mutant resource as an efficient platform for gene discovery in grasses. *Plant Cell* 28(7), 1551–62. doi:org/10.1105/tpc.16.00373.

Jöst, M., Hensel, G., Kappel, C., Druka, A., Sicard, A., Hohmann, U., Beier, S., Himmelbach, A., Waugh, R., Kumlehn, J., Stein, N. and Lenhard, M. 2016a. The INDETERMINATE DOMAIN protein BROAD LEAF1 limits barley leaf width by restricting lateral proliferation. *Curr. Biol.* 26(7), 903–9. doi:10.1016/j.cub.2016.01.047.

Jost, M., Taketa, S., Mascher, M., Himmelbach, A., Yuo, T., Shahinnia, F., Rutten, T., Druka, A., Schmutzer, T., Steuernagel, B., Beier, S., Taudien, S., Scholz, U., Morgante, M., Waugh, R. and Stein, N. 2016b. A homolog of blade-on-petiole 1 and 2 (BOP1/2) controls internode length and homeotic changes of the barley inflorescence. *Plant Physiol.* 171(2), 1113–27. doi:10.1104/pp.16.00124.

Kanyuka, K., Druka, A., Caldwell, D. G., Tymon, A., McCallum, N., Waugh, R. and Adams, M. J. 2005. Evidence that the recessive Bymovirus resistance locus rym4 in barley corresponds to the eukaryotic translation initiation factor 4E gene. *Mol. Plant Pathol.* 6(4), 449–58. doi:10.1111/j.1364-3703.2005.00294.x.

Kapusi, E., Corcuera-Gómez, M., Melnik, S. and Stoger, E. 2017. Heritable genomic fragment deletions and small indels in the putative ENGase gene induced by CRISPR/Cas9 in barley. *Front. Plant Sci.* 8, 540. doi:10.3389/fpls.2017.00540.

Karimi-Ashtiyani, R., Ishii, T., Niessen, M., Stein, N., Heckmann, S., Gurushidze, M., Banaei-Moghaddam, A. M., Fuchs, J., Schubert, V., Koch, K., Weiss, O., Demidov, D., Schmidt, K., Kumlehn, J. and Houben, A. 2015. Point mutation impairs centromeric CENH3 loading and induces haploid plants. *Proc. Natl Acad. Sci. U. S. A.* 112(36), 11211–6. doi:10.1073/pnas.1504333112.

Kim, Y., Schumaker, K. S. and Zhu, J. K. 2006. EMS mutagenesis of *Arabidopsis*. *Methods Mol. Biol.* 323, 101–3. doi:10.1385/1-59745-003-0:101.

Kohata, R., Koitabashi, K., Kitashiba, H. and Nishio, T. 2018. Sensitive mutant detection by concentrating mutant DNA with allele-specific capture and its application to analysis of contaminated grains in rice. *Plant Cell Rep.* 37(6), 865–72. doi:10.1007/s00299-018-2274-9.

Komor, A. C., Kim, Y. B., Packer, M. S., Zuris, J. A. and Liu, D. R. 2016. Programmable editing of a target base in genomic DNA without double-stranded DNA cleavage. *Nature* 533(7603), 420–4. doi:10.1038/nature17946.

Krasileva, K. V., Vasquez-Gross, H. A., Howell, T., Bailey, P., Paraiso, F., Clissold, L., Simmonds, J., Ramirez-Gonzalez, R. H., Wang, X., Borrill, P., Fosker, C., Ayling, S., Phillips, A. L., Uauy, C. and Dubcovsky, J. 2017. Uncovering hidden variation in polyploid wheat. *Proc. Natl Acad. Sci. U. S. A.* 114(6), E913–21. doi:10.1073/pnas.1619268114.

Kumar, A. P. K., McKeown, P. C., Boualem, A., Ryder, P., Brychkova, G., Bendahmane, A., Sarkar, A., Chatterjee, M. and Spillane, C. 2017. TILLING by sequencing (TbyS) for targeted genome mutagenesis in crops. *Mol. Breed.* 37(2). doi:10.1007/s11032-017-0620-1.

Kumlehn, J. 2016. Embryogenesis and plant regeneration from isolated wheat zygotes. *Methods Mol. Biol.* 1359, 503–14. doi:10.1007/978-1-4939-3061-6_29.

Kurowska, M., Labocha-Pawłowska, A., Gnizda, D., Maluszynski, M. and Szarejko, I. 2012. Molecular analysis of point mutations in a barley genome exposed to MNU and gamma rays. *Mutat. Res.* 738–739, 52–70. doi:10.1016/j.mrfmmm.2012.08.008.

Lababidi, S., Mejlhede, N., Rasmussen, S. K., Backes, G., Al-Said, W., Baum, M. and Jahoor, A. 2009. Identification of barley mutants in the cultivar 'Lux' at the Dhn loci through TILLING. *Plant Breed.* 128(4), 332–6. doi:10.1111/j.1439-0523.2009.01640.x.

Landau, A., Lencina, F., Pacheco, M. G. and Prina, A. R. 2016. Plastome mutations and recombination events in barley chloroplast mutator seedlings. *J. Hered.* 107(3), 266–73. doi:10.1093/jhered/esw003.

Lawrenson, T., Shorinola, O., Stacey, N., Li, C., Østergaard, L., Patron, N., Uauy, C. and Harwood, W. 2015. Induction of targeted, heritable mutations in barley and *Brassica oleracea* using RNA-guided Cas9 nuclease. *Genome Biol.* 16, 258. doi:10.1186/s13059-015-0826-7.

Li, M., Hensel, G., Mascher, M., Melzer, M., Budhagatapalli, N., Rutten, T., Himmelbach, A., Beier, S., Korzun, V., Kumlehn, J., Börner, T. and Stein, N. 2019. Leaf variegation and impaired chloroplast development caused by a truncated CCT domain gene in albostrians Barley. *Plant Cell* 31(7), 1430–45. doi:10.1105/tpc.19.00132.

López-González, C., Juárez-Colunga, S., Morales-Elías, N. C. and Tiessen, A. 2019. Exploring regulatory networks in plants: transcription factors of starch metabolism. *Peer J.* 7, e6841. doi:10.7717/peerj.6841.

Lundqvist, U. 2014. Scandinavian mutation research in barley – a historical review. *Hereditas* 151(6), 123–31. doi:10.1111/hrd2.00077.

Lundqvist, U., Meyer, J. and Lundqvist, A. 1991. Mutagen specificity for 71 lines resistant to barley powdery mildew race-D1 and isolated in 4 highbred barley varieties. *Hereditas* 115(3), 227–39. doi:10.1111/j.1601-5223.1992.tb00566.x.

Maluszynski, M. and Szarejko, I. 2003. Induced mutations in the green and gene revolutions. In: Tuberosa, R., Philips, R. L. and Gale, M. (Eds), *In the Wake of the Double Helix: From the Green Revolution to the Gene Revolution*, 27–31 May 2003, Bologna, Italy. Avenue media, Bologna, Italy, pp. 403–25.

Martin, B., Ramiro, M., Martinez-Zapater, J. M. and Alonso-Blanco, C. 2009. A high-density collection of EMS-induced mutations for TILLING in Landsberg erecta genetic background of *Arabidopsis*. *BMC Plant Biol.* 9, 147. doi:10.1186/1471-2229-9-147.

Marzec, M., Gruszka, D., Tylec, P. and Szarejko, I. 2016. Identification and functional analysis of the HvD14 gene involved in strigolactone signaling in Hordeum vulgare. *Physiol. Plant.* 158(3), 341–55. doi:10.1111/ppl.12460.

Mascher, M., Jost, M., Kuon, J. E., Himmelbach, A., Aßfalg, A., Beier, S., Scholz, U., Graner, A. and Stein, N. 2014. Mapping-by-sequencing accelerates forward genetics in barley. *Genome Biol.* 15(6), R78. doi:10.1186/gb-2014-15-6-r78.

Mascher, M., Gundlach, H., Himmelbach, A., Beier, S., Twardziok, S. O., Wicker, T., Radchuk, V., Dockter, C., Hedley, P. E., Russell, J., Bayer, M., Ramsay, L., Liu, H., Haberer, G., Zhang, X. Q., Zhang, Q., Barrero, R. A., Li, L., Taudien, S., Groth, M., Felder, M., Hastie, A., Šimková, H., Staňková, H., Vrána, J., Chan, S., Muñoz-Amatriaín, M., Ounit, R., Wanamaker, S., Bolser, D., Colmsee, C., Schmutzer, T., Aliyeva-Schnorr, L., Grasso, S., Tanskanen, J., Chailyan, A., Sampath, D., Heavens, D., Clissold, L., Cao, S., Chapman, B., Dai, F., Han, Y., Li, H., Li, X., Lin, C., Mccooke, J. K., Tan, C., Wang, P., Wang, S., Yin, S., Zhou, G., Poland, J. A., Bellgard, M. I., Borisjuk, L., Houben, A., Doležel, J., Ayling, S., Lonardi, S., Kersey, P., Langridge, P., Muehlbauer, G. J., Clark, M. D., Caccamo, M., Schulman, A. H., Mayer, K. F. X., Platzer, M., Close, T. J., Scholz, U., Hansson, M., Zhang, G., Braumann, I., Spannagl, M., Li, C., Waugh, R. and Stein, N. 2017. A chromosome conformation capture ordered sequence of the barley genome. *Nature* 544(7651), 427–33. doi:10.1038/nature22043.

McCallum, C. M., Comai, L., Greene, E. A. and Henikoff, S. 2000. Targeted screening for induced mutations. *Nat. Biotechnol.* 18(4), 455–7. doi:10.1038/74542.

Mendiondo, G. M., Gibbs, D. J., Szurman-Zubrzycka, M., Korn, A., Marquez, J., Szarejko, I., Maluszynski, M., King, J., Axcell, B., Smart, K., Corbineau, F. and Holdsworth, M. J. 2016. Enhanced waterlogging tolerance in barley by manipulation of expression of the N-end rule pathway E3 ligase Proteolysis6. *Plant Biotechnol. J.* 14(1), 40–50. doi:10.1111/pbi.12334.

Morita, R., Kusaba, M., Iida, S., Yamaguchi, H., Nishio, T. and Nishimura, M. 2009. Molecular characterization of mutations induced by gamma irradiation in rice. *Genes Genet. Syst.* 84(5), 361–70. doi:10.1266/ggs.84.361.

Nadolska-Orczyk, A., Rajchel, I. K., Orczyk, W. and Gasparis, S. 2017. Major genes determining yield-related traits in wheat and barley. *Theor. Appl. Genet.* 130(6), 1081–98. doi:10.1007/s00122-017-2880-x.

Nakamura, S., Pourkheirandish, M., Morishige, H., Kubo, Y., Nakamura, M., Ichimura, K., Seo, S., Kanamori, H., Wu, J., Ando, T., Hensel, G., Sameri, M., Stein, N., Sato, K., Matsumoto, T., Yano, M. and Komatsuda, T. 2016. Mitogen-activated protein kinase kinase 3 regulates seed dormancy in barley. *Curr. Biol.* 26(6), 775–81. doi:10.1016/j.cub.2016.01.024.

Nilan, R. A., Sideris, E. G., Kleinhofs, A., Sander, C. and Konzak, C. F. 1973. Azide – a potent mutagen. *Mutat. Res.* 17, 142–4.

Owais, W. M. and Kleinhofs, A. 1988. Metabolic activation of the mutagen azide in biological systems. *Mutat. Res.* 197(2), 313–23. doi:10.1016/0027-5107(88)90101-7.

Pourkheirandish, M., Hensel, G., Kilian, B., Senthil, N., Chen, G., Sameri, M., Azhaguvel, P. and Sakuma, S., Dhanagond, S., Sharma, R., Mascher, M., Himmelbach, A., Gottwald, S., Nair, S. K., Tagiri, A., Yukuhiro, F., Nagamura, Y., Kanamori, H., Matsumoto, T., Willcox, G., Middleton, C. P., Wicker, T., Walther, A., Waugh, R., Fincher, G. B., Stein,

N., Kumlehn, J., Sato, K. and Komatsuda, T. 2015. Evolution of the grain dispersal system in barley. *Cell* 162, 527-39.

Poursarebani, N., Seidensticker, T., Koppolu, R., Trautewig, C., Gawroński, P., Bini, F., Govind, G., Rutten, T., Sakuma, S., Tagiri, A., Wolde, G. M., Youssef, H. M., Battal, A., Ciannamea, S., Fusca, T., Nussbaumer, T., Pozzi, C., Börner, A., Lundqvist, U., Komatsuda, T., Salvi, S., Tuberosa, R., Uauy, C., Sreenivasulu, N., Rossini, L. and Schnurbusch, T. 2015. The genetic basis of composite spike form in barley and 'miracle-wheat'. *Genetics* 201(1), 155-65. doi:10.1534/genetics.115.176628.

Puchta, H. 2016. Using CRISPR/Cas in three dimensions: towards synthetic plant genomes, transcriptomes and epigenomes. *Plant J.* 87(1), 5-15. doi:10.1111/tpj.13100.

Raghavan, C., Naredo, M. E. B., Wang, H., Atienza, G., Liu, B., Qiu, F., Mcnally, K. L. and Leung, H. 2007. Rapid method for detecting SNPs on agarose gels and its application in candidate gene mapping. *Mol. Breed.* 19(2), 87-101. doi:10.1007/s11032-006-9046-x.

Rawat, N., Sehgal, S. K., Joshi, A., Rothe, N., Wilson, D. L., Mcgraw, N., Vadlani, P. V., Li, W. and Gill, B. S. 2012. A diploid wheat TILLING resource for wheat functional genomics. *BMC Plant Biol.* 12, 205. doi:10.1186/1471-2229-12-205.

Rawat, N., Schoen, A., Singh, L., Mahlandt, A., Wilson, D. L., Liu, S., Lin, G., Gill, B. S. and Tiwari, V. K. 2018. TILL-D: an *Aegilops tauschii* TILLING resource for wheat improvement. *Front. Plant Sci.* 9, 1665. doi:10.3389/fpls.2018.01665.

Sallaud, C., Gay, C., Larmande, P., Bès, M., Piffanelli, P., Piégu, B., Droc, G., Regad, F., Bourgeois, E., Meynard, D., Périn, C., Sabau, X., Ghesquière, A., Glaszmann, J. C., Delseny, M. and Guiderdoni, E. 2004. High throughput T-DNA insertion mutagenesis in rice: a first step towards in silico reverse genetics. *Plant J.* 39(3), 450-64. doi:10.1111/j.1365-313X.2004.02145.x.

Salvi, S., Piazza, S., Predieri, S., Fuochi, P., Velasco, R. and Malnoy, M. 2015. High frequency of chromosome deletions in regenerated and mutagenized apple (*Malus x domestica* Borkh.) seedlings. *Mol. Breed.* 35(1). doi:10.1007/s11032-015-0199-3.

Sato, Y., Shirasawa, K., Takahashi, Y., Nishimura, M. and Nishio, T. 2006. Mutant selection from progeny of gamma-ray-irradiated rice by DNA heteroduplex cleavage using Brassica petiole extract. *Breed. Sci.* 56(2), 179-83. doi:10.1270/jsbbs.56.179.

Schreiber, M., Barakate, A., Uzrek, N., Macaulay, M., Sourdille, A., Morris, J., Hedley, P. E., Ramsay, L. and Waugh, R. 2019. A highly mutagenised barley (cv. Golden Promise) TILLING population coupled with strategies for screening-by-sequencing. *Plant Methods* 15, 99. doi:10.1186/s13007-019-0486-9.

Sharma, A., Yadav, B. S. and Ritika 2008. Resistant starch: physiological roles and food applications. *Food Rev. Int.* 24(2), 193-234. doi:10.1080/87559120801926237.

Singh, S. and Singh, J. 2017. Sequence-based genetic mapping of Ds-tagged insertions to characterize malting-related traits in barley. *Crop J.* 5(1), 11-20. doi:10.1016/j.cj.2016.07.003.

Slade, A. J., Fuerstenberg, S. I., Loeffler, D., Steine, M. N. and Facciotti, D. 2005. A reverse genetic, nontransgenic approach to wheat crop improvement by TILLING. *Nat. Biotechnol.* 23(1), 75-81. doi:10.1038/nbt1043.

Slamenová, D. and Gabelová, A. 1980. The effects of sodium azide on mammalian cells cultivated in vitro. *Mutat. Res.* 71(2), 253-61. doi:10.1016/0027-5107(80)90077-9.

Sparla, F., Falini, G., Botticella, E., Pirone, C., Talamè, V., Bovina, R., Salvi, S., Tuberosa, R., Sestili, F. and Trost, P. 2014. New starch phenotypes produced by TILLING in barley. *PLoS ONE* 9(10), e107779. doi:10.1371/journal.pone.0107779.

Stolarek, M., Gruszka, D., Braszewska-Zalewska, A. and Maluszynski, M. 2015a. Alleles of newly identified barley gene HvPARP3 exhibit changes in efficiency of DNA repair. *DNA Repair (Amst)* 28, 116–30. doi:10.1016/j.dnarep.2015.02.018.

Stolarek, M., Gruszka, D., Braszewska-Zalewska, A. and Maluszynski, M. 2015b. Functional analysis of the new barley gene HvKu80 indicates that it plays a key role in double-strand DNA break repair and telomere length regulation. *Mutagenesis* 30(6), 785–97. doi:10.1093/mutage/gev033.

Suzuki, T., Eiguchi, M., Kumamaru, T., Satoh, H., Matsusaka, H., Moriguchi, K., Nagato, Y. and Kurata, N. 2008. MNU-induced mutant pools and high performance TILLING enable finding of any gene mutation in rice. *Mol. Genet. Genomics* 279(3), 213–23. doi:10.1007/s00438-007-0293-2.

Szarejko, I. and Maluszynski, M. 1980. Translocations and inversions in barley induced by fast neutrons and N-nitroso-N-methylurea. *Barley Genet. Newsl.* 10, 67–9.

Szarejko, I., Szurman-Zubrzycka, M., Nawrot, M., Marzec, M., Gruszka, D., Kurowska, M., Chmielewska, B., Zbieszczyk, J., Jelonek, J. and Maluszynski, M. 2017. Creation of a TILLING population in barley after chemical mutagenesis with sodium azide and MNU. In: Jankowicz-Cieslak, J., Tai, Th. H., Kumlehn, J. and Till, B. J. (Eds), *Biotechnologies for Plant Mutation Breeding*. Springer International Publishing, Cham, pp. 91–111.

Szurman-Zubrzycka, M. E., Zbieszczyk, J., Marzec, M., Jelonek, J., Chmielewska, B., Kurowska, M. M., Krok, M., Daszkowska-Golec, A., Guzy-Wrobelska, J., Gruszka, D., Gajecka, M., Gajewska, P., Stolarek, M., Tylec, P., Sega, P., Lip, S., Kudełko, M., Lorek, M., Gorniak-Walas, M., Malolepszy, A., Podsiadlo, N., Szyrajew, K. P., Keisa, A., Mbambo, Z., Todorowska, E., Gaj, M., Nita, Z., Orlowska-Job, W., Maluszynski, M. and Szarejko, I. 2018. TILLUS-a rich and renewable source of induced mutations for forward/reverse genetics and pre-breeding programs in barley (*Hordeum vulgare* L.). *Front. Plant Sci.* 9, 216. doi:10.3389/fpls.2018.00216.

Szurman-Zubrzycka, M., Baran, B., Stolarek-Januszkiewicz, M., Kwaśniewska, J., Szarejko, I. and Gruszka, D. 2019. The *dmc1* mutant allows an insight into the DNA double-strand break repair During meiosis in barley (*Hordeum vulgare* L.). *Front. Plant Sci.* 10, 761. doi:10.3389/fpls.2019.00761.

Tadele, Z. 2016. Mutagenesis and TILLING to dissect gene function in plants. *Curr. Genomics* 17(6), 499–508.

Talamè, V., Bovina, R., Sanguineti, M. C., Tuberosa, R., Lundqvist, U. and Salvi, S. 2008. TILLMore, a resource for the discovery of chemically induced mutants in barley. *Plant Biotechnol. J.* 6(5), 477–85. doi:10.1111/j.1467-7652.2008.00341.x.

Talamè, V., Bovina, R., Salvi, S., Sanguineti, M., Piffanelli, P., Lundquist, U. and Tuberosa, R. 2009. *TILLING with TILLMore. Induced Plant Mutations in the Genomics Era*. Food and Agriculture Organization of the United Nations, Rome, pp. 240–2.

Thole, V., Worland, B., Wright, J., Bevan, M. W. and Vain, P. 2010. Distribution and characterization of more than 1000 T-DNA tags in the genome of Brachypodium distachyon community standard line Bd21. *Plant Biotechnol. J.* 8(6), 734–47. doi:10.1111/j.1467-7652.2010.00518.x.

Till, B. J., Colbert, T., Tompa, R., Enns, L. C., Codomo, C. A., Johnson, J. E., Reynolds, S. H., Henikoff, J. G., Greene, E. A., Steine, M. N., Comai, L. and Henikoff, S. 2003. High-throughput Tilling for functional genomics. *Methods Mol. Biol.* 236, 205–20. doi:10.1385/1-59259-413-1:205.

Till, B. J., Burtner, C., Comai, L. and Henikoff, S. 2004a. Mismatch cleavage by single-strand specific nucleases. *Nucleic Acids Res.* 32(8), 2632–41. doi:10.1093/nar/gkh599.

Till, B. J., Reynolds, S. H., Weil, C., Springer, N., Burtner, C., Young, K., Bowers, E., Codomo, C. A., Enns, L. C., Odden, A. R., Greene, E. A., Comai, L. and Henikoff, S. 2004b. Discovery of induced point mutations in maize genes by TILLING. *BMC Plant Biol.* 4, 12. doi:10.1186/1471-2229-4-12.

Till, B. J., Cooper, J., Tai, T. H., Colowit, P., Greene, E. A., Henikoff, S. and Comai, L. 2007. Discovery of chemically induced mutations in rice by TILLING. *BMC Plant Biol.* 7, 19. doi:10.1186/1471-2229-7-19.

Tramontano, A., Jarc, L., Jankowicz-Cieslak, J., Hofinger, B. J., Gajek, K., Szurman-Zubrzycka, M., Szarejko, I., Ingelbrecht, I. and Till, B. J. 2019. Fragmentation of pooled PCR products for highly multiplexed TILLING. *G3 (Bethesda)* 9(8), 2657–66. doi:10.1534/g3.119.400301.

Triques, K., Sturbois, B., Gallais, S., Dalmais, M., Chauvin, S., Clepet, C., Aubourg, S., Rameau, C., Caboche, M. and Bendahmane, A. 2007. Characterization of Arabidopsis thaliana mismatch specific endonucleases: application to mutation discovery by TILLING in pea. *Plant J.* 51(6), 1116–25. doi:10.1111/j.1365-313X.2007.03201.x.

Tsai, H., Howell, T., Nitcher, R., Missirian, V., Watson, B., Ngo, K. J., Lieberman, M., Fass, J., Uauy, C., Tran, R. K., Khan, A. A., Filkov, V., Tai, T. H., Dubcovsky, J. and Comai, L. 2011. Discovery of rare mutations in populations: TILLING by sequencing. *Plant Physiol.* 156(3), 1257–68. doi:10.1104/pp.110.169748.

Tsai, H., Missirian, V., Ngo, K. J., Tran, R. K., Chan, S. R., Sundaresan, V. and Comai, L. 2013. Production of a high-efficiency TILLING population through polyploidization. *Plant Physiol.* 161(4), 1604–14. doi:10.1104/pp.112.213256.

Tsuda, K. and Somssich, I. E. 2015. Transcriptional networks in plant immunity. *New Phytol.* 206(3), 932–47. doi:10.1111/nph.13286.

Vishwakarma, K., Upadhyay, N., Kumar, N., Yadav, G., Singh, J., Mishra, R. K., Kumar, V., Verma, R., Upadhyay, R. G., Pandey, M. and Sharma, S. 2017. Abscisic acid signaling and abiotic stress tolerance in plants: a review on current knowledge and future prospects. *Front. Plant Sci.* 8, 161. doi:10.3389/fpls.2017.00161.

Wang, T. L., Uauy, C., Robson, F. and Till, B. 2012. TILLING in extremis. *Plant Biotechnol. J.* 10(7), 761–72. doi:10.1111/j.1467-7652.2012.00708.x.

Wang, Y., Cheng, X., Shan, Q., Zhang, Y., Liu, J., Gao, C. and Qiu, J. L. 2014. Simultaneous editing of three homoeoalleles in hexaploid bread wheat confers heritable resistance to powdery mildew. *Nat. Biotechnol.* 32(9), 947–51. doi:10.1038/nbt.2969.

Wang, F., Wang, C., Liu, P., Lei, C., Hao, W., Gao, Y., Liu, Y. G. and Zhao, K. 2016. Enhanced rice Blast Resistance by CRISPR/Cas9-targeted mutagenesis of the ERF transcription factor gene OsERF922. *PLoS ONE* 11(4), e0154027. doi:10.1371/journal.pone.0154027.

Weil, C. F. and Monde, R. 2007. Getting the point – mutations in maize. *Crop Sci.* 47(Suppl. 1), S60–7. doi:10.2135/cropsci2006.09.0563tpg.

Xia, J., Zhu, D., Wang, R., Cui, Y. and Yan, Y. 2018. Crop resistant starch and genetic improvement: a review of recent advances. *Theor. Appl. Genet.* 131(12), 2495–511. doi:10.1007/s00122-018-3221-4.

Yang, P., Lüpken, T., Habekuss, A., Hensel, G., Steuernagel, B., Kilian, B., Ariyadasa, R., Himmelbach, A., Kumlehn, J., Scholz, U., Ordon, F. and Stein, N. 2014. Protein disulfide isomerase LIKE 5-1 is a susceptibility factor to plant viruses. *Proc. Natl Acad. Sci. U. S. A.* 111(6), 2104–9. doi:10.1073/pnas.1320362111.

Yang, P., Habekuß, A., Hofinger, B. J., Kanyuka, K., Kilian, B., Graner, A., Ordon, F. and Stein, N. 2017. Sequence diversification in recessive alleles of two host factor genes suggests adaptive selection for Bymovirus resistance in cultivated barley from East Asia. *Theor. Appl. Genet.* 130(2), 331–44. doi:10.1007/s00122-016-2814-z.
Yin, K., Gao, C. and Qiu, J. L. 2017. Progress and prospects in plant genome editing. *Nat. Plants* 3, 17107. doi:10.1038/nplants.2017.107.
Yoshihara, R., Hase, Y., Sato, R., Takimoto, K. and Narumi, I. 2010. Mutational effects of different LET radiations in rpsL transgenic *Arabidopsis*. *Int. J. Radiat. Biol.* 86(2), 125–31. doi:10.3109/09553000903336826.
Yu, Y., Yang, D., Zhou, S., Gu, J., Wang, F., Dong, J. and Huang, R. 2017. The ethylene response factor OsERF109 negatively affects ethylene biosynthesis and drought tolerance in rice. *Protoplasma* 254(1), 401–8. doi:10.1007/s00709-016-0960-4.
Zhao, D., Ferguson, A. A. and Jiang, N. 2016. What makes up plant genomes: the vanishing line between transposable elements and genes. *Biochim. Biophys. Acta* 1859(2), 366–80. doi:10.1016/j.bbagrm.2015.12.005.
Zong, Y., Wang, Y., Li, C., Zhang, R., Chen, K., Ran, Y., Qiu, J. L., Wang, D. and Gao, C. 2017. Precise base editing in rice, wheat and maize with a Cas9-cytidine deaminase fusion. *Nat. Biotechnol.* 35(5), 438–40. doi:10.1038/nbt.3811.

Part 3

Cultivation techniques, pest and disease management

Chapter 9

Advances in postharvest storage and handling of barley: methods to prevent or reduce mycotoxin contamination

Zhao Jin and Paul Schwarz, North Dakota State University, USA

1 Introduction
2 Postharvest handling and storage operations for barley
3 Mycoflora and mycotoxins of barley
4 Prevention or decontamination of mycotoxins in barley storage
5 Post-storage treatment of barley
6 Conclusion and future trends
7 Where to look for further information
8 References

1 Introduction

Barley ranks fourth in terms of global production of cereal grains. Livestock feed is the major use of barley, but feed use has declined in many regions over the years largely due to the lower prices and availability of other feed stuffs. Overall, the food use of barley has also declined over the past 50 years. It does, however, remain an important staple in some regions. Perhaps, the best-known and major value-added use of barley is for the production of malt, which is used as an ingredient in beer, distilled spirits, and some baked products (Izydorczyk and Edney 2017). Malting barley has the strictest quality requirements, and is one of the only uses of cereals, where the grain must be maintained in a living state. Rapid growth of the craft beer and spirit industries is leading a renewed interest in barley, often in new areas or areas that have not produced barley for many years.

As implied by the title of this book, sustainability is becoming increasingly important for barley growers, processors, and researchers. Barley is also important in terms of global food security. Losses, or limitations, in barley production can be caused by abiotic stress factors, disease, poor management or production practices, and by pests. While specific data on barley is limited,

http://dx.doi.org/10.19103/AS.2019.0060.15

it has been estimated that 11–15% of global wheat, rice, and maize production is lost to disease on an annual basis (Pinstrup-Andersen 2001). However, postharvest losses are also important, and these generally occur in storage. Grain losses vary considerably, and can be as high as 50% in less developed countries under poor conditions of storage, but have been reported to be only 1–2% in more developed countries (Jayas 2012). Nevertheless, a 1% loss is still of great significance when one considers that it is millions of metric tons. Losses can be categorized in terms of reductions in weight, quality, nutrition, seed viability, and monetary value (Semple et al. 1992). Insect infestation and fungal growth with the production of mycotoxins are two of the major causes of losses in storage (Kumar and Kalita 2017). Mycotoxins are perhaps the largest threat to the food safety of cereals. Barley, like other cereals, is exposed to microflora from several sources, including the field environment, harvest, postharvest storage and transport. Mycotoxins can be present at the time of harvest, but postharvest storage conditions are a critical factor which can influence the viability and growth of fungi, and potentially the production of additional mycotoxins. In addition, mycotoxins can be produced under some processing operations, such as malting. Finally, one needs to consider that germination is the most important quality requirement for malting barley and seed. Postharvest drying and storage also have a major influence on the maintenance of grain viability and vigor.

The subject of grain storage, and more specifically the storage of malting barley, has been covered by numerous publications. This chapter focuses on postharvest storage and handling practices of barley that can be used to mitigate mycotoxin issues.

2 Postharvest handling and storage operations for barley

In most cases, barley is not consumed or processed immediately after harvesting, and the harvested grain is usually stored for a period of time either on-farm or in commercial granaries. Maltsters generally maintain a certain amount of stored barley, as a buffer against poor crop years. The prevalence of on-farm versus commercial storage of barley can vary considerably between countries, and even between regions of the same country. As an example, malt industry contracts in the upper Midwest of the United States generally have clauses for a certain period of on-farm storage. In all cases, maintenance of quality, and in particular for malting barley, overcoming dormancy, and preservation of germination are of major concern. Protection from the elements and pests, as well as the prevention of microbial spoilage, is also a critical factor. Barley is often stored in dedicated bins with well-managed drying, aeration, and monitoring, but in some cases may be stored in less than ideal conditions, such as unprotected bags and piles (Hellevang 2016; Jayas 2012). Given the importance of drying and storage to malting barley quality, guidelines and recommendations have been published by many academic, governmental, or industry organizations throughout the world. Several examples are listed in Table 1. A particularly rich source of information on the various aspects of barley drying and storage has been prepared by Dr. Kenneth Hellevang of the Department of Agricultural and Biosystems Engineering at North Dakota State University (USA) (Hellevang 2019). Fleurat-Lessard (2017) recently published a review on the integrated management of risks for stored grain. The Home-Grown Cereals Authority (HGCA) has provided an excellent guide on a HACCP approach to grain storage (HGCA 2011).

Goals of storage are to maintain seed viability and quality, and to prevent loss through insect and/or microbial damage. In general, recommendations are similar across regions, with clean and sound grain at the proper moisture content being key concerns as grain enters storage. However, the ambient climatic conditions of a region can have a significant impact on how the grain is managed once in storage.

2.1 General considerations for barley entering storage

The cleanliness and moisture content of incoming harvested grain are immediate concerns when discussing storage. Foreign materials can be a source of both fungal and insect contamination. Skinned, broken, and cracked kernels, as well as immature grains, spoil easier than those that are sound (HGCA 2011; Association American Malting Barley 2016). Immature kernels lose germination more rapidly. As such, efforts should be made to remove these materials prior to storage. While grain cleaning equipment is a fundamental component in commercial grain handling and storage, the availability at the farm level can vary considerably. On the farm, a grain distributer should be used when filling a storage bin, especially when cleaning is not possible. This is to prevent the lighter foreign materials from accumulating non-uniformly at the bin center (Hellevang 2014).

Cleanliness of the storage facility is also important, as residual grain and residue in an empty bin or equipment can serve as a reservoir for insect contamination, and therefore should be thoroughly cleaned prior to new storage. The area surrounding the grain storage should also be clean. If equipment or the bin has been previously infested, they can be treated with an approved fumigant or insecticide.

Grain moisture, along with ambient temperature and relative humidity, is a key factor in maintaining viability and preventing spoilage during the storage. Higher temperatures, moisture, and relative humidity favor the growth of both insects and fungi, as well as contamination with mycotoxins (HGCA 2011; Hellevang 2014). Optimal conditions for fungal growth and insect infestation

Table 1 Examples of resources available on the drying and storage of grains, including malting barley

Country	Organization	Title	Link	Reference
Australia	Grain Research and Development Cooperation	Barley. Section 13. Storage.	https://grdc.com.au/__data/assets/pdf_file/0024/369240/GrowNote-Barley-South-13-Storage.pdf	GRDC (2016)
Canada	Canadian Grain Commission	Manage stored grain.	https://www.grainscanada.gc.ca/en/grain-quality/manage/	CGC (2019a)
United Kingdom	Home Grown Cereal Authority (HGCA)	Grain storage guide for cereals and oilseeds, 3rd ed.	https://cereals.ahdb.org.uk/publications/2011/october/28/grain-storage-guide-for-cereals-and-oilseeds.aspx	HGCA (2011)
United States	American Malting Barley Association	Harvesting, drying, and storing malting barley to get market premiums.	http://ambainc.org/media/AMBA_PDFs/Pubs/Production/Harvesting.pdf	AMBA (2016)
	North Dakota State University	Video presentation. Malt barley drying and storage.	https://www.ag.ndsu.edu/graindrying/presentations-2	Hellevang (2014)
	North Dakota State University	Crop storage management.	https://www.ag.ndsu.edu/graindrying/documents/ae791.pdf	Hellevang (1996)
	North Dakota State University	Grain drying.	https://www.ag.ndsu.edu/graindrying/publications/ae-701-grain-drying	Hellevang (2013)
	North Dakota State University	Estimated allowable storage time for malting barley.	https://www.ag.ndsu.edu/graindrying/documents/BarleyAST.pdf	Hellevang (1990a)

are in the range of about 27°C, 70% relative humidity, and grain moisture above 14.5%. Similarly, one can consider the water activity (a_w) of the grain, which is closely related to fungal growth and metabolism (Fleurat-Lessard 2017). Spore germination and fungal growth are generally inhibited below the threshold of 0.65 a_w. Insect reproduction, as well as fungal growth, stops or slows greatly at a relative humidity of below 65% (HGCA 2011; Hellevang 2014). Insects are killed at temperatures below −1.1°C following a few weeks (Hellevang 2014), while fungi can survive. Usually, 12% is recommended as the maximum moisture content for long-term aerated storage of clean and sound grains (Hellevang 2013). Undried grain with a moisture content of more than 15% should not be stored for more than a week.

Grain that is stored at higher moisture is subject to loss of viability. The estimated allowable storage times for maintaining germination in malting barley, based upon grain moisture and temperature, are listed in Table 2. As an example, when grain at 15% moisture is stored at 26°C, germination will drop to levels unacceptable for malting in less than 2 months. This is a lesson that has been learned in the hard way by a number of new growers and maltsters in the emerging North American craft malt industry. Dedicated grain bins and driers are often not available in these operations, and harvested barley has often been stored in polyethylene/polypropylene flexible bulk containers (super-sacks). Storage in this type of container, with no aeration, exacerbates moisture-related issues. It also needs to be considered that allowable storage times are cumulative. For example, if barley is stored for several weeks at a higher moisture content, before being completely dried to safe storage levels, a portion of the safe storage time shown in Table 2 will be lost (Hellevang 1990a).

The moisture of grain at harvest can be impacted by a number of factors including environmental conditions and management practices, such as straight (direct) combing versus windrowing of the grain. Growers managing multiple fields and grain crops must sometimes harvest at less than ideal moisture due to weather and available resources. In general, barley is considered to be

Table 2 Estimated allowable storage time for malting barley (days)[a]

	Moisture (%)								
Temperature (°C)	11	12	13	14	15	16	17	18	19
26.7	230	175	115	70	40	20	11	9	6
21.1	560	420	270	175	100	50	30	20	15
15.5	*[b]	*	660	430	260	130	65	45	25
10.0	*	*	*	*	630	350	140	100	60

[a] The criterion is germination.
[b] *Allowable storage time exceeds 700 days.
Source: Hellevang (1990a).

physiologically mature at around 35% moisture, and can be cut and windrowed at 20–30% (Association American Malting Barley 2016). Further drying is then dependent upon environmental conditions, but it should be mentioned that the practice of windrowing is becoming less common. Direct combining requires far lower moisture levels, as harvest at moisture levels greater than 18% can result in significant kernel damage.

As harvest moisture is often higher than the safe storage moisture, postharvest drying is often necessary. Drying to a safe moisture level can be accomplished in the bin using aeration and natural air, if moisture levels are relatively low (e.g. 13.5%). Higher moisture levels require grain drying systems using heat. The design of driers can vary considerably in terms of temperature (natural, or low to high heat), operation (batch or continuous), location (in-bin or separate self-contained), and airflow direction (cross-, counter-, and concurrent-flow) (Hellevang 2013). Recommendations will depend to a great degree on the type of drier being used. With in-bin drying, moisture reduction is often accomplished over a period of several days to weeks. The exact time is dependent upon a number of factors including the initial grain moisture and mass (i.e. bin size and amount of water to be removed), weather conditions, and airflow. Drying in a bin with a perforated floor occurs in a zone that moves from the bottom of the bin to the top, with grain at the top staying near the initial moisture content until reached by the drying zone (Fig. 1). Loss of germination in the top layer can occur if moisture reduction is too slow. As such, there is a limit on how much moisture can be removed with this technique.

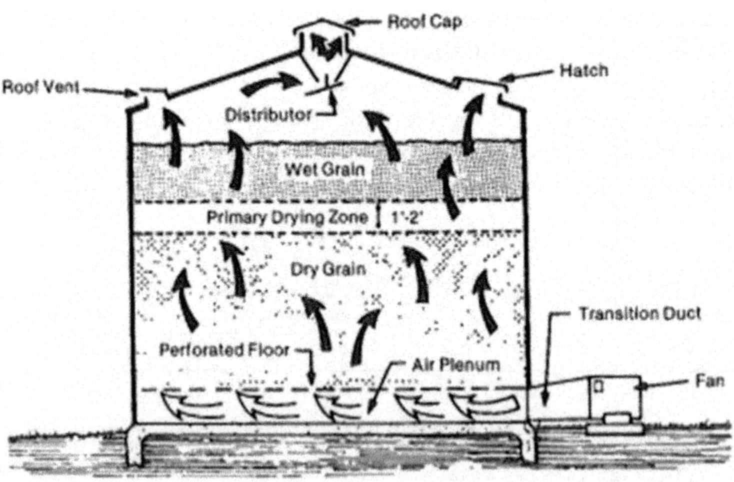

Figure 1 Schematic of a bin dryer utilizing natural air and low temperature drying. Arrows indicate the direction of airflow from fan intake to roof vent. Primary drying zone of 1–2 inches is equivalent to 2.5–5.1 cm. Source: adapted from Hellevang (2013) with permission.

On the contrary, germination loss can also occur when drying temperatures are too high (at a given moisture level). Heated drying can be conducted in the bin, or in a separate dedicated drying unit. Moisture reduction in dedicated driers is generally accomplished in a rapid manner, which requires very good management. The primary risk is heat damage and loss of germination. It is generally recommended that drying temperatures exceeding 40–50°C not be used for malting barley (HGCA 2011; Hellevang 2014). Proper drying requires knowledge of many factors, and it is best to consult recommendations from local authorities or the manufacturer. In addition, malting barley contracts, or guidelines from industry, may specify acceptable drying practices. These may vary between both countries and companies.

2.2 Managing stored barley

As previously stated, spoilage issues with insects and fungi become major concerns at or above 27°C and 70% relative humidity. Thus, prolonged storage can present significant challenges in tropical regions (Mrema et al. 2011). Even though the grain can be dried to an acceptable initial moisture content, it will eventually equilibrate to ambient conditions. While conditions may be more favorable in many of the world's major barley producing areas, the ambient conditions of temperature and relative humidity can vary widely over the seasons. This affects the equilibrium moisture content (EMC) of the grain and necessitates the need for monitoring and management. The EMC is the moisture content of the grain that is equilibrium with air at a specific relative humidity and temperature. This value is used to determine safe storage times, and the grain moisture content should be below the EMC, at 60–65% relative humidity. EMC tables are available from a number of sources (Hellevang 1990b).

Barley harvest times vary between regions, as do the environmental conditions at harvest. As such, it is difficult to present a universal stored barley monitoring scheme. In the northern hemisphere, barley is generally harvested between June and September depending upon latitude and sowing date (i.e. winter vs. spring sown) (Garstang et al. 2010). Harvest times in Australia and Argentina, which are major producers in the southern hemisphere, are between October and December. It must be considered that grain is a good insulator. If the temperature of the barley placed in storage remains higher than ambient, the use of bin fans to cool the grain will extend storage life.

Barley storage facilities need to be monitored across the seasons, and it is recommended that grain temperature and moisture be checked at least every few weeks, and that the temperature and moisture content be recorded at several locations in the bin. Likewise, samples from several locations should be monitored for insects. Frequency of monitoring is dependent upon ambient conditions. As an example, testing can be less frequent during the winter when conditions remain below freezing. Nevertheless, monitoring is still important, and the use of fans may be needed at different times.

A primary concern in storage is that environmental conditions can cause moisture to move or migrate from one part of the stored grain mass to another. This can result in higher moisture ('wet') regions with higher temperature ('hotspots') that become subject to fungal or insect growth. Although moisture migration problems can occur at any time, there are some key factors. For example, when warm grain is stored in cold winter temperatures, the center of storage will be nearly the same temperature as at harvest even when the outdoor temperature has dropped below freezing. This temperature differential causes moisture migration as shown in Fig. 2. The cool air near the bin wall sinks to the bottom, while the inner warm air was pushed up, resulting in moisture condensation on the top of the bin center. The reverse migration can occur when the outdoor temperature increases during late winter and early spring. Here solar energy will warm grains at the top of the bin and near the bin wall to above the average outdoor temperature (Hellevang 1996). The center remains cooler. Again, the use of bin fans for maintaining uniform temperature within

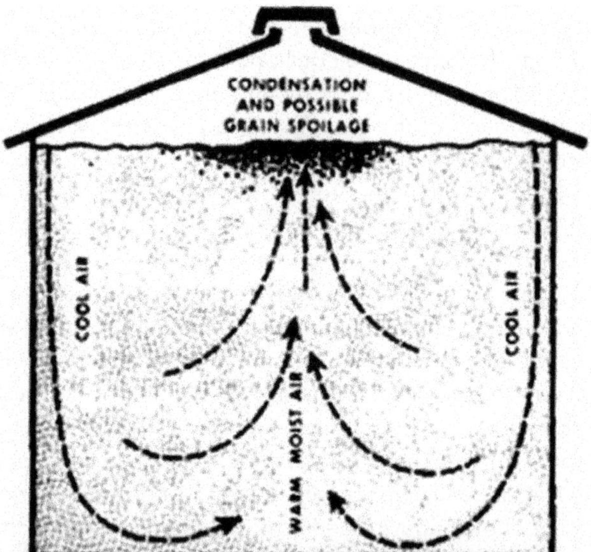

Figure 2 The pattern of moisture migration during fall and winter. The cool air near the bin wall sinks to the bottom of the bin, pushing air up in the center. The rising air is warm by the grain in the center, and then cooled down by the grain near the surface. Moisture in the air condenses and accumulates at the top center of the storage. The condensation indicates possible grain spoilage. Source: adapted from Hellevang (1996) with permission.

the bin is important. Another option is to ventilate the top of the bins with air inlets at the eaves and air outlets at the peak.

In terms of monitoring grain temperature, in-bin cables are an excellent tool, but it must be considered they only detect the temperature of the grain next to the sensor. A hot spot only a few feet from the sensor may not be detected. As such, physical probing might need to check the moisture content and other parameters (Hellevang 2014).

2.3 Storage considerations for dormant and preharvest sprouted barley

Maltsters have long recognized the need to store barley after harvest in order to overcome inherent dormancy, or the inability to geminate, or to geminate in a vigorous and uniform manner. On a physiological basis, dormancy is associated with a number of adaptive advantages such as seasonal detection and spread of germination in time (Barrero et al. 2015). It was a means for the wild ancestors of modern cereals to prevent germination in unfavorable environmental conditions, or in other words, to maximize survivorship. Dormancy in barley is impacted by both the genotype and growing environment, and several studies have shown that the dormancy period can be reduced through warmer storage temperatures (Woods and Fanvier 1994). Storage times to overcome dormancy will be highly dependent upon the cultivar and, to some degree, on the crop year/region and storage conditions.

Dormancy presents a dual problem to the modern malting barley industry, in that it can be either too short, or too long (Barrero et al. 2015). Long-lasting dormancy is an issue as it requires maltsters to maintain larger stocks and storage capacities. However, when there is too little dormancy the grain becomes prone to preharvest sprouting (PHS), which can impact the availability of acceptable grain. Empirical evidence suggests that while some modern malting cultivars that have been bred for high levels of alpha-amylase often have little dormancy, they are also more prone to PHS. This is not surprising as a number of genes associated with dormancy impact PHS (Nakamura 2018) as well as enzyme synthesis.

Grain that has sprouted in the field presents challenges to the maltster, as while it may germinate immediately after harvest, there is a tendency for viability to decrease as the grain is stored. As a consequence, barley initially accepted for malting may be rejected after several months if it loses germination capacity. This loss of viability generally increases with the severity of PHS. The Canadian Grain Commission has compared the measurement of PHS using the Rapid Visco-Analyzer (RVA) to measure the risk of germination loss in storage (Canadian Grain Commission, 2019b) (Table 3). They reported that samples with final viscosity values between 50 and 120 RVU were moderately

Table 3 The risk of rapid germination loss in storage as assessed with the Rapid Visco-Analyzer (RVA)[a]

	RVA Final viscosity	
Risk of rapid germination loss in storage	RVunits	cP
Low	≥120	≥1440
Intermediate	≥50 and <120	≥6000 and <1440
High	<50	<600

[a] Based upon: Using RVA to measure pre-germination in barley and predict germination energy after storage.
Source: Canadian Grain Commission, modified on February 28, 2019 (Canadian Grain Commission 2019b).

pre-germinated. The risk of viability loss is intermediate, but it can be lowered if the barley was stored in cool and dry conditions. Samples with final viscosity values <50 RVU have severe PHS, and should be malted as soon as possible, as there is high risk of viability loss. Again, the conditions of storage are important, and loss of germination will be more rapid at higher temperatures and relative humidity, as well at a higher grain moisture content.

2.4 Management of insects in stored barley

The control of insect pests is extremely important, as when unchecked, they are capable of causing huge losses in stored grains. This subject has been covered by a number of texts (Bhargava and Kumawat 2010) and bulletins from university extension and other agricultural departments (Rees 2007). While somewhat outside the scope of the current review, some basic principles of insect management in stored grain are presented.

As previously mentioned, insect infestation is mainly affected by grain temperature and the cleanliness in and around the storage bin. Insect infestations can occur in grain during storage or in handling equipment, and it is critical to thoroughly clean the storage bins and handling equipment immediately after use, and then again before placing new grains in the storage. However, it can be difficult to clean under a perforated floor or in an aeration duct. Bins can be swept and vacuumed, and then an approved bin spray or fumigant can be applied, especially if an insect infestation occurred during the previous year. Furthermore, seasonal variation and moisture migration to different parts of the grain mass results in moisture accumulation in certain parts of the bulk storage. These areas become a favorable breeding place for insects. Cooling with aeration or by use of a chiller are effective measures in controlling insects, as they can prevent insect feeding and reproduction once the grain temperature is reduced to below 18°C (Navarro 1991).

Pesticides are another means to prevent insect invasion during grain storage, and wide variety of pesticides have been used. These include organo-chlorines, organo-phosphates, carbamates, and pyrethroids (Batey 2017). However, fumigation following a fumigation management plan is often considered as the preferred way to control insects. The main gaseous fumigants are methyl bromide, carbon dioxide, phosphine, and very volatile liquids such as carbon-disulfide and ethyl formate. Nevertheless, there is increasing consumer and legislative pressure on pesticide/fumigant use in stored grains, particularly because of concerns over residues. End users, such as brewers and food processors, may also impose additional restrictions. As such, there is interest of alternative treatments.

The use of diatomaceous earth-based grain protectants has been a popular approach, and control of insects is through desiccation (Athanassiou et al. 2006; Lorini and Beckel 2006). However, this approach is not accepted by some processors due to health and safety concerns involving dust, and also because of potential impacts on processing machinery (HGCA 2011). Modified atmospheres (MA), such as low-oxygen MA and the use of carbon dioxide, are another approach used in the storage of grains to prevent insect pests (Batey 2017). Carbon dioxide has a lethal effect on insects when the concentration is above 40%, and all insects can be killed at 60% concentration with a 4-day exposure at 26°C. Low oxygen is more widely used by holding grains in airtight bins, which can be achieved with the use of nitrogen or exhausting gas burners.

3 Mycoflora and mycotoxins in barley

It has long been understood that harvested grain carries a population of bacteria, yeasts, and filamentous fungi. Environment and geographic origin have a significant influence, and grains from different regions will have different populations (Laitila 2015). Populations also are not static, with some organisms decreasing during storage, while others can develop. In addition, some processing operations, such as germination in malting, provide additional opportunities for growth and shifts in the populations. The presence of microbes can be detrimental, beneficial, or innocuous. While often overlooked as a benefit, grain microorganisms do contribute to the enzyme pool in malting. Negative impacts, which are much more observable, are often associated with damage to grain or product quality, organoleptic characteristics, and grain or product safety. Mycotoxins are the largest food safety concern associated with grains.

3.1 Mycotoxins

Mycotoxins are secondary metabolites produced by a variety of fungal genera, and are capable of causing disease and death in humans, livestock, and other animals. Mycotoxin classes (Bennett and Klich 2003) and their importance to food safety and security (Marroquín-Cardona et al. 2014; Milićević et al. 2010; Moretti et al. 2017) have been reviewed by numerous authors, and only a short synopsis will be presented here. The major classes of mycotoxins have traditionally included aflatoxins (AFs), citrinin, ergot alkaloids, fumonisins, ochratoxin, patulin, trichothecenes, and zearalenone (Bennett and Klich 2003). However, the past decade has seen increased focus on the topics of both modified and emerging mycotoxins. Modified toxins are formed by modification of the parent mycotoxin, often through action of microorganisms or plants. They are sometimes referred to as masked mycotoxins, as they were not commonly detected, and include compounds such as DON-3-β-D-glucoside (DON-3-G), which can be formed in cereals following infection with *Fusarium* spp. The emerging mycotoxins are not strictly defined, but often include the enniatins (ENNs), beauvericin (BEA), moniliformin, fusaproliferin, and others. A more accurate definition might be compounds that have not been routinely determined nor are regulated through legislation (Freire and Sant'Ana 2018). Major mycotoxins, the associated fungi, and their fundamental toxicities/physiological impacts are presented in Table 4.

AFs are considered as the most toxic of mycotoxins, and are mainly produced by *Aspergillus flavus* and *A. parasiticus* (Bennett and Klich 2003). While *A. parasiticus* strains can produce AFB1, AFB2, AFG1, and AFG2, *A. flavus* strains produce only AFB1 and AFB2. AFs have carcinogenic, hepatotoxic, immunosuppressive, mutagenic, and teratogenic effects, and are associated with both acute toxicity and chronic carcinogenicity. AFs can be found on a wide range of materials including cereal grains and other seeds, nuts, dried fruits, and spices. In terms of cereals, AF contamination is commonly associated with maize, but they are found in other grains. Fungal contamination and AF production can occur in the field, but with wheat and barley, AFs are more commonly the result of poor storage (Kumar et al. 2017).

Ergot, caused by the fungus *Claviceps purpurea*, is a disease of many genera of grasses and cereal crops (Milićević et al. 2010). It is commonly seen in wheat, rye, and barley, and the most common sign of ergot are the dark purple to black sclerotia (ergot bodies) that replace kernels in the heads of the grass plant as it develops in the field. The sclerotia consist of a compact mass of hyphae, and contain ergot alkaloid toxins. Ergot is the oldest recognized cause of mycotoxicosis in humans and was endemic during the Middle Ages in much of Europe. Today, ergot mycotoxicosis is more commonly associated with livestock-fed contaminated grain, and symptoms include gangrenous lesions, with the affected tissue eventually being lost. Because of the potential toxicity, strict tolerance limits are enforced within the grain trade for the presence of ergot sclerotia. Modern grain cleaning operations are generally assumed to be quite effective in the removal of sclerotia, but this is likely dependent on a number of factors.

Table 4 Summary of major mycotoxins, associated fungi, examples of optimal conditions for mycotoxin production, and their physiological effects

Mycotoxin	Fungi	Temperature (°C)	Water activity (aw)	Toxicity
Aflatoxins	*Aspergillus flavus* and *A. parasiticus*	33	0.99	Carcinogenic, acute hepatotoxic, immunology suppression
Fumonisins	*Fusarium verticillioides* and *F. proliferatum*	10–30	0.93	Carcinogenic, hepatotoxic
Ochratoxins	*Aspergillus. ochraceus*, *A. carbonarius* and *Penicillium verrucosum*	30	0.98	Carcinogenic, nephrotoxic, hepatotoxic, teratogenic
Trichothecenes (DON)	*Fusarium. sporotrichioides*, *F. graminearum*, *F. culmorum*, *F. roseum*, *F. tricinctum*, *F. acuminatum*	15–25	0.97–0.99	Gastrointestinal hemorrhaging, immunodepressants
Zearalenone	*Fusarium graminearum*	25-30	0.98	Estrogenic activity

Source: Lee and Ryu (2017).

Fumonisins, a group of polyketide-derived mycotoxins, are separated into four main groups: A, B, C, and P. The B group includes the most active, toxic FB1 and its isomers FB2, FB3, and FB4. FB1 accounts for 70–80% of total fumonisins (Alshannaq and Yu 2017). Fumonisin B1 is the cause of considerable toxicological concern and is classified as a probable carcinogen (Ostry et al. 2017). Fumonisins, mainly produced by *Fusarium verticillioides* and *F. proliferatum,* have been detected most frequently in maize, when compared to other cereals.

Ochratoxins are a group of mycotoxins produced by a number of *Aspergillus* species (e.g. *A. ochraceus, A. carbonarius*, and *A. niger*) and some *Penicillium* species (especially *P. verrucosum*) (Bennett and Klich 2003). Ochratoxin A (OTA) is the most prevalent, while ochratoxins B and C are of less importance. This mycotoxin is nephrotoxic, immunosuppressive, teratogenic, and a possible human carcinogen. Ochratoxins can occur in cereals, as well as in coffee, dried fruit, grapes, and red wine. In terms of cereals, ochratoxin contamination is frequently associated with poor storage, although infection with the toxigenic fungi occurs in the field (Miller 1995).

Trichothecenes comprise a vast group of sesquiterpenoid metabolites produced by a number of *Fusarium* species, and several other genera. Trichothecenes are divided into A and B types, mainly based on the presence of different functional groups at the C-8 position of the trichothecene backbone (Shank et al. 2011; Ferrigo et al. 2016). Contamination of cereals with trichothecenes is generally associated with Fusarium Head Blight (FHB) infection in the field, although some can also develop during germination in the malting process. The A types include T-2 and HT-2 toxins, diacetoxy- and monoacetoxy-scirpenol (DAS and MAS) and neosolaniol (NEO). They are potent inducers of oxidative stress and inhibitors of DNA, RNA, protein synthesis, and mitochondrial function (European Food Safety Authority 2011). Type A trichothecenes are mainly produced by *Fusarium langsethiae, F. equiseti, F. poae*, and *F. sporotrichioides* (Desjardins 2006). Type B trichothecenes include DON, nivalenol (NIV), 3-AcetylDON (3-ADON), 15-AcetylDON (15-ADON), and fusarenone X (Shank et al. 2011), and are mainly produced by *F. graminearum* and *F. culmorum* (Desjardins 2006). DON, also known as vomitoxin, is the most frequently occurring mycotoxin in barley and small grains. DON has the potential to cause chronic effects, such as reduced growth and anorexia, as well as acute intoxication leading to vomiting, immunotoxic effects, and changes in brain neurochemicals (Ostry et al. 2017). With respect to toxicity, DON is more toxic against plants, while NIV is more toxic than DON toward animals (Ferrigo et al. 2016). NIV usually occurs with other toxins such as DON and T2 toxin. DON-3-G is a well-known masked mycotoxin which is formed *in planta* through the action of host UDP-glucosyltransferases.

Zearalenone (ZEA), a macrocyclic β-resorcylic acid lactone, is an estrogenic mycotoxin that induces obvious estrogenic effects in humans and animals (Alshannaq and Yu 2017; Ostry et al. 2017). It is mainly produced by *Fusarium graminearum* and a limited number of other species under conditions of high humidity and lower temperature (Desjardins 2006). It is frequently found in cereals, and its contamination simultaneously occurs with DON (Yazar and Omurtag 2008).

Given the importance of mycotoxins to food and feed safety, many national and international public health and governmental authorities such as the US Food and Drug Administration (FDA) and the European Food Safety Authority (EFSA) have adopted strict regulatory (or advisory) guidelines for the major mycotoxin classes in food and feed (FDA 2017; EFSA 2019). Table 5 lists the limits for mycotoxins in cereals or cereal products in a number of countries. In some cases, the regulations specify barley or malt. AFs and DON are commonly regulated, but the regulation of others varies by country. There is limited regulation for fumonisins. While ergot alkaloids are not regulated, *per se*, many countries have specified limits on ergot sclerotia that are permitted in grain for human and animal consumption (http://www.mycotoxins.info/regulations/).

3.2 Fungi and mycotoxins associated with barley

There have been many national to regional surveys of barley for the presence of mycotoxins, some of which are listed in Table 6. Mycotoxins detected in barley grain include AFs, alternariol, BEA, ENNs, OTA, ergot, zearalenone, and a range of trichothecenes including DON and its acetylated derivatives, nivalenol, T-2 and HT-2 toxins, and diacetoxyscirpenol (DAS). It is of interest that many recent studies have detected the cooccurrence of mycotoxins of multiple classes within the same samples. Not surprisingly, geographic origin appears to have a significant influence. Cropping practices and environmental conditions have a tremendous influence on both the mycoflora and development of mycotoxins. OTAs and AFs are most often associated with stored grain, but contamination can be initiated in the field (Ibáñez-Vea et al. 2012a). The presence of other mycotoxins is generally linked directly with field infection.

The trichothecenes, especially DON, have been by far the most widely detected mycotoxins associated with barley (Desjardins 2006). As a consequence, levels are regulated in much of the world, and DON is frequently tested in the grain trade. Levels of DON reported have ranged from the quantitation limit to 9110 μg/kg (Campbell et al. 2000). Frequencies of regional crop contamination as high as 100% of samples have been reported. Levels of DON appear to be higher in North America than Europe, but the presence of HT and T-2 toxins appears to occur more frequently in Europe. T-2 and HT-2 toxins

Table 5 Mycotoxin regulations for cereal grains in different countries and regions

Nation or region	Mycotoxin	Maximum level (μg/kg)	Intended use	Commodity
Brazil	Aflatoxins	5	Human	Cereals and cereal products, including malting barley
	Deoxynivalenol	1000	Human	Cereals except malting barley
		750	Brewing	Malting barley
	Ochratoxin A	10	Human	Cereals and products, including malting barley
	Zearalenone	100	Human	Cereals and products except malting barley
China	Aflatoxin B1	5	Human	Barley, wheat, and maize
	Deoxynivalenol	1000	Human	Barley, wheat, and other cereals
	Ochratoxin A	5	Human	Cereals and products
European Union	Aflatoxin B1	2	Human	Cereals ready to eat (except maize and rice)
	Total aflatoxins (B1+B2+G1+G2)	4	Human	Cereals ready to eat (except maize and rice)
		20	Feed	All feed materials
	Deoxynivalenol	8000	Feed	Cereals and products except maize by products
		1250	Human	Unprocessed cereals except durum wheat, oat, and maize
		750	Human	Cereals intended for direct human consumption
	Ochratoxin A	5	Human	Unprocessed cereals
		3	Human	All products derived from unprocessed cereals
		250	Feed	Cereals and products
	Sum of T-2 and HT-2 toxins	50	Human	Cereals intended for direct human consumption except oat and maize
		200	Human	Unprocessed cereals, including malting barley
		500	Feed	Cereal products
	Zearalenone	2000	Feed	Cereals and products except maize by products
		100	Human	Unprocessed cereals except maize
		75	Human	Cereals intended for direct human consumption
United States	Deoxynivalenol [a]	10000	Feed	Grains and grain by-products destined for ruminating beef and feedlot cattle older than 4 months and for chickens
		5000	Feed	Grains and grain by-products destined for swine
		5000	Feed	Grains and grain by-products destined for other animals

[a] advisory guideline.
Source: micotoxins.info (1998).

were detected in barley grains grown in UK barleys in 2002–2005 (Edwards 2009), as well as in Czech Public and Canada in recent years (Bolechová et al. 2015; Martos et al. 2010).

As previously stated, ZEA is often detected along with DON. In an early survey of barley samples in Korea, ZEA was detected in 51% of samples, with a range of 40–1416 μg/kg (Kim et al. 1993). In a more recent study, ZEA was in 39% of the barley surveyed samples in Spain, but the levels were all below 19 μg/kg (Ibáñez-Vea et al. 2012b).

Ergot can be found in most of the world's cereal production regions, and usually occurs under conditions where the flowering grain is exposed to extended cold and wet periods. Ergot infects the open floret (Coufal-Majewski et al. 2016). Rye is most susceptible followed by wheat and then barley. While the normal incidence in barley is relatively low, widespread occurrence has been reported in some regions during specific crops years. As an example, Schwarz et al. (2006b) reported that ergot was found in 62% of all barley samples (n = 304) surveyed in North Dakota, Minnesota, and Montana in 2005. Levels as high as 1.20% sclerotia were observed, and 90% of ergot contaminated barley samples also contained DON.

The importance of other mycotoxins in barley has sometimes been downplayed given the focus on trichothecenes. It has often been thought that the AFs and fumonisins were primarily problems with maize, but several recent studies have suggested that these mycotoxins can represent a risk to barley in some regions. Gil-Serna et al. (2013) presented the first report on the presence of fumonisins in barley. In this study, fumonisins were detected in 34% of the Spanish barley samples tested, with levels up to 300 μg/kg. FB1 and FB2 have also been reported in barley as part of several other surveys, but the levels were generally relatively low (Bolechová et al. 2015; Martos et al. 2010; Maenetje and Dutton 2007; Juan et al. 2017). The incidence of AFs in barley has generally been believed to be low and was thought to be largely associated with poor storage, often in less developed areas. Maenetje and Dutton (2007) reported low incidence of AFs in barley (levels below 170 μg/kg) grown in South Africa, as well as some malted samples. Ibáñez-Vea et al. (2012b) found AFs in 100% of 123 barley tested from the 2007–08 Spanish crop. AFB1, AFB2, AFG1, and AFG2 were detectable in 100%, 17%, 5%, and 18% of these samples. OTA has been detected in barley samples from the United States (Kuruc et al. 2015), the United Kingdom (Scudamore et al. 2010), Spain (Ibáñez-Vea et al. 2012b), South Africa (Maenetje and and Dutton 2007), Tunisia (Juan et al. 2017), and Egypt (Badr et al. 2017). Kuruc et al. (2015) found OTA in 10% of surveyed stored US barley samples (n = 60), and levels ranged from 0.16 to 185.24 μg/kg. In an earlier survey in the UK, Prickett et al. (2014) found OTA in 19% of barley samples (n = 106). Another study in the France found that the percentage of

Table 6 Examples of surveys for mycotoxins in postharvest barley

			Mycotoxins (%frequency of detection, and level μg/kg)													
Sample numbers	Location	Year	DON	Total 3, 15-ADONs	NIV	T-2 toxin	HT-2 toxin	NEO	ZEA	FB1	FB2	Total aflatoxins	OTA	Total ENNs	BEA	References
39 a	Korea	1990-91	90, 25–1051	18, 13–168	95, 39–6892				51, 40–1416							Kim et al. (1993)
446 b	UK	2002-05	57, 10–1416	<1, 10–40	25, 10–157	12, 10-39	36, 10–105	ND	2, >10				27, 18			Edward (2009), Scudamore et al. (2010)
82	South Africa	2007	78, <3125						Detected in low level	30, <170		AFB1 detected	Detected			Maenetje and Dutton (2007)
123 c	Spain	2007-08	95, 1111	57, <65	20, <142	10, <332	24, <200		39, <19			100, <0.8	58, <4			Ibáñez-Vea et al. (2012b,c)
116	Canada	1991-98	72, <9110	10, <380	34, < 360				15, 30–840				5, 12–26			Campbell et al. (2000)
20	Canada	2010	100, 78–2449	10, 40–41		20, <13	60, 16–67	ND	ND	ND	ND	ND	ND			Martos et al. (2010)
28	Czech Republic	2012	86, 70–602		36, 52–123	35, 15-30	56, 6–57		9, 181–204	5, <20	9, 11–16	ND	ND	100, 25–2079	100, 5–48	Bolechováet al. (2015)
42	Egypt	2014-15							24, <4			5, 11–26	16, <2			Badr et al. (2017)
31 d	Tunisia	2015-16	23, 2–6	ND	10, 16–111			16, 24–35	ND	3, 42		ND	6, 0.3–3	16, 292–855	10, 74–75	Juan et al. (2017)

145 e	USA	2001-12	78, 0.4-63			Schwarz et al. (2014)
60	USA	2011-12			10, 0.2-185	Kuruc et al. (2015)
64	Brazil	2016	91, 46-1155	88, 82-424		Piacentini et al. (2019)

DON is short for deoxynivalenol; total 3, 15-ADONs include 3-acetyl DON and 15-acetyl DON; NIV is nivalenol; NEO is neosolaniol; ZEA is Zearalenone; FB1is fumonisin B1; FB2 is fumonisin B2; total aflatoxins include FB1, FB2, FG1, and FG2; OTA is Ochratoxin; total ENNs include enniatins A, A1, B, and B1; BEA is beauvericin; ND means not detected, Blank means that the mycotoxin was not investigated.

a: 4-acetylnivalenol and 4,15-diacetyldeoxynivalenol were also detected in 44% (12–71 μg/kg) and 13% (15–28 μg/kg) of the investigated barley samples, respectively.

b: diacetoxyscirpenol (DAS) was not detected, moniliformin was detected in 2% (>10 μg/kg), and fusarenone X (FUS-X) and T-2 triol were detected in less than 1% (>10 μg/kg).

c: multi-mycotoxins co-occurred in some samples, and FUS-X and DAS were also found.

d: FUS-X, DAS and sterigmatocystin (STG) were also measured, but not detected in these samples.

e: DON-3-glucoside was present at the levels less than 10 mol% of the DON present on the same samples.

OTA positive samples and mean concentration increases with storage time and moisture content (Malmauret et al. 2002). In fact, OTA is the primary mycotoxin of concern with stored grain in the UK (Edwards 2009; Scudamore et al. 2010; Prickett et al. 2014).

Another concern is recent reports of new or emerging toxins, such as ENNs and BEA, that have been seen in large quantities in barley and other grains (Ferrigo et al. 2016). ENNs are mainly produced by *Fusarium avenaceum*, and consist of more than 20 different six-membered cyclic depsipeptides (Vongvilai et al. 2004). ENN occurrence is high in Northern and Eastern European on both barley and wheat (Bolechová et al. 2015; Lindblad et al. 2013) with incidences up to 100%. An analysis conducted on cereals from Spain indicated that ENNs contaminated 89%, 62%, and 50% of maize, wheat, and barley samples, respectively (Meca et al. 2010). Similar high incidences were found in the Czech Republic and Tunisia (Bolechová et al. 2015; Juan et al. 2017). BEA is a mycotoxin of the cyclohexadepsipeptide family, and is produced by *Fusarium poae, F. avenaceum* and *F. temperatum* (Ferrigo et al. 2016). While BEA has mainly been found in maize, barley grown in both the Czech Republic and Tunisia has been reported to contain BEA, with a maximum content of 75 μg/kg (Bolechová et al. 2015; Juan et al. 2017).

3.3 Stability and development of toxigenic fungi during barley storage

After loading harvested grain into storage facilities, the development of fungal species is highly dependent on environmental conditions. Due to grain respiration, hot spots can be created in some parts of the bulk grain mass. Since the grains are still alive and respiring, CO_2 is evolving. Heat and the heat of respiration increases the temperature of the grain, which can cause microflora which are already present to increase and invade the grain tissue. Fungal-induced hot spots can cause the grain temperature to rise up to about 70°C, and fungal volatiles have been identified in stored grains (Mohapatra et al. 2017). The ecosystem of fungal species is diverse in barley grains. Rabie et al. (1997) enumerated more than 30 fungi in barley under field or storage conditions. The key factors involved in growth of mycotoxigenic fungi in cereal grains during harvest, postharvest processing, and the long-term storage are illustrated in Fig. 3.

Since each species of grain fungi has its own eco-physiological requirements, growth of different species may be in competition, and are dependent on storage conditions, mainly including the temperature and moisture content of the grains (Mohapatra et al. 2017). There are three major types of fungi that can grow in stored grain depending upon prevailing environmental conditions: (i) field fungi, (ii) intermediate microflora, and (iii) storage fungi (Fleurat-Lessard

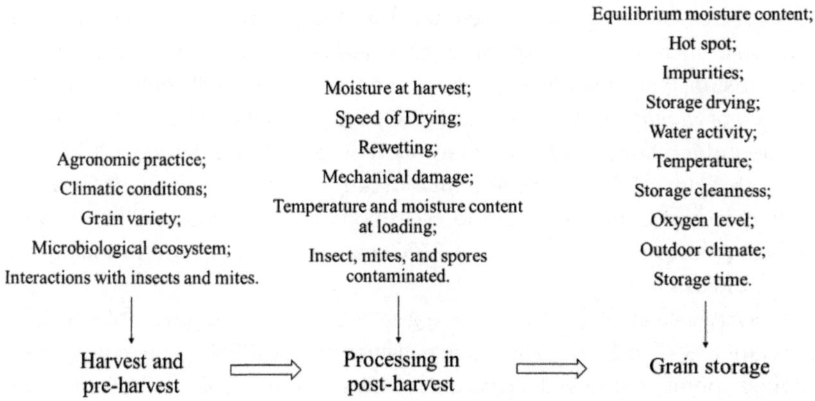

Figure 3 Key factors involved in the germination and growth of mycotoxigenic fungi in cereal grains during a long-term storage. Source: adapted from Fleurat-Lessard (2017) and Magan et al. (2004).

2017). Field fungi are hygrophilic plant pathogenic fungi, such as *Fusarium graminearum*. Intermediate microflora are saprophytic and thermophilic fungi which can contaminate kernels of stressed or injured plants before and after harvest time. Examples are *Fusarium proliferatum* and *Aspergillus flavus*. Storage fungi such as *Eurotium chevalieri*, *Penicillium verrucosum,* and *Aspergillus ochraceus* are usually tolerant under drought conditions, and they colonize grains after the harvest. Field fungi such as *Alternaria* or *Fusarium* species, which have already colonized grain at harvest, are hygrophilic, and are generally expected to disappear after 2–4 months of storage. Some mesophilic species of the intermediate microflora class can invade and replace the hygrophilic mycoflora when the grain moisture content remains high (i.e. more than 18%). If the grain is stored under conditions of more than 20% moisture content, the storage species, such as *Trichoderma viride* Persoon, *Rhizopus nigricans* Ehrenberg, and *Mucor hiemalis* Wehmer would grow after 6–10 months of storage. When grain moisture remains between 15.5% and 17%, the intermediate mesophilic mycoflora are progressively replaced by most xerotolerant species such as some *Aspergillus* spp. after 4–6 months of storage, and secondarily by *Penicillium* spp. with less drought tolerance (Fleurat-Lessard 2017; Mannaa and Kim 2017).

The most irreparable damage caused by the growth of fungi in grain storage is the production of highly toxic and carcinogenic mycotoxins. Ramakrishna et al. (1996) studied the colonization of *A. flavus* on barley grain and formation of AF B1 in the presence of *Penicillium verrucosum*, *Fusarium sporotrichioides*, and *Hyphopichia burtonii* over a 3-week period in combinations of 20°C or 30°C and 0.97, 0.95, or 0.90 a_w. The results showed that the germination of *A. flavus* spores was unaffected by the presence of other species on the grain surface and that the production of AF B1 might decrease or increase, and sometimes changed only slightly in the presence of *P. verrucosum*, *F. sporotrichioides*. Production was temporarily enhanced in the presence of *H. burtonii*. Beattie et al. (1998) evaluated the effect of 7 months of storage under different conditions (ambient outdoor temperature, −20°C, 4°C, 24°C with quiescent air, and 24°C with forced air) on the viability of *Fusarium* infesting stored barley. All storage conditions reduced the viability of both molds, but forced air ventilation at 24°C was most effective in reducing the viability of *Fusarium*.

Wawrzyniak et al. (2013) investigated the effect of unfavorable storage conditions (23°C and 30°C, and water activities of 0.80–0.94 a_w) on mold growth in barley contaminated with *Aspergillus westerdijkiae, Penicillium viridicatum,* and *Fusarium poae*. Water activity turned out to be the main factor affecting the mold growth. The longest lag phase of fungal activation was observed for barley grain at 0.80 a_w, which was not threatened by fungal development for at least 30 days. However, fungal activation occurred within 24–48 h in grains with 0.92 and 0.94 a_w. This is in line with the report on the growth and ochratoxin production by *Aspergillus ochraceus* strains isolated from barley grains (Ramos et al. 1998). The optimal a_w level for *A. ochraceus* was found to be in the range of 0.96–0.98, with an optimal temperature of 30°C. *A. ochraceus* was observed to grow at 10–37°C but only at higher a_w levels. Maximum amounts of ochratoxins were produced at the highest a_w treatment (0.98 a_w) after a 3-week incubation time at 25–30°C.

4 Prevention or decontamination of mycotoxins in barley storage

Avoiding fungal infection or growth is considered the most effective method of controlling mycotoxin problems, since prevention is better than cure. Prevention measures start in the field and continue through postharvest management. Pre-storage measures include the selection of resistant varieties, application of fungicides, harvesting at suitable moisture content, and drying the grains before storage. Good management of stored grain, as previously discussed, is required to prevent the development of conditions which are favorable for fungal growth. However, when necessary, grain can also be treated with physical, chemical, and biological agents, or a combination thereof, to prevent fungi growth and mycotoxin production. Removal of contaminated grains for mycotoxin reduction is an additional option. Applications or treatments that do not leave residues are likely to gain more recognition, because consumers are becoming more concerned about chemical residues in food.

4.1 Removal of mycotoxin-contaminated barley

Removal of broken, thin or shrunken kernels and dust can be used to reduce mycotoxin contamination in grain samples. Results are highly dependent upon the grain type and individual sample. This is because infection can occur at different points during grain development, and that mycotoxins are generally not uniformly distributed within a grain lot. The results of a mycotoxin test represent the average of several hundred individual kernels that would have been ground and extracted (e.g. 5–10 g sample). Again, infection is generally not uniform, and a sample may contain a few heavily contaminated kernels and many with little to no contamination. Any physical factor (e.g. size, density, color) that distinguishes these infected grains can be used as a basis for their removal. For example, a common symptom of FHB in wheat is the presence of shrunken 'tombstone' kernels (FDK, Fusarium damaged kernels). It has been reported that DON in wheat can be reduced 5–20% by cleaning and sorting (Bullerman and Bianchini 2007). The actual effectiveness of cleaning likely depends upon many factors including the toxin and its level, the grain, and the timing of infection during plant development (or storage). Results will be highly dependent upon the individual sample. Barley often flowers prior to head emergence, as opposed to wheat, which flowers after head emergence. As a result, FHB infection of barley often occurs later, and reduction in kernels size is not as common.

The cleaning and sizing of barley has shown variable results in DON reduction. Barley grains that pass through either 2.0 or 2.4 mm (x 19 mm) sieve openings, are generally not used for brewer's malt. Perkowski (1998) separated 48 two-rowed barley samples, with DON levels ranging from 0.1 to 157 μg/g, into four kernel size fractions (>2.8 to <2.2 mm), and found that 80% of the DON was in kernels <2.5 mm. As such, it appeared that it should be possible to remove a significant amount of DON by removing the smallest kernels. However, results from a later study with six-rowed barley showed less promising results. Schwarz et al. (2006a) separated 125 samples of naturally infected barley (0.5–29 μg/g) into four kernel size fractions (>2.8 to <2 mm), and found that a large proportion of DON was present on the thinnest kernels. When samples with DON ranging from 0.5 to 0.9 μg/g were evaluated, removal of thin kernels resulted in reduction of DON levels to <0.5 μg/g in approximately 50% of the samples. However, removal of thins in samples ranging from 1.0 to 2.9 μg/g DON reduced levels to <0.5 μg/g in only 18% of samples. Sizing of samples with ≥3.0 μg/g DON was completely ineffective in reducing DON to acceptable levels.

The technologies of grain dehulling, pearling, and abrasion can also be practically applied for the reduction of mycotoxins in some select situations. This is because fungi and mycotoxins are often located in the husk or bran fractions of the grain (Mohapatra et al. 2017; Gomez-Lopez et al. 2007), which are removed in the pearling process. Pearling has been used in the processing of DON contaminated barley for pet food in the United States.

Grains contaminated with fungi or mycotoxins can also be sorted by color (Fleurat-Lessard 2017). An optical or color sorter inspects grain by means of high-resolution digital cameras (based on near infrared, ultraviolet, and visual light spectra) and removes abnormally colored grain with a short burst of compressed air. As an example, it has been reported that 99% of ergot sclerotia in barley and wheat can be removed by color sorting (Lorenz and Hoseney 1979). Delwiche et al. (2005) reported variable results in DON reduction by optical sorting of FHB-infected wheat. Multiple sorting of 35 samples achieved reductions ranging from 31% to 84%.

4.2 Treatment of stored barley

A number of biological or chemical treatments have been reported for the control of fungi and mycotoxins in stored grain (Fleurat-Lessard 2017). These include biocontrol, application of botanical extracts, ozonation, irradiation, cold plasma, and microwave. Temba et al. (2016) reviewed the effectiveness and drawbacks of nonbiological approaches to remove or inactivate mycotoxins from contaminated food and feed products. Although some novel techniques, such as photosensitization, showed commercial promising, most have not reached the point of practical application, and not all have been evaluated on mycotoxin-contaminated barley.

Perhaps, the most advanced alternative treatment is ozone. Ozone is a strong oxidant, and the use of gaseous ozone as an antimicrobial agent for the treatment, storage, and processing of foods has been formally approved by FDA (Rice and Graham 2001), and there are several companies currently selling ozone generation systems for grain storage applications. The application of gaseous ozone to stored grain is attractive, because if properly used, it does not affect cooking, baking, milling, and germination properties of the grain (Tiwari et al. 2010). Ozone is highly unstable and quickly decomposes to oxygen. The efficacy is also dependent on various factors such as humidity, temperature, and air flow rate (McClurkin et al. 2013). Ozonation also has the added benefit of being effective for the control of some insect pests in stored grain (Bonjour et al. 2011).

Allen et al. (2003) reported that the laboratory application of ozone (0.16 mg/g barley for 5 min at 20°C and 0.98 a_w) during the storage of barley resulted in the inactivation of up to 96% of fungal spores. Increases in temperature and a_w of barley enhanced the fungicidal efficacy of ozone. Mycelia were less resistant to ozone than spores. Higher levels of ozone (0.98 mg/g barley for 15 min), however, were found to damage the germination energy of barley grain. Similar findings were reported by Kottapalli et al. (2005). A fog of ozonated water was recently used to treat barley on an industrial scale, and the results showed 80% disinfection for *Fusarium* spp. and 70% for *Aspergillus* spp. under conditions of 9.8 μg/L of dissolved ozone in water. There was no effect on the germination of barley, but treatment was not effective against *Alternaria* spp. Mendez et al. (2003) studied the application of ozone for stored maize, wheat, and other cereals.

Several studies have also shown that ozone can chemically modify a variety of mycotoxins, such as AFB1, AFB2, AFG1, and AFG2, cyclopiazonic acid, FB1, and DON (McKenzie et al. 1997; Sun et al. 2016).

Sterilization with plasma or gaseous discharges has been proposed as an alternative method to disinfect grain. Plasma is highly energized matter, composed of excited atoms and molecules, ionized gases, radicals, and free electrons (Moisan et al. 2001). The highly active gaseous discharges such as ozone, atomic oxygen, hydroxyl, nitric oxide, and sulfur oxide radicals, as well as other free radicals, are antimicrobial in nature (Laroussi and Leipold 2004). It has been reported that cold atmospheric pressure plasma can effectively degrade pure DON, ZEA, ENNs, FB1, and T2 toxin produced by *Fusarium* spp., sterigmatocystin produced by *Aspergillus* spp., and *Alternaria alternata lycopersici* toxins (AALs) produced by *Alternaria alternata* (Bosch et al. 2017). Selcuk et al. (2008) applied a low-pressure cold plasma (LPCP) system with air gases or SF6 to fungal isolates of *Aspergillus* spp. and *Penicillium* spp., and infected grains and legumes. Sulfur hexafluoride (SF6) plasma treatment decreased the total number of *Aspergillus* spp. from 5.0×10^6 CFU/g to 1.71×10^5 and 1.20×10^4 on barley and wheat, respectively. *Penicillium* spp. on wheat decreased from 4.1×10^7 to 1.6×10^4 CFU/g using LPCP of air or SF6 plasma for 20 min. The effect of plasma treatment on germination was evaluated for wheat and bean seeds. Germination rates were not affected with air gases or SF6 plasma when compared to the control plants. However, N_2 and O_2 gases under the described plasma conditions resulted in the surface discoloration, visible damage, or reduction of germination.

Irradiation is a technology that has been shown to be very effectively used in disinfecting fungi, and reducing mycotoxin production in foods (Mohapatra et al. 2017). However, while the irradiation of grains is approved in some countries, its industrial use is currently limited to higher-value commodities (Maherani et al. 2016). Consumer understanding and acceptance also remains a challenge to acceptance. Research by Aziz et al. (2006a) has demonstrated that irradiation of grains, including barley, wheat, maize, and shelled rice, at doses of 2.0 and 4.0 kGy significantly decreased total fungal counts. After 100 days of storage at room temperature, the unirradiated grains were contaminated with high concentrations of mycotoxins, including AFB1, OTA, cyclopiazonic acid, and citrinin, when compared with irradiated (4.0 kGy) samples. Mycotoxin production in grains decreased with increasing irradiation doses, and the toxins were not detected in samples treated at 6.0 kGy and stored for over 100 days

(Aziz et al. 2007). The dose required for the inhibition of fungi varied in different products. Barley exposed to a Co^{60} gamma irradiation at the dose of 4 kGy eliminated most *Alternaria*, *Fusarium* and *Epicoccum* spp., but a dose of 12 kGy was required to kill *Bacillus* spp. (Ramakrishna et al. 1991). The effect of radiation dose on the quality of stored grains has been evaluated in several studies (Ramakrishna et al. 1991; Lee et al. 2017; Aziz et al. 2006b). The presence of free radicals in seed coat increases with radiation dose, and can affect germination of seeds. The germination energy of barley was improved up to a dose of 8 kGy but gradually decreased with increases in dosage (Ramakrishna et al. 1991).

Biocontrol approaches are considered as an environmentally friendly alternative, in contrast with most physical and chemical techniques of detoxification. These can include application of botanical extracts or microbiological control (Fleurat-Lessard 2017). However, again the use of this method in grain storage has been largely experimental. A large set of bacteria, fungi, and yeasts have shown a capacity to act as bio-competitive agents, microbial antagonists, or 'bio-transformers' of mycotoxins (Pearson et al. 2010). For instance, some lactic acid bacteria and *Geotrichum candidum* can competitively inhibit the growth of *Fusarium* spp., and some bacteria or yeast strains are able to degrade mycotoxins (Leyva Salas et al. 2017). However, at this point the use of bio-competitive agents has been limited to the field or malt processing, and not storage.

5 Post-storage treatment of barley

Regulations in many countries specify mycotoxin limits on grain for various uses including food, feed, and in a few cases, brewing (Table 5). While the ideal situation is the use of mycotoxin-free grain, this is not always possible due to a number of factors, including widespread infection with a field pathogen, such as *Fusarium* spp., within a region. Treatment of mycotoxin-contaminated barley following storage could include the aforementioned processes of cleaning, milling, or abrasion for mycotoxin reduction. However, it must be considered that regulations often specify maximum limits in unprocessed grains, so the utilization of some samples simply may not be possible.

The major uses of barley are for feed and malting. Regulated or advisory limits for animal feed are often higher than those for human food, and as such, the feed market can provide an outlet for some mycotoxin-contaminated barley. Distilled spirits or fuel ethanol is another option as the mycotoxins do no pass to the distillate. However, processors must be cognizant that the mycotoxin may persist in the spent grain, and thus pose a risk to livestock, which is the most common outlet. Most brewing malt is prepared from barley, and the malting process presents some unique challenges as the germination environment can be conducive for the production of some mycotoxins.

5.1 Post-storage treatment of feed barley

The use of additives to prolong the shelf-life, prevent fungal development, or to bind mycotoxins present in animal feed has been the subject of extensive research (Jacela et al. 2010). Additives are often classified as mold inhibitors, mycotoxin binders, and antioxidants. Organic acids, such propionic acid, can be applied to feed to prevent fungal growth. Mycotoxin binders are compounds added to feed rations that are intended to prevent absorption, and can include bentonite clay, zeolites, and yeast cell-wall beta-glucan. Many binders are effective against AFs, but show limited binding of other mycotoxins. Vila-Donat et al. (2018) recently summarized the multi-binding agents for animal feed decontamination by selectively adsorbing different mycotoxins. Some binders, such as carbon/aluminosilicate-based products and grape pomace had the capacity to simultaneously adsorb multi-mycotoxins, with up to 88% for AFB1, followed by ZEN, OTA, and FB1. Treatment with gaseous ammonia is an additional feed treatment that has been used for detoxification of AFs and fumonisins in feed (Fleurat-Lessard 2017).

5.2 Post-storage treatment of malting barley

There have been numerous surveys for the presence of mycotoxins in beer (Schwarz 2017) and DON and D3G are frequently detected. A recent study, focusing on craft beer, confirmed this common occurrence, with about 40% of beers screened (n = 1000) having DON plus D3G levels in excess of 10 μg/L (Peters et al. 2017). AFB1 was detected in several traditional African beers. OTA, fumonisins, T-2 toxin, HT-2 toxin, and ZEN were detected in a few of the samples. The source is either the brewers' malt, or unmalted adjunct grains. The fate of mycotoxins in brewing is dependent upon a number of factors. If present in beer, the toxins are obviously water soluble and stable over the range of temperatures and pH used in brewing. Toxins, which are not water soluble, may remain on the spent grains. Screening grains for mycotoxins is the most direct way to avoid their presence in beer. However, the malt also needs to be screened as the germination environment provides conditions that are favorable for the development of several genera of fungi and production of some mycotoxins. In addition to mycotoxins, some genera of mycotoxigenic fungi also produce compounds that can cause gushing, which is the spontaneous and often violent over-foaming of packaged beer upon opening. Gushing is not caused by mycotoxins, but rather proteins such as hydrophobins (Schwarz 2017).

The development of mycotoxins during malting has been recently reviewed by both Schwarz (2017) and Laitila (2015). The environmental conditions of malting provide favorable opportunities for microbial growth, and today it is generally accepted that the malting of mycotoxin-contaminated grains can present risks for both the growth of fungi, and for additional mycotoxin production. Most work to date has been conducted on the *Fusarium* mycotoxins, especially DON.

Malting of FHB-infected barley can result in dramatic increases in the counts of viable *Fusarium* and cross-contamination of uninfected grains. As previously stated, *Fusarium* spp. generally do not survive long into storage, and empirical evidence and comments from commercial maltsters suggest the general pattern is to see levels of DON decline with steeping, and then to remain low in the malt. However, a dilemma is that germination of lower DON barley occasionally results in malts with DON levels that are considerably higher than the original barley. The reasons for differences in sample behavior are not clear, but it would seem likely that development of DON (and other mycotoxins) during malting is impacted by the host, the pathogen, timing of infection, and distribution of the organism within host tissues. Some maltsters, in fact, refer to these two cases as internal versus external infection.

Avoidance of contaminated grain is obviously the best means of preventing the production of mycotoxins during malting, and/or their transfer to beer. However, this is not always possible. Maltsters and brewers set limits on mycotoxins in the incoming grain, and they will generally utilize cleaning and sizing to further reduce levels. Some research work has also focused on means to control the growth of fungi during malting, particularly *Fusarium*. In terms of the malting process, it has been reported that 70–100% of mycotoxins present in grains can be removed during the steeping phase (Schwarz 2017). The effectiveness of a number of chemical treatments such as hydrogen peroxide, as well as ozonation in the steep, has been investigated.

Conversations with commercial maltsters have also indicated that further reductions of DON and *Fusarium* biomass can be facilitated by the use of increased overflow during the first steep, use of lime ($CaCO_3$) in steeping, and the use of barley washers prior to steeping (Schwarz 2017). Observation from industry also suggests that the ability of *Fusarium* to produce DON declines with grain storage. While barley is generally not malted soon after harvest because of concerns on dormancy, extended storage may also have the added benefit to reduce levels of viable toxigenic field fungi.

Biocontrol has been investigated as a means of controlling the growth of toxigenic fungi during malting, and is an attractive measure as the organisms are often grain-derived and leave no residues (Laitila 2015). A large set of bacteria, fungi, and yeasts have shown a capacity to act as bio-competitive agents, microbial antagonists, or 'bio-transformers' of mycotoxins (Pearson et al. 2010). For instance, some lactic acid bacteria and *Geotrichum candidum* can competitively inhibit the growth of *Fusarium* spp., and some bacteria or yeast strains are able to degrade mycotoxins (Leyva Salas et al. 2017). Using starter culture technology, the microbes can be added during steeping and germination.

6 Conclusion and future trends

Storage is an important component in the sustainable production of cereals, and is a key link in maintaining quality between the producer and processor/consumer. Poor storage conditions and management can result in significant reductions in economic value, to the complete loss of otherwise acceptable grain. Damage through insect and fungal infestation is the most frequent type of loss encountered with storage of barley. Fungal growth can result in the production of several storage-associated mycotoxins, which impact the safety of the grain, and drastically restrict its utilization. Loss of germination can also occur in storage, and while this may not render the sample unfit for consumption, it will downgrade barley from higher- (e.g. malting or seed) to lower-value markets (feed).

The management of stored barley begins following harvest, and cleanliness and moisture content are immediate concerns. The storage unit, location, and associated equipment should be clean. Treatment is advised if insect infestation had been a prior issue. In the future, alternatives to chemical insecticides and fumigants are likely to gain favor because of concerns over toxicity, safety, and environmental impact. These could include treatment with natural plant-derived compounds, ozonation, or controlled atmosphere.

Grain entering storage should ideally be clean, as foreign materials can be a source of both fungal and insect contamination, damaged grains spoil easier, and immature kernels will lose germination more rapidly. Unless stored frozen, higher moisture grain (>12–13%) will be at greater risk of germination loss, insect damage, and mold growth and mycotoxin contamination. As such, it is important that harvested barley be quickly reduced to acceptable moisture levels. This can be performed in the storage unit itself or with external drying equipment. Management of temperature in drying is especially critical for malting barley, as germination must be maintained. Once in storage, grain temperature, moisture/humidity levels need to be frequently, if not continuously, monitored. Stored grain is essentially a man-made ecosystem that is constantly undergoing interactions between abiotic (temperature and relative humidity) and biotic factors (insects and fungi) (Moses et al. 2015). An operator needs to use the tools at their disposal (aeration, cooling and heating) to maintain proper conditions throughout the duration of storage. This is as, environmental conditions change across the seasons, and management practices will be slightly different between regions. However, there is now also convincing evidence that the overall climate is changing on a global basis.

The impacts of climate change on world food production have been documented, and in terms of barley, have already been reported to have had negative effects on yield and production in many of the world's barley-producing regions including much of Europe, North America, and Australia (Ray et al. 2019). However, the impact of rising temperatures and greenhouse gas levels on grain storage have generally been overlooked (Moses et al. 2015). Changing weather patterns will likely impact favorable conditions for insects and fungal pathogens, both in the field and in storage. Geographic regions of crop adaption may change, as will those of pests, pathogens, and the mycotoxins encountered. The amount and frequency of precipitation are expected to increase in some regions, which in turn will also impact drying practices and storage management. As such, individuals managing barley storage will need to adapt practices to new challenges, and will perhaps incur increased costs.

Finally, one also needs to consider that it is often necessary to store barley that was harvested with less than optimal quality. This is particularly true when there is widespread disease or PHS within a region. Sustainability and food security concerns will only emphasize the importance of increasing crop utilization. Improved technologies for the removal of mycotoxin-contaminated kernels from grain lots, such as color sorting, are needed. Treatments for the control mycotoxin development, or the degradation/transformation of mycotoxins are alternatives, but again there will be increasing pressure in terms safety, environmental impact, and economics.

7 Where to look for further information

7.1 Grain storage

A standard introduction to the subject of grain storage is Carl R. Reed's *Managing Stored Grain: To Preserve Quality and Value* (2006. American Association of Cereal Chemists, Inc. St Paul, MN, 235p, ISBN:1891127470).

Specific guidelines for storing barley are often available from academic, industrial, and governmental organizations within individual countries. A partial list of these sources was presented in Table 1.

The Food and Agriculture Organization of the United Nations (FAO) has addressed grain storage techniques in developing countries: http://www.fao.org/3/t1838e/T1838E00.htm

7.2 Fungal Pathogens and Mycotoxins

The USDA GIPSA FGIS (United States Department of Agriculture Grain Inspection, Packers and Stockyards Administration. Federal Grain Inspection Service) has prepared a basic reference on grain fungal diseases and mycotoxins:

- Grain Fungal Diseases and Mycotoxin Reference. 2006. 51 pages. https://www.gipsa.usda.gov/fgis/publication/ref/mycobook.pdf.

Other sources for mycotoxins in grain:

- John F. Leslie and Antonio Logrieco, eds. *Mycotoxin Reduction in Grain Chain*. 2014. Wiley Blackwell. Ames, IA USA, 376p.

8 References

Allen, B., Wu, J. and Doan, H. 2003. Inactivation of fungi associated with barley grain by gaseous ozone. *Journal of Environmental Science and Health – Part B Pesticides, Food Contaminants, and Agricultural Wastes*, 38 (5), 617–30.

Alshannaq, A. and Yu, J. H. 2017. Occurrence, toxicity, and analysis of major mycotoxins in food. *International Journal of Environmental Research and Public Health*, 14 (6), 632.

Association American Malting Barley. 2016. Harvesting, drying, and storing malting barley. AMBA, Milwaukee, WI, 7p. Available at: http://ambainc.org/media/AMBA_PDFs/Pubs/Harvesting_Drying_and_Storing_Barley.pdf.

Athanassiou, C. G., Korunic, Z., Kavallieratos, N. G., Peteinatos, G. G. and Boukouvala, M. C. 2006. New trends in the use of diatomaceous earth against stored-grain insects. In: *9th International Working Conference on Stored Product Protection*, Campinas, Brazil, pp. 730–40.

Aziz, N. H., Mattar, Z. A. and Mahrous, S. R. 2006a. Contamination of grains by mycotoxin-producing molds and mycotoxins and control by gamma irradiation. *Journal of Food Safety*, 26 (3), 184–201.

Aziz, N. H., Souzan, R. M. and Shahin Azza, A. 2006b. Effect of γ-irradiation on the occurrence of pathogenic microorganisms and nutritive value of four principal cereal grains. *Applied Radiation and Isotopes*, 64 (12), 1555–62.

Aziz, N. H., El-Far, F. M., Shahin, A. A. M. and Roushy, S. M. 2007. Control of fusarium moulds and fumonisin B1 in seeds by gamma-irradiation. *Food Control*, 18 (11), 1337–42.

Badr, A. N., Abdel-Fatah, S. M., Sree, Y. H. A. and Amra, H. A. 2017. Mycotoxigenic fungi and mycotoxins in Egyptian barley under climate changes. *Research Journal of Environmental Toxicology*, 11(1), 1-10.

Barrero, M., Corbineau, F., Gubler, F. and Rodrı, V. 2015. Dormancy in cereals (not too much, not so little): About the mechanisms behind this trait. *Seed Science Research*, 25, 99–119.

Batey, I. 2017. Maintaining grain quality during storage and transport. In: *Cereal Grains Assessing and Managing Quality*. Wrigley, C., Batey, I. and Miskelly, D. (Eds). Woodhead Publishing, Duxford, UK, pp. 571–90.

Beattie, S., Schwarz, P. B., Horsley, R., Barr, J. and Casper, H. H. 1998. The effect of grain storage conditions on the viability of fusarium and deoxynivalenol production in infested malting barley. *Journal of Food Protection*, 61 (1), 103–6.

Bennett, J. W. and Klich, M. 2003. Mycotoxins. *Clinical Microbiology Reviews*, 16(3), 497–516.

Bhargava, M. C. and Kumawat, K. C. 2010. *Pests of Stored Grains and Their Management*. New India Publishing, New Delhi, India, 246p.

Bolechová, M., Benešová, K., Běláková, S., Čáslavský, J., Pospíchalová, M. and Mikulíková, R. 2015. Determination of seventeen mycotoxins in barley and malt in the Czech Republic. *Food Control*, 47, 108–13.

Bonjour, E., OPit, G., Hardin, J., Jones, C., Payton, M. and Beeby, R.. 2011. Efficacy of ozone fumigation against the major grain pests in stored wheat. *Journal of Economic Entomology*, 104(1), 308–16.

Bosch, L., Pfohl, K., Avamidis, G., Wieneke, S., Viol, W. and Karlovsky, P. 2017. Plasma-based degradation of mycotoxins produced by *Fusarium*, *Aspergillus* and *Alternaria* species. *Toxins*, 9(3), 97.

Bullerman, L. B. and Bianchini, A. 2007. Stability of mycotoxins during food processing. *International Journal of Food Microbiology*, 119 (1-2), 140–6.

Campbell, H., Choo, T. M., Vigier, B. and Underhill, L. 2000. Mycotoxins in barley and oat samples from Eastern Canada. *Canadian Journal of Plant Science*, 80 (4), 977–80.

Canadian Grain Commission. 2019a. Manage stored grain. Available at: https://www.grainscanada.gc.ca/en/grain-quality/manage/.

Canadian Grain Commission. 2019b. Using RVA to measure pre-germination in barley and predict germination energy after storage. Available at: https://grainscanada.gc.ca/en/grain-research/scientific-reports/rva/.

Coufal-Majewski, S., Stanford, K., McAllister, T., Blakley, B., McKinnon, J., Chaves, A. V. and Wang, Y. 2016. Impacts of cereal ergot in food animal production. *Frontiers in Veterinary Science*, 3, 15.

Delwiche, S. R., Pearson, T. C. and Brabec, D. L. 2005. High-speed optical sorting of soft wheat for reduction of deoxynivalenol. *Plant Disease*, 89 (11), 1214–19.

Desjardins, A. E. 2006. *Fusarium Mycotoxins Chemistry, Genetics, and Biology*. The American Phytopathological Society, St. Paul, MN, pp. 13–76.

Edwards, S. G. 2009. Fusarium mycotoxin content of UK organic and conventional barley. *Food Additives and Contaminants - Part A Chemistry, Analysis, Control, Exposure and Risk Assessment*, 26 (8), 1185–90.

EFSA, 2019. Mycotoxins. https://www.efsa.europa.eu/en/topics/topic/mycotoxins.

European Food Safety Authority. 2011. Scientific opinion on the risks for animal and public health related to the presence of T-2 and HT-2 toxin in food and feed. *European Food Safe*, 9(12), 1–187.

FDA, 2017, Natural toxins and mycotoxins. https://www.fda.gov/food/chemicals-metals-pesticides-food/natural-toxins-and-mycotoxins.

Ferrigo, D., Raiola, A. and Causin, R. 2016. Fusarium toxins in cereals: Occurrence, legislation, factors promoting the appearance and their management. *Molecules*, 21 (5).

Fleurat-Lessard, F. 2017. Integrated management of the risks of stored grain spoilage by seedborne fungi and contamination by storage mould mycotoxins - an update. *Journal of Stored Products Research*, 71, 22–40.

Freire, L. and Sant'Ana, A. S. 2018. Modified mycotoxins: An updated review on their formation, detection, occurrence, and toxic effects. *Food and Chemical Toxicology*, *111*, 189–205.

Garstang, J., Spink, J. Suleimenov, M., Schillinger, W., McKenzie, R., Tanaka, D., Ceccarelli, S., Paynter, B. and Fettell, N 2010. Cultural practices: Focus on major barley producing regions. In: *Barley: Production, Improvement, and Uses*. Ullrich, S. E. (Ed.). Wiley Blackwell, Ames, IA, pp. 221-50.

Gil-Serna, J., Mateo, E. M., González-Jaén, M. T., Jiménez, M., Vázquez, C. and Patiño, B. 2013. Contamination of barley seeds with fusarium species and their toxins in Spain: An integrated approach. *Food Additives and Contaminants - Part A Chemistry, Analysis, Control, Exposure and Risk Assessment*, 30 (2), 372–80.

Gomez-Lopez, V. M., Ragaert, P., Debevere, J. and Devlieghere, F. 2007. Pulsed light for food decontamination: A review. *Trends in Food Science & Technology*, 18 (9), 464–73.

Grains Research & Development Corporation. 2016. GRDC grownnotes: Barley section 13: Storage. GRDC, Canberra, Australia, 14p. Available at: https://grdc.com.au/__data/assets/pdf_file/0024/369240/GrowNote-Barley-South-13-Storage.pdf.

Hellevang, K. 1990a. Estimated allowable storage time for malting barley (Days). *Agricultural and University Extension of North Dakota State University*, Fargo, ND. Available at: https://www.ag.ndsu.edu/graindrying/documents/BarleyAST.pdf.

Hellevang, K. 1990b. Grain equilibrium moisture content charts. *Agricultural and University Extension of North Dakota State University*, Fargo, ND, pp. 1–9. Available at: https://www.ag.ndsu.edu/graindrying/documents/GrainEquilibriumMoistureContentCharts.pdf.

Hellevang, K. 1996. Crop storage management. Agricultural and University Extension of North Dakota State University, Fargo, ND, pp. 1–7. Available at: https://www.ag.ndsu.edu/graindrying/documents/ae791.pdf.

Hellevang, K. 2014. Malt barley dry and storage (video presentation). *Agricultural and University Extension of North Dakota State University*, Fargo, ND. Available at: https://www.ag.ndsu.edu/graindrying/presentations-2.

Hellevang, K. 2016. Consider pros, cons of alternative grain storage methods. *Agricultural and University Extension of North Dakota State University*, Fargo, ND, pp. 1–2. Available at: https://www.ag.ndsu.edu/alerts/soybean-storage/resources/ConsiderProsConsofAlternativeGrainStorageMethods.pdf.

Hellevang, K. 2013. Grain drying. *Agricultural and University Extension of North Dakota State University*, Fargo, ND, pp. 1–23. Available at: https://www.ag.ndsu.edu/publications/crops/grain-drying.

Hellevang, K. 2019. Grain drying and storage. *Agricultural and University Extension of North Dakota State University*, Fargo, ND. Available at: https://www.ag.ndsu.edu/graindrying.

HGCA. 2011. *HGCA Grain Storage Guide for Cereals and Oilseeds*, 3rd Edition. HGCA Publications, Stoneleigh Park, Warwickshire, UK, 28p.

Ibáñez-Vea, M., González-Peñas, E., Lizarraga, E. and De Cerain, A. L. 2012a. Co-occurrence of aflatoxins, ochratoxin A and zearalenone in barley from a Northern Region of Spain. *Food Chemistry*, 132 (1), 35–42.

Ibáñez-Vea, M., González-Peñas, E., Lizarraga, E. and De Cerain, A. L. 2012b. Co-occurrence of mycotoxins in Spanish barley: A statistical overview. *Food Control*, *28* (2), 295–98.

Ibáñez-Vea, M., Lizarraga, E., González-Peñas, E. and De Cerain, A. L. 2012c. Co-occurrence of type-A and type-B trichothecenes in barley from a northern region of Spain. *Food Control*, 25 (1), 81–8.

Izydorczyk, M. S. and Edney, M. 2017. Grain-quality characteristics for specific cereals. In: *Cereal Grains Assessing and Managing Quality*. Wrigley, C., Batey, I. and Miskelly, D. (Eds). Woodhead Publishing, Duxford, pp. 195–228.

Jacela, J. Y., Derouchey, J. M., Tokach, M. D., Goodband, R. D., Nelssen, J. L., David, G. and Dritz, S. S. 2010. Practice tip feed additives for swine: Fact sheets – Flavors and mold inhibitors , mycotoxin binders, and antioxidants. *Journal of Swine Health and Production* 18 (1), 27–32.

Jayas, D. S. 2012. Storing grains for food security and sustainability. *Agricultural Research*, 1(March), 21-4.

Juan, C., Berrada, H., Mañes, J. and Oueslati, S. 2017. Multi-mycotoxin determination in barley and derived products from Tunisia and estimation of their dietary intake. *Food and Chemical Toxicology*, 103, 148–56.

Kim, J. C., Kang, H. J., Lee, D. H., Lee, Y. W. and Yoshizawa, T. 1993. Natural occurrence of fusarium mycotoxins (trichothecenes and zearalenone) in barley and corn in Korea. *Applied and Environmental Microbiology*, *59* (11), 3798–802.

Kottapalli, B., Wolf-Hall, C. E. and Schwarz, P. 2005. Evaluation of gaseous ozone and hydrogen peroxide treatments for reducing fusarium survival in malting barley. *Journal of Food Protection*, 68 (6), 1236–40.

Kumar, D. and Kalita, P. 2017. Reducing postharvest losses during storage of grain crops to strengthen food security in developing countries. *Foods*, 6(1), 8.

Kumar, P., Mahato, D. K., Kamle, M., Mohanta, T. K. and Kang, S. G. 2017. Aflatoxins: A global concern for food safety, human health and their management. *Frontiers in Microbiology*, 7, 2170.

Kuruc, J. A., Schwarz, P. and Wolf-Hall, C. 2015. Ochratoxin A in stored U.S. barley and wheat. *Journal of Food Protection*, 78 (3), 597-601.

Laitila, A. 2015. Toxigenic fungi and mycotoxins in the barley-to-beer chain. In: *Brewing Microbiology*. Hill, A. E. (Ed.). Woodhead Publishing Series in Food Science, Technology and Nutrition. Woodhead Publishing, Oxford, pp. 107-39. Chapter 6.

Laroussi, M. and Leipold, F. 2004. Evaluation of the roles of reactive species, heat, and UV radiation in the inactivation of bacterial cells by air plasmas at atmospheric pressure. *International Journal of Mass Spectrometry*, 233 (1–3), 81–6.

Lee, H. J. and Ryu, D. 2017. Worldwide occurrence of mycotoxins in cereals and cereal-derived food products: Public health perspectives of their co-occurrence. *Journal of Agricultural and Food Chemistry*, 65 (33), 7034–51.

Lee, S. H., Park, S. Y., Byun, K. H., Chun, H. S. and Ha, S. D. 2017. Effects of microwaves on the reduction of aspergillus flavus and aspergillus parasiticus on brown rice (*Oryza Sativa* L.) and barley (*Hordeum Vulgare* L.). *Food Additives and Contaminants – Part A Chemistry, Analysis, Control, Exposure and Risk Assessment*, 34 (7), 1193–200.

Leyva Salas, M., Mounier, J., Valence, F., Coton, M., Thierry, A. and Coton, E. 2017. Antifungal microbial agents for food biopreservation – A review. *Microorganisms*, 5 (3), 37.

Lindblad, M., Gidlund, A., Sulyok, M., Börjesson, T., Krska, R., Olsen, M. and Fredlund, E. 2013. Deoxynivalenol and other selected fusarium toxins in Swedish wheat – occurrence and correlation to specific fusarium species. *International Journal of Food Microbiology*, 167 (2), 284–91.

Lorenz, K. and Hoseney, R. C. 1979. Ergot on cereal grains. *CRC Critical Reviews in Food Science and Nutrition*, 11 (4), 311–54.

Lorini, I. and Beckel, H. 2006. Efficacy of "Diatomaceous Earth" to control the main stored grain pests. In: *9th International Working Conference on Stored Product Protection*, pp. 863-7.

Maenetje, P. W. and Dutton, M. F. 2007. The incidence of fungi and mycotoxins in South African barley and barley products. *Journal of Environmental Science and Health – Part B Pesticides, Food Contaminants, and Agricultural Wastes*, 42 (2), 229–36.

Magan, N., Sanchis, V. and Aldred, D. 2004. The role of spoilage fungi in seed deterioration. In: *Fungal Biotechnology in Agricultural, Food and Environmental Applications*. Arora, D. K. (Ed.). Marcell Dekker, Inc., New York, pp. 311–23.

Maherani, B., Hossain, F., Criado, P., Ben-Fadhel, Y., Salmieri, S. and Lacroix, M. 2016. World market development and consumer acceptance of irradiation technology. *Foods*, 5 (4), 79.

Malmauret, L., Parent-Massin, D., Hardy, J.-L. and Verger, P. 2002. Contaminants in organic and conventional foodstuffs in France. *Food Additives & Contaminants*, 19 (6), 524–32.

Mannaa, M. and Kim, K. D. 2017. Influence of temperature and water activity on deleterious fungi and mycotoxin production during grain storage. *Mycobiology*, 45 (4), 240–54

Maragos, C. M. and Plattner, R. D. 2002. Rapid fluorescence polarization immunoassay for the mycotoxin deoxynivalenol in wheat. *Journal of Agricultural and Food Chemistry*, 50 (7), 1827–32.

Marroquín-Cardona, A. G., Johnson, N. M., Phillips, T. D. and Hayes, A. W. 2014. Mycotoxins in a changing global environment – A review. *Food and Chemical Toxicology*, 69, 220–30.

Martos, P., Thompson, W. and Diaz, G. 2010. Multiresidue Mycotoxin analysis in wheat, barley, oats, rye and maize grain by high-performance liquid chromatography-tandem mass spectrometry. *World Mycotoxin Journal*, 3 (3), 205–23.

McClurkin, J. D., Maier, D. E. and Ileleji, K. E. 2013. Half-life time of ozone as a function of air movement and conditions in a sealed container. *Journal of Stored Products Research*, 55, 41–7.

McKenzie, K., Sarr, A., Mayura, K., Bailey, R., Miller, D., Rogers, T., Norred, W., Voss, K., Plattner, R., Kubena, L. and Phillips, T.. 1997. Oxidative degradation and detoxification of mycotoxins using a novel source of ozone. *Food and Chemical Toxicology*, 35(8): 807–20.

Meca, G., Zinedine, A., Blesa, J., Font, G. and Mañes, J. 2010. Further data on the presence of fusarium emerging mycotoxins enniatins, fusaproliferin and beauvericin in Cereals available on the Spanish markets. *Food and Chemical Toxicology*, 48 (5), 1412–16.

Mendez, F., Maier, D. E., Mason, L. J. and Woloshuk, C. P. 2003. Penetration of ozone into columns of stored grains and effects on chemical composition and processing performance. *Journal of Stored Products Research*, 39 (1), 33–44.

Micotoxins.info. 1998. Regulations. Available at: http://www.mycotoxins.info/regulations/.

Milićević, D. R., Škrinjar, M. and Baltić, T. 2010. Real and perceived risks for mycotoxin contamination in foods and feeds: Challenges for food safety control. *Toxins*, *2* (4), 572–92.

Miller, J. D. 1995. Fungi and mycotoxins in grain: Implications for stored product research. *Journal of Stored Products Research*, 31 (1), 1–16.

Mohapatra, D., Kumar, S., Kotwaliwale, N. and Singh, K. K. 2017. Critical factors responsible for fungi growth in stored food grains and non-chemical approaches for their control. *Industrial Crops and Products*, 108 (June), 162–82.

Moisan, M., Zakrzewski, Z. and Rostaing, J. C. 2001. Waveguide-based single and multiple nozzle plasma torches: The TIAGO concept. *Plasma Sources Science and Technology*, 10 (3), 387–94.

Moretti, A., Logrieco, A. F. and Susca, A. 2017. Mycotoxins: An underhand food problem. *Methods in Molecular Biology* 1542, 3–12.

Moses, J. A., Jayas, D. S. and Alagusundaram, K. 2015. Climate change and its implications on stored food grains. *Agricultural Research*, 4 (1), 21–30.

Mrema, G., Gumbe, L., Chepete, H. and Agullo, J. 2011. Grain crop drying, handling and storage. In: *Rural Structures in the Tropics Design and Development*. FAO, Rome,

Italy, pp. 363–85. Chapter 16. Available at: http://www.fao.org/docrep/015/i2433e/i2433e00.htm.

Nakamura, S. 2018. Grain dormancy genes responsible for preventing pre-harvest sprouting in barley and wheat. *Breeding Science*, 304, 295–304.

Navarro, S. 1991. *Aeration and Cooling for Control of Stored Grain Insects. Conference: International Course on Agricultural Engineering Technology of Grain Storage*, Bet Dagan, Israel, pp. 1–14. Available at: https://www.researchgate.net/publication/235927423.

Ostry, V., Malir, F., Toman, J. and Grosse, Y. 2017. Mycotoxins as human carcinogens – The IARC monographs classification. *Mycotoxin Research*, 33 (1), 65–73.

Pearson, T. C., Wicklow, D. T. and Brabec, D. L. 2010. Characteristics and sorting of white food corn contaminated with mycotoxins. *Applied Engineering in Agriculture*, 26 (1), 109–13.

Perkowski, J. 1998. Distribution of deoxynivalenol in barley kernels infected by *Fusarium*. *Molecular Nutrition and Food Research 42* (2), 81–3.

Peters, J., Dam, R. Van, Doorn, R. Van, Katerere, D. and Berthiller, F., Haasnoot, W., Nielen, M. W. F. 2017. Mycotoxin profiling of 1000 beer samples with a special focus on craft beer. *PLoS ONE*, 1–27.

Piacentini, K. C., Rocha, L. O., Savi, G. D. and Carnielli-Queiroz, L. 2019. Assessment of toxigenic fusarium species and their mycotoxins in brewing barley grains. *Toxins (Basel)*, 11(1), 31.

Pinstrup-Andersen, P. 2001. The future world food situation and the role of plant diseases. *The Plant Health Instructor*. doi:10.1094/PHI-I-2001-0425-01.

Prickett, A. J., Macdonald, S. and Wildey, K. B. 2014. *Survey of Mycotoxins in Stored Grain from the 1999 Harvest in the UK*. Home Grown Cereals Authority, London, UK.

Rabie, C. J., Lübben, A., Marais, G. J. and Jansen Van Vuuren, H. 1997. Enumeration of fungi in barley. *International Journal of Food Microbiology*, 35 (2), 117–27.

Ramakrishna, N., Lacey, J. and Smith, J. E. 1991. Effect of surface sterilization, fumigation and gamma irradiation on the microflora and germination of barley seeds. *International Journal of Food Microbiology*, 13 (1), 47–54.

Ramakrishna, N., Lacey., J. and Smith, J. 1996. Aspergillus flavus colonization and Aflatoxin B1 formation in barley grain during interaction with other fungi. *Mycopathologia* 136(1): 53–63.

Ramos, A. J., Labernia, N., Marín, S., Sanchis, V. and Magan, N. 1998. Effect of water activity and temperature on growth and ochratoxin production by three strains of *Aspergillus ochraceus* on a barley extract medium and on barley grains. *International Journal of Food Microbiology*, 44 (1–2), 133–40.

Ray, D. K., West, P. C., Clark, M., Gerber, J. S., Prishchepov, A. V. and Chatterjee, S. 2019. Climate change has likely already affected global food production. *PLoS ONE*, 14 (5), 1–18.

Rees, D. 2007. *Insects of Stored Grain: A Pocket Reference*, CSIRO Publishing, Collingwood, Victoria Australia, 66p.

Rice, R. P. and Graham, D. M. 2001. U.S. FDA regulatory approval of ozone as an antimicrobial agent – What is allowed and what needs to be understood. Food and Drug Administration. www.ozonia.com/media/pdf/app/FDA-e.pdf.

Schwarz, P. B. 2017. Fusarium head blight and deoxynivalenol in malting and brewing: Successes and future challenges. *Tropical Plant Pathology*, 42 (3), 153–64.

Schwarz, P. B., Horsley, R. D., Steffenson, B. J., Salas, B. and Barr, J. 2006a. Quality risks associated with the utilization of fusarium head blight infected malting barley. *Journal of the American Society of Brewing Chemists*, 64 (1), 1-7.

Schwarz, P. B., Neate, S. M. and Rottinghaus, G. E. 2006b. Widespread occurrence of ergot in upper midwestern US barley, 2005. *Plant Disease*, 90 (4), 527.

Schwarz, P. B., Qian, S. Y., Zhou, B., Xu, Y., Barr, J., Horsley, R. D. and Gillespie, J. 2014. Occurrence of deoxynivalenol-3-glucoside on barley from the Upper Midwestern United States. *Journal of the American Society of Brewing Chemists*, 72 (3), 208–13.

Scudamore, K. A., Patel, S., Breeze, V., Scudamore, K. A. and Patel, S. 2010. Surveillance of stored grain from the 1997 harvest in the United Kingdom for Ochratoxin A. *Food Additives & Contaminants*, *16* (7), 281-90.

Selcuk, M., Oksuz, L. and Basaran, P. 2008. Decontamination of grains and legumes infected with *Aspergillus* Spp. and *Penicillum* Spp. by cold plasma treatment. *Bioresource Technology*, 99 (11), 5104–9.

Semple, R. L., Hicks, P. A., Lozare, J. V. and Castermans, A. 1992. Storage losses and their estimation. In: *Towards Integrated Commodity and Pest Management in Grain Storage*. Semple, R. L., Hicks, P. A., Lozare, J. V. and Castermans, A. (Eds). FAO, Rome, Italy. Available at: http://www.fao.org/3/x5048e/x5048E00.htm#Contents.

Shank, R. A., Foroud, N. A., Hazendonk, P., Eudes, F. and Blackwell, B. A. 2011. Current and future experimental strategies for structural analysis of trichothecene mycotoxins - a prospectus. *Toxins* 3(12), 1518-53.

Sun, C., Ji, J., Wu, S., Sun, C., Pi, F., Zhang, Y., Tang, L., Sun, X. 2016. Saturated aqueous ozone degradation of deoxynivalenol and its application in contaminated grains. *Food Control*, 69, 185–90.

Temba, B. A., Sultanbawa, Y., Kriticos, D. J., Fox, G. P., Harvey, J. J. W. and Fletcher, M. T. 2016. Tools for defusing a major global food and feed safety risk: Nonbiological postharvest procedures to decontaminate mycotoxins in foods and feeds. *Journal of Agricultural and Food Chemistry*, 64(47), 8959–72.

Tiwari, B. K., Brennan, C. S., Curran, T., Gallagher, E., Cullen, P. J. and O'Donnell, C. P. 2010. Application of ozone in grain processing. *Journal of Cereal Science*, 51(3), 248–55.

Vila-Donat, P., Marín, S., Sanchis, V. and Ramos, A. J. A. 2018. A review of the mycotoxin adsorbing agents, with an emphasis on their multi-binding capacity, for animal feed decontamination. *Food and Chemical Toxicology*, 114, 246–59.

Vongvilai, P., Isaka, M., Kittakoop, P., Srikitikulchai, P., Kongsaeree, P., Prabpai, S. and Thebtaranonth, Y. 2004. Isolation and structure elucidation of enniatins L, M1, M2, and N: Novel hydroxy analogs. *Helvetica Chimica Acta*, 87 (8), 2066–73.

Wawrzyniak, J., Ryniecki, A. and Gawrysiak-Witulska, M. 2013. Kinetics of mould growth in the stored barley ecosystem contaminated with *Aspergillus westerdijkiae*, *Penicillium viridicatum* and *Fusarium poae* at 23-30°C. *Journal of the Science of Food and Agriculture*, 93 (4), 895–901.

Woods, J. and Fanvier, J. 1994. Predicting the germinative energy of dormant malting barley during storage. *Journal of Institute of Brewing*, 100, 257-69.

Yazar, S. and Omurtag, G. Z. 2008. Fumonisins, trichothecenes and zearalenone in cereals. *International Journal of Molecular Sciences*, 9 (11), 2062-90.

Zhang, H., van der Lee, T., Waalwijk, C., Chen, W., Xu, J., Xu, J., Zhang, Y. and Feng, J. 2012. Population analysis of the *Fusarium Graminearum* species complex from wheat in China show a shift to more aggressive isolates. *PLoS ONE*, *7* (2), 1–14.

Chapter 10

Fungal diseases affecting barley

Robert S. Brueggeman, Shyam Solanki, Gazala Ameen and Karl Effertz, Washington State University, USA; Roshan Sharma Poudel, North Dakota State University, USA; and Aziz Karakaya, Ankara University, Turkey

1 Introduction
2 Understanding plant genetic resistance to fungal pathogens
3 Biotrophic foliar diseases: stem rust
4 Leaf rust
5 Stripe rust
6 Powdery mildew
7 Necrotrophic diseases: spot blotch
8 Net blotch
9 Ramularia leaf spot
10 Septoria speckled leaf blotch
11 Scald
12 Fusarium head blight
13 A seed-borne disease: barley stripe
14 Conclusion
15 References

1 Introduction

Barley (*Hordeum vulgare* L.) was one of the earliest domesticated crops, and its germplasm subsequently flowed with human migration from five possible loci of domestication, with three in the Fertile Crescent region, one in the Syrian Desert and one in the Tibetan Plateau (Poets et al., 2015). From this mosaic of domestication, the hardy crop was distributed to distant geographies and is now one of the most widely cultivated cereals. On marginal arable farmlands, which include lands with high abiotic stress factors including low moisture, alkaline and saline soils, as well as cool summer temperatures, it has the capability to outperform most other cereals (Mathre, 1997). Barley has been forced onto more marginal lands because of its hardiness and competition from other

http://dx.doi.org/10.19103/AS.2019.0060.16

crops. Thus, profitable barley production on marginal lands or as a rotation alternative requires even more adaptability and further incorporation of abiotic stress traits like water use efficiency and residual herbicide tolerance. Since barley has such a wide range of cultivation covering diverse environments from seacoast low lands to high mountain plateaus across multiple continents, this adaptable cereal is exposed to a wide taxonomy of potential pathogens. These pathogens of barley include fungi, bacteria, viruses and nematodes, but the major class that threaten production are the fungal pathogens. In *Compendium of Barley Diseases*, Mathre (1997) described 73 diseases of barley of which 36 are caused by fungal pathogens. Those that currently cause consistent economic losses are the fungal diseases described in this chapter.

Sustainable barley production is, and more imperatively will be, anchored by breeding programs' primary goal of releasing varieties that are high yielding and produce quality grain across dynamic environments (Horsley et al., 2009). Fungal diseases have negative impacts on these agronomic parameters; thus, in disease epidemic years the selection for resistance equates to the yield and quality enhancement. Due to the low acreage of barley production in many growing regions barley-breeding resources are limited. Varieties are only being developed that contain resistances to the most important diseases, ones that cause losses on an annual basis. However, in the face of climate uncertainty these decisions can become complex. Thus, for barley breeders, it will be even more challenging to meet sustainable barley production goals in the face of climate change. Varieties will need to be stable across uncertain regional environments, which may also require incorporating resistances to a wider diversity of pathogens. Climate change and uncertainty may expand the prevalence and diversity of pathogens that can cause major disease epidemics, but due to the increase of erratic weather patterns across years, different predominant diseases may impact production during any given growing season. Thus, in these uncertain times, breeders and other support scientists must have a thorough understanding of the dynamic disease spectrum and adequate integrated pest management strategies. These strategies must first and foremost take into account the development of disease-resistant varieties and cultural practices. However, for some diseases fungicide application is also necessary for management to meet production demands, especially for the brewing and spirit distilling industries, multi-billion-dollar industries across the world.

Recent studies have determined that climate change presents a major threat to barley production as yield losses are projected to occur due to higher temperatures and water deficiency in some growing regions (Xie et al., 2018). However, as global climate change results in drought in some regions of the world, others will experience excess precipitation. In combination with elevated temperatures, the excess precipitation will make environments more conducive to the development of fungal pathogen epidemics. Thus, without proper disease management practices, climate change will exacerbate disease problems in these regions. This double-edged sword of rising temperatures was not addressed in the recent predictions of future world barley shortages (Xie et al., 2018), suggesting that there could be even greater production shortfalls as a result of enhanced disease pressure in growing regions that experience excess or even adequate precipitation.

Because of the wide distribution of barley, it continuously encounters diverse potential fungal pathogens that have evolved to be specialized on the crop, causing significant disease epidemics. These heterotrophic fungal pathogens are broadly classified into two groups, the biotrophic and necrotrophic pathogens, with both classes obtaining fixed carbon nutrients from their hosts, yet have evolved different strategies for acquisition. The biotrophic pathogens require living host cells to acquire nutrients, typically through feeding structures that require an intimate association with living host cells to complete their lifecycles. On the other hand, the necrotrophic pathogens require dead or dying host tissue to feed and complete their lifecycles.

2 Understanding plant genetic resistance to fungal pathogens

In regard to sustainable barley production, genetic resistance to the fungal pathogens is paramount and is a major focus of this chapter. To understand resistances to the diverse species and different classes of fungal pathogens and to deploy effective resistances, a basic understanding of the strategies employed by both the host and the pathogen for a positive outcome, resistance or virulence, respectively, is important. In the last couple of decades advances in genetic and molecular capabilities have allowed for a better understanding of the plant immune system as well as pathogen virulence.

The plant innate immune system evolved to identify and respond to potential pathogen challengers through receptors that recognize their presence early in the infection process. The first line of active immunity in plants relies on the recognition of highly conserved molecules associated with entire classes of microbes, designated as *m*icrobe-*a*ssociated *m*olecular *p*atterns (MAMPs). These molecules are essential for the microbe's fitness; thus, the microbe population has little plasticity to diversify MAMPs. During the host-pathogen molecular arms race, plants evolved cell-surface pattern recognition receptors (PRRs) to recognize MAMPs, which activate early basal defense mechanisms that are effective against diverse potential pathogens. The major fungal MAMP, chitin, is a structural component of fungal cell walls, and the recognition of this conserved molecule results in physiological reprogramming from normal cellular function toward rapid defense responses, collectively known as pattern-triggered immunity (PTI) (Jones and Dangl,

2006). Because PTI responses are effective at arresting early colonization, fungi evolved virulence effectors to suppress PTI responses, effectively becoming specialized pathogens that could infect and colonize a specific host, which in this chapter are fungi specialized to cause diseases of barley (Couto and Zipfel, 2016; Göhre et al., 2008; Zhang et al., 2017; He et al., 2006; Boch et al., 2014; Read et al., 2016). Plants counter-evolved a second layer of resistance (R) genes that encode predominantly cytoplasmically localized R-proteins, typically containing the nucleotide-binding-leucine-rich repeat (NLR) protein domain architecture. The NLRs effectively recognize pathogen virulence effectors, including those utilized to overcome PTI. Pathogen detection by NLR immunity receptors activates programmed cell death (PCD) responses, known as the hypersensitive response (HR), that are able to sequester and kill biotrophic pathogens in a foci of dead cells (Win et al., 2012; Dodds and Rathjen, 2010). Because biotrophic pathogens produce haustoria or haustoria-like feeding structures to acquire nutrients from living host cells, these dominant resistances typically follow Flor's gene-for-gene hypothesis (Flor, 1956). The gene-for-gene hypothesis follows a functional model where a single dominant resistance gene encodes an immunity receptor protein that recognizes a pathogen effector protein that is encoded from a single dominant avirulence gene. This gene-for-gene interaction elicits a HR that sequesters and kills the invading biotrophic fungal pathogens.

The necrotrophic fungal pathogens evolved to produce host-specific necrotrophic effectors (NEs), including small secreted proteins and low-molecular-weight metabolites that interact with dominant host susceptibility genes, typically encoding immunity receptors including NLRs and PRRs (Shi et al., 2016; Faris et al., 2010). Thus, host genotypes lacking the receptor or a variant that doesn't recognize the NE contain what is considered a recessive resistance to the necrotroph in a genetic context (Wolpert et al., 2002; Faris et al., 2010; Liu et al., 2011, 2015; Stergiopoulos et al., 2013; Shjerve et al., 2014; Shi et al., 2016). These NE-immunity receptor interactions follow the inverse-gene-for-gene model (Friesen et al., 2007), where the necrotrophs evolved NEs to purposefully interact with and activate host immunity receptors to effectively hijack the plant's immune system by triggering PCD to facilitate necrotrophic fungal growth on the resulting dead and dying tissues (Faris et al., 2010; Shi et al., 2016). The resulting compatible or susceptible reactions are called necrotrophic-effector-triggered susceptibility (NETS) (Liu et al., 2015). Thus, when breeding for disease resistances to biotrophic vs. necrotrophic pathogens, two antagonistic approaches must be considered. For biotrophs the focus is to incorporate one or many functional resistance genes. However, for the necrotrophs, it requires the elimination of dominant susceptibility genes/targets, which, based on the dominant susceptibility genes characterized to date, represent functional immunity receptor-like proteins (Faris et al., 2010; Shi et al., 2016). However, from the breeder's point of view, these loci are deployed in the homozygous state so the consideration of dominant resistance or recessive resistance/dominant susceptibility is of little consequence when selecting and deploying the positive alleles that result in disease resistance.

Incorporating genetic resistance is the most sustainable management practice for most fungal diseases of barley, and thus is the major focus of fungal disease management in this chapter. For each disease there is a brief description of the known disease resistance/susceptibility genes or quantitative trait loci (QTL) that have been identified through bi-parental mapping populations. However, association mapping using diverse natural populations and nested association mapping (NAM) population has emerged as a powerful tool to identify genes and/or loci contributing to resistance/susceptibility to many fungal diseases in barley, including stem rust (Sallam et al., 2017; Zhou et al., 2014), net form net blotch (Richards et al., 2017; Amezrou et al., 2018; Adhikari et al., 2019; Daba et al., 2019), spot form net blotch (Tamang et al., 2015; Wang et al., 2013; Burlakoti et al., 2017) and scald (Daba et al., 2019). Since many of these loci are not well characterized, each section focusing on specific diseases will not contain a description of these marker trait associations that have been recently identified via association mapping. However, this does not reflect that the authors do not see the importance of the loci identified by this powerful new genetic mapping technique.

Although genetic resistances are the focus of sustainable management practices of the fungal diseases described here, the management of some fungal diseases, such as fusarium head blight and ramularia leaf spot, is nearly impossible via host genetics because of the lack of effective resistance genes in the primary barley germplasm pool. Thus, adequate disease management to these insidious diseases requires integrated management strategies that include genetic tolerance to toxin accumulation, cultural practices such as rotation to non-host crops to reduce inoculum, as well as fungicide use. However, the management of these and other generalist diseases in other crops, where little effective resistances are available, may be targets for resistance through biotechnological strategies for sustainable disease management and production in the future.

Several previous publications have described important fungal pathogens of barley including *Compendium of Barley Diseases* (Mathre, 1997), *Common Diseases of Small Grain Cereals* (Zillinsky, 1983) and *Barley Production, Improvement and Uses* (Ullrich, 2010). This chapter focuses on the description of major fungal pathogens affecting barley production around the globe and new insights into their mechanisms of virulence and implications on deploying sustainable resistances to these important pathogens.

3 Biotrophic foliar diseases: stem rust

The rusts infect and damage most crops, but those causing disease on the small cereal grains, barley and wheat, historically caused the largest economic losses, food security issues and societal unrest and instability. Historical reports of rust diseases on cereal crops can be traced back to the time of the ancient Greece and Roman empires. Aristotle (384–322 B.C.), the Greek philosopher and scientist, observed and documented that 'the rust was brought on by warm vapors', showing that he made the correlation between a warm moist environment and rust disease onset. On the other hand the Roman sacrificed red animals during the spring festival of Robigallia to satisfy the rust god Robigus in hopes of saving their cereal crops from his wrath (Schumann and Leonard, 2000). Scientific observation and research, beginning with the observations of Aristotle, has compiled a wealth of knowledge about the fungal pathogens, especially those that cause the rust diseases. We now have an extensive list of effective genetic resistances that are used to successfully manage rust diseases. However, these diseases are still feared and considered a threat to food security due to the pathogen's adaptability and its potential to quickly evolve to overcome deployed resistances. The following sections review rust diseases, starting with stem rust.

Stem rust, caused by the obligate biotrophic fungal pathogen *Puccinia graminis*, is a notorious pathogen of the small grain cereals (Leonard and Szabo, 2005) and has a special place in the history of cereal disease epidemics and crop loss. *P. graminis* spores have been identified in archaeological sites in Israel containing grain bins dating back to 1300 B.C. (Kislev, 1982). Thus, historically and prehistorically, dating back to the dawn of barley and wheat domestication in the Fertile Crescent, stem rust is believed to have played a role in crop failure and famine.

Depending upon primary host specificity, many '*forma specialis*' (f. sp.) of *P. graminis* were described in the literature including wheat stem rust caused by *P. graminis* f. sp. *tritici* (*Pgt*), which infects wheat and barley; rye stem rust caused by *P. graminis* f. sp. *secalis*, which infects rye and barley; and oat stem rust caused by *P. graminis* f. sp. *avenae*, which infects oats and barley. The secondary host for *Pgt* is barberry or *Mahonia* spp., which shelters pycniospores and aeciospores, whereas the primary hosts (barley and wheat) shelters urediniospores, teliospores and basidiospores (Aurthur, 1934). *P. graminis* is macrocyclic and heteroecious, producing five different types of spores to complete its lifecycle. The onset of yearly infections starts early in the spring with the alternate host via basidiospores, when environmental conditions are conducive for pathogen infection, warm (27–30°C) and moist conditions that allow for ample dew formation on leaf surfaces or precipitation during the night. Sexual recombination on the secondary host occurs, allowing the pathogen to generate recombinant progeny with new virulence gene combinations. Aecia and aeciospores form on the abaxial surface of the host leaf, acting as a primary source of inoculum for the cereal host. When favorable conditions repeat, asexual spores, dikaryotic urediniospores (n + n), act as a secondary source of inoculum in the fields (Schumann and Leonard, 2000). Urediniospores are produced in large amounts and can travel across very long distances on wind currents, exemplified by the annual phenomenon across the ocean from Australia to New Zealand (Luig, 1985) and from the South to North up the North American Great Plains, popularly known as the 'Puccinia Pathway' (Schumann and Leonard, 2000). *P. graminis* urediniospores germinate on barley leaf surfaces and then grow perpendicularly to the leaf veins until they encounter a stomata and form an appressorium over the guard cells. Historically, it was concluded that until the daylight induces the stomatal opening, *P. graminis* appressoria remained quiescent and did not penetrate through stomatal guard cells (Yirgou and Caldwell, 1963). However, recent evidence suggested that invasion is light independent (Solanki, 2017; Solanki et al., 2019). Upon successful penetration of the stomatal guard cells, the tube between the guard cells initiates substomatal growth with primary infection hyphae further branching into secondary infection hyphae that grow intercellularly until contacting mesophyll cells, which trigger differentiation into haustorial mother cells. The haustorial mother cells invaginate into the host mesophyll cells forming haustoria, which act as a nutrient conduit and colonization powerhouse for the pathogen. Apart from the haustoria acting as a feeding structure, it facilitates pathogen manipulation of the host. The pathogen hijacks host cell physiology by utilizing its effector repertoire that it secretes from haustoria to establish a nutrient sink, which leads to efficient pathogen feeding and further colonization. Eventually a lifecycle shift results in the formation of uredinia, with millions of urediniospores erupting from the epidermis of the stem and leaf surfaces approximately 10–14 days after spore germination, fulfilling the pathogen's main goal of reproduction (Leonard and Szabo, 2005). Urediniospores act as the secondary inoculum causing polycyclic disease when the environmental conditions are conducive and a suitable host is available. Upon the conclusion of the favorable conditions near the end of the growing season, mature uredinia form telia, producing teliospores. Stem rust is also known as black rust due to the physical appearance of infected barley plant during senescence, since telia are black in color. Teliospores overwinter through harsh climatic conditions and germinate to produce basidiospores and restart the *Puccinia* lifecycle.

Signs of pathogen infection occur primarily on the stem and leaf sheaths but may also be found on leaves and glumes. Typical signs of *Pgt* are characterized by small chlorotic flecks, which occur at 4–5 days after inoculation, progressing into round to elongated diamond-shaped, brick red uredinial lesions on its cereal host at ~8–10 days post spore germination (Roelfs, 1985; Roelfs and Bushnell, 1985), rupturing the epidermal cells on the stem and abaxial and adaxial leaf surfaces (Fig. 1). Severe disease results in heavy yield losses due to the reduction in the photosynthetic area, loss of water through epidermal rupture at the point of pathogen spore production, and channeling of plant nutrients toward the pathogen infection sight. Thus, the pathogen colonization results in the overall reduction of plant vigor and stem breakage that ultimately leads to further losses due to lodging (Roelfs and Bushnell, 1985; Leonard and Szabo, 2005). Historically stem rust caused devastating epidemics to cereal crops worldwide (Singh et al., 2015), notably in the regions of the northern Great Plains of the USA (Eversmeyer and Kramer, 2000), western Canadian Prairies (Steffenson, 1992b), Australia, eastern Europe (Zadoks, 1967), Indian subcontinent (Nagarajan and Joshi, 1975) and northeast Africa (Kolmer, 2005).

In barley, the single dominant R-gene, *Rpg1*, has effectively managed stem rust in the upper Great Plains of the USA and western Canadian Prairies since 1942 (Steffenson, 1992a). However, a race of stem rust, *Pgt* race QCCJB,

Figure 1 Stem rust field infection with the upper-left panel showing a close-up photo of a barley stem with urediniospores erupting through the epidermis.

was identified in the early 1990s in North Dakota that was virulent on *Rpg1* (Steffenson, 1992b; Brueggeman et al., 2002). *Pgt* race QCCJB caused minor epidemic and yield loss in northern Great Plains barley until the race QCCJB-susceptible wheat cultivar TAM 105 was removed from production in the lower Great Plains. The removal of TAM 105 alleviated the buildup of race QCCJB inoculum that was spreading the *Rpg1* virulent pathogen to barley fields (Eversmeyer and Kramer, 2000).

The emergence of the virulent *Pgt* race TTKSK (A.K.A. Ug99) and its lineage in Uganda, Africa, in 1998, which was virulent on barley containing *Rpg1* and multiple wheat stem rust resistance genes including *Sr34,* posed a major threat to world food security due to its potential to cause major yield losses (Saari and Prescott, 1985; Singh et al., 2008; Bhattacharya, 2017). *Pgt* race TTKSK was shown to be virulent on 95% of the barley accessions assayed, including those carrying *Rpg1,* threatening barley production in North America if or when it was introduced (Steffenson et al., 2013). This new race and its variants spread very rapidly across the African continent and the Middle East, causing crop failure when conducive conditions and a susceptible host were present.

In barley, the *rpg4/Rpg5*-mediated resistance locus (RMRL) provides broad resistance to most *P. graminis* races/isolates including race TTKSK (Steffenson et al., 2016). The genes at the locus required for resistance at the RMRL have been identified (Brueggeman et al., 2008; Wang et al., 2013). This locus was also shown to provide dominant resistance to races of rye stem rust caused by *P. graminis* f. sp. *secalis* (Steffenson et al., 1995) and oat stem rust caused by *P. graminis* f. sp. *avenae* (Dracatos et al., 2015). Although isolates or races have been reported as virulent against both barley-stem rust resistance genes, no isolates have been identified that are virulent against both. Recent attempts to pyramid *Rpg1* and RMRL into elite malting barley backgrounds, which together provide resistance to all known races of stem rust, resulted in non-function of both genes when they were introgressed into certain backgrounds together, suggesting that at some level the two resistance mechanisms interact. It was determined that to function together a third gene designated *Rrr1* (required for *rpg4* resistance 1) was needed (Sharma Poudel et al., 2018). Thus, this gene, closely linked to the RMRL locus, must also be selected for when pyramiding these two resistance sources. To identify new sources of stem rust resistance in the *Hordeum* primary germplasm pool, exhaustive screening and genome-wide association mapping efforts have also recently yielded novel stem rust resistance genes/QTL (Sallam et al., 2017; Zhou et al., 2014) that still require further characterization. These continued efforts are warranted by the dynamic evolution of this pathogen that will require gene pyramiding from different sources to achieve and maintain sustainable stem rust resistance in barley.

Interestingly, it must be noted that when we refer to *Pgt* isolates that have been classified into diverse pathogen 'races', it is based on their virulence/

avirulence to specific wheat stem rust resistance genes (Jin et al., 2008). Yet it must be pointed out that these race designations appear to have little relevance in regard to barley interactions that result in virulence/avirulence as to date only two effective *Pgt* resistance genes, *Rpg1* and *rpg4*, have been characterized in barley (Brueggeman et al., 2002, 2008). Considering the very limited number of specific resistance genes identified in barley and the fact that the currently known resistances resemble broad nonspecific resistances, it can be speculated that barley is a near non-host or recent host of *Pgt*. This is the most plausible explanation for the limited numbers of specific resistance genes that have evolved in barley, likely due to the putative short history of coevolution of this pathosystem.

4 Leaf rust

Leaf rust on barley, also known as brown rust, is caused by the obligate biotrophic fungal pathogen *Puccinia hordei* (*Ph*). Leaf rust is the most common and widely occurring rust disease of barley, yet the pathogen does not consistently cause severe epidemics. It is only a concern in growing regions when the crop matures late in the season and the environment is conducive for disease epidemics. In Australia, leaf rust is considered one of the five major barley diseases (Murray and Brennan, 2010), causing potential yield losses up to 60% when susceptible varieties are grown under environmental conditions conducive for disease development. Depending upon disease severity, it was determined that for each 1% increase in disease severity on the flag leaf, penultimate leaf or whole plant, potential yield losses of 0.77% (Melville et al., 1976), 0.40% (King and Polley, 1976) and 0.60% were shown, respectively (Park et al., 2015). Late season epidemics can also result in decreased kernel weight and higher protein content, affecting the acceptance for malting, which is where producers receive premium prices that make barley production much more profitable.

P. hordei is a heteroecious and macrocyclic pathogen that infects its primary host barley and various wild *Hordeum* species in its uredinial and telial spore stages, whereas the pycnial and aecial stages reside on the alternate hosts of several genera (*Ornithogalum*, *Leopoldia* and *Dipcadia*) in the Liliaceae family (Anikster, 1984; Clifford, 1985). As the lifecycle of leaf rust mirrors that of stem rust, which was previously described in detail, it will not be described again here. Airborne urediniospores are the primary source of inoculum of barley, and initial signs of infection are small green islands on the leaf surfaces. Successful infection and colonization, primarily through the stomatal openings of a susceptible host, results in the eruption of uredinia through the leaf epidermis and urediniospore release. Uredinia of *P. hordei* emerge mainly on the adaxial leaf surface but can also occur on the abaxial surface and leaf blades, appearing as orange to orange-brown minute circular to oval pustules encompassed by chlorotic halos (Fig. 2). During conditions conducive for disease epidemics, glumes, awns and stems can also be infected. Upon host maturation and senescence, leaf rust uredinia transition to dark brown, resting spore telia, giving the disease its alternate name 'brown rust'. Leaf rust has become a concern in recent decades in temperate barley-growing environments where optimum temperatures throughout the year and intensive agricultural practices provided optimal conditions for leaf rust epidemics in these barley-growing regions.

The deployment of genetic resistance provides the most cost effective and environmentally friendly option for the management of leaf rust. Genetic resistances to *P. hordei* can be seedling, adult plant or incomplete resistances conferred by the *Rph* (Resistance to *P. hordei*) genes or QTL. However, virulence has been reported for most of the characterized seedling *Rph* genes. To date, 22 resistance loci for seedling resistance and two adult plant resistance genes have been identified from cultivated and wild barley lines (Park et al., 2015).

Figure 2 Leaf rust field infection with the upper-left panel showing a close-up photo of a barley leaf with urediniospores erupting through the epidermis.

Apart from the *Rph8* locus, all the other seedling resistance genes have been genetically mapped to regions of the barley genome. The identification of these resistance sources is important for broad-spectrum resistance breeding and gene pyramiding for constructing durable resistance against leaf rust. Clifford (1985) first suggested using a set of ten differential genotypes to monitor the virulence of *Ph*. Thus, unlike the barley-stem rust pathosystem, there are barley-specific leaf rust isolates that can be race-typed with barley differential resistance genes.

Aside from genetic resistance, cultural practices can be adopted to reduce sources of the primary inoculum by removing alternate hosts and volunteer plants. This effectively removes the green bridge. Alternate sowing dates can also be a useful tactics to avoid diseases by limiting exposure to high inoculum loads. Disease forecasting models have also been developed to assist decisions on the timing of foliar fungicide application (Clifford, 1985). However, the use of expensive chemical applications in regions such as the northern Great Plains of the USA and Australia where barley production demands low-input costs can prove uneconomical.

5 Stripe rust

Stripe rust of barley is caused by the obligate biotrophic fungal pathogen *Puccinia striiformis* Westend. f. sp. *hordei* Eriks. & Henn. *P. striiformis* f. sp. *hordei* is a destructive disease in many growing regions of the world and is considered one of the most important diseases of barley in the western USA (Line, 2002). Barley stripe rust was first described in Europe in the late nineteenth century, and minor epidemics causing significant yield losses occurred in the UK and Europe through the 1960s (Dantuma, 1964; Nover and Scholz, 1969). A large-scale epidemic occurred in Columbia due to *P. striiformis* f. sp. *hordei* race 24 in 1975, a race that was introduced from Europe (Dubin and Stubbs, 1986; Marshall and Sutton, 1995). The pathogen spread in South America and into North America where it was detected in Mexico by 1987, the southern USA (Texas) by 1991 and the northern USA (Washington) by 1995 (Chen and Line, 1999; Line, 2002; Marshall and Sutton, 1995; Wellings, 2011). Establishment of *P. striiformis* f. sp. *hordei* led to epidemics in South America in 1988 (Sandoval-Islas et al., 1998) and the Pacific Northwest in the USA in 1996 (Line, 2002). Stripe rust is the most commonly occurring rust disease in the Pacific Northwest of the USA because of the relatively cool climatic conditions that favor stripe rust infection during the barley-growing season. Today, the pathogen is endemic in Asia, Africa, Europe, and North and South America (Wellings, 2011). It is also considered a potential threat to Australia (Dracatos et al., 2016). The presence of *P. striiformis* f. sp. *hordei* has not been recorded in Australia, but it represents a future threat when it arrives on the continent, as a study undertaken by CIMMYT, in Mexico, determined that 70% of Australian barley varieties are susceptible to the aggressive *P. striiformis* f. sp. *hordei* race 24, indicating a preemptive need to incorporate genetic resistance(s) into certain Australian adapted materials or varieties (Dracatos et al., 2016).

Stripe rust infection effectively reduces the photosynthetic area or capability of the leaves, severely affecting the plant and causing a significant reduction in the weight and numbers of kernels. The losses due to stripe rust can be 40 to 70% (Marshall and Sutton, 1995). In the Transhimalayan region of India, 66% losses due to stripe rust were reported from fields sown with susceptible barley cultivars (Vaish et al., 2011).

Unlike stem rust and leaf rust, stripe rust is favored by cool temperatures ranging from 11 to 15°C, coupled with intermittent rain and dew for effective infection and disease epidemic development. However, the disease can continue to progress at higher temperatures and in relatively drier conditions (Milus et al., 2009). Recently, Barberry species were discovered as an alternate host for *P. striiformis* (Jin et al., 2010), indicating that it can also be a source of early infection and cause inoculum buildup as well as contribute to genetic diversity within the pathogen population. Yet wind-borne urediniospores remain the primary means of dispersal and infection of the primary host, barley. Initial symptoms appear in seven to ten days post inoculation on susceptible barley in the form of visible chlorotic spots at the sight of infection. Later, yellow-orange-colored elongated uredinia develop in the form of stripes along the leaf veins and glumes, characteristic of the disease name. On severely infected plants the stripe pattern may not be evident and a random pattern can be observed. Late in the growing season dark-colored teliospores are produced on the senesced barley plants, serving the purpose of survival and overwintering. Depending upon the cropping system and environment, the stripe rust lifecycle can adapt. In Asia and warmer areas of Europe where a continuous cycle of cereals are grown or volunteers are present, a year-round green bridge can provide a continuous source of stripe rust inoculum. However, in regions that experience harsh winters, the pathogen can survive as urediniomycelium and teliospores, which will overwinter and sporulate upon the arrival of warmer temperatures.

Considerable variation is present in the *P. striiformis* f. sp. *hordei* isolates collected in the USA, yet barley genotypes containing resistance to all evaluated races have been identified. Currently ~70 different *Psh* races have been characterized, indicating a large amount of variability present in the pathogen population. In 1998, another form of stripe rust specialized to infect wild barley grasses was discovered in Australia (Wellings et al., 2000), now known as barley grass yellow rust (*P. striiformis* f. sp. *pseudo-hordei*). Later it was found that it can also cause disease on certain Australian commercial cultivars, and thus needs to be monitored.

Genetic, cultural and chemical approaches can be used to control stripe rust; however, an integrated approach is advocated. Barley stripe rust is an obligate fungal pathogen; thus, a green bridge between growing seasons facilitates pathogen survival and contributes to the early season inoculum the following season. Minimizing the green bridge reduces the initial inoculum load for the next cropping season, thus reducing the incidence and severity of disease. The use of animal grazing or herbicides is recommended to limit overwintering plants. Early planting is also recommended to reduce disease severity especially in the western USA, where stripe rust frequently occurs in late planted barley. Irrigation scheduling to reduce the length of moist periods in the field also reduces infection.

Foliar fungicides are effective against stripe rust, and different applications are recommended depending upon disease severity. Seed treatments in conjunction with foliar applications later in the season are also suggested. In regions where stripe rust occurs regularly, the use of foliar systemic fungicides containing triazoles is recommended.

Although cultural practices and fungicide use can be utilized to help manage the disease, breeding for resistance remains the most economically viable and sustainable method of management. In barley, 26 *Psh* resistance loci (Chen and Line, 1999) have been analyzed, but only five *Rps* (Resistance to *P. striiformis)* loci have been mapped (Chen and Line, 1999; Chełkowski et al., 2003). In a recent study, 45 new barley genotypes showing adult plant resistance for Asian *Psh* races in the field were described (Verma et al., 2018). Interestingly, many of the genes identified in barley for stripe rust resistance are recessive, deviating from the classic gene-for-gene, dominant, race-specific resistances commonly identified for rusts. Thus, it is not known if these resistance mechanisms are as prone to breakdown as typical rust resistance genes. However, the continued identification of novel sources of effective resistances is important for the pyramiding of both seedling and adult plant resistance in barley.

6 Powdery mildew

Powdery mildew is the widest spread foliar disease of barley, producing the most consistent and damaging disease epidemics in northern Europe and Japan (Mathre, 1997). It causes considerable damage in temperate zones with continuous cereal cropping systems, but is endemic in subtropical and tropical areas as well (Oerke et al., 2012; Mathre, 1997). In temperate regions, powdery mildew can cause barley yield losses of 15–20% and sporadically up to 40% (Chaure et al., 2000). Powdery mildew is a major disease of barley in central Europe, with losses reported up to 14% (Oerke et al., 2012; Mathre, 1997). Yield losses due to powdery mildew are typically accompanied by a reduction of kernel weight, tiller number, grain protein and root growth. A study determined that powdery mildew infection showed yield losses of 26% that were accompanied by kernel weight loss of 11% and a 9% grain protein reduction (Mathre, 1997). This same study also determined that resistant varieties still experienced up to 7% yield losses.

Powdery mildew in barley is caused by the ascomycete fungus *Blumeria graminis* (DC.) E. O. Speer, f. sp. *hordei* emend. É. J. Marchal (anamorph: *Oidium monilioides* Link) (= *Bgh*) (Mathre, 1997). *B. graminis* f. sp. *hordei* is an obligate biotrophic fungal pathogen and can survive only on a live host. Its host range is restricted to *H. vulgare* and wild *Hordeum* species. The pathogen can infect all aerial tissues but is mainly observed on the adaxial surface of the leaves. Disease development is optimal between 15 and 22°C and is reduced dramatically above 25°C. Severe disease epidemics are more common in cooler regions with humid climates. The early signs of *B. graminis* f. sp. *hordei* infection initially appear as small fluffy white powdery patches of mycelia and conidia on the surface of the infected leaf (Fig. 3). As the disease progresses, it moves from the lower to upper leaves and eventually to the stem and head of the crop. The fungal colonies eventually expand across the leaf surface, impairing photosynthesis and leading to yellowing and chlorosis of the infected leaves. Moderate to severe disease can cause the leaves to curl and die. The fungal colonies on the older leaves and heads develop fruiting bodies (chasmothecia, formerly known as cleistothecia), which appear as small black or brown specks on the mycelial mat (Braun et al., 2002; Heffer et al., 2006).

B. graminis f. sp. *hordei* can survive as mycelium on volunteer grasses or straw in milder climates, or overwinters as chasmothecia on cereal straw and stubble in regions that experience harsh winters. It can spread by both windborne ascospores (sexual spores from chasmothecia) and conidia (asexual spores) (Ridout, 2009). These spores are unique because they germinate on barley in the absence of free moisture, but the optimum germination occurs at 100% relative humidity and over a wider temperature range (1–30°C). The disease cycle starts with the germination of spores (ascospore and/or conidium) upon landing on a susceptible host. The spore elongates to form a primary germ tube (PGT) and penetrates directly into the host epidermal cuticle. The formation of the PGT is a distinct characteristic of *B. graminis* f. sp. *hordei* infection compared to other powdery mildew. Within 3–4 hours of infection, a second appressorial germ tube (AGT) is produced that elongates and differentiates into an appressorium. The appressorium attaches to the leaf surface and produces an infection peg to breach the host cuticle and cell wall. A specialized multidigitate haustoria is developed within host epidermal cells that mediates the transfer of nutrients from host to pathogen. The infection progresses to give rise to conidiophores from which conidia, the asexual spores, form (Heffer et al., 2006; Ridout, 2009). Within 4–6 days of infection, the fungus

Figure 3 Powdery mildew greenhouse infection with the upper-left panel showing a close-up photo of a barley leaf with lesions having the darkly pigmented chasmothecia formed within the mycelial mat.

gives rise to large numbers of conidia that become wind borne and initiate the secondary cycle of infection. As the plant matures, a darkly pigmented globose-shaped chasmothecia containing sexual spores (ascospores) are formed within the mycelial mat (Fig. 3). The chasmothecia remain dormant during the winter or summer in the absence of a viable host. Later, under favorable conditions, the chasmothecia break open by soaking up water and discharge ascospores from asci. The ascospores can be spread by wind and may serve as a source of the primary inoculum to initiate barley infection.

Powdery mildew in barley can be effectively controlled by an integrated approach of cultural practices and the use of resistant varieties and/or fungicides. Cultural practices such as crop rotation, removal of crop residues and volunteer plants can eliminate the green bridge, reducing the density of overwintering chasmothecia or mycelia and minimizing the local inoculum. However, exogenous spores travelling with the prevailing wind can cause devastating effects in the absence of proper resistant varieties and/or fungicide use. The use of fungicides to control powdery mildew has been a common practice in Europe and other regions. Several classes of fungicides have been utilized to effectively control powdery mildew (Coghlan et al., 1991; Brown et al., 1996). However, additional cost associated with fungicide applications and several reports of fungicide-resistant *B. graminis* f. sp. *hordei* isolates require an integrated approach to effectively manage this disease. The use of biological control agents, including compounds isolated from microorganisms like mycelial extracts from *Bipolaris oryzae*, *Pythium ultimum* and *Rhizopus stolonifer*, may serve as environmentally friendly alternatives to fungicides in controlling *B. graminis* f. sp. *hordei* (Haugaard et al., 2002).

The development and deployment of *B. graminis* f. sp. *hordei*–resistant barley cultivars is the most sustainable approach to managing powdery mildew (Dreiseitl, 2011; Ames et al., 2015). Substantial efforts were invested in the identification of powdery mildew resistance from both cultivated (Jørgensen, 1992) and wild barley (Dreiseitl and Dinoor, 2004; Ames et al., 2015). However, the deployment of single R-genes without broad-spectrum resistance suffered the fate of the 'boom-and-bust' cycle during the exerted selection for new virulence types in the pathogen population. The most effective and extensively deployed powdery mildew resistance genes in barley are *mlo* and *Mla*. The basal defense resistance gene *mlo* has provided broad-spectrum resistance against powdery mildew for nearly 40 years (Jørgensen, 1992). Unfortunately, the use of the *mlo* allele in barley has been attributed to yield penalty and, surprisingly, enhanced susceptibility to the necrotophic pathogens *Magnaporthe oryzae*, *Cochliobolus sativus*, *Fusarium graminearum*, *Pyrenophora teres* and *Ramularia collo-cygni* (McGrann et al., 2014).

The knowledge of the molecular mechanisms of resistance genes, effector biology and non-host resistance is and will continue to significantly contribute to the development of more durable disease control against powdery mildew (Ridout, 2009). *B. graminis* f. sp. *hordei* is a rapidly evolving and adaptive pathogen; thus, the identification of non-host resistances and selective breeding can help achieve adequate levels of durable resistance. Likewise, R-genes are known to guard pathogen virulence targets in a host; thus, studies focused on the characterization of *B. graminis* f. sp. *hordei* effectors could help to identify virulence targets. Such findings can lead to the identification of alleles of the virulence targets with natural polymorphism that can resist pathogen effector manipulation. The deployment of these genes would reduce susceptibility as well as the reliance on race-specific R-genes (Ridout, 2009). On the pathogen side, the understanding of effectors will allow for the selection for R-genes that target conserved effectors and could be deployed in barley for more durable resistance (Avrova and Knogge, 2012).

7 Necrotrophic diseases: spot blotch

The necrotrophic foliar diseases of barley make up a large proportion of the fungal diseases that threaten barley production worldwide. Research of the molecular mechanisms governing the host-pathogen interactions of this class of fungal pathogens has recently expanded our knowledge into how these pathogens manipulate the host to induce disease. The following paragraphs on the necrotrophic pathogens will describe each pathogen, but will also focus on resistance to the pathogens, which typically follow the inverse gene-for-gene interaction model. The discussion starts with spot blotch.

The disease spot blotch caused by the necrotrophic pathogen *Bipolaris sorokiniana* (Sacc.) Shoem. is one of the most common and economically devastating diseases of barley (Mathre, 1982). Spot blotch is prevalent worldwide, covering all five continents wherever barley and wheat are cultivated, including the upper Midwest region of the USA and the Canadian Prairie provinces (Acharya et al., 2011; Gupta et al., 2018). It was reported that this disease was identified from every sample collected from barley fields of Manitoba, Canada, in the year 2002 (Tekauz, 2003). The disease is severe in the warm and humid environment found in subtropical regions (Acharya et al., 2011; Gupta et al., 2018). Yield losses of about 30% are common and can exceed this level under favorable disease development conditions (Mathre, 1982; Wilcoxson et al., 1990). Spot blotch was reported to cause more than 30% yield losses, which accounted for average annual losses of $1 to $2 million in northern New South Wales and Queensland, Australia, across multiple growing seasons (Murray and Brennan, 2010).

B. sorokiniana (teleomorph: *C. sativus* (S. Ito & Kurib.) Drechsler ex Dastur) is an ascomycete fungal pathogen belonging to the class Dothideomycetes. Shoemaker proposed the name '*Bipolaris*' to the *Helminthosporium* species, which produces fusoid, straight or curved conidia with three to ten septations that germinate to produce two germ tubes, one from each end, hence the name *Bipolaris*. Spot blotch is one of the most destructive diseases of both barley and wheat (*Triticum aestivum*). It can also infect wild emmer (*Triticum dicoccoides*), triticale, oats (*Avena sativa*), rye (*Secale cereale*), *Brassica campestris*, maize (*Zea mays*), sorghum (*Sorghum bicolor*), soybean (*Glycine max*) and many grasses including switch grass (*Panicum virgatum*) and *Brachypodium distachyon*. The characteristic symptoms of spot blotch are brown necrotic lesions on the leaves with or without chlorosis, depending on the pathotype of the pathogen and host genotype (Fig. 4) (Fetch and Steffenson, 1994, 1999; Mathre, 1997). *B. sorokiniana* is capable of causing common root rot of barley and wheat, and can also infect the spikes, rendering the seeds shriveled with black discoloration, known as black point or kernel blight. In 1999, the leaf spot disease severity scale was developed by Fetch and Steffenson, ranging from 1 to 9 based on

Figure 4 Spot blotch-infected barley plants in the field. The upper-left panel shows the typical spot blotch necrotic lesions on a susceptible barley variety.

the relative size of the necrotic lesions and the chlorotic margins on barley seedlings. The 1 to 9 rating scale was summarized as 1–3 as resistant, 4–5 as moderately resistant, 6–7 as moderately susceptible and 8–9 as susceptible.

B. sorokiniana is a polycyclic plant pathogen that tends to infect the older leaves in the canopy first and progresses up the plant toward the flag leaves (Brandle et al., 1987; Raemaekers, 1988). Long durations of leaf wetness (>12 h) with high relative humidity and temperatures from 18 to 32°C, favor conidia germination and infection (White and Rodriguez-Aguilar, 2001). The disease cycles begin when the asexual conidia land on the surface of the leaf, germinate from both ends of the spore and directly penetrate the cuticle of the leaves by appressorium (Braun and Howard, 1994; Kumar et al., 2002). The biotrophic phase is confined to the first host epidermal cell penetrated, and then invades the neighboring cells by multiple branching of infection hyphae, which marks the switch to the pathogen's necrotrophic phase (Kumar et al., 2002). The necrotrophic phase is associated with the cell death (Santén et al., 2005), with necrotic and/or chlorotic lesions expanding on the leaves. The pathogen then sporulates from the lesions, producing multiple secondary asexual cycles and giving rise to multiple infection cycles by conidia. *B. sorokiniana* can survive as a saprophyte and can overwinter as mycelia or conidia on crop residue or infected seed (Kiesling, 1985). During the next growing season, the overwintering mycelia or conidia germinate to produce the asexual spores as a source of the primary inoculum (Duczek, 1984). The saprophytic stage, shared with other pathogenic fungi, has provided them with the opportunity for horizontal gene transfer of virulence factors. For example, the ToxA NE has been reported in *Parastagonospora nodorum*, in *Pyrenophora tritici-repentis* and recently in *B. sorokiniana*. Recently, it has been shown that a horizontal gene transfer event between these wheat necrotrophic pathogens gave rise to the spread of ToxA (McDonald et al., 2018; Friesen et al., 2018).

In the upper Midwest regions of the USA, three pathotypes were designated based on their virulence pattern on the three barley differential lines, ND5883, Bowman and NDB112. Isolates having low virulence on all three are pathotype 0, whereas isolates having high virulence on the ND5883 and low virulence on the other two are pathotype 1 and high virulence on Bowman and low virulence on the other two differentials are pathotype 2 (Valjavec-Gratian and Steffenson, 1997). Recently, an isolate was identified as having high virulence on all three differentials and was denoted as pathotype 3 (Zhong et al., 2015). The high virulence of pathotype 2 isolate ND90Pr was mapped on a bi-parental population created between ND90Pr and ND93-1 to the *VHv1* locus (Valjavec-Gratian and Steffenson, 1997; Zhong et al., 2002). The *VHv1* locus had 43 predicted genes, out of which two were secondary metabolites belonging to nonribosomal peptide synthetases (NRPSs) and were unique to the pathotype 2 isolate. The deletion of one of the two genes effectively reduces virulence on barley cultivar Bowman (Condon et al., 2013).

Spot blotch can be managed by cultural practices, such as the destruction of the previous year's crop residues, which serve as a primary source of inoculum (Duczek et al., 1999). Burying, burning or removal of the residues can achieve the destruction of crop residues. Crop rotation with a non-host crop or with a fallow can also decrease the primary inoculum in the field (Mathre, 1997). Two- or more-year rotations with flax (*Linum usitatissimum*) have been reported to decrease viable inoculum of *B. sorokiniana* in the soil (Conner et al., 1996). Other strategies include seed treatment by systemic fungicides or foliar fungicide application. However, fungicide applications incur extra input cost for the farmer, making host resistance the most desirable means of effectively controlling this disease.

Resistance to spot blotch in North American barley varieties was deployed in the mid-1960s and has remained effective for more than 55 years (Steffenson et al., 1996; Bilgic et al., 2006; Zhou and Steffenson, 2013). This remarkably durable resistance was derived from the line NDB112 and has been successfully deployed into six-row malting barley cultivars in the region (Wilcoxson et al., 1990; Zhou and Steffenson, 2013). NDB112 was selected from the cross CI 7117-77 x Kindred at North Dakota State University (Wilcoxson et al., 1990a). The major source of this resistance was mapped in a doubled haploid population of Steptoe (susceptible) by Morex (resistant) to a 3.3 cM region on the short arm of chromosome 7H (Steffenson et al., 1996) designated as Reaction to *C. sativus* 5 (*Rcs5*). The *Rcs5* region was saturated with markers and further refined to a 2.8 cM region, encompassing an approximate 2 Mb physical region (Drader et al., 2009; Drader and Kleinhofs, 2010). The *Rcs5* gene (now referred to as *rcs5*) was recently shown to be a recessive resistance gene or more appropriately functions as a dominant susceptibility gene present in spot blotch susceptible cultivars of barley (Ameen et al., 2016). The *rcs5* resistance is provided by the absence of the induction of the two tightly linked wall-associated kinases, designated as Susceptibility to *B. sorokiniana* 1 and 2 (*Sbs1* and *Sbs2*) in barley (Ameen et al., unpublished). *Rcs5* has remained responsible for the majority of seedling and some adult plant resistance in both six-row and two-row upper Midwestern USA and Canadian Prairie province barley cultivars (Steffenson et al., 1996; Bilgic et al., 2005, 2006; Zhou and Steffenson, 2013). Although, the *rcs5* locus has remained effective for a remarkable 55 years, reports of *B. sorokiniana* pathotypes virulent on the NDB112 pose a threat to this resistance. Therefore, screening barley for new resistance sources against the evolving *B. sorokiniana* pathogen is imperative for future sustainable barley production in growing regions where *B. sorokiniana* can cause disease epidemics.

In a study conducted by Çelik Oğuz et al. (2016), the seedling reactions of 25 advanced barley lines to 5 Turkish isolates of *C. sativus* were evaluated with no line exhibiting low infection responses. Two barley lines showed an intermediate infection response to all five isolates, and one barley line showed an intermediate infection response to four isolates. However, differential virulence was reported among the isolates, suggesting diversity within virulence effectors in the isolate collection used. Balcı et al. (2018) assessed the seedling reactions of 22 hulless barley cultivars and genotypes to two isolates of *C. sativus*. One hulless barley cultivar showed low infection response, whereas the hulless barley cultivars Yalın and Özen showed intermediate infection responses. Virulence difference between the two isolates was reported. Çelik Oğuz and Karakaya (2017) evaluated the reactions of 39 barley cultivars. Cultivar Vamıkhoca 98 exhibited low infection response to all three isolates used. Cultivar Zafer 160 showed low infection responses to two isolates and showed intermediate infection response to one isolate. Cultivar Fahrettinbey showed low infection response to one isolate and showed intermediate infection responses to two isolates. Cultivars Balkan 96 and Erginel 90 showed intermediate infection responses to all three isolates used. These studies screening for new resistances within Turkish lines that are effective against Turkish *C. sativus* isolates showed that virulence differences among the isolates was evident yet resistance was present within the advances barley lines and cultivars tested.

8 Net blotch

The necrotrophic fungal pathogen *Pyrenophora teres* Drechsler (anamorph *Drechslera teres* [Sacc.] Shoem.) is the causal agent of the disease net blotch of barley. Net blotch is an economically important barley disease around the globe causing yield losses ranging from 10 to 100%, with typical epidemics causing between 10 and 40% yield loss when susceptible varieties are grown under conditions conducive to disease development (Mathre, 1997; Murray and Brennan, 2010). Foliar infection of *P. teres* has the largest impact on yield, but the infection of other aerial structures (kernels, leaf sheaths and stems) also impacts yield and quality (Liu et al., 2011). Because malt-quality-grade barley brings the producer premium prices, a major concern is the negative effect that the disease has on kernel plumpness, size and bulk density (Grewal et al., 2008).

The net blotch pathogen has evolved into two very closely related forms, *P. teres* f. *teres* and *P. teres* f. *maculata*, which cause distinct symptoms. *P. teres* f. *teres* is the causal agent of net form net blotch (NFNB), which is characterized by the longitudinal and transverse necrotic striations produced on foliar tissue, forming net pattern lesions (Fig. 5) typically surrounded by chlorosis (Shipton et al., 1974). NFNB was first described by Atanasoff and Johnson in 1920. Symptoms develop on the leaves within 24 hours of infection, starting as pinpoint lesions. On susceptible genotypes, the lesions expand to eventually form the typical netting symptoms (Liu et al., 2011). Smedegård-Petersen (1971) suggested the intraspecific classification of a new form of *P. teres,* through his observation of spot-type symptom on barley leaves. This new form was named *P. teres* f. *maculata*, which is the causal agent of spot form net blotch (SFNB). Typical symptoms of SFNB include the presence of dark brown elliptical necrotic lesions surrounded by chlorosis (Fig. 5). The symptoms of SFNB begin as pinpoint lesions and expand on susceptible barley genotypes (McLean et al., 2009). These expanded lesions are indistinguishable from spot blotch necrotic lesions caused by the pathogen *B. sorokiniana*. Besides the distinct lesions produced by the two forms of *P. teres*, they cannot be differentiated based on morphological characteristics including spore morphology. However, molecular markers specific to the mating-type genes of the pathogens can be used for the diagnosis of the forms causing disease epidemics (Keiper et al., 2008; Lu et al., 2010).

P. teres f. *teres* and *P. teres* f. *maculata* can be hybridized and produce fertile recombinant progeny under laboratory conditions (Smedegård-Petersen, 1971; Campbell et al., 1999) that produce intermediate circular spot symptoms

Figure 5 Barley field infected with a mix of both net form net blotch and spot form net blotch where nearly all the leaf tissue senesced due to pathogen infection. The upper-right panel shows the typical net form net blotch lesions, and the upper-right panel shows the typical spot form net blotch lesions.

with jagged edges. It has been shown that *P. teres* f. *teres* and *P. teres* f. *maculata* do not commonly mate in nature (Rau et al., 2003, 2005); however, isolates collected in South Africa and Eastern Europe were shown to be naturally occurring hybrids (Campbell et al., 2002; Leišova et al., 2005).

The lifecycles of *P. teres* f. *teres* and *P. teres* f. *maculata* are essentially identical with both overwintering on barley residue as sexual fruiting bodies known as pseudothecia. At the beginning of the following growing season, ascospores are actively released from the pseudothecia, become wind borne and act as the primary inoculum. Alternatively, mycelium from infected seed may become the primary inoculum, although it may not be an important factor in disease development (Shipton et al., 1974; Jordan, 1981). Following the primary infection, the production of conidia is initiated and they act as the secondary inoculum. Multiple generations of conidia occur throughout the growing season, which can be disseminated via rain splash or wind and, unlike ascospores, are not forcibly discharged (Shipton et al., 1974; Mathre, 1997). Conidia have an important role in the infections (Karakaya et al., 2004). To complete the cycle, senescent tissue is then colonized by the pathogen, whereby pseudothecia may be formed to produce the primary inoculum for the following growing season.

Several methods are employed to manage net blotch, but an integrated pest management strategy including fungicide and genetic resistances is the most effective. As net blotch is a stubble-borne disease, crop rotation is important to reduce the primary inoculum (Liu et al., 2011). Additionally, seed treatment with the fungicides triadimenol or maneb resulted in 12.6% and 12.4% yield increase, respectively (Martin, 1985). However, as with all fungal diseases, the deployment of effective genetic resistance is considered the most sustainable and environmentally friendly means of disease management (Mathre, 1997).

For the effective deployment of genetic resistance to *P. teres*, a thorough understanding of the dynamic pathogen population is essential. Because the diverse fungal pathogen populations lead to a high level of adaptability, which allows resistance genes to be overcome relatively quickly, care and stewardship of resistance genes must be considered when deployed. Because *P. teres* reproduce both sexually and asexually, and have a high level of population diversity (Liu et al., 2011; Çelik Oğuz and Karakaya, 2017; Çelik Oğuz et al., 2018, 2019a), it is considered to have rapid evolutionary potential (McDonald and Linde, 2002). Extensive research has been conducted to determine the genetic variability of *P. teres* populations, including the diversity of virulence (Khan and Tekauz, 1982; Steffenson and Webster, 1992; Wu et al., 2003; Arabi et al., 2003; McLean et al., 2014; Leišová-Svobodová et al., 2014; Akhavan et al., 2016; Çelik Oğuz and Karakaya, 2017; Çelik Oğuz et al., 2018, 2019a). Since this pathogen has the ability to adapt quickly and overcome host genetic resistances or fungicides that exert selective pressure on the populations, the knowledge of the genetic structures of pathogen populations is important in resistance gene deployment and stewardship (McDonald and Linde, 2002). Thus, pathogen populations need to be continually monitored to intelligently deploy effective net blotch resistance in barley.

NEs previously referred to as host-selective toxins (HSTs) are typically small secreted proteins or low-molecular-weight secondary metabolite molecules produced by necrotrophic pathogens that function in pathogenicity by targeting host sensitivity proteins, facilitating the induction of PCD (Wolpert et al., 2002). The infiltration of NEs into a plant containing the susceptibility target gene results in PCD responses (Liu et al., 2015). It has been shown that *P. teres* f. *teres* produces both proteinaceous and non-proteinaceous NEs or HSTs (Smedegård-Peterson, 1971; Sarpeleh et al., 2007); however, the isolated NE proteins induced a necrotic response, whereas the low-molecular-weight molecules elicited general chlorosis. The necrosis induced by the proteinaceous NEs was host specific, as determined by the induction of PCD/necrosis on susceptible genotypes. However, chlorosis induced by the non-proteinaceous HSTs was not specific, as the symptoms appeared on susceptible and resistant barley lines as well as other non-host species (Sarpeleh et al., 2007). The characterization of these NE proteins determined that they had properties, temperature stability and light dependency, similar to NEs isolated from related fungal pathogens, such as *Ptr* ToxA from *Pyrenophora tritici-repentis* (Ballance et al., 1989; Manning and Ciuffetti, 2005; Sarpeleh et al., 2007) and Snn1 from *Parastagonospora nodorum* (Shi et al., 2016).

A proteinaceous NE was identified from *P. teres* f. *teres* designated PttNE1. Intracellular wash fluids (IWFs) from the susceptible barley variety Hector inoculated with *P. teres* f. *teres* isolate 0-1 were infiltrated into the susceptible variety Hector producing necrosis, whereas the same IWFs did not elicit symptoms in the resistant line NDB112. The IWF sensitivity was mapped on a Hector x NDB112 recombinant inbred line (RIL) population and localized to the centromeric region of barley chromosome 6H (Liu et al., 2015).

Ongoing efforts to identify *P. teres* f. *teres* NEs by pathogen genetics utilized a bi-parental population developed by crossing the isolates 15A and 6A, virulent on barley varieties Kombar and Rika, respectively, identified two virulence QTL from each isolate, designated VK1, VK2, VR1 and VR2. Mendelized progeny isolates with single virulence loci were used to phenotype a Rika x Kombar double haploid population and susceptibility to each of the four independently segregating virulence loci mapped to the centromere of chromosome 6H. This indicated that *P. teres* f. *teres* isolates 15A and 6A each harbor two NEs that target or interact with multiple genes or alleles of the same gene at the centromeric region of barley chromosome 6H (Shjerve et al., 2014).

The first documented resistance to NFNB was determined to be a single dominant R-gene (Geschele, 1928), yet subsequent analysis of genetic resistance to *P. teres* f. *teres* has uncovered a remarkably complex interaction with dominant, incomplete, and recessive resistances identified (Koladia et al., 2017; Abu Qamar et al., 2008; Friesen et al., 2006a,b; Douiyssi et al., 1996; Mode and Schaller, 1958). The quantitative nature of NFNB resistance and different nature of resistances residing at closely linked loci, for example the centromeric region of barley chromosome 6H, complicates breeding. Dominant resistance genes need to be incorporated into the genome while recessive resistance or dominant susceptibility targets need to be removed from elite backgrounds (Liu et al., 2011). At least 20 studies have been conducted using bi-parental populations to identify QTL associated with resistance or susceptibility to NFNB (Grewal et al., 2008; Steffenson et al., 1996; Richter et al., 1998; Manninen et al., 2000, 2006; Raman et al., 2003; Cakir et al., 2003; Ma et al., 2004 Emebiri et al., 2005; Yun et al., 2005; Friesen et al., 2006a,b; Abu Qamar et al., 2008; St. Pierre et al., 2010; König et al., 2014; Liu et al., 2015; Koladia et al., 2017) and QTL have been identified on all seven barley chromosomes. In some cases the same sources of resistance are effective at both the adult plant and seedling stages, including the chromosome 6H locus.

The centromeric region of barley chromosome 6H is an important resistance/susceptibility locus for NFNB. QTL analyses utilizing diverse barley populations and diverse pathogen isolates have identified both dominant resistances and dominant susceptibilities co-localizing to the centromeric region of chromosome 6H (Steffenson et al., 1996; Richter et al., 1998; Raman et al., 2003; Cakir et al., 2003; Ma et al., 2004; Emebiri et al., 2005; Yun et al., 2005; Manninen et al., 2000, 2006; Friesen et al., 2006a,b; Grewal et al., 2008; Abu Qamar et al., 2008; St. Pierre et al., 2010; Liu et al., 2015; Koladia et al., 2017). Since many studies resulted in the identification of dominant resistance and susceptibility genes at the same locus, a cluster of tightly linked host sensitivity or resistance genes reside within the region, which interact with a diverse repertoire of NE or avirulence proteins present within the pathogen populations.

Efforts to fine-map and clone NFNB resistance/susceptibility genes have made relatively slow progress due to the complexity involved. High-resolution mapping of the 6H locus utilizing 2976 Rika x Kombar F_2 recombinant gametes was used to fine-map the recessive resistance genes *rpt.r* and *rpt.k*, hereafter referred to as the dominant susceptibility gene *Spt1*, to the two NFNB isolates, 6A and 15A. It was observed that Rika was susceptible to isolate 6A, while Kombar was resistant. The opposite was true with isolate 15A, where Rika was resistant and Kombar was susceptible. F_2 segregation analysis resulted in a 3:1 (S:R), confirming the nature of the susceptibility as dominant (Abu Qamar et al., 2008; Liu et al., 2010). The F_2 population was genotyped with the sequence tag site (STS), and single nucleotide polymorphism (SNP) markers and the dominant susceptibility gene to both isolates were mapped to a ~0.24 cM interval in the centromeric region of chromosome 6H (Richards et al., 2017), with the gene(s) being tightly linked in repulsion. These mapping efforts greatly advance the progress toward the positional cloning of the *P. teres* f. *teres* isolate 6A and 15A dominant susceptibility factor *Spt1*. Recently, a candidate gene has been identified that represents a single receptor-like protein gene with unprecedented allelic variation that corresponds to pathogen virulence specificities. The eventual cloning and characterization of this gene(s) will grant the scientific community a greater understanding of the intricate molecular mechanisms at work within this pathosystem and enhance the ability of breeders to effectively develop NFNB-resistant barley cultivars.

Several studies have reported major and minor QTL conferring SFNB resistance and/or susceptibility. These mapping studies utilizing bi-parental populations, developed from diverse sources of resistance assayed with diverse *P. teres* f. *maculata* isolates, identified QTL located on all seven barley chromosomes (Ho et al., 1996; Steffenson et al., 1996; Manninen et al., 2000, 2006; Molnar et al., 2000; Friesen et al., 2006a,b; Grewal et al., 2008; Cakir et al., 2011). These studies and those characterizing pathogen populations (Arabi et al., 2003; Karki and Sharp, 1986; Tekauz, 1990; McLean et al., 2014) show that SFNB resistance or susceptibility is complex and polygenic in nature. Although many SFNB-resistance QTL have been identified, they are just beginning to be utilized in breeding programs; thus, most barley cultivars grown are still susceptible to the regional population of *P. teres* f. *maculata*.

Compared to NFNB, relatively few SFNB-resistance QTL have been characterized to date; thus, the understanding of SFNB-resistance/susceptibility mechanisms and the deployment of these genes is lagging. Although these two pathogens are very closely related, it appears that they have distinct host-pathogen genetic interactions, and thus must be treated as separate diseases when breeding for resistance. However, since both diseases typically occupy the same environmental niche, with ebb and flow between predominate forms throughout time (Louw, 1996; Arabi et al., 1992; McLean et al., 2009; Liu et al., 2010), resistance to both must be considered in most growing regions.

It has been determined that barley landraces and wild barley (*Hordeum spontaneum*) genotypes are good sources of resistance to both forms of net blotch (Çelik Oğuz et al., 2017, 2019b,c). Yet resistance to SFNB was more common among the landraces and genotypes tested.

Dikilitas et al. (2018) compared *P. teres* f. *maculata* and *P. teres* f. *teres* isolates differing in virulence in terms of protease activity and glucose production. Virulent *P. teres* f. *maculata* isolates produced significantly more protease and glucose than the less virulent isolates. Also, virulent *P. teres* f. *teres* isolates produced significantly more protease than the less virulent isolates. Virulence mechanism in *P. teres* was closely associated with both protease and glucose production. The authors concluded that the protease enzyme secreted by these fungi could be the determining factor for aggressiveness and symptom expression.

9 Ramularia leaf spot

Ramularia leaf spot (RLS) is a late season disease in barley that is emerging as a serious threat to production across the temperate regions of Europe, New Zealand and South America (Walters et al., 2008; Havis et al., 2015). Though the first symptoms of RLS were reported more than a century ago in northern Italy (Cavara, 1893), it was not recognized as a severe problem in barley until the early 1980s (Walters et al., 2008). The pathogen causing RLS was initially described as the fungus *Ophiocladium hordei*, but was later named as *Ramularia collo-cygni* (*Rcc*) because of the special swan's neck-like shape of the fungal condiophores (collum = neck, cygnus = swan) (Sutton and Waller, 1988; Sachs, 2006). Currently this disease has been reported across European countries including Austria, Denmark, England, France, Ireland, Scotland, Switzerland and the Czech Republic, as well North and South American countries (Argentina, Chile, Columbia, Mexico and the USA) and New Zealand (Walters et al., 2008; Havis et al., 2015). More recently, there have been reports of this disease on barley in Estonia, Iceland, Poland, Russia, Spain and Slovakia (Gubiš et al., 2008; Afanasenko et al., 2012; Sooväli et al., 2014).

Typically, RLS can lead to the yield losses between 20 and 35%, but has been reported to reduce yield up to 70% in susceptible varieties (Pinnschmidt and Jorgensen; Havis et al., 2015). It can also reduce malt quality. It is unclear if the changes in host and pathogen genetics and/or agronomic practices have contributed to the sudden uprising of RLS as a threat to the global barley production (McGrann et al., 2014). The lack of recognition of this disease in past years could be partly due to the symptoms imparted by RLS that are confused as physiological leaf spots, or other diseases and rapid senescence (Havis et al., 2015). Reliable methods of fungal isolation and improved molecular diagnostic tools to detect these pathogens are now available (Havis et al., 2006; McGrann et al., 2014). With these novel tools, it is now possible to accurately distinguish RLS from other disease symptoms (*Pyrenophora teres* f. *maculata,* and senescing *B. graminis* lesions) that can resemble RLS symptoms in the field. Recently there have been reports of *R. collo-cygni* infection on wheat, rye, oats, maize and other classes of grasses including the model grass *B. distachyon* (Peraldi et al., 2014).

RLS can infect all the aerial tissues of barley, but infection chiefly occurs on the leaves and leaf sheath. The symptoms of RLS typically occur later in the growing season, after heading (Walters et al., 2008). The infection commonly starts from conidial spores landing on young leaves, requiring moisture for germination and subsequent infection. The penetration of the host is through open stomata and occurs rapidly, within 24 hours of the spore landing on the leaf surface (Sutton and Waller, 1988). The pathogen grows intracellular by forming branched hyphae to colonize the mesophyll tissues. The early symptoms can be observed within 1–2 weeks of infection and appear as brown to dark brown spots with a pale halo, typically producing 1–2 mm long lesions that are delimited by the leaf veins. The local lesions can eventually coalesce to form larger necrotic regions that may progress over the entire leaf surface. Later, the entire leaf can become chlorotic and necrotic, starting from the leaf tips and margins. These lesions resemble physiological leaf spot and thus can be misdiagnosed (Paulitz and Steffenson, 2011). However, RLS lesions can be distinguished from spot blotch and SFNB lesions because of their more spindle to ellipsoid shape. The conidiophores of *R. collo-cygni* break out of the necrotic tissue on the abaxial leaf surface and appear as white cluster in rows (Miethbauer et al., 2003). *Ramularia collo-cygni* produces a non-host-specific anthraquinoid toxin (rubellin) that causes the necrotic cell death in leaves (Miethbauer et al., 2003). Under heavy inoculum pressure, RLS can progress rapidly and severely impact the yield and quality of grains.

Though some symptoms are visible occasionally in the early growth stage of barley, they are typically observed after heading and thus RLS is considered a late season disease. The transition from the symptomless to symptom-causing phase is still not completely understood, but may depend on the developmental stage of the plant or light conditions (Walters et al., 2008; Kaczmarek et al., 2017). The pathogen is thought to have a latent endophytic phase before the symptom develops and then gradually shifts toward necrotrophy, depending on the host developmental stage and possibly light intensity (Salamati and Reitan, 2006; Walters et al., 2008). The development of symptoms and sporulation was shown to be more apparent during the transition of the host from the vegetative to reproductive stage (Salamati and Reitan, 2006). In some cases, *R. collo-cygni* can complete its lifecycle without developing any symptoms in barley (Kaczmarek et al., 2017). PCR-based investigation of the lifecycle of *R. collo-cygni* indicated that the pathogen grows systemically and asymptomatically in plants and can be detected even without the visual symptoms, depending upon the season and the barley variety (Havis et al., 2006).

The primary source of inoculum for RLS in barley is infected seed (Havis et al., 2014; Kaczmarek et al., 2017). The other source of inoculum is airborne conidia that are released from infected barley, other cereal crops (oat, wheat, rye) and wild grasses. Frei et al. (2007) demonstrated that the *R. collo-cygni* overwinters on winter barley and volunteer plants to infect spring barley. Salamati and Reitan (2006) described a secondary spore type of *R. collo-cygni* known as the asteromella stage that develops late in the season and can possibly act as a source of secondary inoculum. The asteromella-type spermogonial stage is not fully understood but is speculated as a potential site of sexual recombination (Havis et al., 2015).

RLS can be effectively managed by an integrated disease management strategy involving proper diagnosis, sanitation practices and fungicide treatment (Paulitz and Steffenson, 2011). Since the symptoms of this disease develop later in the season and could be confused with other diseases and physiological stress, an accurate diagnosis is important to manage RLS. Because *R. collo-cygni* is a seed-borne pathogen, it is imperative to obtain disease-free seeds to avoid potential epidemics (Walters et al., 2008). *R. collo-cygni* is also a wind-borne pathogen that can survive as a saprophyte in volunteer crop and stubble. Thus, cultural practices involving crop rotation with non-host crops and sanitation measures to remove the green bridge can significantly contribute to disease management. Currently, the most effective measure of RLS management is the use of fungicides with efficacy depending on the application timing that coincides with the onset of the disease (Paulitz and Steffenson, 2011; Havis et al., 2015). Strobulin-based fungicides (QoI) were effective for years to control RLS but quickly became less effective after the development and spread of several QoI-resistant *R. collo-cygni* populations across different countries (Fountaine and Fraaije, 2009; Matusinsky et al., 2010; Havis et al., 2015).

As with most fungal diseases of barley, genetic resistance would be a sustainable alternative to the costly use of fungicides and the development of fungicide-resistant *R. collo-cygni* populations; however, screening of a large collection of winter and spring barley lines did not identify any sources of effective or even substantial RLS resistance (Havis et al., 2015). These analyses provided no clear divisions between resistance and susceptibility, suggesting that the low levels of resistance are determined by small-effect QTL, which confound breeding efforts, and determining the mechanisms underlying RLS resistance (Sjokvist et al., 2019). A recent study utilizing transcriptome profiling of both the host and the pathogen recently suggested that the upregulation of the barley *Mlo* gene may mediate defenses against *R. collo-cygni,* but the pathogen has evolved effectors to suppress *Mlo* (Lewis et al., 2012), increasing susceptibility (Sjokvist et al., 2019). These studies and work by others are slowly unraveling the mystery of this pathogen's infection process and virulence, but it appears that this is another fungal pathogen of barley that is difficult to manage.

Currently the only genetic factor in barley that is known to contribute to RLS disease severity is the barley *Mlo* gene (Sjokvist et al., 2019). Non-functional mutant *mlo-11* alleles have been effectively deployed to confer powdery mildew resistance in European barley cultivars for more than 35 years (Acevedo-Garcia et al., 2014), yet the *mlo* mutant allele has been associated with enhanced susceptibility to *R. collo-cygni* and other necrotrophic fungal pathogens. Thus, previous studies have suggested that the widespread deployment of *mlo*-11 in Europe to manage powdery mildew may have resulted in the emergence of RLS as a major fungal pathogen over the past 15 years (McGrann et al., 2014). It was determined that *mlo*-11- and *Mlo*-containing barley lines had significantly different levels of susceptibility to RLS, suggesting minor RLS QTL could be selected for developing lines that are resistant to both powdery mildew and RLS (McGrann et al., 2014). However, for the sustainable management of RLS hopefully wide screens of diverse germplasm will uncover RLS-resistance sources that are effective, yet maintain resistance against other major diseases.

10 Septoria speckled leaf blotch

Septoria speckled leaf blotch (SSLB) is an economically important disease in major barley regions of Australia, Europe, North America and North Africa (Green and Bendelow, 1961; Mathre, 1997). Although SSLB has been reported to cause severe and extensive epidemics in some regions of North America, they occur sporadically (Toubia-Rahme and Steffenson, 2004; Paulitz and Steffenson, 2011). SSLB is commonly caused by the fungal pathogens *Septoria passerini* Sacc., although *Stagonospora avenae* Bisset f. sp. *triticea* T. Johnson (teleomorph: *Phaeosphaeria avenaria* (G. F. Weber) O. Eriksson. f. sp. *triticea* T. Johnson) is also occasionally observed in SSLB-infected barley tissue (Toubia-Rahme and Steffenson, 2004). The pathogen requires rain and extended moist conditions for its dispersal and germination; thus the requirement of these specific conditions results in sporadic and somewhat infrequent disease epidemics. During the 1950s, SSLB was endemic in the upper Midwestern region of the USA and in the Prairie of Canada, but remained contained until 1993 when the disease reemerged as an epidemic due to the inoculum buildup from extended wet and cool conditions (Toubia-Rahme and Steffenson, 2004). The SSLB epidemics typically cause yield losses of 20%, but have been reported up to 40% under severe conditions (Green and Bendelow, 1961; Toubia-Rahme and Steffenson, 2004). Apart from its impact on yield, this disease can significantly reduce several malting attributes like kernel plumpness and malt extract.

The symptoms of SSLB in barley are different depending on the growth stage of the host, but this can also be dependent on the causal pathogen (*S. passerini* or *S. avenae* f. sp. *triticea*) (Cunfer, 2000; Paulitz and Steffenson, 2011). The pathogen generally infects the leaves and leaf sheath of barley, and symptoms can be observed as early as the seedling stage. The early symptoms from *S. passerini* initially appear as small gray spots on the leaf that gradually turn straw colored, elongate and coalesce. As these lesions mature, dark brown pycnidia are observed inside the lesion margins, but in some cases lesions on the young plant may not produce pycnidia. The appearances of the lesions vary between the two causal agents of SSLB at the adult plant stage. *S. passerini* develops long and rectangular lesions delimited by veins and appear as light gray to white, but eventually turns tan or straw colored. On the contrary, *S. avenae* f. sp. *triticea* develops lens-shaped yellow-brown lesions that expand across the leaf veins. However, under severe infection, the lesions of both pathogens coalesce and kill the entire leaf or leaf sheath. Barley plants severely infected with SSLB become weak and experience straw breakage and lodging. As the barley matures, numerous pycnidia form on the senescent leaves and leaf sheaths, even when the disease severity is low (Cunfer, 2000).

The primary source of *S. passerini* inoculum is pycnidia that remain viable in the infected host debris/straw on the soil surface or below ground (Cunfer, 2000; Paulitz and Steffenson, 2011). The pycnidiospores are dispersed locally over short distances by rain and/or wind that carry infested debris. *S. passerini* requires an extended period of moisture (about 48 hours) for germination and leaf penetration, which is prolonged compared to the majority of foliar pathogens of barley. Disease epidemics were reported to be severe in seasons with high rainfall coupled with low temperatures, conditions that are also conducive for an extended barley vegetative phase (Cunfer, 2000). *S. passerini* is chiefly a monocyclic pathogen because of the newly developed pycnidia that form late in the season. The pathogen does not commonly infect healthy plants in the same season (Paulitz and Steffenson, 2011). *S. passerini* has a narrow host range that is limited to barley and some wild *Hordeum* species. In contrast, *S. avenae* f. sp. *triticea* can infect barley, wheat, wild *Hordeum* species and other grasses. Unlike *S. passerini, S. avenae* f. sp. *triticea* ascospores, along with pycnidiospores, act as a primary inoculum source. The infection process of *S. passerini* starts with the germination of the spores on the leaf surface that elongate to form germ tubes. On the surface of the leaf, the germ tubes branch repeatedly and produce swelling at multiple points, possibly due to the unsuccessful attempts of penetrating the host (Cunfer, 2000). Several germ tubes can be observed over the stomata at 48 hours post germination, followed by the stomatal penetration and hyphal growth in leaf mesophyll by 72 hours. The intracellular hyphae grow through the mesophyll, clustered at substomatal cavities where the pycnidia are formed.

SSLB can be effectively managed by cultural practices like plowing and burning of debris and crop rotation with non-host crops to reduce the reservoir of pycnidia in the subsequent season. Crop rotation with commercial crops like crambe, canola and dry pea was effective in reducing the severity of SSLB in barley. Fungicide applications were also effective at managing the disease; however, the additional cost incurred with such application makes this approach unattractive.

The most practical and environmental friendly alternative to managing SSLB is the deployment of genetic resistance. To date, four major genes (*Rsp1, Rsp2, Rsp3* and *Rsp4*) conferring resistance to *S. passerini* have been identified and introgressed into barley lines (St Pierre et al., 2010; Paulitz and Steffenson, 2011). Apart from the major R-genes, QTL conferring partial resistance to SSLB have also been mapped in several studies (Yu et al., 2010; Luckert et al., 2012; Zhou and Steffenson, 2013). However, as with most fungal diseases, the reliance on a limited number of major *R*-genes leads to the boom-and-bust cycle due to the strong selection pressure exerted on the pathogen. Extensive study on the diverse population of *S. passerini* is lacking; thus, the knowledge of virulence on known SSLB *R*-genes and QTL within populations is limited (Paulitz and Steffenson, 2011). There is the possibility of a sexual stage for this pathogen that would allow it to rapidly evolve new virulence spectra through sexual recombination. Additionally, SSLB is also caused by *S. avenae* f. sp. *triticea,* and no reports are available on whether the genes providing resistance to *S. passerinii* are effective on *S. avenae*. This highlights a need for further studies to map and identify genetic resistances that are effective on both of these SSLB-causing pathogens. In the meantime, the pyramiding of multiple major genes, along with QTL providing partial resistance, could contribute to durable resistance against SSLB.

11 Scald

Scald (leaf blotch) is a common foliar disease of barley in most growing regions, but is a major concern in temperate regions with cool and moist winters (Caldwell, 1937; Shipton et al., 1974). It can cause significant yield losses and reduce grain quality, especially for malting. In areas with intensive barley production and conducive environment for diseases, scald can cause severe damage up to 40%, although losses under 10% are common. Scald typically results in yield loss; however, in severe epidemics, the number of spikes per plant and the number of kernels per spike are also reduced (Mathre, 1997).

Barley scald is caused by the fungal pathogen *Rhynchosporium commune* Zaffarano, McDonald and Linde sp. nov (formerly known as *Rhynchosporium secalis* (Oudem.) J.J. Davis) (Caldwell, 1937; Shipton et al., 1974; Zaffarano et al., 2011). The first case of *Rhynchosporium* was discovered in rye (*Secale cereale*) in the year 1897 in the Netherlands by Oudemans who reported it as *Marsonia secalis* (Oudemans, 1897; Caldwell, 1937). In the same year Frank (1897) recognized this fungus as a causal agent of a disease in barley and rye. In 1901, Heinsen reclassified this fungus as a novel genus *Rhynchosporium* because of its typical single septated, beak-shaped spores and named it *Rhynchosporium graminicola*. Later in 1919, it was renamed as *R. secalis* (Oud.) by J.J. Davis to comply with the International Rules of Nomenclature. However, the evidences of host specialization in *Rhynchosporium* isolates originating from rye, barley and wild barley (*H. spontaneum*) and phylogenetic analysis of multilocus DNA sequences of these isolates led to the subdivision of *Rhynchosporium* into three species (Zaffarano et al., 2011; Avrova and Knogge, 2012). The species colonizing rye and triticale belongs to *R. secalis*, *Hordeum* spp. including barley and *B. diandrus* represents *R. commune* and *Agropyron* spp. belongs to *R. agropyri*.

The early symptoms of barley scald appear as oval, water-soaked, greyish lesion on the leaves, leaf sheath and head. The lesions gradually elongate, often to a diamond shape, and turn light gray, tan or white in the center with distinctive dark brown margins (Fig. 6). Under severe conditions, the lesions coalesce to form large necrotic areas from which the leaf withers and dies. These symptoms frequently occur in the leaf axil but can move to all aerial tissues except the upper stem. The disease symptoms can be seen on seeds as typical lesions with a dark brown margin and light center at the base of the awn (Avrova and Knogge, 2012). Although infected seeds may not exhibit

Figure 6 Barley field infected with scald. The upper-left panel shows the typical lesions formed on the leaves.

symptoms, they can still harbor *R. commune* and barley grown from the seed may develop scald (Lee et al., 2001).

Barley scald is a wind-borne and seed-borne pathogen. The primary source of inoculum is conidia produced from the stroma on the surface of locally infected debris, infected seeds and possibly yet unidentified ascospores (Fountaine and Fraaije, 2009; Avrova and Knogge, 2012). The stroma, but not the conidia, can survive extreme climates, both very hot and cold temperatures (Paulitz and Steffenson, 2011). Unlike other pathogens, wind alone cannot properly disseminate *R. commune*, but it requires splashing rain along with wind for dispersal. The infection of *R. commune* from the conidia starts with its germination on the leaf surface, producing hyphae that directly penetrate the cuticle above the epidermal cells (Jones and Ayres, 1974; Davis and Fitt, 1990). The growth of the fungus is restricted to the subcuticular region of epidermis where the hypha forms a linked network along the anticlinal walls (right angle to the surface of cells). This early stage is considered the biotrophic phase of the pathogen, during which it grows asymptomatically and absorbs nutrients from the host (Thirugnanasambandam et al., 2011; Avrova and Knogge, 2012). This initial biotrophic phase is succeeded by a necrotrophic phase, which is characterized by the collapse of epidermal and mesophyll cells and the eruption of conidia from leaf cuticles. Because of this long asymptomatic phase followed by the appearance of necrotic lesions in infected areas, this pathogen should be considered a hemibiotroph rather than a necrotroph (Zhan et al., 2008). The sporulation of new conidia is typically observed in healthy-looking leaves, but also in necrotic areas (Avrova and Knogge, 2012). These new conidia act as a source of inoculum for the secondary infection, and thus *R. commune* is considered a polycyclic pathogen. The healthy plants can become infected from contact with the wetted conidia harbored in neighboring infected plants (Paulitz and Steffenson, 2011). The disease cycle from infected seeds starts when the weather is cool and moist after planting. The hyphae present in the pericarp or hull of the seed infect the coleoptile emerging from an embryo. The infected seed acts as a major source of the long-distance transmission of the disease and spread of new physiological races.

Barley scald can be sustainably controlled by cultural practices, fungicide application and the use of resistant cultivars. The removal of volunteer crops, alternate hosts as well as susceptible barley that acts as a reservoir of the inoculum can control the spread of this pathogen. Sanitation practices like burning and plowing under debris, rotation with non-host crops, planting crop mixtures and modification of sowing date can help in reducing the inoculum buildup for the next season (Zhan et al., 2008). The elimination of seed-borne inoculum by planting healthy seeds or fungicide-treated seeds in areas conducive to disease development is also an effective management measure. Foliar fungicides are also effective to manage scald (Avrova and Knogge, 2012), but several compounds such as methyl benzimidazole carbamates (MBCs) and demethylation inhibitors (DMIs; triazoles) were effective until fungicide resistance evolved within pathogen populations. The use of fungicide with the single mode of action exerts a strong selection pressure and increases the opportunity for the pathogen to gain resistance over it. Such strong selection pressure could be reduced by implementing integrated crop protection (ICP) system that recommends a proper dosage, timing and fungicide mix with different modes of action. Still, the use of fungicides faces the public concern regarding the threat to environments, as well as increased cost associated with the application. Thus, host resistance is the most economically and environmentally friendly approach to control this disease (Zhan et al., 2008).

Genetic resistance to *R. commune* in barley is conferred by both quantitative (complete resistance) and qualitative genes (partial resistance) (Zhan et al., 2008). To date seven host-specific R-genes and their alleles, *Rrs1* (11 alleles), *Rrs2* (two alleles), *Rrs3*, *Rrs4* (two alleles), *Rrs12*, *Rrs13* and *Rrs14*, and four unconfirmed *R*-genes (*Rh5*, *rh8*, *Rh10* and *rh11*), have been reported for barley scald (Zhan et al., 2008; Avrova and Knogge, 2012). Several clusters of QTL for scald resistance have also been reported. The pyramiding of major genes along with QTL can contribute toward obtaining durable resistance against scald in barley. However, the continued evaluation of barley collections is needed to identify and characterize new resistance sources. Düşünceli et al. (2008) evaluated the reactions of 36 barley cultivars and 683 barley genotypes to scald under greenhouse (seedling) and field (adult plant) conditions. Among the 683 barley genotypes tested 44% and 39% were found to be resistant to moderately resistant at the seedling and adult plant stages, respectively, with some adapted barley cultivars containing both seedling and adult plant resistance under greenhouse and field conditions, respectively. Barley landraces and wild barley (*H. spontaneum*) genotypes are also good sources of scald resistance, but a recent report showed that wild barley populations contained better resistance sources (Azamparsa et al., 2019).

Apart from the resistance genes, several morphological traits like plant height, early stem elongation, maturity and canopy structures have also been shown to be correlated with disease development, and, thus, breeding for such traits could potentially contribute to controlling this disease. *R. commune* is a highly variable pathogen that can rapidly adapt and evolve to a given selection pressure (Zhan et al., 2008). Since pathogen populations contain numerous pathotypes (Ceoloni, 1980; Tekauz, 1991; Goodwin et al., 1992; Salamati and Tronsmo, 1997; Meles et al., 2004; Bouajila et al., 2006; Azamparsa, 2015), the virulence present within pathogen populations must be considered when breeding for scald resistance, which requires monitoring pathogen populations with differential barley lines.

Because of the variability within *R. commune* populations and the adaptability of this pathogen reliance on fungicide or genetic resistance alone could lead to an outbreak of new and highly virulent/aggressive pathogen genotypes. Thus, the understanding of effector biology of this pathogen could contribute toward developing better-informed approaches to manage this disease. The availability of genome sequences of *R. commune* and studies leading to the identification and characterization of effectors can aid in the rapid monitoring of evolving pathogen populations, as well as select for R-genes that target conserved effectors to attain more durable scald resistance (Avrova and Knogge, 2012).

12 Fusarium head blight

Fusarium head blight (FHB) or scab is the most destructive disease of barley. It has a potential to destroy a high-yielding crop within a single week (McMullen et al., 1997). It occurs in almost every wheat- and barley-growing region that experiences frequent rainfall, excessive humidity and moisture during the flowering and grain filling period of the crops (McMullen et al., 1997). For many years, FHB was not a major problem in barley grown in the USA and other parts of the continent (Steffenson, 2003). However, it has globally reemerged as a major disease in barley and has caused a significant impact on the yield and quality of barley. FHB was first reported in 1884 in England but was not considered a major threat to wheat and barley until the early twentieth century (Goswami and Kistler, 2004). FHB is endemic to eastern Asia and has caused several epidemics in these areas with yield losses up to 98% (Steffenson, 2003). The USA also suffered multiple FHB epidemics of barley and wheat in the twentieth century. The earliest, in 1919, FHB caused a severe reduction in barley production. The disease then became a rare occurrence in barley for several decades in the upper Midwest of the USA, but reemerged in an epidemic proportion in 1990 that shook the entire barley industry in the Red River Valley of the USA and Canada with losses beyond comparison to any known epidemics of barley in the region. Following these epidemic years, barley acreage was significantly reduced in these areas and has yet to recover. The severity of these unprecedented epidemics forced barley and wheat breeders, pathologists and extension scientists to focus on research efforts in combating this insidious disease (Steffenson, 2003).

FHB infection typically results in the reduction of the number of grains per spike and seed weight that leads to yield losses up to 40% (Perkowski et al., 1995). It can have a huge impact on the malt industry because of its adverse impact on malting qualities like kernel plumpness and poor germination that reduces malt extract (Schwartz, 2003). Additionally, the infected grain may contain a significant level of toxic trichothecene mycotoxins such as deoxynivalenol (DON) that exacerbates the losses from FHB and makes the grain inedible for human or animal consumption (Gardiner et al., 2010). The consumption of mycotoxin-contaminated grain can often cause acute toxicosis symptoms in humans and vomiting and hyperestrogenism in swine (Steffenson, 2003; Paulitz and Steffenson, 2011). Due to the acute nature of these FHB mycotoxins on humans and animals, even healthy-looking grains that contain mycotoxins at low concentrations (1.0 ppm) are not accepted at the elevator. The impact of head blight on mycotoxin accumulation is more severe in barley compared to wheat (Steffenson, 2003). This is due to presence of an intact lemma and palea on barley kernels as the lemma and palea of the florets have higher accumulation of mycotoxins.

FHB is a fungal disease caused by several *Fusarium* species including *Fusarium acuminatum, F. avenaceum, F. culmorum, F. crookwellense, F. equiseti, F. graminearum, F. poae* and *F. sporotrichioides* (Perkowski et al., 1995; Xue et al., 2006). Among these *Fusarium* species, *F. crookwellense, F. culmorum* and *F. graminearum* exhibited a higher pathogenicity on barley grown in the greenhouse condition (Xue et al., 2006). Of all these species, *F. graminearum*

Schwabe (teleomorph *Gibberella zeae* (Schweinitz) Petch) is the most common species causing FHB in North America. *F. graminearum* is commonly found in the areas that grow maize (Paulitz and Steffenson, 2011), with the other species of *Fusarium* being common in areas with cool climates and cultivation practices that favor species other than corn (McMullen et al., 1997).

F. graminearum infects barley spikes immediately after the emergence from the flag leaf sheath (Paulitz and Steffenson, 2011). The most noticeable distinction between FHB infection in barley and wheat is the appearance of distinct and sporadic infection of florets in barley compared to the continuous infection seen on wheat spikes (Fig. 7). The infection of FHB is favored by the moist weather and spreads rapidly at temperatures higher than 25°C. Under favorable conditions, the infection starts with the entry of the pathogen into the floret through the crevices between lemma and palea, and occasionally through the apex of a floret (Bushnell et al., 2003). The pathogen enters the floret tissue either by penetrating the epidermal wall by infectious hyphae

Figure 7 A two-rowed barley field with a severe fusarium head blight infection. The panel in the upper-left corner shows a barley head with the distinct tan to dark brown lesions on the kernels with the typical sporadic FHB infection of florets in barley.

or through natural openings like stomata. The first visible symptoms of FHB appears as tan to dark brown lesions at the base of the kernel (Paulitz and Steffenson, 2011). During prolonged warm and moist conditions, the infection spreads across kernels and can result in the complete sterility of spikes. Such early infection can lead to high yield loss. Late infection has a lower impact on the yield but can induce high DON accumulation in the grain. Other visual symptoms of diseases are salmon-orange masses of *F. graminearum* mycelia and conidia on the infected kernels, sometimes having black perithecia of the sexual stage *Gibberella zea*. If moist conditions persist longer, the fungal mycelia can be expanded to a length long enough to infect kernels from outside while spreading through the spike. The conidia produced from the infected kernels can serve as a source of the secondary inoculum, though its impact on overall disease development is not considered significantly important. The blackish perithecia on the infected kernels are the overwintering structures that can survive in crop debris. The ascospores released from these perithecia as well as the conidia from overwintering mycelia act as a primary source of inoculum for the next growing season.

FHB is a notoriously difficult disease to manage and requires an integrated management strategy to prevent yield loss and mycotoxin accumulation (McMullen et al., 1997; Paulitz and Steffenson, 2011). The use of appropriate cultural practices, chemical control and genetic resistance together are needed to ensure the management of this disease and in major epidemic years can still fall short. The rotation of non-susceptible host as well as non-host, combined with the tillage cultivation and spreading of crop residues, can assist in reducing the inoculum loads for the next season. In the areas with a high risk of scab infection, staggered planting at multiple dates can avoid entire crop loss. Moisture is a key factor in developing FHB, and, thus, cultivation practices in more arid regions can reduce disease incidence. Over-fertilization leading to lodging should be avoided for proper FHB management (Steffenson, 2003). Timely application of fungicides has been reported to control FHB and DON accumulation, but results have been inconsistent in the past and, thus, could not be relied on completely (McMullen et al., 1997; Paulitz and Steffenson, 2011). The most attractive means of effectively controlling this disease is genetic resistance that is durable, economic and environmentally friendly. Several large-scale screenings of barley accessions and genetic mapping efforts have led to the identification of barley accession and QTL for FHB resistance and low DON accumulation, but explain a low amount of the variability (Steffenson, 2003; Horsley et al., 2006; Paulitz and Steffenson, 2011; Dahleen et al., 2012). FHB resistance and DON accumulation are complex and confounded by agronomic traits (heights, heading date) and spike architecture. This impedes the accurate identification of QTL that are truly related to FHB resistance and DON accumulation. However, QTL explaining variation up to 60% for FHB resistance and 14% for DON accumulation have been identified and can be used in conjunction with other management strategies for better disease control (Horsley et al., 2006).

13 A seed-borne disease: barley stripe

Barley stripe disease is caused by the fungus *Pyrenophora graminea* S. Ito & Kuribay (anamorphic stage: *Drechslera* teres (Rab.) Shoem., syn: *Helminthosporium gramineum* Rabh.). This disease can cause significant losses in areas where seed treatment fungicides are not used (Mathre, 1982; Karakaya et al., 2016).

Barley stripe is a single-cycle seed-borne disease where the fungal pathogen systemically moves within the plant. The disease symptoms first appear as characteristic yellow stripes along the whole leaves that later become necrotic, and sporulation occurs. The fungus impairs head formation, and heads frequently are not formed on heavily infected plants. If heads form, they typically show distortions and frequently become trapped in the boot. The diseased plants also show stunting, and premature death commonly occurs. Abundant conidia form on the early and advanced leaf stripes, producing conidia spores that serve as the inoculum that infects nearby healthy heads and seeds. Perithecia, the sexual stage of the fungus, have also been reported (Mathre, 1982). The use of pathogen-free seed is the best method of control; however, in regions with a high incidence of the disease, seed fungicides should be used. Barley-stripe-resistant cultivars do exist (Ulus and Karakaya, 2007; Bayraktar and Akan, 2012), and the use of resistant cultivars is a good management practice. Differential virulence between fungal isolates has been reported (Ulus and Karakaya, 2007; Çelik et al., 2016; Karakaya et al., 2017). It has also been determined that barley landraces and wild barley genotypes (*H. spontaneum*) have differential reactions to the pathogen and are valuable sources of resistance (Çelik et al., 2016; Karakaya et al., 2017; Çelik Oğuz, 2019).

14 Conclusion

Barley diseases are important to producers and end users because they significantly decrease the yield and quality when an epidemic occurs. In the presence of the disease, resistance genes become important yield and quality loci for selection and are the most sustainable tool for management. However, deploying effective and durable resistance into adapted backgrounds takes considerable effort and resources and, depending on the disease and pathogen population, can be very complex. For some fungal diseases like FHB, deploying resistance is nearly impossible because adequate resistance is not present within the primary germplasm pool. With these diseases it may require biotech approaches to achieve adequate genetic resistance but industry and end users' acceptance and perception will need to shift before this option could be translated to the field.

Barley breeders make decisions and prioritize their efforts based on the frequency and severity of disease epidemics and the effects these diseases have on end users. These decisions are becoming more complicated by the dynamic effects of a warming planet, which result in unpredictable weather patterns from year to year. Because of the importance of environmental influence on which diseases develop over a growing season, different predominant diseases can occur from year to year because weather conditions determine how widespread and severe epidemics become. In production regions where multiple diseases can cause significant yield and quality reduction, breeders have to prioritize their resources to those that present a perennial or consistent problem; however, when multiple diseases are present it can be very difficult to set up field-screening nurseries because the effect on the yield and quality can be the result of a complex of diseases. Yet the alternative screening of prebreeding materials for multiple diseases under controlled environments is timely and expensive, and there is also the concern that growth chamber and greenhouse experiments may not translate to the field. Thus, in many cases disease phenotyping is a major bottleneck in the selection.

As the expense of genotyping breeding materials goes down via DNA sequencing technologies, we are beginning to see that the use of genome-wide association mapping and genomic selection will become much more common when making early generation selections, avoiding costly and time-consuming phenotyping. Yet these methodologies require robust genetic mapping and marker development. The best markers are generated when gene discovery efforts identify the actual resistance gene and causal polymorphisms that differentiate compatible (susceptibility) from incompatible (resistance) reactions, allowing for the development of perfect markers for disease evaluation. As we move into the next generation of breeding, we can envision a not-so-distant time where association mapping becomes so powerful that causal SNPs could be identified at numerous loci across the genome and genome-wide selection will be utilized in early generation selection to identify the best combinations of resistance loci to multiple diseases. With the acceleration in DNA sequencing technologies and rapid gene discovery techniques coming to fruition, these capabilities in plant breeding are not far on the horizon.

15 References

Abu Qamar, M., Liu, Z. H., Faris, J. D., Chao, S., Edwards, M. C., Lai, Z., Franckowiak, J. D. and Friesen, T. L. 2008. A region of barley chromosome 6H harbors multiple major genes associated with net type net blotch resistance. *Theoretical and Applied Genetics* 117(8), 1261–70. doi:10.1007/s00122-008-0860-x.

Acevedo-Garcia, J., Kusch, S. and Panstruga, R. 2014. Magical mystery tour: MLO proteins in plant immunity and beyond. *New Phytologist* 204(2), 273–81. doi:10.1111/nph.12889.

Acharya, K., Dutta, A. K. and Pradhan, P. 2011. *Bipolaris sorokiniana* (Sacc.) Shoem.: the most destructive wheat fungal pathogen in the warmer areas. *Australian Journal of Crop Science* 5(9), 1064.

Adhikari, A., Steffenson, B. J., Smith, M. J. and Dill-Macky, R. 2019. Genome-wide association mapping of seedling net form net blotch resistance in an Ethiopian and Eritrean barley collection. *Crop Science* 59(4). doi:10.2135/cropsci2019.01.0003.

Afanasenko, O. S., Havis, N. D., Bespalova, L. A., Ablova, I. B. and Marienko, V. I. 2012. Ramularia leaf spot is a new barley disease in Russia. *Plant Protection Quarantine* 1, 11–3.

Akhavan, A. T., Turkington, T. K., Askarian, H., Tekauz, A., Xi, K., Tucker, J. R., Kutcher, H. R. and Strelkov, S. E. 2016. Virulence of *Pyrenophora teres* populations in western Canada. *Canadian Journal of Plant Pathology* 38(2), 183–96. doi:10.1080/07060661.2016.1159617.

Ameen, G., Drader, T., Sager, L., Steffenson, B., Kleinhofs, A. and Brueggeman, R. S. 2016. Rcs5 is a wall associated kinase gene that putatively functions as a dominant susceptibility factor in the barley- *Cochliobolus sativus* interaction. *Proceedings of 12th International Barley Genetics Symposium*, Minneapolis, MN.

Ames, N., Dreiseitl, A., Steffenson, B. J. and Muehlbauer, G. J. 2015. Mining wild barley for powdery mildew resistance. *Plant Pathology* 64(6), 1396–406. doi:10.1111/ppa.12384.

Amezrou, R., Verma, R. P. S., Chao, S., Brueggeman, R. S., Belqadi, L., Arbaoui, M., Rehman, S. and Gyawali, S. 2018. Genome-wide association studies of net form of net blotch resistance at seedling and adult plant stages in spring barley collection. *Molecular Breeding* 38, 1–14.

Anikster, Y. 1984. Parasitic specialization of *Puccinia hordei* in Israel. *Phytopathology* 74(9), 1061–4. doi:10.1094/Phyto-74-1061.

Arabi, M. I. E., Barrault, G., Sarrafi, A. and Albertini, L. 1992. Variation in the resistance of barley cultivars and in the pathogenicity of *Drechslera teres* f. sp. *maculata* and *D. teres* f. sp. *teres* isolates from France. *Plant Pathology* 41(2), 180–6. doi:10.1111/j.1365-3059.1992.tb02336.x.

Arabi, M. I. E., Al-Safadi, B. and Charbaji, T. 2003. Pathogenic variation among isolates of *Pyrenophora teres*, the causal agent of barley net blotch. *Journal of Phytopathology* 151(7–8), 376–82. doi:10.1046/j.1439-0434.2003.00734.x.

Aurthur, J. C. 1934. *Manual of Rusts in the United States and Canada*. The Science Press Printing Co, Lancaster, PA.

Avrova, A. and Knogge, W. 2012. *Rhynchosporium commune*: a persistent threat to barley cultivation. *Molecular Plant Pathology* 13(9), 986–97. doi:10.1111/j.1364-3703.2012.00811.x.

Azamparsa, M. R. 2015. Determination of the pathotypes of *Rhynchosporium commune* in Turkey and assessment of the reactions of some barley landraces and wild barley (*Hordeum spontaneum*) populations to scald. Ph.D. thesis. Ankara University, Graduate School of Natural and Applied Sciences, Department of Plant Protection, Ankara, Turkey (in Turkish).

Azamparsa, M. R., Karakaya, A., Ergün, N., Sayim, İ., Duran, R. M. and Özbek, K. 2019. Identification of barley landraces and wild barley (*Hordeum spontaneum*) genotypes resistant to *Rhynchosporium commune*. *Tarım Bilimleri Dergisi* (in press).

Balcı, S., Karakaya, A., Çelik Oğuz, A., Ergün, N., Sayim, İ. and Aydoğan, S. 2018. Assessment of the seedling reactions of some hulless barley cultivars and genotypes to Cochliobolus spot blotch disease. *Bitki Koruma Bülteni* 58(4), 221-6. (in Turkish).

Ballance, G. M., Lamari, L. and Bernier, C. C. 1989. Purification and characterization of a host-selective necrosis toxin from *Pyrenophora tritici-repentis*. *Physiological and Molecular Plant Pathology* 35(3), 203-13. doi:10.1016/0885-5765(89)90051-9.

Bayraktar, H. and Akan, K. 2012. Genetic characterization of *Pyrenophora graminea* isolates and the reactions of some barley cultivars to leaf stripe disease under greenhouse conditions. *Turkish Journal of Agriculture and Forestry* 36, 329–39.

Bhattacharya, S. 2017. Deadly new wheat disease threatens Europe's crops. *Nature News* 542(7640), 145–6. doi:10.1038/nature.2017.21424.

Bilgic, H., Steffenson, B. J. and Hayes, P. M. 2005. Comprehensive genetic analyses reveal differential expression of spot blotch resistance in four populations of barley. *Theoretical and Applied Genetics* 111(7), 1238–50. doi:10.1007/s00122-005-0023-2.

Bilgic, H., Steffenson, B. J. and Hayes, P. M. 2006. Molecular mapping of loci conferring resistance to different pathotypes of the spot blotch pathogen in barley. *Phytopathology* 96(7), 699–708. doi:10.1094/PHYTO-96-0699.

Boch, J., Bonas, U. and Lahaye, T. 2014. TAL effectors-pathogen strategies and plant resistance engineering. *New Phytologist* 204(4), 823–32. doi:10.1111/nph.13015.

Bouajila, A., Haouas, S., Fakhfakh, M., Rezgui, S., El Ahmed, M. and Yahyaoui, A. 2006. Pathotypic diversity of *Rhynchosporium secalis* (Oudem) in Tunisia. *African Journal of Biotechnology* 5(8), 570–9.

Brandle, J. E., Namwilla, J. C. P. and Little, R. 1987. Effect of plant architecture on levels of *Helminthosporium sativum* infection on spring wheat grown in Zambia. *Crop Protection* 6(3), 153–6. doi:10.1016/0261-2194(87)90003-2.

Braun, E. J. and Howard, R. J. 1994. Adhesion of fungal spores and germlings to host plant surfaces. *Protoplasma* 181(1-4), 202–12. doi:10.1007/BF01666396.

Braun, U., Cook, R. T. A., Inman, A. J. and Shin, H. D. 2002. The taxonomy of the powdery mildew fungi. In: *The Powdery Mildews: a Comprehensive Treatise*, pp. 13–55.

Brown, J. K. M., Le Boulaire, S. and Evans, N. 1996. Genetics of responses to morpholine-type fungicides and of avirulences in *Erysiphe graminis* f. sp. *hordei*. *European Journal of Plant Pathology* 102(5), 479–90. doi:10.1007/BF01877142.

Brueggeman, R., Rostoks, N., Kudrna, D., Kilian, A., Han, F., Chen, J., Druka, A., Steffenson, B. and Kleinhofs, A. 2002. The barley stem rust-resistance gene Rpg1 is a novel disease-resistance gene with homology to receptor kinases. *Proceedings of the National Academy of Sciences of the United States of America* 99(14), 9328–33. doi:10.1073/pnas.142284999.

Brueggeman, R., Druka, A., Nirmala, J., Cavileer, T., Drader, T., Rostoks, N., Mirlohi, A., Bennypaul, H., Gill, U., Kudrna, D., Whitelaw, C., Kilian, A., Han, F., Sun, Y., Gill, K., Steffenson, B. and Kleinhofs, A. 2008. The stem rust resistance gene Rpg5 encodes a protein with nucleotide-binding-site, leucine-rich, and protein kinase domains. *Proceedings of the National Academy of Sciences of the United States of America* 105(39), 14970–5. doi:10.1073/pnas.0807270105.

Burlakoti, R. R., Gyawali, S., Chao, S., Smith, K. P., Horsley, R. D., Cooper, B., Muehlbauer, G. J. and Neate, S. M. 2017. Genome-wide association study of spot form of net blotch

resistance in the upper midwest barley breeding programs. *Phytopathology* 107(1), 100–8. doi:10.1094/PHYTO-03-16-0136-R.
Bushnell, W. R., Hazen, B. E., Pritsch, C. and Leonard, K. 2003. *Histology and Physiology of Fusarium Head Blight: Fusarium Head Blight of Wheat and Barley*. Leonard, K. J. and Bushnell, W. R. (Eds). The American Phytopathological Society, St. Paul, MN, pp. 44–83.
Cakir, M., Gupta, S., Platz, G. J., Ablett, G. A., Loughman, R., Emebiri, L. C., Poulsen, D., Li, C. D., Lance, R. C. M., Galwey, N. W., Jones, M. G. K. and Appels, R. 2003. Mapping and validation of the genes for resistance to *Pyrenophora teres* f. *teres* in barley (*Hordeum vulgare* L.). *Australian Journal of Agricultural Research* 54(12), 1369–77. doi:10.1071/AR02229.
Cakir, M., Gupta, S., Li, C., Hayden, M., Mather, D. E., Ablett, G. A., Platz, G. J., Broughton, S., Chalmers, K. J., Loughman, R., Jones, M. G. K. and Lance, R. C. M. 2011. Genetic mapping and QTL analysis of disease resistance traits in the barley population Baudin × AC Metcalfe. *Crop and Pasture Science* 62(2), 152–61. doi:10.1071/CP10154.
Caldwell, R. M. 1937. *Rhynchosporium secalis* of barley, rye, and other grasses. *Journal of Agricultural Research* 55, 175–98.
Campbell, G. F., Crous, P. W. and Lucas, J. A. 1999. *Pyrenophora teres* f. *maculata*, the cause of Pyrenophora leaf spot of barley in South Africa. *Mycological Research* 103(3), 257–67. doi:10.1017/S0953756298007114.
Campbell, G. F., Lucas, J. A. and Crous, P. W. 2002. Evidence of recombination between net-and spot-type populations of *Pyrenophora teres* as determined by RAPD analysis. *Mycological Research* 106(5), 602–8. doi:10.1017/S0953756202005853.
Cavara, F. 1893. Über einige parasitische Pilze auf dem Getreide. *Zeitschrift für Pflanzenkrankheiten* 3, 16–26.
Çelik, Y., Karakaya, A., Oğuz, A. Ç., Mert, Z., Akan, K., Ergün, N. and Sayim, İ. 2016. Determination of the reactions of some barley landraces and cultivars to *Drechslera graminea*. *Mediterranean Agricultural Sciences* 29(2), 43–7.
Çelik Oğuz, A. 2019. Resistance of wild barley (*Hordeum spontaneum* C. Koch) and barley landraces to barley leaf stripe disease. *Phytopathologia Mediterranea* (in press).
Çelik Oğuz, A. and Karakaya, A. 2017. Seedling response of commonly grown barley cultivars in Turkey to spot blotch disease. *Fresenius Environmental Bulletin* 26(11), 6734–8.
Çelik Oğuz, A., Karakaya, A., Mert, Z., Ergün, N., Sayim, İ. and Aydoğan, S. 2016. Determination of the seedling reactions of advanced barley lines to spot blotch disease caused by *Cochliobolus sativus*. *Works of the Faculty of Agriculture and Food Sciences, University of Sarajevo* 61(66/1), 244–6.
Çelik Oğuz, A., Karakaya, A., Ergün, N. and Sayim, İ. 2017. Turkish barley landraces resistant to net and spot forms of *Pyrenophora teres*. *Phytopathologia Mediterranea* 56, 217–23.
Çelik Oğuz, A., Ölmez, F. and Karakaya, A. 2018. Mating type idiomorphs of *Pyrenophora teres* in Turkey. *Zemdirbyste-Agriculture* 105(3), 271–8. doi:10.13080/z-a.2018.105.035.
Çelik Oğuz, A., Ölmez, F. and Karakaya, A. 2019a. Genetic diversity of net blotch pathogens of barley in Turkey. *International Journal of Agriculture and Biology* 21, 1089–96.

Çelik Oğuz, A., Karakaya, A., Duran, R. M. and Özbek, K. 2019b. Identification of *Hordeum spontaneum* genotypes resistant to net blotch disease. *Tarım Bilimleri Dergisi* 25, 15-122.

Çelik Oğuz, A., Rahimi, A. and Karakaya, A. 2019c. Seedling response of Iranian barley landraces to *Pyrenophora teres* f. *teres* and *Pyrenophora teres* f. *maculata*. *Tarım Bilimleri Dergisi* 25, 86-92.

Ceoloni, C. 1980. Race differentiation and searches for sources of resistance to *Rhynchosporium secalis* in barley in Italy. *Euphytica* 29(3), 547-53. doi:10.1007/BF00023201.

Chaure, P., Gurr, S. J. and Spanu, P. 2000. Stable transformation of *Erysiphe graminis*, an obligate biotrophic pathogen of barley. *Nature Biotechnology* 18(2), 205-7. doi:10.1038/72666.

Chelkowski, J., Tyrka, M. and Sobkiewicz, A. 2003. Resistance genes in barley (*Hordeum vulgare* L.) and their identification with molecular markers. *Journal of Applied Genetics* 44(3), 291-309.

Chen, X. M. 2007. Challenges and solutions for stripe rust control in the United States. *Australian Journal of Agricultural Research* 58(6), 648-55. doi:10.1071/AR07045.

Chen, X. and Line, R. F. 1999. Recessive genes for resistance to *Puccinia striiformis* f. sp. *hordei* in barley. *Phytopathology* 89(3), 226-32. doi:10.1094/PHYTO.1999.89.3.226.

Clifford, B. C. 1985. Barley leaf rust. In: *Diseases, Distribution, Epidemiology, and Control*. Elsevier, pp. 173-205. doi:10.1016/B978-0-12-148402-6.50014-6.

Coghlan, M. J., Krumkalns, E. V., Caley, B. A., Hall, H. R. and Arnold, W. R. 1991. Novel agents for the control of cereal and grape powdery mildew: synthesis and investigation of 4-phenoxyquinolines. In: *ACS Symposium Series*. American Chemical Society, USA, pp. 538-52. doi:10.1021/bk-1991-0443.ch043.

Condon, B. J., Leng, Y., Wu, D., Bushley, K. E., Ohm, R. A., Otillar, R., Martin, J., Schackwitz, W., Grimwood, J., MohdZainudin, N., Xue, C., Wang, R., Manning, V. A., Dhillon, B., Tu, Z. J., Steffenson, B. J., Salamov, A., Sun, H., Lowry, S., LaButti, K., Han, J., Copeland, A., Lindquist, E., Barry, K., Schmutz, J., Baker, S. E., Ciuffetti, L. M., Grigoriev, I. V., Zhong, S. and Turgeon, B. G. 2013. Comparative genome structure, secondary metabolite, and effector coding capacity across *Cochliobolus* pathogens. *PLoS Genetics* 9(1), e1003233. doi:10.1371/journal.pgen.1003233.

Conner, R. L., Duczek, L. J., Kozub, G. C. and Kuzyk, A. D. 1996. Influence of crop rotation on common root rot of wheat and barley. *Canadian Journal of Plant Pathology* 18(3), 247-54. doi:10.1080/07060669609500620.

Couto, D. and Zipfel, C. 2016. Regulation of pattern recognition receptor signalling in plants. *Nature Reviews. Immunology* 16(9), 537-52. doi:10.1038/nri.2016.77.

Cunfer, B. M. 2000. Stagonospora and Septoria diseases of barley, oat, and rye. *Canadian Journal of Plant Pathology* 22(4), 332-48. doi:10.1080/07060660009500452.

Daba, S. D., Horsley, R., Brueggeman, R., Chao, S. and Mohammadi, M. 2019. Genome-wide association study and candidate gene identification for leaf scald and net blotch in barley (*Hordeum vulgare* L.). *Plant Disease* 103(5), 880-9. doi:10.1094/PDIS-07-18-1190-RE.

Dahleen, L. S., Morgan, W., Mittal, S., Bregitzer, P., Brown, R. H. and Hill, N. S. 2012. Quantitative trait loci (QTL) for Fusarium ELISA compared to QTL for Fusarium head blight resistance and deoxynivalenol content in barley. *Plant Breeding* 131(2), 237-43. doi:10.1111/j.1439-0523.2012.01952.x.

Dantuma, G. 1964. Sources of resistance to mildew and stripe rust in breeding spring barley. *Euphytica* 13(3), 245–9.

Davis, H. and Fitt, B. D. L. 1990. Symptomless infection of *Rhynchosporium secalis* on leaves of winter barley. *Mycological Research* 94(4), 557–60. doi:10.1016/S0953-7562(10)80023-1.

Dikilitas, M., Çelik Oğuz, A. and Karakaya, A. 2018. Extracellular protease activity and glucose production in isolates of net blotch pathogens differing in virulence. *Zemdirbyste-Agriculture* 105(1), 89-94. doi:10.13080/z-a.2018.105.012.

Dodds, P. N. and Rathjen, J. P. 2010. Plant immunity: towards an integrated view of plant-pathogen interactions. *Nature Reviews. Genetics* 11(8), 539–48. doi:10.1038/nrg2812.

Douiyssi, A., Rasmusson, D. C. and Wilcoxson, R. D. 1996. Inheritance of resistance to net blotch in barley in Morocco. *Plant Disease* 80(11), 1269–72. doi:10.1094/PD-80-1269.

Dracatos, P., Singh, D., Fetch, T. and Park, R. 2015 Resistance to *Puccinia graminis* f. sp. avenae in Barley Is Associated with the Rpg5 Locus. *Phytopathology* 105(4), 490–4. doi:10.1094/PHYTO-08-14-0224-R.

Dracatos, P. M., Zhang, P., Park, R. F., McIntosh, R. A. and Wellings, C. R. 2016. Complementary resistance genes in wheat selection 'Avocet R' confer resistance to stripe rust. *Theoretical and Applied Genetics* 129(1), 65–76. doi:10.1007/s00122-015-2609-7.

Drader, T. and Kleinhofs, A. 2010. A synteny map and disease resistance gene comparison between barley and the model monocot *Brachypodium distachyon*. *Genome* 53(5), 406–17. doi:10.1139/g10-014.

Drader, T., Johnson, K., Brueggeman, R., Kudrna, D. and Kleinhofs, A. 2009. Genetic and physical mapping of a high recombination region on chromosome 7H(1) in barley. *Theoretical and Applied Genetics* 118(4), 811–20. doi:10.1007/s00122-008-0941-x.

Dreiseitl, A. 2011. Changes in the population of *Blumeria graminis* f. sp. *hordei* in the Czech Republic from 2009 to 2010. *Plant Protection Science* 47(2), 43–51.

Dreiseitl, A. and Dinoor, A. 2004. Phenotypic diversity of barley powdery mildew resistance sources. *Genetic Resources and Crop Evolution* 51(3), 251–7. doi:10.1023/B:GRES.0000024010.12369.b3.

Dubin, H. J. and Stubbs, R. W. 1986. Epidemic spread of barley stripe rust in South America. *Plant Disease* 70(2), 141–4. doi:10.1094/PD-70-141.

Duczek, L. J. 1984. Comparison of the common root rot reaction of barley lines and cultivars in northwestern Alberta and central Saskatchewan. *Canadian Journal of Plant Pathology* 6(1), 81–9. doi:10.1080/07060668409501596.

Duczek, L. J., Sutherland, K. A., Reed, S. L., Bailey, K. L. and Lafond, G. P. 1999. Survival of leaf spot pathogens on crop residues of wheat and barley in Saskatchewan. *Canadian Journal of Plant Pathology* 21(2), 165–73. doi:10.1080/07060669909501208.

Düşünceli, F., Çetin, L., Albustan, S., Mert, Z., Akan, K. and Karakaya, A. 2008. Determination of the reactions of some barley cultivars and genotypes to scald under greenhouse and field conditions. *Tarım Bilimleri Dergisi* 14(1), 46–50.

Emebiri, L. C., Platz, G. and Moody, D. B. 2005. Disease resistance genes in a doubled haploid population of two-rowed barley segregating for malting quality attributes. *Australian Journal of Agricultural Research* 56(1), 49–56. doi:10.1071/AR04102.

Eversmeyer, M. G. and Kramer, C. L. 2000. Epidemiology of wheat leaf and stem rust in the central great plains of the USA. *Annual Review of Phytopathology* 38(1), 491-513. doi:10.1146/annurev.phyto.38.1.491.

Faris, J. D., Zhang, Z., Lu, H., Lu, S., Reddy, L., Cloutier, S., Fellers, J. P., Meinhardt, S. W., Rasmussen, J. B., Xu, S. S., Oliver, R. P., Simons, K. J. and Friesen, T. L. 2010. A unique wheat disease resistance-like gene governs effector-triggered susceptibility to necrotrophic pathogens. *Proceedings of the National Academy of Sciences of the United States of America* 107(30), 13544-9. doi:10.1073/pnas.1004090107.

Fetch Jr., T. G. and Steffenson, B. J. 1994. Identification of Cochliobolus sativus isolates expressing differential virulence on two-row barley genotypes from North Dakota. *Canadian Journal of Plant Pathology* 16(3), 202-6. doi:10.1080/07060669409500754.

Fetch Jr., T. G. and Steffenson, B. J. 1999. Rating scales for assessing infection responses of barley infected with *Cochliobolus sativus*. *Plant Disease* 83(3), 213-7. doi:10.1094/PDIS.1999.83.3.213.

Flor, H. H. 1956. The complementary genic systems in flax and flax rust. *Advances in Genetics* 8, 29-54. doi:10.1016/S0065-2660(08)60498-8.

Fountaine, J. M. and Fraaije, B. A. 2009. Development of QoI resistant alleles in populations of *Ramularia collo-cygni*. Asp. *Applied Biology* 92, 123-6.

Frank, A. B. 1897. Über die Zerstörung der Gerste durch einen neuen Getreidepilz. *Wochenschrift für Brauerei* 42, 518-20.

Frei, P., Gindro, K., Richter, H. and Schürch, S. 2007. Direct-PCR detection and epidemiology of *Ramularia collo-cygni* associated with barley necrotic leaf spots. *Journal of Phytopathology* 155(5), 281-8. doi:10.1111/j.1439-0434.2007.01228.x.

Friesen, T. L., Faris, J. D., Lai, Z. and Steffenson, B. J. 2006a. Identification and chromosomal location of major genes for resistance to *Pyrenophora teres* in a doubled-haploid barley population. *Genome* 49(7), 855-9. doi:10.1139/g06-024.

Friesen, T. L., Stukenbrock, E. H., Liu, Z., Meinhardt, S., Ling, H., Faris, J. D., Rasmussen, J. B., Solomon, P. S., McDonald, B. A. and Oliver, R. P. 2006b. Emergence of a new disease as a result of interspecific virulence gene transfer. *Nature Genetics* 38(8), 953-6. doi:10.1038/ng1839.

Friesen, T. L., Meinhardt, S. W. and Faris, J. D. 2007. The *Stagonospora nodorum*-wheat pathosystem involves multiple proteinaceous host-selective toxins and corresponding host sensitivity genes that interact in an inverse gene-for-gene manner. *The Plant Journal: for Cell and Molecular Biology* 51(4), 681-92. doi:10.1111/j.1365-313X.2007.03166.x.

Friesen, T. L., Holmes, D. J., Bowden, R. L. and Faris, J. D. 2018. ToxA is present in the U.S. *Bipolaris sorokiniana* population and is a significant virulence factor on wheat harboring *Tsn1*. *Plant Disease* 102(12), 2446-52. doi:10.1094/PDIS-03-18-0521-RE.

Gardiner, D. M., Kazan, K., Praud, S., Torney, F. J., Rusu, A. and Manners, J. M. 2010. Early activation of wheat polyamine biosynthesis during Fusarium head blight implicates putrescine as an inducer of trichothecene mycotoxin production. *BMC Plant Biology* 10(1), 289. doi:10.1186/1471-2229-10-289.

Geschele, E. E. 1928. The response of barley to parasitic fungi Helminthosporium teres Sacc. *Bull. Appl. Bot. Genet. Plant Breed* 19, 371-84.

Göhre, V., Spallek, T., Häweker, H., Mersmann, S., Mentzel, T., Boller, T., de Torres, M., Mansfield, J. W. and Robatzek, S. 2008. Plant pattern-recognition receptor FLS2 is directed for degradation by the bacterial ubiquitin ligase AvrPtoB. *Current Biology: CB* 18(23), 1824–32. doi:10.1016/j.cub.2008.10.063.

Goodwin, S. B., Allard, R. W., Hardy, S. A. and Webster, R. K. 1992. Hierarchical structure of pathogenic variation among *Rhynchosporium secalis* populations in Idaho and Oregon. *Canadian Journal of Botany* 70(4), 810-7. doi:10.1139/b92-103.

Goswami, R. S. and Kistler, H. C. 2004. Heading for disaster: *Fusarium graminearum* on cereal crops. *Molecular Plant Pathology* 5(6), 515–25. doi:10.1111/j.1364-3703.2004.00252.x.

Green, G. J. and Bendelow, V. M. 1961. Effect of speckled leaf blotch, *Septoria passerinii* Sacc., on the yield and malting quality of barley. *Canadian Journal of Plant Science* 41(2), 431–5. doi:10.4141/cjps61-058.

Grewal, T. S., Rossnagel, B. G., Pozniak, C. J. and Scoles, G. J. 2008. Mapping quantitative trait loci associated with barley net blotch resistance. *Theoretical and Applied Genetics* 116(4), 529–39. doi:10.1007/s00122-007-0688-9.

Gubiš, J., Hudcovicova, M. and Klčová, L. 2008. First report of *Ramularia collo-cygni* in Slovakia. *Journal of Plant Pathology* 90, 149.

Gupta, P. K., Chand, R., Vasistha, N. K., Pandey, S. P., Kumar, U., Mishra, V. K. and Joshi, A. K. 2018. Spot blotch disease of wheat: the current status of research on genetics and breeding. *Plant Pathology* 67(3), 508–31. doi:10.1111/ppa.12781.

Haugaard, H., Collinge, D. B. and Lyngkjær, M. F. 2002. Mechanisms involved in control of *Blumeria graminis* f. sp. *hordei* in barley treated with mycelial extracts from cultured fungi. *Plant Pathology* 51(5), 612–20. doi:10.1046/j.1365-3059.2002.00748.x.

Havis, N. D., Oxley, S. P., Piper, S. R. and Langrell, S. R. 2006. Rapid nested PCR-based detection of *Ramularia collo-cygni* direct from barley. *FEMS Microbiology Letters* 256(2), 217–23. doi:10.1111/j.1574-6968.2006.00121.x..

Havis, N. D., Nyman, M. and Oxley, S. J. P. 2014. Evidence for seed transmission and symptomless growth of *Ramularia collo-cygni* in barley (*Hordeum vulgare*). *Plant Pathology* 63(4), 929–36. doi:10.1111/ppa.12162.

Havis, N. D., Brown, J. K., Clemente, G., Frei, P., Jedryczka, M., Kaczmarek, J., Kaczmarek, M., Matusinsky, P., McGrann, G. R., Pereyra, S., Piotrowska, M., Sghyer, H., Tellier, A. and Hess, M. 2015. *Ramularia collo-cygni* an emerging pathogen of barley crops. *Phytopathology* 105(7), 895–904. doi:10.1094/PHYTO-11-14-0337-FI.

He, P., Shan, L., Lin, N. C., Martin, G. B., Kemmerling, B., Nürnberger, T. and Sheen, J. 2006. Specific bacterial suppressors of MAMP signaling upstream of MAPKKK in Arabidopsis innate immunity. *Cell* 125(3), 563–75. doi:10.1016/j.cell.2006.02.047.

Heffer, V., Powelson, M. L., Johnson, K. B. and Shishkoff, N. 2006. Identification of powdery mildew fungi anno 2006. *The Plant Health Instructor*. doi:10.1094/PHI-I-2006-0706-01.

Heinsen, E. 1901. Beobachtungen über den neuen Getreidepilz *Rhynchosporium graminicola*. *Jahrb. wissenschaft. Anstalten* 3, 43.

Ho, K. M., Choo, T. M., Tekauz, A. and Martin, R. A. 1996. Genetic studies on net blotch resistance in a barley cross. *Canadian Journal of Plant Science* 76(4), 715–9. doi:10.4141/cjps96-123.

Horsley, R. D., Schmierer, D., Maier, C., Kudrna, D., Urrea, C. A., Steffenson, B. J., Schwarz, P. B., Franckowiak, J. D., Green, M. J., Zhang, B. and Kleinhofs, A. 2006. Identification of QTLs associated with Fusarium head blight resistance in barley accession CIho 4196. *Crop Science* 46(1), 145–56. doi:10.2135/cropsci2005.0247.

Horsley, R. D., Franckowiak, J. D. and Schwarz, P. B. 2009. Barley. In: *Cereals*. Springer, New York, NY, pp. 227–50.

Jin, Y., Szabo, L. J., Pretorius, Z. A., Singh, R. P., Ward, R. and Fetch, T. 2008. Detection of virulence to resistance gene Sr24 within race TTKS of *Puccinia graminis* f. sp. *tritici*. *Plant Disease* 92(6), 923–6. doi:10.1094/PDIS-92-6-0923.

Jin, Y., Szabo, L. J. and Carson, M. 2010. Century-old mystery of *Puccinia striiformis* life history solved with the identification of *Berberis* as an alternate host. *Phytopathology* 100(5), 432–5. doi:10.1094/PHYTO-100-5-0432.

Jones, P. and Ayres, P. G. 1974. *Rhynchosporium* leaf blotch of barley studied during the subcuticular phase by electron microscopy. *Physiological Plant Pathology* 4(2), 229–33. doi:10.1016/0048-4059(74)90011-3.

Jones, J. D. and Dangl, J. L. 2006. The plant immune system. *Nature* 444(7117), 323–9. doi:10.1038/nature05286.

Jordan, V. W. L. 1981. Aetiology of barley net blotch caused by *Pyrenophora teres* and some effects on yield. *Plant Pathology* 30(2), 77–87. doi:10.1111/j.1365-3059.1981.tb01232.x.

Jørgensen, I. H. 1992. Discovery, characterization and exploitation of Mlo powdery mildew resistance in barley. *Euphytica* 63(1–2), 141–52. doi:10.1007/BF00023919.

Kaczmarek, M., Piotrowska, M. J., Fountaine, J. M., Gorniak, K., McGrann, G. R. D., Armstrong, A., Wright, K. M., Newton, A. C. and Havis, N. D. 2017. Infection strategy of *Ramularia collo-cygni* and development of Ramularia leaf spot on barley and alternative graminaceous hosts. *Plant Pathology* 66(1), 45–55. doi:10.1111/ppa.12552.

Karakaya, A., Katırcıoğlu, Y. Z. and Aktaş, H. 2004. Studies on the biology of *Drechslera teres* under Ankara conditions. *Tarım Bilimleri Dergisi* 10, 33–135.

Karakaya, A., Mert, Z., Oğuz, A. Ç. and Çetin, L. 2016. Distribution of barley stripe disease in Central Anatolia, Turkey. *Selcuk Journal of Agriculture and Food Sciences* 30(2), 59–61.

Karakaya, A., Çelik Oğuz, A. and Rahimi, A. 2017. Response of Iranian barley landraces to *Drechslera graminea*. *Works of the Faculty of Agriculture and Food Sciences University of Sarajevo* XLII(67/2), 225–30.

Karki, C. B. and Sharp, E. L. 1986. Pathogenic variation in some isolates of Pyrenophora teres f. sp. maculata on barley. *Plant Disease* 70(7), 684–7. doi:10.1094/PD-70-684.

Keiper, F. J., Grcic, M., Capio, E. and Wallwork, H. 2008. Diagnostic microsatellite markers for the barley net blotch pathogens, *Pyrenophora teres* f. *maculata* and *Pyrenophora teres* f. *teres*. *Australasian Plant Pathology* 37(4), 428–30. doi:10.1071/AP08035.

Khan, T. N. and Tekauz, A. 1982. Occurrence and pathogenicity of *Drechslera teres* isolates causing spot-type symptoms on barley in Western Australia. *Plant Disease* 66(1), 423–5. doi:10.1094/PD-66-423.

Kiesling, R. L. 1985. The diseases of barley. In: Rasmusso, D. C. (Ed.), *Barley*. American Society of Agronomy, Madison, WI, pp.269–312.

King, J. E. and Polley, R. W. 1976. Observations on the epidemiology and effect on grain yield of brown rust in spring barley. *Plant Pathology* 25(2), 63–73. doi:10.1111/j.1365-3059.1976.tb01926.x.

Kislev, M. E. 1982. Stem rust of wheat 3300 years old found in Israel. *Science* 216(4549), 993–4. doi:10.1126/science.216.4549.993.

Koladia, V. M., Richards, J. K., Wyatt, N. A., Faris, J. D., Brueggeman, R. S. and Friesen, T. L. 2017. Genetic analysis of virulence in the *Pyrenophora teres* f. *teres* population

BB25× FGOH04Ptt-21. *Fungal Genetics and Biology* 107, 12–9. doi:10.1016/j.fgb.2017.07.003.

Kolmer, J. A. 2005. Tracking wheat rust on a continental scale. *Current Opinion in Plant Biology* 8(4), 441–9. doi:10.1016/j.pbi.2005.05.001.

König, J., Perovic, D., Kopahnke, D. and Ordon, F. 2014. Mapping seedling resistance to net form of net blotch (*Pyrenophora teres* f. *teres*) in barley using detached leaf assay. *Plant Breeding* 133(3), 356–65. doi:10.1111/pbr.12147.

Krattinger, S. G., Lagudah, E. S., Spielmeyer, W., Singh, R. P., Huerta-Espino, J., McFadden, H., Bossolini, E., Selter, L. L. and Keller, B. 2009. A putative ABC transporter confers durable resistance to multiple fungal pathogens in wheat. *Science* 323(5919), 1360–3. doi:10.1126/science.1166453.

Kumar, J., Schäfer, P., Hückelhoven, R., Langen, G., Baltruschat, H., Stein, E., Nagarajan, S. and Kogel, K. H. 2002. *Bipolaris sorokiniana*, a cereal pathogen of global concern: cytological and molecular approaches towards better control. *Molecular Plant Pathology* 3(4), 185–95. doi:10.1046/j.1364-3703.2002.00120.x.

Lee, H., Tewan, J. and Turkington, T. 2001. Symptomless infection of barley seed to Rhynchosporium secalis. *Canadian Journal of Plant Pathology* 23(3): 1–10.

Leišová-Svobodová, L., Minaříková, V., Kučera, L. and Ovesna, J. 2005. Genetic diversity of *Pyrenophora teres* isolates as detected by AFLP analysis. *Journal of Phytopathology* 153(10), 569–78. doi:10.1111/j.1439-0434.2005.01019.x.

Leišová-Svobodová, L., Minaříková, V., Matušinsky, P., Hudcovicová, M., Ondreičková, K. and Gubiš, J. 2014. Genetic structure of *Pyrenophora teres* net and spot populations as revealed by microsatellite analysis. *Fungal Biology* 118(2), 180–92. doi:10.1016/j.funbio.2013.11.008.

Leonard, K. J. and Szabo, L. J. 2005. Stem rust of small grains and grasses caused by *Puccinia graminis*. *Molecular Plant Pathology* 6(2), 99–111. doi:10.1111/j.1364-3703.2005.00273.x.

Lewis, J., Wan, J., Ford, R., Gong, Y., Fung, P., Nahal, H., Wang, P. W., Desveaux, D. and Guttman, D. S. 2012. Quantitative interactor screening with next-generation sequencing (QIS-Seq) identifies Arabidopsis thaliana MLO2 as a target of the Pseudomonas syringae type III effector HopZ2. *BMC Genomics* 13(8): 1–13.

Line, R. F. 2002. Stripe rust of wheat and barley in North America: a retrospective historical review. *Annual Review of Phytopathology* 40(1), 75–118. doi:10.1146/annurev.phyto.40.020102.111645.

Liu, Z., Faris, J. D., Edwards, M. C. and Friesen, T. L. 2010. Development of expressed sequence tag (EST)–based markers for genomic analysis of a barley 6H Region harboring multiple net form net blotch resistance genes. *The Plant Genome* 3(1), 41–52. doi:10.3835/plantgenome2010.02.0002.

Liu, Z., Ellwood, S. R., Oliver, R. P. and Friesen, T. L. 2011. *Pyrenophora teres*: profile of an increasingly damaging barley pathogen. *Molecular Plant Pathology* 12(1), 1–19. doi:10.1111/j.1364-3703.2010.00649.x.

Liu, Z., Holmes, D. J., Faris, J. D., Chao, S., Brueggeman, R. S., Edwards, M. C. and Friesen, T. L. 2015. Necrotrophic effector-triggered susceptibility (NETS) underlies the barley-*Pyrenophora teres* f. *teres* interaction specific to chromosome 6H. *Molecular Plant Pathology* 16(2), 188–200. doi:10.1111/mpp.12172.

Louw, J. P. J. 1996. Relative importance of the barley net blotch pathogens *Pyrenophora teres* f. *teres* (net-type) and *P. teres* f. *maculata*. *African Plant Protection* 2(2).

Lu, S., Platz, G. J., Edwards, M. C. and Friesen, T. L. 2010. Mating type locus-specific polymerase chain reaction markers for differentiation of *Pyrenophora teres* f. *teres* and *P. teres* f. *maculata*, the causal agents of barley net blotch. *Phytopathology* 100(12), 1298–306. doi:10.1094/PHYTO-05-10-0135.

Luckert, D., Toubia-Rahme, H., Steffenson, B. J., Choo, T. M. and Molnar, S. J. 2012. Novel septoria speckled leaf blotch resistance loci in a barley doubled-haploid population. *Phytopathology* 102(7), 683–91. doi:10.1094/PHYTO-06-11-0181.

Luig, N. H. 1985. Epidemiology in Australia and New Zealand. In: *Diseases, Distribution, Epidemiology, and Control*. Academic Press, pp. 301–28.

Ma, Z. Q., Lalpitan, N. L. V. and Steffenson, B. 2004. QTL mapping of net blotch resistance genes in a doubled-haploid population of six-rowed barley. *Euphytica* 137(3), 291–6. doi:10.1023/B:EUPH.0000040441.36990.58.

Manninen, O., Kalendar, R., Robinson, J. and Schulman, A. H. 2000. Application of BARE-1 retrotransposon markers to the mapping of a major resistance gene for net blotch in barley. *Molecular and General Genetics: MGG* 264(3), 325–34. doi:10.1007/s004380000326.

Manninen, O. M., Jalli, M., Kalendar, R., Schulman, A., Afanasenko, O. and Robinson, J. 2006. Mapping of major spot-type and net-type net-blotch resistance genes in the Ethiopian barley line CI 9819. *Genome* 49(12), 1564–71. doi:10.1139/g06-119.

Manning, V. A. and Ciuffetti, L. M. 2005. Localization of Ptr ToxA produced *by Pyrenophora tritici-repentis* reveals protein import into wheat mesophyll cells. *The Plant Cell* 17(11), 3203–12. doi:10.1105/tpc.105.035063.

Marshall, D. and Sutton, R. L. 1995. Epidemiology of stripe rust, virulence of *Puccinia striiformis* f. sp. hordei, and yield loss in barley. *Plant Disease* 79(7), 732–7. doi:10.1094/PD-79-0732.

Martin, R. A. 1985. Disease progression and yield loss in barley associated with net blotch, as influenced by fungicide seed treatment. *Canadian Journal of Plant Pathology* 7(1), 83–90. doi:10.1080/07060668509501520.

Mathre, D. E. 1982. *Compendium of Barley Diseases*. American Phytopathological Society (APS) Press. St. Paul, MN.

Mathre, D. E. 1997. *Compendium of Barley Diseases* (2nd edn.). American Phytopathological Society (APS) Press, St. Paul, MN.

Matusinsky, P., Svobodova-Leisova, L., Mariks, P., Tvaruzek, L., Stemberkova, L., Hanusova, M., Minarikova, V., Vysohlidova, M. and Spitzer, T. 2010. Frequency of a mutant allele of cytochrome b conferring resistance to QoI fungicides in the Czech population of *Ramularia collo-cygni*. *Journal of Plant Diseases and Protection* 117(6), 248–52. doi:10.1007/BF03356369.

McDonald, B. A. and Linde, C. 2002. Pathogen population genetics, evolutionary potential, and durable resistance. *Annual Review of Phytopathology* 40(1), 349–79. doi:10.1146/annurev.phyto.40.120501.101443.

McDonald, M. C., Ahren, D., Simpfendorfer, S., Milgate, A. and Solomon, P. S. 2018. The discovery of the virulence gene *ToxA* in the wheat and barley pathogen *Bipolaris sorokiniana*. *Molecular Plant Pathology* 19(2), 432–9. doi:10.1111/mpp.12535.

McGrann, G. R., Stavrinides, A., Russell, J., Corbitt, M. M., Booth, A., Chartrain, L., Thomas, W. T. and Brown, J. K. 2014. A trade off between *mlo* resistance to powdery mildew and increased susceptibility of barley to a newly important disease, Ramularia leaf spot. *Journal of Experimental Botany* 65(4), 1025–37. doi:10.1093/jxb/ert452.

McLean, M. S., Howlett, B. J. and Hollaway, G. J. 2009. Erratum to: epidemiology and control of spot form of net blotch (*Pyrenophora teres* f. *maculata*) of barley: a review. *Crop and Pasture Science* 60(5), 499-. doi:10.1071/CP08173_ER.

McLean, M. S., Martin, A., Gupta, S., Sutherland, M. W., Hollaway, G. J. and Platz, G. J. 2014. Validation of a new spot form of net blotch differential set and evidence for hybridisation between the spot and net forms of net blotch in Australia. *Australasian Plant Pathology* 43(3), 223–33. doi:10.1007/s13313-014-0285-8.

McMullen, M., Jones, R. and Gallenberg, D. 1997. Scab of wheat and barley: a re-emerging disease of devastating impact. *Plant Disease* 81(12), 1340–8. doi:10.1094/PDIS.1997.81.12.1340.

Meles, K. A., Hulluka, M. and Abang, M. M. 2004. Phenotypic diversity in *Rhynchosporium secalis* from Ethiopia and host response to barley scald. *Plant Pathology Journal* 3(1), 26–34. doi:10.3923/ppj.2004.26.34.

Melville, S. C., Griffin, G. W. and Jemmett, J. L. 1976. Effects of fungicide spraying on brown rust and yield in spring barley. *Plant Pathology* 25(2), 99–107. doi:10.1111/j.1365-3059.1976.tb01932.x.

Miethbauer, S., Heiser, I. and Liebermann, B. 2003. The phytopathogenic fungus Ramularia collo-cygni produces biologically active rubellins on infected barley leaves. *Journal of Phytopathology* 151(11-12), 665–8. doi:10.1046/j.1439-0434.2003.00783.x.

Milus, E. A., Kristensen, K. and Hovmøller, M. S. 2009. Evidence for increased aggressiveness in a recent widespread strain of *Puccinia striiformis* f. sp. *tritici* causing stripe rust of wheat. *Phytopathology* 99(1), 89–94. doi:10.1094/PHYTO-99-1-0089.

Mode, C. J. and Schaller, C. W. 1958. Two Additional factors for host resistance to net blotch in barley. *Agronomy Journal* 50(1), 15–8. doi:10.2134/agronj1958.00021962005000010005x.

Molnar, S. J., James, L. E. and Kasha, K. J. 2000. Inheritance and RAPD tagging of multiple genes for resistance to net blotch in barley. *Genome* 43(2), 224–31. doi:10.1139/g99-111.

Murray, G. M. and Brennan, J. P. 2010. Estimating disease losses to the Australian barley industry. *Australasian Plant Pathology* 39(1), 85–96. doi:10.1071/AP09064.

Nagarajan, S. and Joshi, L. M. 1975. Historical account of wheat rust epidemics in India, and their significance. *Cereal Rusts Bulletin*.

Nover, I. and Scholz, F. 1969. Genetic studies on the resistance of barley to yellow rust (*Puccinia striiformis* West.). *Theoretical and Applied Genetics* 39(4), 150–5. doi:10.1007/BF00272523.

Oerke, E. C., Dehne, H. W., Schönbeck, F. and Weber, A. 2012. *Crop Production and Crop Protection: Estimated Losses in Major Food and Cash Crops*. Elsevier.

Oudemans, C. A. J. 1897. Observations mycologiques. *Konink. Akad. Wetensch. Amsterdam*. 6, 86–92.

Park, R. F., Golegaonkar, P. G., Derevnina, L., Sandhu, K. S., Karaoglu, H., Elmansour, H. M., Dracatos, P. M. and Singh, D. 2015. Leaf rust of cultivated barley: pathology and control. *Annual Review of Phytopathology* 53, 565–89. doi:10.1146/annurev-phyto-080614-120324.

Paulitz, T. C. and Steffenson, B. J. 2011. Biotic stress in barley: disease problems and solutions. In: *Barley Production, Improvement, and Uses*, pp. 307–54.

Peraldi, A., Griffe, L. L., Burt, C., McGrann, G. R. D. and Nicholson, P. 2014. *Brachypodium distachyon* exhibits compatible interactions with *Oculimacula* spp. and *Ramularia collo-cygni*, providing the first pathosystem model to study eyespot and Ramularia leaf spot diseases. *Plant Pathology* 63(3), 554–62. doi:10.1111/ppa.12114.

Perkowski, J., Kiecana, I. and Chelkowski, J. 1995. Susceptibility of barley cultivars and lines to Fusarium infection and mycotoxin accumulation in kernels. *Journal of Phytopathology* 143(9), 547–51. doi:10.1111/j.1439-0434.1995.tb00661.x.

Pinnschmidt, H. and Jorgensen, L. N. 2009. Yield effects of Ramularia leaf spot on spring barley. The 2nd European Ramularia Workshop – A new disease and challenge in barley production. *Aspects of Applied Biology* 92.

Poets, A. M., Fang, Z., Clegg, M. T. and Morrell, P. L. 2015. Barley landraces are characterized by geographically heterogeneous genomic origins. *Genome Biology* 16, 173. doi:10.1186/s13059-015-0712-3.

Raemaekers, R. H. 1988. *Helminthosporium sativum*: disease complex on wheat and sources of resistance in Zambia. In: Klatt, A. R. (Ed.), *Wheat Production Constraints in Tropical Environments*. CIMMYT, Mexico D.F., pp. 175–86.

Raman, H., Platz, G. J., Chalmers, K. J., Raman, R., Read, B. J., Barr, A. R. and Moody, D. B. 2003. Mapping of genomic regions associated with net form of net blotch resistance in barley. *Australian Journal of Agricultural Research* 54(12), 1359–67. doi:10.1071/AR03026.

Rau, D., Brown, A. H., Brubaker, C. L., Attene, G., Balmas, V., Saba, E. and Papa, R. 2003. Population genetic structure *of Pyrenophora teres* Drechs. the causal agent of net blotch in Sardinian landraces of barley (*Hordeum vulgare* L.). *Theoretical and Applied Genetics. Theoretische und Angewandte Genetik* 106(5), 947–59. doi:10.1007/s00122-002-1173-0.

Rau, D., Maier, F. J., Papa, R., Brown, A. H., Balmas, V., Saba, E., Schaefer, W. and Attene, G. 2005. Isolation and characterization of the mating-type locus of the barley pathogen *Pyrenophora teres* and frequencies of mating-type idiomorphs within and among fungal populations collected from barley landraces. *Genome* 48(5), 855–69. doi:10.1139/g05-046.

Read, A. C., Rinaldi, F. C., Hutin, M., He, Y. Q., Triplett, L. R. and Bogdanove, A. J. 2016. Suppression of Xo1-mediated disease resistance in rice by a truncated, non-DNA-binding TAL effector of *Xanthomonas oryzae*. *Frontiers in Plant Science* 7, 1516. doi:10.3389/fpls.2016.01516.

Richards, J. K., Friesen, T. L. and Brueggeman, R. S. 2017. Association mapping utilizing diverse barley lines reveals net form net blotch seedling resistance/susceptibility loci. *Theoretical and Applied Genetics* 130(5), 915–27. doi:10.1007/s00122-017-2860-1.

Richter, K., Schondelmaier, J. and Jung, C. 1998. Mapping of quantitative trait loci affecting *Drechslera teres* resistance in barley with molecular markers. *Theoretical and Applied Genetics* 97(8), 1225–34. doi:10.1007/s001220051014.

Ridout, C. J. 2009. Profiles in pathogenesis and mutualism: powdery mildews. In: *Plant Relationships*. Springer, Berlin, Heidelberg, pp. 51–68.

Roelfs, A. P. 1985. Epidemiology in North America. *The Cereal Rusts* 2, 403–34.

Roelfs, A. P. and Bushnell, W. R. 1985. *The Cereal Rusts, Volume II: Diseases, Distribution, Epidemiology, and Control*. Academic Press, Orlando.

Saari, E. E. and Prescott, J. M. 1985. World distribution in relation to economic losses. In: *Diseases, Distribution, Epidemiology, and Control*. Academic Press, pp. 259–98.

Sabelleck, B. and Panstruga, R. 2018. Novel jack-in-the-box effector of the barley powdery mildew pathogen? *Journal of Experimental Botany* 69(15), 3511–4. doi:10.1093/jxb/ery192.

Sachs, E. 2006. The history of research into Ramularia leaf spot on barley. In: *Proceedings of the 1st European Ramularia Workshop*, March 2006, pp. 9–15.

Saintenac, C., Zhang, W., Salcedo, A., Rouse, M. N., Trick, H. N., Akhunov, E. and Dubcovsky, J. 2013. Identification of wheat gene *Sr35* that confers resistance to Ug99 stem rust race group. *Science* 341(6147), 783–6. doi:10.1126/science.1239022.

Salamati, S. and Reitan, L. 2006. *Ramularia collo-cygni* on spring barley, an overview of its biology and epidemiology. In: *Proceedings 1st European Ramularia Workshop*, March 2006, pp. 19–35.

Salamati, S. and Tronsmo, A. M. 1997. Pathogenicity of *Rhynchosporium secalis* isolates from Norway on 30 cultivars of barley. *Plant Pathology* 46(3), 416–24. doi:10.1046/j.1365-3059.1997.d01-20.x.

Salcedo, A., Rutter, W., Wang, S., Akhunova, A., Bolus, S., Chao, S., Anderson, N., De Soto, M. F., Rouse, M., Szabo, L., Bowden, R. L., Dubcovsky, J. and Akhunov, E. 2017. Variation in the *AvrSr35* gene determines *Sr35* resistance against wheat stem rust race Ug99. *Science* 358(6370), 1604–6. doi:10.1126/science.aao7294.

Sallam, A. H., Tyagi, P., Brown-Guedira, G., Muehlbauer, G. J., Hulsa, A. and Steffenson, B. J. 2017. Genome-wide association mapping of stem rust resistance in *Hordeum vulgare* subsp. *spontaneum*. *G3* 7(10), 3491–507. doi:10.1534/g3.117.300222.

Sandoval-Islas, J. S., Broers, L. H. M., Vivar, H. and Osada, K. S. 1998. Evaluation of quantitative resistance to yellow rust (*Puccinia striiformis* f. sp. *hordei*) in the ICARDA/CIMMYT barley breeding programme. *Plant Breeding* 117(2), 127–30. doi:10.1111/j.1439-0523.1998.tb01465.x.

Santén, K., Marttila, S., Liljeroth, E. and Bryngelsson, T. 2005. Immunocytochemical localization of the pathogenesis-related PR-1 protein in barley leaves after infection by *Bipolaris sorokiniana*. *Physiological and Molecular Plant Pathology* 66(1–2), 45–54. doi:10.1016/j.pmpp.2005.04.006.

Sarpeleh, A., Wallwork, H., Catcheside, D. E., Tate, M. E. and Able, A. J. 2007. Proteinaceous metabolites from *Pyrenophora teres* contribute to symptom development of barley net blotch. *Phytopathology* 97(8), 907–15. doi:10.1094/PHYTO-97-8-0907.

Schumann, G. L. and Leonard, K. J. 2000. Stem rust of wheat (black rust). *The Plant Health Instructor*. doi:10.1094/PHI-I-2000-0721-01.

Schwartz, P. B. 2003. Impact of fusarium head blight on malting and brewing quality barley. In: Leonard, K. J. and Bushnell, W. R. (Eds), *Fusarium Head Blight of Wheat and Barley*. The American Phytopathological Society, St. Paul, MN, pp. 395–419.

Sharma Poudel, R., Al-Hashel, A. F., Gross, T., Gross, P. and Brueggeman, R. 2018. Pyramiding *rpg4*-and *Rpg1*-mediated stem rust resistance in barley requires the *Rrr1* gene for both to function. *Frontiers in Plant Science* 9, 1789. doi:10.3389/fpls.2018.01789.

Shi, G., Zhang, Z., Friesen, T. L., Raats, D., Fahima, T., Brueggeman, R. S., Lu, S., Trick, H. N., Liu, Z., Chao, W., Frenkel, Z., Xu, S. S., Rasmussen, J. B. and Faris, J. D. 2016. The hijacking of a receptor kinase-driven pathway by a wheat fungal pathogen leads to disease. *Science Advances* 2(10), e1600822. doi:10.1126/sciadv.1600822.

Shipton, W. A., Boyd, J. R. and Ali, S. M. 1974. Scald of barley. *Review of Plant Pathology* 53, 839–61.

Shjerve, R. A., Faris, J. D., Brueggeman, R. S., Yan, C., Zhu, Y., Koladia, V. and Friesen, T. L. 2014. Evaluation of a *Pyrenophora teres* f. *teres* mapping population reveals multiple independent interactions with a region of barley chromosome 6H. *Fungal Genetics and Biology* 70, 104–12. doi:10.1016/j.fgb.2014.07.012.

Singh, R. P., Hodson, D. P., Huerta-Espino, J., Jin, Y., Njau, P., Wanyera, R., Herrera-Foessel, S. A. and Ward, R. W. 2008. Will stem rust destroy the world's wheat crop? *Advances in Agronomy* 98, 271–309. doi:10.1016/S0065-2113(08)00205-8.

Singh, R. P., Hodson, D. P., Jin, Y., Lagudah, E. S., Ayliffe, M. A., Bhavani, S., Rouse, M. N., Pretorius, Z. A., Szabo, L. J., Huerta-Espino, J., Basnet, B. R., Lan, C. and Hovmøller, M. S. 2015. Emergence and spread of new races of wheat stem rust fungus: continued threat to food security and prospects of genetic control. *Phytopathology* 105(7), 872–84. doi:10.1094/PHYTO-01-15-0030-FI.

Sjokvist, E., Lemcke, R., Kamble, M., Turner, F., Blaxter, M., Havis, N. H. D., Lyngkjær, M. F. and Radutoiu, S. 2019. Dissection of Ramularia leaf spot disease by integrated analysis of barley and *Ramularia collo-cygni* transcriptome responses. *Molecular Plant-Microbe Interactions : MPMI* 32(2), 176–93. doi:10.1094/MPMI-05-18-0113-R.

Smedegård-Petersen, V. 1971. *Pyrenophora teres* f. *maculata* f. nov. and *Pyrenophora teres* f. *teres* on barley in Denmark. *Yearbook of the Royal Veterinary and Agricultural University (Copenhagen)* 1971, 124–44.

Solanki, S. 2017. Dissecting the mystery behind *Rpg5* mediated *Puccinia graminis* resistance in barley using genetics, molecular and bioinformatics approaches. Doctoral dissertation. North Dakota State University.

Solanki, S., Ameen, G., Borowicz, P. and Brueggeman, R. S. 2019. Shedding light on penetration of cereal host stomata by wheat stem rust using improved methodology. *Scientific Reports* 9(1), 7939. doi:10.1038/s41598-019-44280-6.

Sooväli, P., Tikhonova, M. and Matušinsky, P. 2014. First report of Ramularia leaf spot caused by *Ramularia collo-cygni* on leaves and seeds of barley in Estonia. *Plant Disease* 98(7), 997–. doi:10.1094/PDIS-10-13-1035-PDN.

St. Pierre, S., Gustus, C., Steffenson, B., Dill-Macky, R. and Smith, K. P. 2010. Mapping net form net blotch and Septoria speckled leaf blotch resistance loci in barley. *Phytopathology* 100(1), 80–4. doi:10.1094/PHYTO-100-1-0080.

Steffenson, B. J. 1992a. Analysis of durable resistance to stem rust in barley. *Euphytica* 63(1–2), 153–67. doi:10.1007/BF00023920.

Steffenson, B. J. 1992b. Analysis of durable resistance to stem rust in barley. *Euphytica* 63, 153–67.

Steffenson, B. J. 2003. Fusarium head blight of barley: Impact, epidemics, management, and strategies for identifying and utilizing genetic resistance. In: Leonard, K. J. and Bushnell, W. R. (Eds), *Fusarium Head Blight of Wheat and Barley*. The American Phytopathological Society, St. Paul, MN, pp. 241–95.

Steffenson, B. J. and Webster, R. K. 1992. Pathotype diversity of *Pyrenophora teres* f. *teres* on barley. *Phytopathology* 82(2), 170–7. doi:10.1094/Phyto-82-170.

Steffenson, B. J., Jin, Y., Rossnagel, B. G., Rasmussen, J. B. and Kao, K. 1995. Genetics of multiple disease resistance in a doubled-haploid population of barley. *Plant Breeding* 114(1), 50–4. doi:10.1111/j.1439-0523.1995.tb00758.x.

Steffenson, B. J., Hayes, P. M. and Kleinhofs, A. 1996. Genetics of seedling and adult plant resistance to net blotch (*Pyrenophora teres f. teres*) and spot blotch (*Cochliobolus sativus*) in barley. *Theoretical and Applied Genetics* 92(5), 552–8. doi:10.1007/BF00224557.

Steffenson, B. J., Zhou, H., Chai, Y. and Grando, S. 2013. Vulnerability of cultivated and wild barley to African stem rust race TTKSK. In: *Advance in Barley Sciences*. Springer, Dordrecht, pp. 243–55.

Steffenson, B. J., Solanki, S. and Brueggeman, R. S. 2016. Landraces from mountainous regions of Switzerland are sources of important genes for stem rust resistance in barley. *Alpine Botany* 126(1), 23–33. doi:10.1007/s00035-015-0161-3.

Stergiopoulos, I., Collemare, J., Mehrabi, R. and De Wit, P. J. 2013. Phytotoxic secondary metabolites and peptides produced by plant pathogenic *Dothideomycete* fungi. *FEMS Microbiology Reviews* 37(1), 67–93. doi:10.1111/j.1574-6976.2012.00349.x.

Steuernagel, B., Periyannan, S. K., Hernández-Pinzón, I., Witek, K., Rouse, M. N., Yu, G., Hatta, A., Ayliffe, M., Bariana, H., Jones, J. D., Lagudah, E. S. and Wulff, B. B. 2016. Rapid cloning of disease-resistance genes in plants using mutagenesis and sequence capture. *Nature Biotechnology* 34(6), 652–5. doi:10.1038/nbt.3543.

Sutton, B. C. and Waller, J. M. 1988. Taxonomy of *Ophiocladium hordei*, causing leaf lesions on triticale and other Gramineae. *Transactions of the British Mycological Society* 90(1), 55–61. doi:10.1016/S0007-1536(88)80180-3.

Tamang, P., Neupane, A., Mamidi, S., Friesen, T. and Brueggeman, R. 2015. Association mapping of seedling resistance to spot form net blotch in a worldwide collection of barley. *Phytopathology* 105(4), 500–8. doi:10.1094/PHYTO-04-14-0106-R.

Tekauz, A. 1990. Characterization and distribution of pathogenic variation in *Pyrenophora teres* f. *teres* and *P. teres* f. *maculata* from western Canada. *Canadian Journal of Plant Pathology* 12(2), 141–8. doi:10.1080/07060669009501017.

Tekauz, A. 1991. Pathogenic variation in *Rhynchosporium secalis* on barley in Canada. *Canadian Journal of Plant Pathology* 13(4), 298–304. doi:10.1080/07060669109500915.

Tekauz, A. 2003. Spot blotch caused by *Bipolaris sorokiniana*: an emerging problem in barley in western Canada. In: *Proceedings of the of 8th International Congress of Plant Pathology* (vol. II), Christchurch, New Zealand, p. 103.

Thirugnanasambandam, A., Wright, K. M., Atkins, S. D., Whisson, S. C. and Newton, A. C. 2011. Infection of *Rrs1* barley by an incompatible race of the fungus *Rhynchosporium secalis* expressing the green fluorescent protein. *Plant Pathology* 60(3), 513–21. doi:10.1111/j.1365-3059.2010.02393.x.

Toubia-Rahme, H. and Steffenson, B. J. 2004. Sources of resistance to septoria speckled leaf blotch caused by *Septoria passerinii* in barley. *Canadian Journal of Plant Pathology* 26(3), 358–64. doi:10.1080/07060660409507153.

Ullrich, S. E. 2010. *Barley: Production, Improvement, and Uses* (vol. 12). John Wiley & Sons.

Ulus, C. and Karakaya, A. 2007. Assessment of the seedling reactions of some Turkish barley cultivars to barley stripe. *Tarım Bilimleri Dergisi* 13(4), 409–12. doi:10.1501/Tarimbil_0000000383.

Vaish, S. S., Ahmed, S. B. and Prakash, K. 2011. First documentation on status of barley diseases from the high altitude cold arid trans-Himalayan Ladakh region of India. *Crop Protection* 30(9), 1129–37. doi:10.1016/j.cropro.2011.04.015.

Valjavec-Gratian, M. and Steffenson, B. J. 1997. Pathotypes of *Cochliobolus sativus* on barley in North Dakota. *Plant Disease* 81(11), 1275–8. doi:10.1094/PDIS.1997.81.11.1275.

Verma, R. P. S., Selvakumar, R., Gangwar, O. P., Shekhawat, P. S., Bhardwaj, S. C., Rehman, S., Sharma-Poudyal, D. and Gyawali, S. 2018. Identification of additional sources of resistance to *Puccinia striiformis* f. sp. *hordei* (*PSH*) in a collection of barley genotypes adapted to the high input condition. *Journal of Phytopathology* 166(5), 355–64. doi:10.1111/jph.12693.

Walters, D. R., Havis, N. D. and Oxley, S. J. 2008. *Ramularia collo-cygni*: the biology of an emerging pathogen of barley. *FEMS Microbiology Letters* 279(1), 1-7. doi:10.1111/j.1574-6968.2007.00986.x.

Wang, X., Richards, J., Gross, T., Druka, A., Kleinhofs, A., Steffenson, B., Acevedo, M. and Brueggeman, R. 2013. The *rpg4*-mediated resistance to wheat stem rust (*Puccinia graminis*) in barley (*Hordeum vulgare*) requires *Rpg5*, a second NBS-LRR gene, and an actin depolymerization factor. *Molecular Plant-Microbe Interactions* 26(4), 407–18. doi:10.1094/MPMI-06-12-0146-R.

Wellings, C. R. 2011. Global status of stripe rust: a review of historical and current threats. *Euphytica* 179(1), 129–41. doi:10.1007/s10681-011-0360-y.

Wellings, C. R., Burdon, J. J., McIntosh, R. A., Wallwork, H., Raman, H. and Murray, G. M. 2000. A new variant of *Puccinia striiformis* causing stripe rust on barley and wild *Hordeum* species in Australia. *Plant Pathology* 49(6), 803-. doi:10.1046/j.1365-3059.2000.00506.x.

White, J. W. and Rodriguez-Aguilar, A. 2001. An agroclimatological characterization of the Indo-Gangetic Plains. *Journal of Crop Production* 3(2), 53–65. doi:10.1300/J144v03n02_03.

Wilcoxson, R. D., Rasmusson, D. C. and Miles, M. R. 1990. Development of barley resistant to spot blotch and genetics of resistance. *Plant Disease* 74(3), 207–10. doi:10.1094/PD-74-0207.

Win, J., Chaparro-Garcia, A., Belhaj, K., Saunders, D. G. O., Yoshida, K., Dong, S., Schornack, S., Zipfel, C., Robatzek, S., Hogenhout, S. A. and Kamoun, S. 2012. Effector biology of plant-associated organisms: concepts and perspectives. In: *Cold Spring Harbor Symposia on Quantitative Biology* (vol. 77), pp. 235–47. doi:10.1101/sqb.2012.77.015933.

Wolpert, T. J., Dunkle, L. D. and Ciuffetti, L. M. 2002. Host-selective toxins and avirulence determinants: what's in a name? *Annual Review of Phytopathology* 40(1), 251–85. doi:10.1146/annurev.phyto.40.011402.114210.

Wu, H.-L., Steffenson, B. J., Zhong, S., Li, Y. and Oleson, A. E. 2003. Genetic variation for virulence and RFLP markers in *Pyrenophora teres*. *Canadian Journal of Plant Pathology* 25(1), 82–90. doi:10.1080/07060660309507052.

Xie, W., Xiong, W., Pan, J., Ali, T., Cui, Q., Guan, D., Meng, J., Mueller, N. D., Lin, E. and Davis, S. J. 2018. Decreases in global beer supply due to extreme drought and heat. *Nature Plants* 4(11), 964–73. doi:10.1038/s41477-018-0263-1.

Xue, A. G., Ho, K. M., Butler, G., Vigier, B. J. and Babcock, C. 2006. Pathogenicity of Fusarium species causing head blight in barley. *Phytoprotection* 87(2), 55–61. doi:10.7202/013973ar.

Yirgou, D. and Caldwell, R. M. 1963. Stomatal penetration of wheat seedlings by stem and leaf rust: effect of light and carbon dioxide. *Science* 141(3577), 272–3. doi:10.1126/science.141.3577.272.

Yu, G. T., Franckowiak, J. D., Lee, S. H., Horsley, R. D. and Neate, S. M. 2010. A novel QTL for Septoria speckled leaf blotch resistance in barley (*Hordeum vulgare* L.) accession PI 643302 by whole-genome QTL mapping. *Genome* 53(8), 630–6. doi:10.1139/g10-032.

Yun, S. J., Gyenis, L., Hayes, P. M., Matus, I., Smith, K. P., Steffenson, B. J. and Muehlbauer, G. J. 2005. Quantitative trait loci for multiple disease resistance in wild barley. *Crop Science* 45(6), 2563–72. doi:10.2135/cropsci2005.0236.

Zadoks, T. C. 1967. Epidemiology of wheat rust in Europe. *International Journal of Pest Management: Part B* 13(1), 29–46. doi:10.1080/05331846709432232.

Zaffarano, P. L., McDonald, B. A. and Linde, C. C. 2011. Two new species of *Rhynchosporium*. *Mycologia* 103(1), 195–202. doi:10.3852/10-119.

Zhan, J., Fitt, B. D. L., Pinnschmidt, H. O., Oxley, S. J. P. and Newton, A. C. 2008. Resistance, epidemiology and sustainable management of *Rhynchosporium secalis* populations on barley. *Plant Pathology* 57(1), 070918211612002–???. doi:10.1111/j.1365-3059.2007.01691.x.

Zhang, J., Shao, F., Li, Y., Cui, H., Chen, L., Li, H., Zou, Y., Long, C., Lan, L., Chai, J., Chen, S., Tang, X. and Zhou, J. M. 2007. A *Pseudomonas syringae* effector inactivates MAPKs to suppress PAMP-induced immunity in plants. *Cell Host and Microbe* 1(3), 175–85. doi:10.1016/j.chom.2007.03.006.

Zhang, W., Chen, S., Abate, Z., Nirmala, J., Rouse, M. N. and Dubcovsky, J. 2017. Identification and characterization of *Sr13*, a tetraploid wheat gene that confers resistance to the Ug99 stem rust race group. *Proceedings of the National Academy of Sciences of the United States of America* 114(45), E9483–92. doi:10.1073/pnas.1706277114.

Zhong, S., Steffenson, B. J., Martinez, J. P. and Ciuffetti, L. M. 2002. A molecular genetic map and electrophoretic karyotype of the plant pathogenic fungus *Cochliobolus sativus*. *Molecular Plant-Microbe Interactions* 15(5), 481–92. doi:10.1094/MPMI.2002.15.5.481.

Zhong, S., Ali, S., Leng, Y., Wang, R. and Garvin, D. F. 2015. *Brachypodium distachyon-Cochliobolus sativus* pathosystem is a new model for studying plant-fungal interactions in cereal crops. *Phytopathology* 105(4), 482–9. doi:10.1094/PHYTO-08-14-0214-R.

Zhou, H. and Steffenson, B. 2013. Genome-wide association mapping reveals genetic architecture of durable spot blotch resistance in US barley breeding germplasm. *Molecular Breeding* 32(1), 139–54. doi:10.1007/s11032-013-9858-4.

Zhou, H., Steffenson, B. J., Muehlbauer, G., Wanyera, R., Njau, P. and Ndeda, S. 2014. Association mapping of stem rust race TTKSK resistance in US barley breeding germplasm. *Theoretical and Applied Genetics* 127(6), 1293–304. doi:10.1007/s00122-014-2297-8.

Zillinsky, F. J. 1983. *Common Diseases of Small Grain Cereals. A Guide to Identification.*

Chapter 11

Integrated disease management of barley

Adrian C. Newton, James Hutton Institute and SRUC, UK; and Henry E. Creissen, Neil D. Havis, and Fiona J. Burnett, SRUC, UK

1 Introduction
2 Barley production context: requirements and constraints
3 Diseases overview
4 Inoculum management: sources and epidemiological conditions
5 Varietal resistance
6 Crop protectants
7 Agronomy
8 IPM knowledge sources and tools
9 Uptake and communication of IPM
10 Farming systems, soil and research platforms
11 Conclusion and future trends
12 Acknowledgements
13 Where to look for further information
14 References

1 Introduction

It is widely recommended and required by EU Directive law (2009/128/EC, article 14) that member states should promote low pesticide input pest management; in effect that all farmers and growers have an integrated pest management (IPM) plan. This is no less important for broad acre cereal crops like barley than it is for, for example, tomato crops in glasshouses. However, the perception is that the concept applies more in the latter type of crop as it requires the use of carefully managed biocontrol agents to eliminate fungicides and their residues from the system. Whilst it may be true that the latter represents very successful application of IPM principles, the same principles can and should be applied to the former, and many growers do to some degree. Nevertheless, in many

http://dx.doi.org/10.19103/AS.2019.0060.18

countries most farmers have no IPM plan as such (76% in Scotland: Scottish Government, 2017) but to effectively manage diseases in barley, this must be carried out in an IPM context to be sustainable.

A definition of IPM is 'the careful consideration of all available pest control techniques and subsequent integration of appropriate measures that discourage the development of pest populations and keep pesticides and other interventions to levels that are economically justified and reduce or minimize risks to human health and the environment. IPM emphasizes the growth of a healthy crop with the least possible disruption to agro-ecosystems and encourages natural pest control mechanisms'; or more succinctly: 'Integrated Pest Management (IPM) is an ecosystem approach to crop production and protection that combines different management strategies and practices to grow healthy crops and minimize the use of pesticides' (FAO, 2018).

How do these principles apply to disease management in barley production? We will describe the main options available in the sections below and consider them in the key context of integration. They may also be considered in the context of management decisions under the headings of (1) seed-based, (2) mechanical/growing medium, (3) cropping/biological control, (4) timing and (5) input efficiency (Table 1). Integration can mean simply additive effects of the components, but more than likely is that some components will interact synergistically while others may be antagonistic. Furthermore, the outcomes will be different whether considered from an economic, ecological or sociological context. However, in many cases we do not know how they interact in any of these contexts and research is needed to determine the mechanisms and develop tools to aid their integration in practice.

IPM can include all possible biotic and abiotic stress factors or influences but as the title implies, we focus on disease and generally exclude for example, weeds, nematodes and insects as these usually require different approaches and are therefore best considered separately. Furthermore, diseases represent the biggest threats to barley yield for much of the barley growing area not already constrained by abiotic stress, and for which there are many management options.

2 Barley production context: requirements and constraints

Barley is grown for animal feed, human food, as a substrate for grain or malt distilling or for brewing, and as whole crop for feed, silage or anaerobic digestion (Newton et al., 2011). For each of these uses different cultivars, mixtures or even populations will be grown in very different environments with different agronomies. Management and inputs can be from minimal to

Table 1 IPM categories for consideration in an ecological, economic or sociological context

1) *Seed based*
Using certified seed
Selecting cultivars with resistance to key risks in your area
Testing home-saved seed
Use appropriate seed treatments on affected or early drilled crops
Stale seedbeds
Cultivar diversification
Use of cultivar mixtures where markets allow
Alteration of crop/plant population density
2) *Mechanical/growing medium*
Direct drilling
Minimum tillage
Regular ploughing
Rotational ploughing
Under-sowing of crops
Increasing soil organic matter
Choice of cover crops
3) *Cropping/biological control*
Summer fallow
Encourage beneficial organisms
Use of biological controls
Catch/cover crops to manage soil pests (e.g. PCN)
4) *Timing*
Planning crop nutrition
Planning drilling date
Planning drainage requirements
An agri-environment scheme requirements
5) *Input efficiency*
Use of treated seed
Spot treatment
Drift reduction technology
Precision application technology
Field mapping
Attention to growth stages, spray timings and intervals
Technical updates on product efficacy
Crops monitored pre- and post-treatment
Thresholds, risk warnings or weather data used
Where feasible to do so reduce pesticide dosage and/or frequency of pesticide use
Incorporate resistance elicitor-based products in crop protection programme where available
Calibration and maintenance of sprayers
Pesticide resistance stewardship advice followed

intensive and therefore outcomes in economic, ecological or sociological contexts may be very contrasting. For example, a malting barley crop for the Scotch whisky industry must be of a single, pure, often-specified cultivar and achieve specific grain size, grain nitrogen and germination standards. In contrast, a whole crop of barley for silage may need to produce the highest quantity of digestible dry matter from negligible inputs. The consequences for pests and pathogens as well as the options available will be very different. In the former case herbicide and fungicide inputs will be required and there is scope for minimising them with monitoring, forecasting and interpretation of crop intelligence data. In the latter case herbicides and fungicides may not be required because the risks and consequences of not applying them are low due to the inherent qualities of the system. However, there may be other options that cannot be considered in the malting crop such as companion- or inter-cropping.

Winter and spring barley systems are another contrast where the presence or absence of a cold period required for vernalisation also has consequences for pests and pathogens. However, other barley crops are winter sown to better utilise rain and some crops are grazed before then going on to grain harvest. Such contrasting systems cannot be considered for all the components of IPM so below we will consider the options available in the systems where they have potential to contribute to beneficial outcomes.

3 Diseases overview

Barley is plagued by a great diversity of diseases caused by a wide range of fungal, viral and bacterial pathogens, many of which can seriously impact upon the crop's ability to produce high yields and quality grains. When considering the management of barley disease it is important to understand which diseases pose the greatest threat and the management strategies that are effective in controlling them. Due to differences in life cycles it may be that different diseases require different management strategies which may sometimes require conflicting approaches, for example reduced tillage regime to control common eyespot (Burnett et al., 2012) or ploughing to control tan spot (Bockus and Claassen, 1992). Table 2 summarises information on many of the current economically important barley diseases, their causal agents, suitable management strategies and issues to consider when attempting to control them. It's important to understand that over-reliance on a single control measure puts a selection pressure on the pathogen to evolve and overcome the control measure, thus an integrated approach to disease management is required if the disease is to be managed in a sustainable fashion.

Table 2 Summary of the most economically important barley diseases, the biology of their causal agents and potential control methods

Disease	Pathogen	Biotroph/ hemibiotroph/ necrotroph	Target (foliar/ ear/stem base/ root)	Dispersal method	Occurrence	Economic importance	Genetic resistance (mg, pr)	Pesticide resistance (1→3 low→high)	Cultural control	References
Fungi										
Powdery mildew	*Blumeria graminis* f. sp. *hordei*	Biotroph	Foliar	Wind	Very common	Medium	mg, pr	2	Control volunteers, crop residue removal, don't overfertilise	Lyngkjaer et al. (2000), Kusch and Panstruga (2017)
Brown rust	*Puccinia hordei*	Biotroph	Foliar	Wind	Common	Medium	mg, pr	1	Control volunteers	Grasso et al. (2006)
Stem rust	*Puccinia graminis* f. sp. *secalis*, *Puccinia graminis* f. sp. *tritici*	Biotroph	Foliar	Wind	Very rare	High	mg, pr	1	Control volunteers	Grasso et al. (2006), Kleinhofs et al. (2009)
Yellow/ stripe rust	*Puccinia striiformis* f. sp. *hordei*	Biotroph	Foliar	Wind	Rare	Medium	mg, pr	1	Control volunteers	Chen et al. (1994), Yan and Chen 2006
Leaf scald	*Rhynchosporium commune*	Hemibiotroph	Foliar	Seed, trash	Very common	High	mg, pr	2	Rotation, crop residue removal, delay sowing	Kendall et al. (1994), Oxley and Burnett (2009), Zhan et al. (2008), Cooke et al. (2004)
Ramularia leaf spot	*Ramularia collo-cygni*	Hemibiotroph	Foliar	Seed, wind	Common	Medium	Unidentified	3	None known	Walters et al. (2008), Havis et al. (2015)
Net blotch (net form)	*Pyrenophora teres* f. *teres* (net form)	Necrotroph	Foliar	Trash, seed	Rare	High	mg, pr	3	Crop rotation, crop residue removal	Semar et al. (2007)
Net blotch (spot form)	*Pyrenophora teres* f. *maculata* (spot form)	Necrotroph	Foliar	Trash, seed	Rare	Medium	mg, pr	3	Crop rotation, crop residue removal	Semar et al. (2007)

(Continued)

Table 2 (*Continued*)

Disease	Pathogen	Biotroph/ hemibiotroph/ necrotroph	Target (foliar/ ear/stem base/ root)	Dispersal method	Occurrence	Economic importance	Genetic resistance (mg, pr)	Pesticide resistance (1→3 low→high)	Cultural control	References
Tan spot	*Pyrenophora tritici-repentis*	Necrotroph	Foliar	Trash, seed	Rare	Medium	mg, pr	NA	Crop rotation, crop residue removal	Bockus and Claassen (1992), Carignano et al. (2008), Duczek et al. (1999)
Stagonospora blotch	*Parastagonospora nodorum*	Hemibiotroph	Foliar	Trash	Rare	Low	mg, pr	2	Crop rotation, crop residue removal	Duczek et al. (1999), Caten and Newton (2000)
Head/ear/ seedling blight	*Microdochium nivale, Fusarium graminearum* but also *F. avenaceum, F. culmorum, F. poae*	Necrotroph	Ear/stem base	Soil	Common	Medium	pr	2	Plough, rotation not including maize	Bateman et al. (2007), Cromey et al. (2006), Browne and Cooke (2005), Bai and Shaner (2004), Scherm et al. (2013)
Ergot	*Claviceps purpurea*	Necrotroph	Ear	Seed	Very rare	Low	Unidentified	3	Rotation, plough, weed control	Orlando et al. (2017)
Cephalosporium stripe	*Cephalosporium gramineum*	Necrotroph	Foliar	Trash/soil	Rare	Low	Unidentified	1	Rotation, plough, weed control	Havis and Gorniak (2004)
Take all	*Gaeumannomyces graminis* var *tritici*	Necrotroph	Root	Soil	Rare	Low	Unidentified	NA	Rotation, delay sowing	Oxley and Burnett (2009), Cromey et al. (2006)
Common eyespot	*Oculimacula yallundae, O. acuformis*	Necrotroph	Stem base	Trash, soil	Rare	Low	mg, pr	2	Control volunteers, rotation, delay sowing	Burnett and Hughes (2004), Burnett et al. (2012)
Sharp eyespot	*Ceratobasidium cereale*	Necrotroph	Stem	Trash, soil	Rare	Low	pr	2	Control volunteers, control weeds, plough	Colbach et al. (1997), Cromey et al. (2006)

Loose smut	*Ustilago nuda* f. sp. *hordei*	Biotroph	Ear	Seed	Very rare	Low	pr	2	Clean seed	Menzies et al. (2009), Menzies et al. (2010)
Covered smut	*Ustilago hordei*	Biotroph	Ear	Seed	Very rare	Low	pr	1	Clean seed	Menzies et al. (2009)
Leaf stripe	*Pyrenophora graminea*	Necrotroph	Foliar	Seed	Rare	Low	pr	1	Rotation	Pecchioni et al. (1999)
Bacteria										
Bacterial leaf streak	*Xanthomonas campestris* pv. *hordei*	Necrotroph	Foliar	Seed, splash	Very rare	Low	pr	NA	Plough, remove weeds	Alizadeh et al. (1994), El Attari et al. (1998)
Bacterial leaf blight /basal kernel blight	*Pseudomonas syringae* pv. *syringae*	Necrotroph	Foliar	Wind, splash, seed	Very rare	Low	pr	NA	Plough, clean seed	Braun-Kiewnick et al. (2000)
Viral										
Barley yellow mosaic disease	Barley yellow mosaic virus (BaYMV), Barley mild mosaic virus (BaMMV)	Biotroph	Foliar	Soil (*Polymyxa graminis*)	Rare	Low	Unidentified	NA	None known	
Barley yellow dwarf virus (BYDV)	Barley yellow dwarf virus (BYDV)	Biotroph	Foliar	Aphid vector	Common	High	Unidentified	NA with respect to virus. 3 with respect to aphid vector	Ploughing/destroying crop residues several weeks before drilling, rotation, altering sowing dates to limit contact with vector	Miller and Rasochova (1997)

mg = major gene, pr = partial resistance.
NA = Not applicable as pesticides rarely used.

4 Inoculum management: sources and epidemiological conditions

Two major considerations determine whether a disease has the potential to become problematic: (1) whether the inoculum is present and (2) whether favourable epidemiological conditions exist, for both host and environment. In addition, not all pathogens cause enough yield or quality losses to be economically viable to treat even in their most severe cases. IPM choices should be determined by this knowledge. For inoculum that is brought predominantly by wind such as the rusts and mildews, crop intelligence reports can be very useful for determining risk and therefore choice of appropriate treatments. Knowledge of outbreak locations, virulence specificity of the inoculum and fungicide resistance traits can all be used to aid decisions, although some of these data will be interpolated from trends determined by full characterisation of previous years' isolates. Examples of this include the UK Cereal Pathogen Virulence Survey (UKCPVS, 2018) where the previous year's data is disseminated before the growing season starts and updates about outbreaks are released during the season for powdery mildew and yellow rust.

A second source of incoming inoculum is seed-borne. For barley pathogens this largely affects rhynchosporium (barley leaf blotch or scald) and ramularia (Ramularia leaf spot), though it can be important for net blotch too. Here clean seed or adequate seed treatment with fungicide, heat, resistance elicitor or biocontrol agent may reduce or eliminate the inoculum source. It can also control other inoculum such as the smuts and ergot. Inoculum that is predominantly on the seed surface is effectively treated in this way but ramularia inoculum is often present throughout the seed including the embryo, so effective treatment is very difficult without risking seed viability or vigour loss (Havis et al., 2014b). Indeed, ramularia colonises the whole plant endophytically, and we will consider the implications of endophytes in more detail later.

As pathogens that are seed-borne are generally non-obligate fungi they are usually already present in a third source, the crop environment, surviving on crop debris or soil depending on their saprophytic growth characteristics. Therefore, agronomic factors and general phytosanitation factors can be manipulated to reduce risk. Soil cultivation, particularly inversion tillage (ploughing) where the plant debris is buried, compared with non-inversion tillage (minimum tillage or direct drilling) can make a difference particularly to early season disease.

A fourth source is volunteer plants and reservoirs in adjacent crops or crop margins. This is a potential increasing problem for non-specialised inoculum such as ergot, but where host specificity even at a species level is important, such as for rhynchosporium (King et al., 2013; Penselin et al., 2016), inoculum and symptoms on neighbouring plants are low risk. The risk from volunteer and other reservoir plants as hosts to any type of pathogen can be reduced only by appropriate agronomic decisions and phytosanitary measures. This might include rotation, probably the most effective tool for rhizoctonia and take-all control.

Another source of disease that could be from any of the original inoculum sources is asymptomatic infection, that is they are not causing disease but infection can be extensive. Such infected plants require signals that cause the transition to pathogenicity. Such signals could be developmental, abiotic stress or temporal (Newton et al., 2010). However, in an IPM context we should consider host–microbe interactions in a broader context by managing our crops in such a way that benign and beneficial associations are favoured and that pathogenic or parasitic associations are disadvantaged (Looseley and Newton, 2014). An example of the former would be to encourage mycorrhizal associations which can be achieved by minimising soil disturbance in our agronomic management. Knowledge of the source of inoculum can help to target treatment too, that is whether seed-borne, soil, trash, foliar for wind or splash-dispersed transmission.

Having considered (1) whether inoculum is present, if it is then we must consider (2) whether favourable epidemiological conditions exist. A good example of this is the effect of seed heavily infected with rhynchosporium versus clean seed on epidemic development. Here it was found that in some seasons the seed-borne inoculum had a big impact on epidemic development, but in others it was the later epidemiological conditions that were dominant and the extra seed-borne inoculum had no impact (Havis et al., 2014a). The general nature of the favourable epidemiological conditions can be deduced for each disease. These include wet conditions, particularly heavy rain for splash-dispersed pathogens such as rhynchosporium, especially during stem extension which promotes transmission up the canopy. Other conducive conditions include dew on the leaves during cool nights, which favours rust infection; and warm, humid conditions without free water which promotes mildew infection. However, for some pathogens such as ramularia and rhynchosporium there are also stress factors that are less clearly understood making epidemic prediction less certain (Newton et al., 2010).

5 Varietal resistance

Use of durable resistance is the obvious and most effective strategy to deploy in IPM. However, durable resistance in practice may take many forms and should be a product of its context, that is considering all the environment factors too, both natural and managed. There are very few examples of single major genes that have proved durable and the general strategy should be to not use them as the main component of a resistance strategy. To do so would be to expose them to such strong selection that the pathogen will mutate to virulence rendering the resistance ineffective. The exceptions would be where other components of the IPM strategy are more effective resulting in very little selection pressure, or an unusual type of genetic resistance such as the *mlo* gene in barley that has remained durable despite widespread use in spring barley (Kusch and Panstruga, 2017). However, the more likely genetic basis of durable resistance is variously referred to as polygenic, partial resistance or field resistance. Crucially, it comprises multiple expressions of resistance mechanisms that together amount to effective, although rarely total, resistance. The components separately are likely expressions of different genes each of which has a small effect on resistance, the net effect of which is additive or even synergistic. On resistance ratings such as one (susceptible) to nine (fully resistant), AHDB Recommended lists (AHDB, 2018) these are likely to be seven or eight but rarely nine.

Partial resistance is not necessarily either polygenic or race-non-specific and there are examples of where it is neither (Newton, 1989). Therefore, not relying on a single genotype whatever its resistance phenotype is a good strategy for reducing risk. Expressed another way, diversification is key to durable or effective resistance in practice. This can be from crop types through region and fields to within a crop or even as intercrops. The deployment of the barley *mlo* gene is an example of crop diversification between crop types as it is deployed in spring barley but not in winter barley at present (Lyngkjær et al., 2000). Diversification schemes have been published in the past to encourage farmers to grow cultivars with different resistance genes in neighbouring fields. These schemes are devised based on knowledge of the virulence frequencies or race composition of the current pathogen population (UKCPVS, 2018). Within-crop diversity is effective in cultivar mixtures or blends, or more rarely where multilines have been developed, that is a common cultivar with different resistance genes back-crossed into it (Mundt, 2002). Again, ideally these need to be designed with respect to the composition of the pathogen populations that affect the crop, and complex strategies can be very effective (Newton and Guy, 2011). Mixtures must be designed with all other aspects of the interaction for the end use in mind too (Newton et al., 2009). However, multiple components and patchiness are the key to enhancing the effectiveness of these strategies (Newton et al., 1997; Newton and Guy, 2009). Intercrops can be regarded as like cultivar mixtures but where the trait differences are much larger. Nevertheless, the principles of more diversity and heterogeneity prevail (Newton et al., 2009).

Disease tolerance is a concept little exploited in IPM as it is hard to define (Newton, 2016). A simple definition is less yield loss than might be expected from the disease severity observed but this begs many questions about disease assessment and the disease-yield relationship of crops with respect to resources

(Bingham and Newton, 2009; Bingham and Topp, 2009; Bingham et al., 2009), that is resource use efficiency. However, cultivar mixtures too have complex relationships with resource use efficiency, disease and thereby tolerance and in all these considerations we find that it is a balance between the negative effects of competition and the positive effects of facilitation between plants, be it in monocultures or mixtures (Brooker et al., 2016).

Disease escape is a simpler concept to understand and potentially to exploit as it is mechanistically based. In splash-dispersed pathogens height is a major component. A mapping population of barley segregating for two dwarfing genes shows two strong quantitative trait loci (QTL) for rhynchosporium resistance. These are in fact the loci of the dwarfing genes as shorter plants get more splash onto the upper foliage that then infect and cause disease (Looseley et al., 2012). Erectoid leaf habit also reduces infection as fewer leaves are exposed to splash. Other leaf habit attributes affect the microclimate of the canopy and thus affect likelihood of successful infection. All of these are escape mechanisms rather than resistance but can be deployed in a complementary way.

Breeders exploit molecular marker technology increasingly to enhance the likelihood of desirable traits being expressed in advanced selections under field conditions. However, these are used effectively only on genes with strong expression and many effective resistance expressions are the product of many genes each of small effect. Some virus resistance such as Barley Mild Mosaic Virus and Barley Yellow Mosaic Virus are routinely screened for in breeding. However, among the fungal diseases only the *mlo* gene for powdery mildew resistance is both oligogenic, strongly expressed and durable, although other genes with useful levels of resistance such as alleles *Rrs1* for resistance to rhynchosporium could be selected. While genetic modification might appear to offer a potential fast-track to resistance deployment in some crops, in barley not only are there so few candidates, but also most varietal backgrounds with the exception of Golden Promise have proved to be intransigent. Gene editing is likely to offer considerably enhanced opportunities, particularly with the advent of the barley Pan-Genome projects such as SHAPE at IPK Gatersleben in Germany which aims to unravel the complexity of the core genome shared in all barley haplotypes and the 'dispensable' part of the genome found only in individual haplotypes. Such technologies with these bioinformatic resources could lead to strategies for pyramiding or cassettes of resistance components that should be durable and therefore very valuable. However, commercial varieties exploiting these resources still have to pass DUS (Distinctiveness, Uniformity and Stability) and VCU (Value for Cultivation and Use) regulations in most markets for widespread availability and exploitation, and must still be managed as components of IPM to maximise their effectiveness and maintain their durability.

6 Crop protectants

One of the core principles of IPM systems is to promote crop management strategies which reduce reliance on pesticides (Lamichhane et al., 2016). The application of crop protectant agents in targeted and sustainable ways is therefore a component of all IPM systems. In practical terms most crop protectant agents applied to barley are fungistatic and classed as pesticides, although alternatives such as elicitors, biologicals and biostimulants are also the subject of research effort, with some very limited uptake in practice (Chandler et al., 2011). The use of pesticides carries cost implications as well as benefits and can be contentious (Geiger et al., 2010). They are heavily regulated, both in terms of their inherent safety/toxicity to operators, consumers and the wider environment and also in terms of the dosage, timings, skills and equipment required. In addition, specific criteria need to be met to satisfy regulators that efficacy claims are justified and to demonstrate that the crop protection agents are not phytotoxic on the target crop (Regulation (EC) No 1107/2009 of the European Parliament). There are many practical considerations to be overcome too, such as ease of manufacture and stability and shelf life.

Fungicides are defined as chemical agents that kill a fungal pathogen or limit its growth and they can be synthetic or natural compounds. Their mode of action (MOA) is not always known or understood prior to launch but usually they are effective by interfering with key metabolic process or pathways within the target pathogen and can be grouped accordingly by their MOA. Knowledge of their MOA is useful for inferring their likely field of efficacy, for example strobilurin fungicides which target energy production are far more effective in inhibiting spore germination than in preventing later mycelial growth (Hamdy, 2007). MOA is also useful in terms of developing stewardship and anti-resistance strategies which aim to incorporate a diversity of modes of actions in control programmes, and also to protect non-target species such as aquatic life and bees.

Other crop protectant agents may not be fungistatic as such but can be effective by switching on the host plant's own defence mechanisms so that it is better able to detect and recognise a pathogen early in the infection process (Walters et al., 2014). These are often termed resistance elicitors. Biological control agents are more common in terms of control of insect pests than they are for fungal pests, for example nematodes that kill leatherjackets (Kergunteuil et al., 2016), although these are costly and only used in high-value contexts such as sports turf and airports and not in the arable context to protect barley.

Crop protection agents can have both fungistatic effects and also effects on the host plant, which may be as obvious as the growth regulatory or greening effects seen with some fungicide groups. Biostimulant is a term used to cover products applied to the crop which have greening or nutritional benefit, some of these have potential to allow a crop to yield well even in the presence of disease, and some biostimulants may also elicit host defence mechanisms (Xu et al., 2011).

Key principles in the integrated use of fungicides in managing a barley disease is that their use is targeted, not just at the key pathogen risks but also targeted in terms of timings (Jørgensen et al., 2017). Control programmes should be based on the identification of the key disease targets, but also predicated on the key timings of application and what their purpose is. Barley is generally recognised as being sink limited – in other words that yield potential is more driven by the number of grain sites than it is by the source, that is the amount of photosynthates captured by green leaf (Kennedy et al., 2017). This means that retaining tillers, and hence grain sites, is crucial early in crop development and why late tillering/start of stem extension is a key timing for maximising yield response to fungicides in both the winter and spring crop. Prior to this in an IPM system any earlier disease would only be targeted if indicated by risk – for example the very early ingress of mildew, brown rust or other foliar diseases in the autumn or early spring. Later sprays to retain green leaf would be used to manage established disease not well controlled by this first timing and to retain green leaf, which is particularly important in the spring sown crop. The spring crop has less capacity to compensate for periods of poor growth so protection of later green leaf is often required where the site or cultivar need dictates. Protection against ramularia is an example of a late season treatment often applied to barley crops and timing is dictated by the need to protect crops prior to the stress of flowering, and also by last application dates for many fungicides. Fungicide applied after ear emergence gives an example of a timing where yield responses are very small and seldom exceed the cost of application, and which in an integrated management programme are easily dropped.

Crop protection programmes should be reactive to risk but largely built around these key crop timings where benefits to yield are maximised. The fungicide choice, dose rate and mixing partners should target the main disease threats for the barley crop in question. The key disease risks for each crop should be assessed prior to designing crop protection programmes and then modified in response to emerging in-season intelligence as described above (Hughes et al., 1999). This can be formal knowledge of risk systems and crop monitoring, but also informal knowledge of the endemic disease risks in an area – for example rhynchosporium may be an annual threat in wet areas where there is a history of growing susceptible cultivars. Programmes should be tailored to the risk in individual fields, set within the context of a sustainable farm system and rotation. The growth of continuous spring barley has been discouraged through farmer direct support payments which incentivise diverse rotations by making payments conditional on cross-compliance with good practice (Council Regulation 1306/2013), but is an example of poor practice which can build-up trash-borne issue such as Cephalosporium and stem base diseases like eyespot. The targeting of inputs can extend as far as precision surveying and application techniques allow – so spot treatment or applications to areas of fields identified as at higher risk can be part of an IPM system (O'Grady and O'Hare, 2017). Spot liming barley crops is already practised and as technological solutions present themselves this could encompass remote detection and application to specific areas of fields. Early identification of risk allows for pre-planting control measures to be implemented such as a different crop choice or the use of a cultivar with resistance to the main threat.

6.1 Fungicide methods of application

Fungicides can be applied as seed treatments, foliar sprays or soil treatments (Matthews, 2000). The latter tend to have high costs and significant negative environmental impacts and so there are no current examples used in barley. Seed treatments are a targeted and effective way of addressing seed-borne problems. Seed treatments can also have some effect in reducing trash and soil-borne diseases such as take-all. Seed treatments are an effective means of managing seed-borne barley health issues such as loose smut, leaf stripe and seed-borne net blotch. They can also help to manage Fusarium and Microdochium species in the context in which these pathogens can lead to seedling blights. They should always be used in conjunction with seed testing and with the purchase and drilling of seed of a high health status, for example as purchased through official certification schemes. Seed treatments should be used, as with other fungicide inputs, as part of an integrated package and should not be used to pull seed with known health issues up to acceptable levels of seed health. Some seed treatments will also move into emerging leaf tissue and protect against early season foliar diseases so used to be targeted at early sown and high-risk crops. However, as these foliar-acting seed treatments tend to be applied at higher doses they present a greater potential hazard and so have been removed from the European market, leaving what are often termed the 'single-purpose' seed treatments, that is those purely targeting the classic seed-borne health issues. The addition of silthiofam (to manage take-all) is an example of a seed treatment that can be added to extend the range of the 'single-purpose' treatments where take-all is deemed a risk. Other measures to reduce take-all such as extended rotation and late drilling should of course be practised as a first resort.

Fungicides are also applied as foliar treatments to protect against foliar pathogens. Fungicides are usually applied to barley crops in programmes which aim to protect the crop over the growing season and, as indicated above, should be targeted with the key timings and the key disease risk in mind (Young et al., 2006).

6.2 Fungicide modes of action

Knowledge of the characteristic of each fungicide is useful in targeting their use to the situations where they will be most effective. Some fungicides are termed protectant and will act early in fungal life cycles, such as preventing spore germination, but will be less effective as eradicants where disease is established. Some fungicides are more eradicant in nature and can be used to target established infections – for example the morpholines. Many fungicide groups sit somewhere on this continuum – for example the demethylation inhibitors (DMIs) or succinate dehydrogenase inhibitors (SDHIs) are fungicide groups both containing fungicide actives that give both eradicant or protectant activity (Oliver and Hewitt, 2014). There is a potential conflict in IPM in terms of applying fungicides prophylactically and getting maximum efficacy using lower doses or waiting until thresholds are reached and disease is established, and using fungicides eradicantly (also termed curatively) but at higher doses. This is why as broad a consideration of risk as possible is helpful in identifying historic and regional risk and allowing key risks to be targeted in a timely manner. Using a diversity of fungicide MOAs allows for a range of crop growth stages and diseases to be targeted in a crop and for varied programmes to be developed and tailored to the crops in question. In addition, diversifying MOAs in barley programmes reduces reliance on any one component with potential benefits to the environment and with benefits to fungicide resistance management strategies (FRAC Code List, 2018). MOAs with activity against barley pathogens include quinone outside inhibitors (QoIs), anilino-pyrimidines, methyl benzimidazole carbamates (MBCs), morpholines, a phthalonitrile and dithio-carbamates, as well as the DMI or SDHI fungicide groups mentioned above.

6.3 Fungicide stewardship

Although there are strict testing regimes before pesticides are registered for use and statutory limitations on their usage, the use of fungicides can still be contentious, and the health and environmental impacts are kept under constant review (European Union, 2009). Many pathways are shared between fungal pathogens and other organisms so fungicide groups under review currently include the DMI group of actives where endocrine disruption is a health concern for some actives within this group, or chlorothalonil which has been linked to bee declines in the United States (McArt et al., 2017) which breaks down into multiple metabolites, not all of which can be assessed accurately. Statutory limitations such as buffer strips around fields and restrictions around water courses, maximum dosages, harvest intervals and other restrictions on use are commonly applied. Adherence is a legal requirement.

Some countries, such as Denmark have further disincentivised fungicide use by imposing measures such as pesticide taxes. Other countries, such as the United Kingdom have opted for voluntary initiatives (VI) to reduce reliance on pesticides, and market drivers such as supermarket protocols or quality assurance schemes also promote sustainable, responsible and integrated usage. VI measures include such measures as skills and training for sprayer operators, the use of skilled and qualified agronomists (BASIS registered) and other stewardship measures such as testing sprayers, use of low drift nozzles and other methods which reduce the risk drift or damage to non-target organisms.

Fungicide resistance management is a further example of stewardship measures that are applied to barley crops. The selective advantage conferred on resistant individuals in a population, particularly where individual MOAs are intensively used, drives resistance development. Anti-resistance strategies which aim to reduce selection pressure are based on principles of reduced reliance (van den Bosch et al., 2014a). The risk of resistance developing to fungicides that act at a single point in the metabolic chain, termed single-site fungicides, is very high. It is much harder for pathogens to develop resistance against fungicides that act at multiple points in a cell's metabolism and they tend to carry fitness costs for the pathogen and also a higher risk of side effects in non-target species. For this reason, there are few examples of new multi-site actives that have made it through the development and registration processes recently. In barley folpet and chlorothalonil give examples that have been used in the multi-site context, but other multi-sites include older products such as mancozeb and sulphur. To reduce the risk of fungicide resistance, reliance on any single active must be reduced and in practical terms this means incorporating other IPM measures which reduce reliance on chemistry and, when planning a fungicide programme, using a diversity of chemistry. This can be implemented either as mixtures and/or as alternations, and ideally as both. The inclusion of multi-sites is beneficial in reducing risk to the higher risk actives (van den Bosch et al., 2014b). Annual updates and guidance on key crops are provided in the United Kingdom by FRAG-UK and by other groups such as NOROBAG in Nordic countries. FRAC is an agrochemical industry initiative which is pan global, to monitor resistance and formulate strategies.

Barley pathogens have a track record of developing resistance problems with most key pathogens carrying some form of reduced efficacy in the field. Ramularia is a well-documented example with field resistance to all the main single-site fungicides which were previously effective including MBCs, DMIs, SDHIs and QoIs. Other examples of barley pathogens with resistance issues include loose smut (historic examples of SDHI resistance), rhynchosporium (older DMIs), mildew (QoIs, older DMIs), tan spot (SDHIs, QoIs) and net blotch (QoIs and SDHIs). Low doses generally reduce the risk of resistance development (Grimmer et al., 2015). Anti-resistance stewardship is currently a mix of statutory advice on labels (maximum application number) plus 'softer' on label guidance such as the advice to follow FRAG-UK guidance or include a suitable mixing partner, and voluntary guidance such as that advocated through FRAG-UK. Some consideration of resistance risk has to be made prior to registration so that this can be incorporated into statutory controls.

7 Agronomy

Several agronomic components of IPM have already been mentioned, namely reduced cultivation leading to potential enhanced mycorrhizal colonisation but also to enhanced pathogen inoculum, and also rotation length to minimise inoculum, drilling date (spring crop the ultimate example), sowing density, ploughing (increases eyespot) and use of growth regulators. The former is likely to enhance nutrition and prime resistance mechanisms, whereas the latter may enhance disease. Intercropping was mentioned above as a way of exploiting diversity and it will promote facilitation between species if well designed. It can also limit the crop protectants that can be used due to their MOAs, for example herbicides that affect dicots but not monocots, or for regulatory or product approval reasons. Crops can also be sown later before winter to avoid pathogen build-up before the dormant period in colder regions. The crop sequence can be important with respect to soil microbes in general, pathogen inoculum in particular and nutrition and soil structure. An example would be the effect of maize (corn) debris on fusarium inoculum causing problems in subsequent barley and wheat crops. However, here again there are many other factors that affect risk and these can be evaluated in an IPM plan (Edwards, 2013).

Resistance elicitors were referenced above but they can be components of products regulated as biostimulants. Although disease control cannot be claimed as an attribute of biostimulants without them being classified and regulated as pesticides, plant health claims are made and these effects can be of value in IPM protocols. Likewise, many nutritional products that target micronutrients can be beneficial to plant health and resistance expression. However, it is rarely individual nutrients that have an effect irrespective of the status of other major and minor nutrients (Walters and Bingham, 2007) so there are few rules that can be used to optimise the contribution of these. Indeed, the effect of soil pH alone on disease can be considerable as it affects nutrient availability differentially (Holland et al., 2017).

8 IPM knowledge sources and tools

The successful implementation of IPM by growers is dependent on the availability of information on the health status of crops, potential disease threats and changes in pathogen variation. Monitoring of disease is carried out at a regional level in many countries to give an accurate record of annual epidemics and an indication to growers of disease threats. Disease outbreaks are reported via online information platforms such as 'Adopt-a-Crop' in Scotland, 'Crop Monitor' in England and Wales, and Mississippi Crop Situation in the United States (Mississippi State University, 2018). In addition to intelligence on crop disease levels, growers require updates on changes in pathogen variation and fungicide sensitivity. National bodies often undertake monitoring of pathogen variation, for example the National Institute of Agricultural Botany in the United Kingdom monitors changes in the race structure of the yellow rust pathogen of wheat (*Puccinia striiformis*). In the United Kingdom there was a rapid change in race structure of the yellow rust pathogen in 2016 which rendered many previously resistant cultivars sensitive to the disease. Research institutes and agrochemical companies also monitor changes in pathogen sensitivity to fungicides and changes on a larger scale are reported by bodies like the Fungicide Resistance Action Group UK (FRAG-UK). Recent mutations in the barley pathogen *Ramularia collo-cygni* have reduced the efficacy of two major groups of fungicides to control disease symptoms (Havis et al., 2018). This information should be interpreted in the context of on-farm issues such as the historical risk, the recorded effects of diseases on yield and quality and the likely threats locally, all important components of planning IPM strategies on-farm.

Many of the principles, concepts and practical information discussed above are implemented in the AHDB Barley Disease Management Guide for the United Kingdom (AHDB, 2016) and many other countries have similar sources of applied information and advice. A risk forecast has been elucidated for a number of barley diseases such as Fusarium head blight and Ramularia leaf spot. However, the ramularia forecast was based on a limited number of environmental parameters and requires more research to become a fully robust forecast (Havis et al., 2018). A risk assessment has been produced for barley yellow dwarf virus based on monitoring of aphid populations (Fabre et al., 2006). The model uses a single assessment of the proportion of plants infested with aphids and then a temperature-dependent simulation of aphid dynamics.

The assessment of risk in any barley crop requires the integration of a number of factors including previous cropping, cultivation methods, crop nutrition, varietal susceptibility, fungicide efficacy and availability of alternative control measures. Each of the major barley pathogens has an optimal temperature for growth and development such as 15–20°C for *Rhynchosporium commune* and 20–25°C for *Pyrenophora teres* (Mathre, 1997). Therefore, the use of weather forecasts combined with actual data should provide growers with an indication of the likelihood of disease development

9 Uptake and communication of IPM

IPM is a knowledge-intensive process in which evidence-supported approaches are selected based on the biotic threats threatening the crop (Byerlee, 1996). It is therefore important to consider the ways in which famers gather and decipher information that ultimately results in change in practice. Farmers constantly receive floods of information and plant protection advice from various sources. Knowledge transfer messages via farming press, advisory body publication, radio broadcasts and so on do not always constitute a consensus, which can confuse the receiver who is asked to digest information and consider advice that can sometimes be contradictory and even inaccurate. It is therefore important that details regarding specific information sources are clearly provided and properly referenced so that the grower is in a position to access and assess the quality of the information and act on it accordingly. A neat solution to these issues would be a coordinated and coherent set of simple IPM messages presented through established and respected communication bodies. Whereas this would be an ideal solution it is not likely to become a reality because different stakeholders communicate different messages to suit their specific needs, whether that be the selling of pesticides or seed, a reduction in pesticide contamination of waterways or an increase in local biodiversity. This point is illustrated by the fact that there are now multiple different definitions of IPM (Prokopsy, 2003), each abiding by the same principle of a holistic approach to pest management that attempts to minimise the negative impacts of pest management programmes on the environment (Parsa et al., 2014). However, slight variations in definition are employed to suit the end goal of the authors. Farmers too may also disagree as to what constitutes best IPM practices as they have different drivers to uptake of IPM practices which may be a combination of economic, environmental and social (Sherman and Gent, 2014). Despite such differences in opinion, the main components of IPM programmes and the potential benefits of adopting them for the grower are often fairly well conserved among practitioners, which is important as crop protection decisions often involve multiple decision makers (Burnett and Hughes, 2015). The question is whether adequate evidence-based advice supporting IPM adoption currently exists, and if it does, whether the messages encouraging adoption are being communicated effectively.

Recent research has shown that growers who perceive themselves to be more familiar with IPM are more actively seeking IPM information citing knowledge exchange activities, via open days, crop walks, discussion groups, exchanges with independent agronomists, as being their most valued sources of pest management advice (Creissen et al., 2018). Those growers who attend knowledge exchange events are, in turn, more likely to adopt IPM practices on farm (Creissen et al., 2018). The concern is that those individuals who are actively engaging with experts and peers to seek and understand current evidence-based advice are not in the majority. That presents an interesting question, how can we incentivise growers to engage in knowledge exchange activities? There are various 'carrots' and 'sticks' that can be used to this end. A 'carrot' may be in the form of financial rewards for attending knowledge exchange events. In

Ireland growers are rewarded to the sum of €750 per annum for their attendance at five discussion group meetings per annum and completion of associated tasks such as a farm improvement plan (https://www.teagasc.ie/about/farm-advisory/advisory-services/discussion-groups/). 'Sticks' to incentivise uptake of IPM include subsidy reductions and fines for those not deemed to be practising IPM. This is a less desirable route as it requires the development of system to assess whether IPM is being adopted and IPM approaches will vary greatly according to the farm business, farming system and the associated biotic threats to crop production. Many potential 'stick or 'carrot' systems would require an effective way of measuring the complex technology that is IPM to be established. Recent efforts to establish a universal metric for assessing IPM uptake have been made (Creissen et al. 2019). If accepted by the industry this could pave the way for a system that could be employed to encourage farmers to increase their level of IPM adoption.

When considering malting barley production we must also be aware of the high level of influence of those involved in the supply chain (Barnes et al., 2018). Distillers aim to acquire malt that provides the greatest alcohol spirit yield, brewers may also consider the flavour profile of the malt. Maltsters specify grain quality parameters which often relate to uniformity of germination rates and biochemistry of the grains in relation to malting ability, such traits include specific grain weight and degree of grain skinning (Hoad et al., 2016). To meet these requirements growers look to maximise grain fill and minimise skinning which are both affected by fungicides and growth regulators (Wilson et al., 2018). For a high-value crop this may reduce the likelihood of growers accepting potential risks associated with a reduction in inputs. Due to the need to grow for the market, farmers are also heavily restricted in the cultivars they may grow as the end users have already pre-selected a few cultivars they would be willing to accept. Removal of choices associated with this key and fundamental component of any disease management programme means that growers must alleviate risks associated with disease by other means, including the use of fungicides. This presents a considerable barrier to uptake of IPM in the malting barley sector. However, for winter barley the issues surrounding heterogeneity of the end produce are far less significant as nearly all of the grain produced is intended for livestock feed. Research into the potential for, and barriers to, IPM uptake along the entire supply chain is critical to improving our understanding of the most effective approaches to communicate the IPM message.

10 Farming systems, soil and research platforms

As an integrated approach, both the crop and soil environment are the product of multiple components spread through time. Factorial experimentation can be used only to evaluate the possible contribution of a few components and their interactions. IPM must be assessed as a system in comparison with other systems. Therefore, it is not possible to do this for barley alone as an IPM system will have barley as a crop in a sequence. Nevertheless, IPM schemes have been compared in sequences not only across multiple sites but also across years in long-term experiment platforms which reveals where they work well and where their weaknesses are (Lechenet et al., 2016; Vasileiadis et al., 2017).

Such research platforms are helpful and while few of these focus on barley, the principles still apply. However, in common with cultivar testing and recommendation trials, they are seldom representative of on-farm conditions. On the other hand, on-farm conditions are highly variable so it is difficult to conceive what representative trial characteristics should comprise. Fundamentally the requirement for robust IPM on-farm is resilience, so trialling systems that express biotic and abiotic stress factors as a proxy for on-farm conditions to identify reliable IPM protocols are what is needed. Borrowing concepts used particularly in disciplines like molecular biology, an indication of the key components of a system, can be gained by overexpression or knock-out of that factor (Newton, 2016). Much criticised by systems biologists, nevertheless such approaches can indicate the sensitive or rate-limiting factors in IPM systems that should be the focus of research systems comparisons. For example, zero tillage and substantially reduced nutrients could be used as an overexpression/knock-out proxy for on-farm conditions (Newton et al., 2018) to identify cultivars likely to be resilient. Tolerance, essentially the expression of yield resilience with respect to disease, can be characterised in this way (Newton, 2016).

We have discussed briefly plant-microbe management in the broad as well as IPM-applied context including endophytes and mycorrhizal colonisation. We must also reference the microbiome in general and perhaps the rhizosphere biome in particular. The ability of many soils to be disease suppressive is well known, as is the ability of bacteria in particular to induce resistance or induced systemic resistance (ISR) (Bakker et al., 2013) and plant growth-promoting rhizobacteria. With modern sequencing technologies we are beginning to obtain data on the importance of microbiome characteristics, but we have yet to understand sufficiently their function and how we might manage and manipulate them for beneficial outcomes. Crucial to any microbial activity in the soil, however, is soil management and cultivation in particular as soil disturbance can, for example, disrupt mycorrhizal networks while enhancing bacterial oxidative processes.

Implied in the above discussion of uptake of IPM is the incremental adoption of measures that should improve crop health. A high risk but potentially much more rewarding strategy is the adoption of multiple measures together in a systems approach. Although there is a transition phase to successful implementation, the outcome cannot be achieved using an incremental approach. An example is intercropping in a conservation tillage system. Here several technologies must be adopted at once. Inversion soil tillage is replaced by non-inversion tillage (minimum or zero tillage/direct drilling). Soil structure and fertility is managed by using cover crops. Disease is managed by a combination of resistance in the crop cultivars, diversity and deployment strategies. Nutrient management is further managed in this context taking due regard for each component of the crop system. The barley crop itself is grown together with another species, often a legume such as peas. Harvest dates must be matched and subsequent yield separation may be needed if used in different markets such as for malting. This can be achieved and examples of an approach as the one described have been achieved on farm even in challenging winter environments in Scotland and demonstrated to other farmers to encourage adoption. Therefore, to properly implement IPM in barley the whole system should be considered before decisions are made about adoption to determine the best approach or direction of travel for any given situation.

11 Conclusion and future trends

For increased IPM adoption in barley production we need continued research and innovation, but equally important is translation of the research outcomes into practical implementation and communication of these outcomes to farmers. However, to both ensure the practicality and that the solutions are needed, all IPM research and development should be carried out in a co-production/co-innovation framework. Indeed, many practical implementations of IPM components are being developed by farmers both deliberately and inadvertently through a practical understanding of what works in practice. This needs to inform more formal experimentation and innovation. Discussion groups, field labs, monitor farms, farm and trial visits, open days and other groupings of farmers together with researchers and advisors can all help achieve these outcomes, but coordinated and collaborative trialling and demonstration, especially if facilitated by funding, can be particularly effective.

IPM involves multiple decision makers that includes farmers, agronomists and other farm business experts, and all have different attitudes towards risk driven by their personal nature, circumstances and their business environment. However, IPM has a major public goods driver too in the form of the environment and long-term sustainability. This needs to be separately incentivised, as does the development of all the tools needed to implement effective IPM to deliver these goods.

12 Acknowledgements

We are grateful for the financial support for this work from the Rural and Environmental Science and Analytical Services Division of the Scottish Government under its Environmental Change and Food, Land and People Research Programmes.

13 Where to look for further information

13.1 Further reading

- A good review of control of foliar diseases of barley is Walters et al. (2012).
- The broader context of barley is reviewed by Newton et al. (2011).

13.2 Key conferences

- International Congress for Plant Pathology, Lyon, France, 2023.
- Many countries host national agronomy and pathology meetings on IPM, for example check the Association of Applied Biologists website.

13.3 Major international research projects

- The EU hosted several large IPM projects that have legacy websites, ENDURE http://www.endure-network.eu/, and PURE, http://www.pure-ipm.eu/being the major ones.

14 References

AHDB. 2016. Agriculture and Horticulture Development Board barley disease management guide. Available at: https://cereals.ahdb.org.uk/publications/2016/february/23/barley-disease-management-guide.aspx (accessed November 2018).

AHDB. 2018. Agriculture and Horticulture Development Board recommended lists. Available at: https://cereals.ahdb.org.uk/varieties/ahdb-recommended-lists.aspx (accessed November 2018).

Alizadeh, A., Benetti, V., Sarrafi, A., Barrault, G. and Albertini, L. 1994. Genetic-analysis for partial resistance to an Iranian strain of bacterial leaf streak (*Xanthomonas campestris* pv *hordei*) in barley. *Plant Breed.* 113, 323–6.

Bai, G. H. and Shaner, G. 2004. Management and resistance in wheat and barley to Fusarium head blight. *Annu. Rev. Phytopathol.* 42, 135–61. doi:10.1146/annurev.phyto.42.040803.140340.

Bakker, P. A. H. M., Doornbos, R. F., Zamioudis, C., Berendsen, R. L. and Pieterse, C. M. J. 2013. Induced systemic resistance and the rhizosphere microbiome. *Plant Pathol. J.* 29(2), 136–43. doi:10.5423/PPJ.SI.07.2012.0111.

Balba, H. 2007. Review of strobilurin fungicide chemicals. *J. Environ. Sci. Health B* 42(4), 441–51. doi:10.1080/03601230701316465.

Barnes, A., Burnett, F. and Creissen, H. 2018. Assessing the barriers to uptake of Integrated Pest Management in Scotland. Available at: https://www.sruc.ac.uk/downloads/file/3764/assessing_the_barriers_to_uptake_of_integrated_pest_management_in_scotland (accessed November 2018).

Bateman, G. L., Gutteridge, R. J., Gherbawy, Y., Thomsett, M. A. and Nicholson, P. 2007. Infection of stem bases and grains of winter wheat by *Fusarium culmorum* and *F. graminearum* and effects of tillage method and maize-stalk residues. *Plant Pathol.* 56(4), 604–15. doi:10.1111/j.1365-3059.2007.01577.x.

Bingham, I. J. and Newton, A. C. 2009. Crop tolerance of foliar pathogens: possible mechanisms and potential for exploitation. In: Walters, D. (Ed.), *Disease Control in Crops-Biological and Environmentally Friendly Approaches*. Wiley-Blackwell, Chichester, UK, pp. 142–61.

Bingham, I. J. and Topp, C. F. E. 2009. Potential contribution of selected canopy traits to the tolerance of foliar disease by spring barley. *Plant Pathol.* 58(6), 1010–20. doi:10.1111/j.1365-3059.2009.02137.x.

Bingham, I. J., Walters, D. R., Foulkes, M. J. and Paveley, N. D. 2009. Crop traits and the tolerance of wheat and barley to foliar disease. *Ann. Appl. Biol.* 154(2), 159–73. doi:10.1111/j.1744-7348.2008.00291.x.

Bockus, W. W. and Claassen, M. M. 1992. Effects of crop-rotation and residue management-practices on severity of tan spot of winter-wheat. *Plant Dis.* 76(6), 633–6. doi:10.1094/PD-76-0633.

Braun-Kiewnick, A., Jacobsen, B. J. and Sands, D. C. 2000. Biological control of *Pseudomonas syringae* pv. *syringae*, the causal agent of basal kernel blight of barley, by antagonistic *Pantoea agglomerans*. *Phytopathology* 90(4), 368–75. doi:10.1094/PHYTO.2000.90.4.368.

Brooker, R. W., Karley, A. J., Newton, A. C., Pakeman, R. J. and Schöb, C. 2016. Facilitation and sustainable agriculture: a mechanistic approach to reconciling crop production and conservation. *Funct. Ecol.* 30(1), 98–107. doi:10.1111/1365-2435.12496.

Browne, R. A. and Cooke, B. M. 2005. A comparative assessment of potential components of partial disease resistance to Fusarium head blight using a detached leaf assay of wheat, barley and oats. *Eur. J. Plant Pathol.* 112(3), 247–58. doi:10.1007/s10658-005-2077-z.

Burnett, F. and Hughes, G. 2004. The development of a risk assessment method to identify wheat crops at risk from eyespot. AHDB project report 347. AHDB, Stoneleigh Park, Kenilworth, Warwickshire, UK.

Burnett, F., Butler-Ellis, C., Hughes, G., Knight, S. and Rumiana, R. 2012. Forecasting eyespot development and yield losses in winter wheat. AHDB project report 491. AHDB, Stoneleigh Park, Kenilworth, Warwickshire, UK.

Byerlee, D. 1996. Modern varieties, productivity, and sustainability–recent experience and emerging challenges. *World Dev.* 24(4), 697–718. doi:10.1016/0305-750X(95)00162-6.

Carignano, M., Staggenborg, S. A. and Shroyer, J. P. 2008. Management practices to minimize tan spot in a continuous wheat rotation. *Agron. J.* 100, 145–53.

Caten, C. E. and Newton, A. C. 2000. Variation in cultural characteristics, pathogenicity, vegetative compatibility and electrophoretic karyotype within field populations of *Stagonospora nodorum*. *Plant Pathol.* 49(2), 219–26. doi:10.1046/j.1365-3059.2000.00441.x.

Chandler, D., Bailey, A. S., Tatchell, G. M., Davidson, G., Greaves, J. and Grant, W. P. 2011. The development, regulation and use of biopesticides for integrated pest management. *Philos. Trans. R. Soc. Lond., B, Biol. Sci.* 366(1573), 1987–98. doi:10.1098/rstb.2010.0390.

Chen, F. Q., Prehn, D., Hayes, P. M., Mulrooney, D., Corey, A. and Vivar, H. 1994. Mapping genes for resistance to barley stripe rust (*Puccinia striiformis* f. sp. *hordei*). *Theor. Appl. Genet.* 88(2), 215–9. doi:10.1007/BF00225900.

Colbach, N., Lucas, P., Cavelier, N. and Cavelier, A. 1997. Influence of cropping system on sharp eyespot in winter wheat. *Crop Prot.* 16(5), 415–22. doi:10.1016/S0261-2194(97)00018-5.

Cooke, L. R., Locke, T., Lockley, K. D., Phillips, A. N., Sadiq, M. D. S., Coll, R., Black, L., Taggart, P. J. and Mercer, P. C. 2004. The effect of fungicide programmes based on epoxiconazole on the control and DMI sensitivity of *Rhynchosporium secalis* in winter barley. *Crop Prot.* 23(5), 393–406. doi:10.1016/j.cropro.2003.09.009.

Creissen, H., Tranter, R., Jones, P., Burnett, F., Jess, J., Girling, R., Gaffney, M., Thorne, F. and Kildea, S. 2018. IPM practices on arable farms in the UK and Ireland. In: *Crop Production in Northern Britain*, Dundee Conference, UK, 2018.

Creissen, H. E., Jones, P. J., Tranter, R. B., Girling, R. D., Jess, S., Burnett, F. J., Gaffney, M., Thorne, F. S. and Kildea, S. 2019. Measuring the unmeasurable? A method to quantify adoption of Integrated Pest Management practices in temperate arable farming systems. *Pest. Man. Sci.* doi: 10.1002/ps.5428

Cromey, M. G., Parkes, R. A. and Fraser, P. M. 2006. Factors associated with stem base and root diseases of New Zealand wheat and barley crops. *Aust. Plant Pathol.* 35, 391–400.

Duczek, L. J., Sutherland, K. A., Reed, S. L., Bailey, K. L. and Lanford, G. P. 1999. Survival of leaf spot pathogens on crop residues of wheat and barley in Saskatchewan. *Can. J. Plant Pathol.* 21, 165–73.

Edwards, S. 2013. HGCA risk assessment for Fusarium mycotoxins in wheat. Caledonia House: Home Grown Cereals Authority Topic Sheet 121. AHDB, Stoneleigh Park, Kenilworth, Warwickshire, UK.

El Attari, H., Hayes, P. M., Rebai, A., Barrault, G., Dechamp-Guillaume, G. and Sarrafi, A. 1998. Potential of doubled-haploid lines and localization of quantitative trait loci (QTL) for partial resistance to bacterial leaf streak (*Xanthomonas campestris* pv. *hordei*) in barley. *Theor. Appl. Gen.* 96(1), 95–100. doi:10.1007/s001220050714.

European Union. 2009. Council Directive 2009/128/EC of the European Parliament and of the Council of 21 October 2009 establishing a framework for Community action to achieve the sustainable use of pesticides. *Off. J. Eur. Comm.* L309/71, 71–86.

Fabre, F., Pierre, J. S., Dedryver, C. A. and Plantegenest, M. 2006. Barley yellow dwarf virus disease risk assessment based on Bayesian modelling of aphid population dynamics. *Ecol. Modell.* 193(3–4), 457–66. doi:10.1016/j.ecolmodel.2005.08.021.

FAO. 2018. Food and Agriculture Organization of the United Nations (FAO) definition of Integrated Pest Management (IPM). Available at: http://www.fao.org/agriculture/crops/thematic-sitemap/theme/pests/ipm/en/(accessed on 4 August 2018).

FERA. 2018. Crop monitor. Available at: http://www.cropmonitor.co.uk/ (accessed on 16 September 2018).

FRAC Code List. 2018. Fungicides sorted by mode of action (including FRAC Code numbering). Available at: http://www.frac.info/docs/default-source/publications/frac-code-list/frac_code_list_2018-final.pdf?sfvrsn=6144b9a_2.

Geiger, F., Bengtsson, J., Berendse, F., Weisser, W. W., Emmerson, M., Morales, M. B., Ceryngier, P., Liira, J., Tscharntke, T., Winqvist, C., et al. 2010. Persistent negative effects of pesticides on biodiversity and biological control potential on European farmland. *Basic Appl. Ecol.* 11(2), 97–105. doi:10.1016/j.baae.2009.12.001.

Grasso, V., Sierotzki, H., Garibaldi, A. and Gisi, U. 2006. Characterization of the cytochrome *b* gene fragment of *Puccinia* species responsible for the binding site of QoI fungicides. *Pest. Biochem. Physiol.* 84(2), 72–82. doi:10.1016/j.pestbp.2005.05.005.

Grimmer, M. K., van den Bosch, F., Powers, S. J. and Paveley, N. D. 2015. Fungicide resistance risk assessment based on traits associated with the rate of pathogen evolution. *Pest Manag. Sci.* 71(2), 207–15. doi:10.1002/ps.3781.

Havis, N. D. and Gorniak, K. 2014. Cephalosporium leaf stripe - an emerging threat to wheat crops in short rotations. AHDB project report 542. AHDB, Stoneleigh Park, Kenilworth, Warwickshire, UK.

Havis, N. D., Burnett, F. J., Hughes, G., Mercer, P. C., Cooke, L. R., Fraaije, B. A., Hunter, E. A. and Oxley, S. J. P. 2014a. *Rhynchosporium commune* – understanding the effects of variety, fungicide resistance and seed-borne infection on disease levels in barley. In: *The Dundee Conference. Crop Protection in Northern Britain 2014*, Dundee, UK, pp. 167–72.

Havis, N. D., Nyman, M. and Oxley, S. J. P. 2014b. Evidence for seed transmission and symptomless growth of *Ramularia collo-cygni* in barley (*Hordeum vulgare*). *Plant Pathol.* 63(4), 929–36. doi:10.1111/ppa.12162.

Havis, N. D., Brown, J. K. M., Clemente, G., Frei, P., Jedryczka, M., Kaczmarek, J., Kaczmarek, M., Matusinsky, P., McGrann, G. R. D., Pereyra, S., et al. 2015. *Ramularia collo-cygni* – an emerging pathogen of barley crops. *Phytopathology* 105(7), 895–904. doi:10.1094/PHYTO-11-14-0337-FI.

Havis, N. D., Gorniak, K., Taylor, J., Stanisz-Migal, M. and Burnett, F. J. 2018. Controlling Ramularia leaf spot in barley. In: *Therapeutic Dundee Conference. Crop Production in Northern Britain 2018*, Dundee, UK, pp. 91–6.

Hoad, S. P., Brennan, M., Wilson, G. W. and Cochrane, P. M. 2016. Hull to caryopsis adhesion and grain skinning in malting barley: identification of key growth stages in the adhesion process. *J. Cer. Sci.* 68, 8–15. doi:10.1016/j.jcs.2015.10.007.

Holland, J. E., Bennett, A., Newton, A. C., White, P., McKenzie, B., George, T., Pakeman, R., Bailey, J., Fornara, D. and Hayes, R. 2017. Liming impacts on soils, plants and biodiversity in the UK: a review. *Sci. Total Environ.* 610-611, 316–32.

Hughes, G., McRoberts, N. and Burnett, F. J. 1999. Decision making and diagnosis in disease management. *Plant Pathol.* 48(2), 147–53. doi:10.1046/j.1365-3059.1999.00327.x.

Hughes, G. and Burnett, F., J. 2015. Integrating experience, evidence and expertise in the crop protection decision process. *Plant Disease*, Vol 99 No 9, 1197–1203.

Jørgensen, L. N., van den Bosch, F., Oliver, R. P., Heick, T. M. and Paveley, N. D. 2017. Targeting fungicide inputs according to need. *Annu. Rev. Phytopathol.* 55, 181–203. doi:10.1146/annurev-phyto-080516-035357.

Kendall, S., Hollomon, D. W., Ishii, H. and Heaney, S. P. 1994. Characterization of benzimidazole-resistant strains of *Rhynchosporium secalis*. *Pestic. Sci.* 40(3), 175–81. doi:10.1002/ps.2780400302.

Kennedy, S. P., Bingham, I. J. and Spink, J. H. 2017. Determinants of spring barley yield in a high yield potential environment. *J. Agric. Sci.* 155(1), 60–80. doi:10.1017/S0021859616000289.

Kergunteuil, A., Bakhtiari, M., Formenti, L., Xiao, Z., Defossez, E. and Rasmann, S. 2016. Biological control beneath the feet: a review of crop protection against insect root herbivores. *Insects* 29(4). pii:E70. doi:10.3390/insects7040070.

King, K. M., West, J. S., Brunner, P. C., Dyer, P. S. and Fitt, B. D. L. 2013. Evolutionary relationships between *Rhynchosporium lolii* sp. nov. and other *Rhynchosporium* species on grasses. *PLOS ONE* 8(10), e72536. doi:10.1371/journal.pone.0072536.

Kleinhofs, A., Brueggeman, R., Nirmala, J., Zhang, L., Mirlohi, A., Druka, A., Rostoks, N. and Steffenson, B. J. 2009. Barley stem rust resistance genes: structure and function. *Plant Genome* 2(2), 109–20. doi:10.3835/plantgenome2009.02.0011.

Kusch, S. and Panstruga, R. 2017. *mlo*-based resistance: an apparently universal 'weapon' to defeat powdery mildew disease. *Mol. Plant Microbe Int.* 30, 179–89.

Lamichhane, J. R., Dachbrodt-Saaydeh, S., Kudsk, P. and Messean, A. 2016. Toward a reduced reliance on conventional pesticides in European agriculture. *Plant Dis.* 100(1), 10–24. doi:10.1094/PDIS-05-15-0574-FE.

Lechenet, M., Deytieux, V., Antichic, D., Aubertot, J.-N., Bàrberi, B., Bertrand, M., Cellier, V., Charles, R., Colnenne-David, C., Dachbrodt-Saaydeh, S., et al. 2016. Diversity of methodologies to experiment Integrated Pest Management in arable cropping systems: analysis and reflections based on a European network. *Eur. J. Agron.* 83, 86–99.

Looseley, M. E. and Newton, A. C. 2014. Assessing the consequences of microbial infection in field trials: seen, and unseen, beneficial, parasitic and pathogenic. *Agronomy* 4(2), 302–21. doi:10.3390/agronomy4020302.

Looseley, M. E., Newton, A. C., Atkins, S. D., Fitt, B. D. L., Fraaije, B. A., Thomas, W. T. B., Keith, R., Macaulay, M., Lynott, J. and Harrap, D. 2012. Genetic basis of control of *Rhynchosporium secalis* infection and symptom expression in barley. *Euphytica* 184(1), 47–56. doi:10.1007/s10681-011-0485-z.

Lyngkjaer, M. F., Newton, A. C., Atzema, J. L. and Baker, S. J. 2000. The barley *mlo*-gene: an important powdery mildew resistance source. *Agronomie* 20(7), 745–56. doi:10.1051/agro:2000173.

Mathre, D. E. 1997. *Compendium of Barley Diseases* (2nd edn.). The American Phytopathological Society Press, St Paul, MN.

Matthews, G. A. 2000. *Pesticide Application Methods* (3rd edn.). Blackwell Publishing Science, Oxford, UK.

McArt, S. H., Urbanowicz, C., McCoshum, S., Irwin, R. E. and Adler, L. S. 2017. Landscape predictors of pathogen prevalence and range contractions in US bumblebees. *Proc. Biol. Sci.* 284(1867). doi:10.1098/rspb.2017.2181:29142119.

Menzies, J. G., Turkington, T. K. and Knox, R. E. 2009. Testing for resistance to smut diseases of barley, oats and wheat in western Canada. *Can. J. Plant Pathol.* 31(3), 265–79. doi:10.1080/07060660909507601.

Menzies, J. G., Steffenson, B. J. and Kleinhofs, A. 2010. A resistance gene to *Ustilago nuda* in barley is located on chromosome 3H. *Can. J. Plant Pathol.* 32(2), 247-51. doi:10.1080/07060661003739977.

Miller, W. A. and Rasochova, L. 1997. Barley yellow dwarf viruses. *Ann. Rev. Phytopathol.* 35, 167–90.

Mississippi State University. 2018. Mississippi crop situation. Available at: http://www.mississippi-crops.com/disease-monitoring/.

Mundt, C. C. 2002. Use of multiline cultivars and cultivar mixtures for disease management. *Annu. Rev. Phytopathol.* 40, 381–410. doi:10.1146/annurev.phyto.40.011402.113723.

Newton, A. C. 1989. Genetic adaptation of *Erysiphe graminis* f.sp. *hordei* to barley with partial resistance. *J. Phytopathol.* 126(2), 133–48. doi:10.1111/j.1439-0434.1989.tb01097.x.

Newton, A. C. 2016. Exploitation of diversity within crops - the key to disease tolerance? *Front. Plant Sci.* 7, 665. doi:10.3389/fpls.2016.00665.

Newton, A. C. and Guy, D. C. 2009. The effects of uneven, patchy cultivar mixtures on disease control and yield in winter barley. *Field Crops Res.* 110(3), 225–8. doi:10.1016/j.fcr.2008.09.002.

Newton, A. C. and Guy, D. C. 2011. Scale and spatial structure effects on the outcome of barley cultivar mixture trials for disease control. *Field Crops Res.* 123(2), 74–9. doi:10.1016/j.fcr.2011.05.002.

Newton, A. C., Ellis, R. P., Hackett, C. A. and Guy, D. C. 1997. The effect of component number on *Rhynchosporium secalis* infection and yield in mixtures of winter barley cultivars. *Plant Pathol.* 46(6), 930–8. doi:10.1046/j.1365-3059.1997.d01-83.x.

Newton, A. C., Begg, G. S. and Swanston, J. S. 2009. Deployment of diversity for enhanced crop function. *Ann. Appl. Biol.* 154(3), 309–22. doi:10.1111/j.1744-7348.2008.00303.x.

Newton, A. C., Fitt, B. D. L., Atkins, S. D., Walters, D. R. and Daniell, T. 2010. Pathogenesis, mutualism and parasitism in the trophic space of microbe-plant interactions. *Trends Microbiol.* 18, 365–73.

Newton, A. C., Flavell, A. J., George, T. S., Leat, P., Mullholland, B., Ramsay, L., Revoredo-Giha, C., Russell, J., Steffenson, B. J., Swanston, J. S., et al. 2011. Crops that feed the world 4. Barley: a resilient crop? Strengths and weaknesses in the context of food security. *Food Sec.* 3(2), 141–78. doi:10.1007/s12571-011-0126-3.

Newton, A. C., Guy, D. C., Valentine, T., George, T. S., McKenzie, B. M. and Hackett, C. A. 2018. Differential adaptation of spring barley cultivars to inversion and non-inversion tillage. In: *Proceedings, Crop Production in Northern Britain 2018: The Dundee Conference, Environmental Management and Crop Production*, pp. 109–14.

O'Grady, M. J. and O'Hare, G. M. P. 2017. Modelling the smart farm Information. *Proc. Agric.* 4, 179–87.

Oliver, R. P. and Hewitt, H. G. 2014. *Fungicides in Crop Protection* (2nd edn.). CABI Publishing, CAB International, Wallingford, UK.

Orlando, B., Maumene, C. and Piraux, F. 2017. Ergot and ergot alkaloids in French cereals: occurrence, pattern and agronomic practices for managing the risk. *World Mycotoxin J.* 10(4), 327–38. doi:10.3920/WMJ2017.2183.

Oxley, S. J. P. and Burnett, F. 2009. *Barley Disease Control*. SAC Technical Note TN619. SAC, Edinburgh, UK. ISBN: 1 85482 873 8.

Parsa, S., Morse, S., Bonifacio, A., Chancellor, T. C. B., Condori, B., Crespo-Perez, V., Hobbs, S. L. A., Kroschel, J., Ba, M. N., Rebaudo, F., et al. 2014. Obstacles to integrated pest management adoption in developing countries. *Proc. Natl. Acad. Sci. U.S.A.* 111(10), 3889–94. doi:10.1073/pnas.1312693111.

Pecchioni, N., Vale, G., Toubia-Rahme, H., Faccioli, P., Terzi, V. and Delogu, G. 1999. Barley-*Pyrenophora graminea* interaction: QTL analysis and gene mapping. *Plant Breed.* 118(1), 29–35. doi:10.1046/j.1439-0523.1999.118001029.x.

Penselin, D., Münsterkötter, M., Kirsten, S., Felder, M., Taudien, S., Platzer, M., Ashelford, K., Paskiewicz, K. H., Harrison, R. J., Hughes, D. J., et al. 2016. Comparative genomics to explore phylogenetic relationship, cryptic sexual potential and host specificity of *Rhynchosporium* species on grasses. *BMC Genomics* 17(1), 953, doi:10.1186/s12864-016-3299-5.

Prokopy, R. J. 2003. Two decades of bottom-up, ecologically based pest management in a small commercial apple orchard in Massachusetts. *Agric. Ecosyst. Environ.* 94(3), 299–309. doi:10.1016/S0167-8809(02)00036-1.

Scherm, B., Balmas, V., Spanu, F., Pani, G., Delogu, G., Pasquali, M. and Migheli, Q. 2013. *Fusarium culmorum*: causal agent of foot and root rot and head blight on wheat. *Mol. Plant Pathol.* 14(4), 323–41. doi:10.1111/mpp.12011.

Scottish Government. 2017. Pesticide usage in Scotland: Arable crops and potato stores 2016. Appendix 6 - Integrated pest management. Available at: https://www.gov.scot/Publications/2017/10/7179/59 (accessed November 2018).

Semar, M., Strobel, D., Koch, A., Klappach, K. and Stammler, G. 2007. Field efficacy of pyraclostrobin against populations of *Pyrenophora teres* containing the F129L mutation in the cytochrome *b* gene. *J. Plant Dis. Prot.* 114(3), 117–9. doi:10.1007/BF03356718.

Sherman, J. and Gent, D. H. 2014. Concepts of sustainability, motivations for pest management approaches, and implications for communicating change. *Plant Dis.* 98(8), 1024–35. doi:10.1094/PDIS-03-14-0313-FE.

UKCPVS. 2018. The United Kingdom cereal pathogen virulence survey. Available at: https://cereals.ahdb.org.uk/ukcpvs.

van den Bosch, F., Oliver, R., van den Berg, F. and Paveley, N. 2014a. Governing principles can guide fungicide resistance management tactics. *Annu. Rev. Phytopathol.* 52, 175–95. doi:10.1146/annurev-phyto-102313-050158.

van den Bosch, F., Paveley, N., van den Berg, F., Hobbelen, P. and Oliver, R. 2014b. Mixtures as a fungicide resistance management tactic. *Phytopathology* 104(12), 1264–73. doi:10.1094/PHYTO-04-14-0121-RVW.

Vasileiadis, V. P., Dachbrodt-Saaydeh, S., Kudsk, P., Colnenne-David, C., Leprince, F., Holb, I. J., Kierzek, R., Furlan, L., Loddo, D., Melander, B., et al. 2017. Sustainability of European winter wheat- and maize-based cropping systems: economic, environmental and social *ex-post* assessment of current and IPM-based systems. *Crop Prot.* 97, 60–9.

Walters, D. R. and Bingham, I. J. 2007. Influence of nutrition on disease development caused by fungal pathogens: implications for plant disease control. *Ann. Appl. Biol.* 151(3), 307–24. doi:10.1111/j.1744-7348.2007.00176.x.

Walters, D. R., Havis, N. D. and Oxley, S. J. P. 2008. *Ramularia collo-cygni*: the biology of an emerging pathogen of barley. *FEMS Microbiol. Lett.* 279(1), 1–7. doi:10.1111/j.1574-6968.2007.00986.x.

Walters, D. R., Avrova, A., Bingham, I. J., Burnett, F. J., Fountaine, J., Havis, N. D., Hoad, S. P., Hughes, G., Looseley, M., Oxley, S. J. P., et al. 2012. Control of foliar diseases in barley: towards an integrated approach. *Eur. J. Plant Pathol.* 133(1), 33-73. doi:10.1007/s10658-012-9948-x.

Walters, D. R., Lyon, G. D. and Newton, A. C. 2014. *Induced Resistance for Plant Defence: a Sustainable Approach to Crop Protection* (2nd edn.). Blackwell Publishing Science, Oxford, UK.

Wilson, W., Dahl, B. and Nganje, W. 2018. Economic costs of Fusarium Head Blight, scab and deoxynivalenol. *World Mycotoxin J.* 11(2), 291–302. doi:10.3920/WMJ2017.2204.

Xu, X. M., Jeffries, P., Pautasso, M. and Jeger, M. J. 2011. Combined use of biocontrol agents to manage plant diseases in theory and practice. *Phytopathology* 101(9), 1024–31. doi:10.1094/PHYTO-08-10-0216.

Young, C. S., Thomas, J. M., Parker, S. R. and Paveley, N. D. 2006. Relationship between leaf emergence and optimum spray timing for leaf blotch (*Rhynchosporium secalis*) control on winter barley. *Plant Pathol.* 55(3), 413–20. doi:10.1111/j.1365-3059.2006.01361.x.

Yan, G. P. and Chen, X. M. 2006. Molecular mapping of a recessive gene for resistance to stripe rust in barley. *Theor. Appl. Genet.* 113(3), 529–37. doi:10.1007/s00122-006-0319-x.

Zhan, J., Fitt, B. D. L., Pinnschmidt, H. O., Oxley, S. J. P. and Newton, A. C. 2008. Resistance, epidemiology and sustainable management of *Rhynchosporium secalis* populations on barley. *Plant Pathol.* 57, 1–14.

Chapter 12

Integrated weed management in barley cultivation

Michael Widderick, Queensland Department of Agriculture and Fisheries, Australia

1 Introduction
2 Integrated Weed Management
3 Weed control tactics
4 IWM in practice
5 Examples of IWM in barley
6 Conclusion
7 Where to look for further information
8 References

1 Introduction

A wide variety of weed species can infest barley crops, with broadleaf and grass weeds being the most common. A weed survey of cereal crops in central southern England found 16 grass weed species in winter barley and 8 in spring barley (Froud-Williams and Chancellor, 1982). A second survey of the same region found 21 and 22 broadleaf weeds in winter and spring barley, respectively (Chancellor and Froud-Williams, 1984). A survey of weeds in Canadian barley fields also found a variety of grass and broadleaf weeds (Harker, 2001).

Weed infestation in barley can cause yield reductions through competition for resources including water, light, nutrients and space. A survey of the impact of weeds on crop losses due to weeds in Canada found a 7–15% loss in barley (Swanton et al., 1993). In the United States, *Avena fatua* L. (wild oat) reduced barley yield by 6.5% and 25.9% when present at 83 and 199 seedlings m^{-2} (Bell and Nalewaja, 1968). In a UK study of 51 spring barley crops infested with *A. fatua*, Wilson and Peters (1982) found infestations ranging from 200 to 350 plants m^{-2} reduced barley yields by between 16% and 20%. Infestation by the weed *Galium aparine* L. (cleavers, goosegrass) in a UK study, reduced the yield of two barley cultivars by 53-61% (Seavers and Wright, 1999). In a Western Australian study, increasing *Lolium rigidum* Gaud. (annual ryegrass,

http://dx.doi.org/10.19103/AS.2019.0060.19

rigid ryegrass) plant population from 16 to 125 plants m^{-2} decreased barley grain yield through a reduction in barley biomass, tiller number and grain number (Paynter and Hills, 2009).

Weed infestations in barley can also reduce harvested grain quality. Wilson and Peters (1982) found the proportion of small barley seeds (below 2 mm in size) was increased from 13% in weed-free barley to 20% in the presence of wild oats.

The impact of weeds on barley crop production can differ depending upon the time and duration of weed infestation. In most cases, weed infestation at or close to crop establishment will cause greatest yield reductions (Sardana et al., 2017; Swanton et al., 2008). Therefore, weed control prior to planting and early in the crop's life is critical. However, weed infestation later in the crop, although it may not cause major reduction in crop production, can interfere with harvest and contaminate the harvested product. Also, uncontrolled weeds can produce large quantities of seed that will return to the soil's seed bank to infest future crops. At densities of 8–1100 *A. fatua* plants m^{-2} in a barley crop, *A. fatua* seed production was 180–9950 seeds m^{-2} respectively, in the absence of herbicides (Wille et al., 1998). Therefore, weed management at all stages of the barley crop is important.

Weed infestation in a field is the product of farming practices used and the environment (Buhler, 2002). The primary goal in barley weed management should be to reduce the impact of weeds on the productivity of current and future crops. To provide maximum, long-term control of weeds in barley, it is necessary to apply a range of weed management tactics. In this chapter, we outline the various tactics that can be applied for weed management in barley cropping and will highlight the need to integrate or use a combination of weed management tactics as part of an integrated weed management (IWM) strategy.

2 Integrated Weed Management

IWM is a strategic approach to weed management used over the long-term that combines diverse weed control tactics in an integrated approach. IWM considers the causes of weed problems rather than reacting to existing weed populations (Buhler, 2002) and combines chemical, non-chemical and agronomic weed control tactics to reduce weed numbers, stop seed bank replenishment and minimise the risk of selection for species and biotype shift, including herbicide resistance (Storrie, 2014). Integrated systems diversify the selection pressure on weed populations by applying a broad range of management tactics (Buhler, 2002).

A very basic example of an IWM approach is the combination of two tactics, growing a competitive crop (agronomic) combined with the application of in-crop herbicides. By combining these two tactics in barley, weed control can be improved compared to control achieved by only using one tactic alone (Wille et al., 1998; O'Donovan et al., 2001; Harker et al., 2009).

No one weed management tactic will consistently provide 100% control of weeds in the long term, and without complete weed control, weeds are allowed to reproduce and spread. However, weed management tactics used in combination can have an additive or multiplying effect and provide improved weed control, thereby reducing weed spread and proliferation (Storrie, 2014).

Furthermore, relying on one weed control tactic will result in species or biotype shift. Weed populations are genetically diverse and are able to adapt to changes in weed management and farming systems. The rate of population shift will depend on the selection pressure imposed, the genetic variability in the population and the weeds' ecology and environmental factors (Murphy and Lemerle, 2006). Weed populations can change rapidly in response to adding or removing tillage. A 6-year study by Tuesca et al. (2001) found grassy annual weeds increased rapidly in no-tillage treatments. Weed surveys have shown the introduction of glyphosate-resistant cotton cropping has resulted in an increase in small-seeded, wind-dispersed species *Conyza bonariensis* and *Sonchus oleraceus* after only 7–9 years (Werth et al., 2013). A sustained reliance on any one weed management tactic will drive the weed population towards weed species and biotypes that are able to survive the tactic.

Conservation Agriculture systems have largely removed cultivation from the farming system and placed a heavy reliance on herbicides for weed control. The removal of cultivation has favoured the proliferation of small-seeded weed species in Australia by retaining larger quantities of weed propagules in the soil surface layers where their germination is favoured (Wu et al., 2007; Widderick et al., 2010). The reliance on herbicides has resulted in the widespread occurrence of herbicide resistance in weeds worldwide (Heap, 2018).

Herbicide resistance has been a key driver behind the development and adoption of IWM (Norsworthy et al., 2012). As herbicide resistance deems a herbicide or herbicides ineffective, other tactics are required to control these weeds.

IWM systems provide improved weed control in the long-term by reducing selection pressure for species shift and herbicide resistance and by preventing reproduction on surviving plants (Llewellyn et al., 2007). In a 6-year French study, IWM cropping systems with combinations of crop rotations (including barley), soil tillage and mechanical and chemical weed control tactics were compared (Chikowo et al., 2009). The study found the IWM systems were successful in reducing reliance on herbicides, reducing potential environmental impacts from herbicides, including herbicide resistance potential, and were able to control weed infestations in the long-term.

An IWM system will better control a weed infestation that is made up of a number of different weed species. Rarely will only one weed species or weed type be present in crop. Often a variety of weeds (grasses and broadleaf weeds) will be present and commonly at a range of growth stages. Relying on one weed management tactic is unlikely to provide control of a range of weed types and ages.

An improved consistency in weed management can also be achieved through IWM. Where one tactic may fail or provide poor weed control, another tactic used in combination or in addition to the first tactic may provide control where the first tactic has failed.

IWM is a system that requires an increase in diligence and planning and an increase in inputs. As such, IWM systems may cost more in the short-term. However, the long-term benefits of avoiding additional costs associated with herbicide resistance are clear (Norsworthy et al., 2012).

3 Weed control tactics

3.1 Herbicides

The cornerstone of Conservation Agriculture is the minimisation of soil disturbance by tillage (Dumanski et al., 2006). Such systems are becoming common across the globe and are heavily reliant on herbicides for weed control. In Australia, >90% of grain production is under Conservation Agriculture systems (minimum or zero cultivation) and 98% of farmers use herbicides for weed control (Llewellyn et al., 2016).

Herbicides have been an effective means of weed control in crop production for many years, greatly reducing yield losses and reducing weed populations to low levels (Chikowo et al., 2009). Herbicides are an effective, easy-to-use, and often safe and affordable means of weed control. Herbicide-based weed management has allowed the development of relatively simple cropping systems (Buhler, 2002). As a result, herbicides have been widely adopted in agricultural production systems worldwide. Although the dependence upon herbicides has caused widespread herbicide resistance, herbicides will continue to dominate weed control. Their preservation through judicious use is therefore recommended and should include control of survivors of herbicide treatment to stop seed set and the potential spread of herbicide resistance.

Weed control in barley starts prior to the crop being planted. Starting with a weed-free planting bed is important for the establishment of a uniform barley crop and enables the barley crop to grow in the absence of competition with weeds. Knock-down or post-emergent herbicides are commonly used to obtain a weed-free planting bed and commonly a single, non-selective herbicide will be applied (e.g. glyphosate, paraquat, gramoxone) (Chauhan et al., 2012). However, using a single herbicide alone applies a high selection pressure for resistance towards these herbicides (Neve et al., 2003).

An alternative, integrated approach, that improves weed control and reduces the selection pressure for herbicide resistance, is the application of knock-down herbicides as part of a double knock approach. The double knock herbicide tactic is the sequential application of two herbicides with different chemical modes of action (Borger and Hashem, 2007). The purpose of a double knock is for the second herbicide application to control survivors of the first application and thereby minimise seed production.

A double knock does not have to be herbicide-based and can be a combination of herbicide and mechanical weed control (e.g. a knock-down herbicide followed by cultivation to control survivors). However, current double knock strategies are commonly herbicide based. The most effective herbicide double knocks use a translocated herbicide for the first knock and a contact herbicide for the second knock. A common example is the application of glyphosate followed by paraquat (Werth et al., 2010).

The efficacy of the double knock is determined by the weed growth stage (with better control on pre-tillering or rosette stage weeds), the moisture stress of the weed (non-stressed weeds are better controlled), the duration between the two herbicide treatments (generally 7–14 days provides maximum efficacy) and the herbicides used (Werth et al., 2010; Borger and Hashem, 2007; Widderick and McLean, 2018).

Pre-emergent herbicides are becoming more commonly used for the control of weeds at planting and for prolonged control during the crop. Pre-emergent herbicides allow the crop to grow without competition from weeds and can be applied alone or in mixture with a knock-down herbicide, or as part of a double knock strategy. Applying a pre-emergent herbicide in combination with a knock-down herbicide will provide control of weeds that have already emerged and future emergences while reducing the number and cost of herbicide applications.

Effective broadleaf weed control was achieved by the pre-emergent herbicide prosulfuron when applied in barley crops in Ontario, Canada, without any adverse effect on the barley crop (Soltani et al., 2014). A study by Sikkema et al. (2008) evaluated the pre-emergent herbicide saflufenacil in barley and found no negative impact on barley growth or yield when applied pre-emergence (3 days after crop planting). However, application post crop emergence reduced yield of barley by 24%.

The efficacy of pre-emergent herbicides can be variable as it is affected by many factors including stubble and crop interception (potentially limiting the amount of herbicide reaching the soil), photochemical decomposition, volatilisation, run off, microbial decomposition, chemical breakdown and leaching (Congreve and Cameron, 2018). Each of these factors is different for each pre-emergent herbicide and each environment (soil type, rainfall, temperature) and will determine the duration for which the herbicide remains active in the soil. Due to the variability in control achieved via pre-emergent herbicides, they should always be used in combination with other weed control tactics to improve control, for example in combination with a competitive crop. Many pre-emergent herbicides can negatively impact the growth and production of subsequent susceptible crops if there is a high enough concentration of the herbicide in the soil at their planting time. As such, the use of residual herbicides in farming systems with crop rotation requires careful planning.

Weed control in a barley crop can also be achieved via the application of in-crop post-emergent herbicides (Wille et al., 1998; O'Donovan et al., 2001; Harker et al., 2009). It is important to consider the growth stage of both the weed and the crop in applying in-crop herbicides. Applying herbicides at the incorrect weed growth stage can result in ineffective control. Applying herbicides at the incorrect crop growth stage can cause crop damage and yield reductions. In an Indian study, pinoxaden was evaluated for the post-emergence control of grass weeds in barley (Chhokar et al., 2008). The herbicide was found to effectively control *Phalaris minor* Retz. (canary grass), *Avena ludoviciana* Durieu (ludo wild oats) and *Polypogon monspeliensis* (L.) Desf. (beard grass) without causing any toxicity to the 18 barley cultivars tested.

It is important to note there can be varietal differences in barley tolerance towards in-crop herbicides. Research in Australia through the National Variety Trials (2015) programme has evaluated the susceptibility of barley varieties to commonly used in-crop herbicides across Australian cropping regions and years. For example, some barley cultivars were found to be highly sensitive to the herbicide dicamba with up to 15% yield reduction, while other cultivars had no yield reduction. This information is published annually to inform farmers about herbicides that can be safely used in crop.

The efficacy of in-crop herbicides will also be determined by the weed density. Wille et al. (1998) found in their US study that as *A. fatua* densities increased from 8 to 1100 plants m^{-2}, *A. fatua* seed production increased from 180 to 9950 seeds m^{-2} in the absence of herbicides. However, when the herbicide imazamethabenz was applied, *A. fatua* seed production was 0–2810 seeds m^{-2} respectively. This study shows in-crop post-emergence herbicides were not able to provide 100% *A. fatua* control at a high weed density.

Herbicide-tolerant (HT) barley varieties now exist and allow for in-crop application of herbicides that non-HT barley varieties are susceptible to. For example, some barley varieties enable the in-crop application of the herbicides imazamox and imazapyr for the post-emergent knock-down and residual control of a range of grass and broadleaf weeds (Moody, 2015). However, HT crops encourage reliance on one herbicide for in-crop weed control and are therefore expected to increase the likelihood of herbicide resistance development (Vencill et al., 2012). In addition, if not used according to label instructions, there can be excessive chemical residues in barley grain products, limiting use and export potential.

If control measures have failed before and during the crop, there are herbicide options that can be used as a salvage operation to control weeds and stop their seed set. Unfortunately, these options provide little or no gain for crop production of the current crop, but can greatly reduce the number of weed seeds produced, having implications for future cropping. Salvage operations that may be suitable for use in barely include selective spray topping (the application of post-emergent selective herbicides to prevent weed seed set) and wiper technology (wipe herbicide onto weed plants that are growing above the crop canopy) (Storrie, 2014). There are risks associated with the late application of herbicide in crop including chemical contamination of the harvested grain, which can eliminate the option of export or use in food products.

While there are many herbicide options for the control of weeds in barley cropping, very rarely will these tactics provide 100% control of weeds when used in isolation. They should therefore be used in combination with other herbicide and non-chemical tactics. In addition, a reliance on herbicides for weed control is likely to increase the risk for herbicide resistance development.

3.1.1 Herbicide resistance

The widespread adoption of Conservation Agriculture has resulted in a heavy reliance on herbicides for weed control. This reliance on herbicides has been in response to the exclusion of non-chemical weed management tactics, including cultivation (Norsworthy et al., 2012). In addition, herbicides have proven to be very effective in controlling weeds and they are often convenient, safe and easy to apply. However, as a result of an over-reliance on herbicides, widespread herbicide resistance has developed worldwide, weed populations have shifted towards difficult-to-control and there can commonly be off-site movement of herbicides (Buhler, 2002).

Herbicide resistance is the inherited ability of a plant to survive and reproduce following the application of a herbicide lethal to normal individuals of that plant (Vencill et al., 2012). In practical terms, herbicide resistance will deem a once effective herbicide ineffective on a weed population.

There have been 495 unique cases of herbicide resistance in weeds documented through the International Survey of Herbicide Resistant Weeds (Heap, 2018). Of these, there are 255 weed species in 92 crops and 70 countries. Of the confirmed cases, 75 have been identified in fields under barley production and include both grass and broadleaf weeds.

Herbicide resistance is genetically controlled and the genes for resistance naturally exist in most weed populations. Although rare, these genes are selected for when the same herbicide mode of action (MOA) is used repeatedly and in isolation of other weed control tactics. Each time the herbicide is applied, susceptible plants die and resistant individuals survive. If these resistant weeds are not controlled, they set seed and over time the proportion of resistant individuals increases. Eventually, the majority of the population becomes resistant.

The development of resistance will be affected by several factors including:

- Initial frequency of the resistance gene(s),
- Size of the weed population,
- Herbicide efficacy and selection pressure and
- Weed biology (Diggle and Neve, 2001; Norsworthy et al., 2012).

The frequency of the resistance gene in an untreated weed population will differ for different herbicide MOA groups and weed species. For example, the frequency of *L. rigidum* individuals resistant to ALS-inhibiting herbicides ranged from 2.2×10^{-5} to 1.2×10^{-4} for three unselected populations (Preston and Powles, 2002). The same resistance has been estimated to occur at a frequency of 1×10^{-9} in *Arabidopsis thaliana* (L.) Heynh. (thale cress, mouse-ear cress) (Haughn and Somerville, 1987).

The selection pressure imposed on a population, through herbicide application, will affect resistance development. A highly effective herbicide will apply a strong selection pressure for resistance by controlling a large proportion of susceptible individuals (Diggle and Neve, 2001). Applying herbicides at a low, sublethal rate will allow resistant individuals to survive, but will also allow some susceptible individuals to survive. Therefore, weed control will be poor and more likely to result in yield loss. A robust herbicide rate should always be applied, but surviving weeds should be controlled.

The size of a weed population will also change the likelihood of the resistance gene being present. In a large and dense weed population, there is a greater likelihood of the resistance gene being present, compared with a sparse and small population (see Table 1). A greater initial frequency of the resistance gene will also increase the presence of resistant individuals in a weed population.

A weed's biology can also influence the likelihood and speed of herbicide resistance development. A weed species that produces a large number of seeds will produce a large number of resistant progeny that can quickly increase and spread. If a weed species has seed that is highly dormant and therefore not able to readily germinate, this could delay the onset of resistance as the susceptible seed in the soil seed bank will dilute the proportion of resistant individuals.

The spread of resistance will be more rapid in cross-pollinating weed species (Diggle and Neve, 2001). In cross-pollinated weed species, resistant individuals that survive herbicide application can cross with susceptible weeds.

Table 1 Predicted number of resistant individuals in weed populations of varying density, size and initial gene frequencies

Weed density (seeds m^{-2})	Population size (m^{-2})	Initial gene frequency	Number of resistant individuals
10	10 000	1×10^{-8}	0.0001
		1×10^{-6}	0.01
	1 000 000	1×10^{-8}	0.1
		1×10^{-6}	10
100	10 000	1×10^{-8}	0.01
		1×10^{-6}	1
	1 000 000	1×10^{-8}	1
		1×10^{-6}	100
1000	10 000	1×10^{-8}	0.1
		1×10^{-6}	10
	1 000 000	1×10^{-8}	10
		1×10^{-6}	1000

Source: Diggle and Neve (2001).

For self-pollinated weed species, a herbicide-resistant parent plant will have identical herbicide-resistant progeny.

Herbicide resistance can also be introduced onto a field through machinery, feed for stock and through wind and water dispersal. While introduction of seed and propagules can be minimised, more important is how these individuals are controlled, as this will determine whether they proliferate and spread.

It is important to identify herbicide resistance as early as possible to gain knowledge of which herbicides are effective and to ensure other control measures can be put in place for weed control. Testing for herbicide resistance can be done in a laboratory or on farm. In a laboratory setting either seed or live plant samples can be used and tested for survival once treated with herbicides. In a field situation a farmer can run a test strip of a range of herbicides to see if they are effective in controlling the weeds.

The likelihood of herbicide resistance is increased by using the same herbicide MOA in the absence of other herbicide modes of action and other, non-chemical weed control options. Also, not controlling seed set on weeds that survive herbicide treatment will increase the likelihood of spread of herbicide resistance.

The only guaranteed way to prevent and manage herbicide resistance is to achieve 100% weed seed set control. In stopping seed set or propagule production, the resistance gene is stopped. No one weed control tactic, used in isolation of other tactics, is able to provide 100% weed control for a sustained period of time. This is why IWM is so important in effective and sustained weed management.

Herbicide-resistant weeds pose a major threat to agricultural food production. Therefore the adoption of IWM practices is critical in order to preserve the useful life of effective herbicides.

3.2 Harvest weed seed control

Uncontrolled weeds in barley have the potential to set large quantities of viable seed, replenish the soil seed bank and spread herbicide resistance. In a standard crop harvest, weed seeds present and mature at harvest time are collected and dispersed along with the chaff fraction back into the field.

Harvest weed seed control (HWSC) was developed in Western Australian cropping systems in response to widespread multiple herbicide resistance, particularly in the weeds *L. rigidum* and *Raphanus raphanistrum* L. (wild radish), to minimise weed seed inputs into the seed bank (Walsh et al., 2013). HWSC is now being examined for the control of other widespread herbicide-resistant weeds in the United States (*Amaranthus palmeri* S. Wats. – Palmer amaranth) (Norsworthy et al., 2016) and Canada (*Avena* spp.) (Tidemann et al., 2016).

Through HWSC, weed seeds in the chaff fraction are collected and destroyed during or after grain harvest. HWSC exploits two common biological features of annual weeds, namely their ability to retain seed at maturity and their relatively short-lived seed bank (Walsh et al., 2013). HWSC is very effective in reducing seed return on weed species that retain a large proportion of their seed at harvest time and above harvest height. However, not all weed species are suited for control using HWSC. Weed species with indeterminate flowering will not have all their seed maturing at the same time and weeds with wind-dispersed seeds or shattering pods are not likely to retain a large proportion of their seeds at harvest.

HWSC systems include chaff carts where the chaff fraction is collected behind the harvester before being dumped and burnt or grazed; bale-direct where the chaff and straw are collected and baled directly behind the harvester, usually for stock food; narrow windrow burning where the chaff and straw are directed into narrow windrows to later be burnt; chaff tramlining where the chaff fraction is directed on to permanent wheel tracks to reduce weed emergence and speed up weed seed degradation; and the HSD that destroys chaff and weed seeds though a rotating mill system (Walsh et al., 2017). These HWSC systems have been shown to remove up to 99% of weed seeds that enter the harvester (Walsh et al., 2013).

HWSC can greatly reduce weed seeds in the soil seed bank over time when used in combination with in-crop herbicides. In a Western Australian study, weed management systems of herbicide versus herbicides + HWSC were compared for *L. rigidum* populations in 25 fields over a 10-year period (Walsh et al., 2013). In fields that combined herbicides with HWSC, the *L. rigidum* populations reduced from an average of 35 plants m^{-2} in 2002 to just 0.5 plants m^{-2} in 2011. In contrast, when only herbicides were applied, average annual *L. rigidum* populations remained above 4 plants m^{-2}. Furthermore, modelling has identified herbicides were better protected from future resistance evolution when their use was combined with annual HWSC (Somerville et al., 2018).

However, HWSC used on its own may not be sufficient to reduce the weed seed bank over time. A Canadian study (Tidemann et al., 2016) on seed retention of *A. fatua* found that early shattering of *A. fatua* seed, especially in later planted winter cereal crops, demonstrated HWSC was unlikely to be effective on this weed. They recommend HWSC should be used in combination with other weed management techniques for improved control of *A. fatua*. Similar results were found in an Australian study where *A. fatua* seed retention declined over time from 84% at the start of crop maturity to 39% after 28 days (Walsh and Powles, 2014).

Overuse of HWSC in the absence of other weed control tactics imposes a selection pressure on weed populations to shift towards weed species and biotypes that escape treatment. These may be earlier maturing (i.e. seed shatters or is dropped from the plant prior to harvest) or more prostrate (weeds that grow and produce seed below the height of harvest). However, a survey of nine Australian wheat fields found no evidence that *L. rigidum* had adapted to escape HWSC via change in seed retention height following 5–10 years of HWSC (Walsh et al., 2018). This study suggests weed adaptation in the field may not take place or may be a slow process.

3.3 Crop rotation

Weeds tend to associate with crops that have similar life cycles and times of growth (Blackshaw et al., 2007). Rotating crops with different life cycles can disrupt the development of weed-crop associations and will also enable the application of different weed control tactics. Choosing a different crop type can enable the inclusion of additional tactics and expand the opportunities for weed control (Storrie, 2014).

A cropping monoculture of barley will be heavily reliant on the same weed control tactics, most commonly the same suite of herbicides for weed control. Such farming systems are at high risk for the selection of a specific weed or suite of weeds and also for the development of herbicide resistance.

In regions where cropping is only possible in one season, rotating crop species between years can enable the use of different herbicide modes of action. For example, growing barley (a grass crop) in rotation with a broadleaved crop (e.g. chickpea).

Some agricultural regions are able to grow and rotate crops across seasons. In these regions where crops can be grown all year round, there are more options for crop rotation and therefore more opportunities to apply herbicides with different modes of action. In addition to crop rotation, having a crop-free period will enable other alternative weed control tactics to be applied. A fallow period enables the application of non-selective herbicides and residual herbicides, strategic cultivation and so on, and can assist in preventing herbicide resistance.

3.4 Crop competition

Weeds compete with crops for resources necessary for plant growth including water, sunlight, nutrients and space. An uncontrolled weed will use these resources and can cause reductions in crop growth and production. Growing a competitive barley crop can reduce the impact of weeds on crop production and increase the effectiveness of weed control tactics used in crop (e.g. in crop herbicides).

Barley is well known as a competitive crop species. In a Swedish study by Bertholdsson (2005), barley's early crop biomass production and potential allelopathic potential were identified as parameters that contributed towards the crops' competitiveness.

In a 1998 trial in Western Australia, *L. rigidum* dry matter in barley was half that in wheat and triticale (*Triticale hexaploide* Lart.) (Peltzer, 1999). On the Darling Downs in Queensland, Australia, barley was found to be more competitive than wheat against the broadleaf weeds *Rapistrum rugosum* (L.) All. (turnip weed) and *Fallopia convolvulus* (L.) A. Löve. (black-bindweed, wild buckwheat) (Marley and Robinson, 1990). An additional study in NSW showed barley to have a more rapid early growth rate than both wheat and *L. rigidum*, and as a result, barley was able to suppress ryegrass more than wheat and maintain crop yield (Cousens, 1996).

The competitiveness of a barley crop can be increased by growing a more competitive variety, by sowing at a narrower row spacing and higher density, by sowing in an East–West row orientation and through fertiliser placement.

Different barley cultivars have different competitive ability. A study by Watson et al. (2006) compared the growth and weed suppressive ability of 29 barley cultivars grown on the Canadian prairies. They found there was considerable variability in competitive ability; however, the trend was for semi-dwarf and hull-less cultivars to be less competitive against *Avena sativa* L. (common oat). A further study identified differences in the competitive ability of barley cultivars grown in Denmark (Christensen, 1995). In this study weed suppression measured via dry weight biomass was correlated with the rate of canopy height development and light interception, with greater biomass reduction in quicker growing and taller cultivars.

Several studies have shown growing barley at narrow row spacing can increase competitiveness, reduce weed growth and increase barley yield.

In a Western Australian study, barley planted in wide rows (36–50 cm) compared with narrow rows (18–25 cm) resulted in lower barley yield and higher *L. rigidum* tiller number at three of four sites (Paynter, 2010). Wide rows favoured the growth of ryegrass dry matter and tiller numbers in the order of a 50–100% increase. In a Canadian study (Kirkland, 1993), biomass production of *A. fatua*, *Brassica kaber* (DC.) L.C. Wheeler (wild mustard) and volunteer (*Brassica campestris* L.) (canola, rapeseed) were significantly reduced by narrower row spacings. Additionally, adjusting barley rows from 46 to 11 cm produced yield increases of between 67% and 77% in 1990 and 1991, respectively. Generally, wide row spacing will reduce competition with weeds and reduce crop yield in some situations, while closer row spacing will improve crop competition for light and soil moisture (Blackshaw et al., 2007).

The competitiveness of barley can also be increased by increasing the sowing rate or crop density. Increasing barley density (75, 150 and 300 plants m^{-2}) reduced ryegrass dry matter production and tiller number in a Western Australian study (Paynter and Hills, 2009). Similarly, an increased barley sowing density (50, 90, 150 and 220 kg ha^{-1}) significantly reduced the biomass production of *A. fatua*, *B. kaber* and *B. campestris* (Kirkland, 1993). Additionally, increasing the seeding density from 50 to 220 kg ha^{-1} resulted in yield increases of between 80% and 93%. A study by Lafond (1994) found increasing seeding rate of barley from 27 to 161 kg ha^{-1} increased barley yield by 32%.

Planting barley crop rows at right angles to sunlight direction has been shown to increase inter-row shading and reduce weed biomass. In a Western Australian study (Borger et al., 2010) light interception by the barley crop canopy was 18% greater in crops oriented East–West, compared with North–South. The resulting weed biomass was 37% less and grain yield increased by 26%. Location and latitude will determine in which direction the crops should be planted for best utilisation of sunlight.

Several studies have shown barley can reduce the growth of weeds through allelopathy via the production of over 44 chemicals (Kremer and Ben-Hammouda, 2009) including the secondary metabolites gramine and hordenine, common in barley (Lovett et al., 1994). In a 1966 study, Overland showed a reduction in *Stellaria media* L. (chickweed) and *Capsella bursa-pastoris* (L.) Medic. (shepherd's purse) growth in response to living plants and aqueous leachates of barley. The allelopathic potential of barley was confirmed by Liu and Lovett (1990) when their research showed barley delayed germination and inhibited the growth of *Sinapis alba* L. (white mustard). A recent study identified the metabolite gramine had a greater level of toxicity than hordenine in reducing the growth of *Matricaria recutita* L. (chamomile) (Lebecque et al., 2018).

3.5 On-farm hygiene

The unnecessary introduction of weeds and their seeds can be avoided and managed to reduce future weed problems and prevent the introduction of herbicide resistance.

Weed seeds can readily be imported along with crop seed for planting and in feed for livestock. In these cases, endeavour to use certified seed and feed. For planting seed, make sure to check the seed lot to ensure it is not contaminated with weed seeds. Most commercially available crop seed will have an analysis of weed contamination and there will be allowable limits. Crop seed and feed purchased from another farmer may contain higher quantities of weed seeds. In these cases, know where the seed or feed is coming from, examine a sample for contamination and also evaluate whether the weed seed is herbicide resistant. To limit the spread of weeds from stock feed, feed stock in one location to limit spread and to be able to closely monitor for weed emergence.

Weed seeds can also be introduced via machinery. In some cases contracted machinery is used for planting and harvesting of barley. Be sure machinery is clean of weed seeds before entering and exiting your property. If using your own machinery, try to harvest the cleaner paddocks first to limit weed seed spread.

It is important to control weeds in non-crop areas such as along fence lines, field edges and roadsides as weed incursions and spread can come from these areas and into cropping fields. This is particularly important for weed species that spread via wind-dispersed seeds. Commonly pre-emergent and post-emergent herbicides can be used along with mowing and cultivation, or a combination of these.

Weed seeds can also be introduced via wind or floods and these are not preventable. However, if there has been a flood event, diligence is required in monitoring for new weed incursions. These incursions should be controlled prior to weed seed set to ensure the incursion does not persist or spread.

4 IWM in practice

Adopting an IWM approach can be a challenge. Ultimately, each weed population and environment is different so decisions regarding which tactics to use in combination need to be site- and time-specific (Buhler, 2002).

IWM is more complex and less convenient than relying on individual weed management tactics. IWM requires forward planning and consideration of the whole farming system and practices and how these interact with weed populations. In addition, IWM can be more costly, in the short-term. However, simulation modelling has shown IWM practices have long-term economic benefits. A simulation model of a New South Wales (Australian) continuous cropping system over a 20-year period determined the economic benefit of using IWM for key weeds such as *Avena* spp., *R. raphanistrum* and *L. rigidum* (Storrie, 2014). Weed management that relied on herbicides (no IWM) provided lower economic return than systems utilising IWM, for each weed species (see Table 2). The negative economic return for *R. raphanistrum* was attributed to the large seed production and seed build up over time due to poor herbicidal control due to staggered emergence of this species.

Weeds can commonly occur in patches or only in certain parts of a field (Clay et al., 1999). As such, it can be more cost-effective to apply weed management only to these specific areas. This approach to weed management is commonly referred to as site-specific weed control. In an IWM system, site-specific weed control will help to minimise the additional short-term expenses of applying multiple weed control tactics and reduce herbicide use in agricultural production. Recent advances in weed detection and mapping technologies (e.g. GPS (global positioning systems), GIS (geographic information systems), remote sensing and weed detection) have enabled weeds to be mapped and strategically targeted. When combined with precision agriculture techniques, a 4-year study in barley found a 60% and 90% reduction in herbicides used to control broadleaf and grass weeds, respectively (Gerhards and Christensen, 2003).

5 Examples of IWM in barley

In a Canadian study, Harker et al. (2009) determined the influence of repeated chemical and cultural practices on *A. fatua* populations and barley production. In the study, cultural practices included crop competition through growing tall barley cultivars and via doubling seeding rate, combined with crop rotation with canola and peas. These tactics were used in combination with recommended herbicide rates of 25%, 50% and 100%, respectively. The study showed the cultural practices interacted synergistically to reduce *A. fatua* emergence, biomass and most importantly seed production. As an example, *A. fatua* biomass was reduced 2- to 3-, 6- to 7- or 19-fold when single, double or triple

Table 2 Economic return (AUD\$ ha^{-1}) of no IWM (herbicide-based) vs IWM systems for *Avena* spp., *Raphanus raphanistrum* and *Lolium rigidum* in a southern NSW continuous cropping system as determined through simulation modelling over a 20-year period

	Economic return (AUD\$ ha^{-1})					
	Avena spp.		*R. raphanistrum*		*L. rigidum*	
No IWM	268	±35	−9	±27	284	±34
IWM	332	±38	315	±37	335	±38

The values following ± are the standard deviation.
Source: Storrie (2014).

cultural treatments were combined. In addition, there was a proportionate increase in barley yield.

Another Canadian study combined different barley row spacing (20 and 30 cm) and seeding rates (75, 125 and 175 kg ha^{-1}) and rates of the herbicide mixture tralkoxydim + bromoxynil + MCPA (0%, 25%, 50% and 100% recommended rate) and measured the effect on *A. fatua* seed production (O'Donovan et al., 2001). The study found there was a positive interaction between these factors. At each herbicide rate, *A. fatua* seed production decreased as seeding rate increased. The combination of higher herbicide and seeding rate reduced the number of *A. fatua* seed in the soil bank whereas at the lowest seeding rate (75 kg ha^{-1}) large amounts of *A. fatua* seed were produced when the herbicide mixture was applied at 25% or 50% recommended rate. In this study row spacing had no effect on *A. fatua* seed production and seeding rate and row spacing had little effect on barley yield.

6 Conclusion

Weeds are an ever-present constraint in barley production and can cause major production losses and increased production costs. However, there is a range of weed control tactics that when applied correctly and used in combination can provide effective weed control in current and future barley crops. The tactics used and the application strategy will differ for different weed species, weed growth stages, crop growth stages and herbicide resistance status. The cost of the tactics and the ability to practically apply the tactics will also influence which tactics are used. Correct application of weed control tactics within an IWM approach can successfully reduce the impact of weeds on barley production.

7 Where to look for further information

There are many publications available with useful information on weeds and their control in agriculture. In addition to the references used in this chapter, the following books provide a thorough overview of weeds and their impact on agricultural production and strategies for their control.

Recent Advances in Weed Management: Eds. Chauhan, B.S. and Mahajan, G. (2014). This publication provides up-to-date information on IWM in cropping systems including organic farming, conservation agriculture systems and horticultural crops. There is also information on weed management systems in specific crops including rice, wheat, maize, cotton and soybean. Additionally, the authors have explored challenges and opportunities in weed management under changing systems of increased herbicide resistance.

Australian Weed Management Systems: Ed. Sindel, B. (2000). This publication explains weed ecology and population dynamics and how these characteristics contribute towards a weed's success. Having a better understanding of these ecological drivers can help in making a more informed decision on which control tactics may be effective. The authors include information on a range of different weed control tactics including tillage, cultural control, grazing management and biological control. In addition, there is material on devising weed control strategies in agricultural, environmental and amenity systems.

8 References

Bell, A. R. and Nalewaja, J. D. 1968. Competition of wild oat in wheat and barley. *Weed Sci.* 16(4), 505–8.

Bertholdsson, N. O. 2005. Early vigour and allelopathy - two useful traits for enhanced barley and wheat competitiveness against weeds. *Weed Res.* 45(2), 94–102. doi:10.1111/j.1365-3180.2004.00442.x.

Blackshaw, R. E., Anderson, R. L. and Lemerle, D. 2007. Cultural weed management. In: Upadhyaya, M. K. and Blackshaw, R. E. (Eds), *Non-Chemical Weed Management - Principles, Concepts and Technology*. CABI, Wallingford, UK, pp. 35–48. Chapter 3.

Borger, C. P. and Hashem, A. 2007. Evaluating the double knockdown technique: sequence, application interval, and annual ryegrass growth stage. *Aust. J. Agric. Res.* 58, 265–71.

Borger, C. P. D., Hashem, A. and Pathan, S. 2010. Manipulating crop row orientation to suppress weeds and increase crop yield. *Weed Sci.* 58(2), 174–8. doi:10.1614/WS-09-094.1.

Buhler, D. D. 2002. 50th Anniversary–Invited Article: Challenges and opportunities for integrated weed management. *Weed Sci.* 50(3), 273–80. doi:10.1614/0043-1745(2002)050[0273:AIAAOF]2.0.CO;2.

Chancellor, R. J. and Froud-Williams, R. J. 1984. A second survey of cereal weeds in central southern England. *Weed Res.* 24(1), 29–36. doi:10.1111/j.1365-3180.1984.tb00568.x.

Chauhan, B. S. and Mahajan, G. (Eds). 2014. *Recent Advances in Weed Management*. Springer-Verlag, New York, NY.

Chauhan, B. S., Singh, R. G. and Mahajan, G. 2012. Ecology and management of weeds under conservation agriculture: a review. *Crop Prot.* 38, 57–65. doi:10.1016/j.cropro.2012.03.010.

Chhokar, R. S., Sharma, R. K. and Verma, R. P. S. 2008. Pinoxaden for controlling grass weeds in wheat and barley. *Indian J. Weed Sci.* 40, 41–6.

Chikowo, R., Faloya, V., Petit, S. and Munier-Jolain, N. M. 2009. Integrated weed management systems allow reduced reliance on herbicides and long-term weed control. *Agric. Ecosyst. Environ.* 132(3–4), 237–42. doi:10.1016/j.agee.2009.04.009.

Christensen, S. 1995. Weed suppression ability of spring barley varieties. *Weed Res.* 35(4), 241–7. doi:10.1111/j.1365-3180.1995.tb01786.x.

Clay, S. A., Lems, G. J., Clay, D. E., Forcella, F., Ellsbury, M. M. and Carlson, C. G. 1999. Sampling weed spatial variability on a fieldwide scale. *Weed Sci.* 47, 674–81.

Congreve, M. and Cameron, J. (Eds). 2018. *Soil Behaviour of Pre-Emergent Herbicides in Australian Farming Systems - a National Reference Manual for Advisers* (2nd edn.). GRDC publication, Australia.

Cousens, R. D. 1996. Comparative growth of wheat, barley, and annual ryegrass (*Lolium rigidum*) in monoculture and mixture. *Aust. J. Agric. Res.* 47(3), 449–64. doi:10.1071/AR9960449.

Diggle, A. J. and Neve, P. 2001. The population dynamics and genetics of herbicide resistance – a modeling approach. In: Powles, S. B. and Shaner, D. L. (Eds), *Herbicide Resistance and World Grains*. CRC Press, Boca Raton, FL, pp. 62–96.

Dumanski, J., Peiretti, R., Benetis, J., McGarry, D. and Pieri, C. 2006. The paradigm of conservation tillage. In: *Proc. World Assoc. Soil and Water Conserv.*, pp. 58–64.

Froud-Williams, R. J. and Chancellor, R. J. 1982. A survey of grass weeds in cereals in central southern England. *Weed Res.* 22(3), 163–71. doi:10.1111/j.1365-3180.1982.tb00160.x.

Gerhards, R. and Christensen, S. 2003. Real-time weed detection, decision making and patch spraying in maize, sugarbeet, winter wheat and winter barley. *Weed Res.* 43(6), 385–92. doi:10.1046/j.1365-3180.2003.00349.x.

Harker, K. N. 2001. Survey of yield losses due to weeds in central Alberta. *Can. J. Plant Sci.* 81(2), 339–42.

Harker, K. N., O'Donovan, J. T., Irvine, R. B., Turkington, T. K. and Clayton, G. W. 2009. Integrating cropping systems with cultural techniques augments wild oat (*Avena fatua*) management in barley. *Weed Sci.* 57(3), 326–37. doi:10.1614/WS-08-165.1.

Haughn, G. and Somerville, C. R. 1987. Selection for herbicide resistance at the whole plant level. In: LeBaron, H. M., Mumma, R. O., Honeycutt, R. C. and Duesing, J. H. (Eds), *Biotechnology in Agricultural Chemistry*. American Chemical Society, Washington DC, pp. 98–107. Chapter 7.

Heap, I. 2018. The International Survey of Herbicide Resistant Weeds. Online. 25 October 2018. Available at: www.weedscience.org.

Kirkland, K. J. 1993. Weed management in spring barley (*Hordeum vulgare*) in the absence of herbicides. *J. Sustain. Agr.* 3(3–4), 95–104. doi:10.1300/J064v03n03_07.

Kremer, R. J. and Ben-Hammouda, M. 2009. Allelopathic plants. 19. Barley (*Hordeum vulgare* L.). *Allelopathy J.* 24, 225–42.

Lafond, G. P. 1994. Effects of row spacing, seeding rate and nitrogen on yield of barley and wheat under zero-till management. *Can. J. Plant Sci.* 74(4), 703–11. doi:10.4141/cjps94-127.

Lebecque, S., Crowet, J., Laurence, L., Delory, B. M., du Jardin, P., Fauconnier, M. and Deleu, M. 2018. Interaction between the barley allelochemical compounds gramine and hordenine and artificial lipid bilayers mimicking the plant plasma membrane. *Nature* 8, 9784.

Liu, D. L. and Lovett, J. V. 1990. Assessment of barley allelopathy. In: *Proceedings of the 9th Australian Weeds Conference*, Adelaide, Australia, pp. 440–5.

Llewellyn, R. S., Lidner, R. K., Pannell, D. J. and Powles, S. B. 2007. Herbicide resistance and the adoption of integrated weed management by Western Australian grain growers. *Agric. Econ.* 36, 121–30.

Llewellyn, R. S., Ronning, D., Ouzman, J., Walker, S., Mayfield, A. and Clarke, M. 2016. Impact of weeds on Australian grain production: the cost of weeds to Australian grain growers and the adoption of weed management and tillage practices. Report for GRDC. CSIRO, Australia.

Lovett, J. V., Hoult, A. H. C. and Christen, O. 1994. Biologically active secondary metabolites of barley. IV. Hordenine production by different barley lines. *J. Chem. Ecol.* 20(8), 1945–54. doi:10.1007/BF02066235.

Marley, J. M. and Robinson, G. R. 1990. Strategies for broadleaf control in barley. Final report. Barley Research Committee for Queensland, Australia.

Moody, D. 2015. Breeding for imidazolinone tolerant barley varieties: industry issues and concerns. GRDC. Available at: https://grdc.com.au (accessed on 25 October 2018).

Murphy, C. E. and Lemerle, D. 2006. Continuous cropping systems and weed selection. *Euphytica* 148(1-2), 61-73. doi:10.1007/s10681-006-5941-9.

National Variety Trials. 2015. Barley cultivars response to herbicides in the Northern Region 1999-2015. GRDC. Available at: https://www.nvtonline.com.au/herbicide-tolerance/(accessed 25 October 2018).

Neve, P., Diggle, A. J., Smith, F. P. and Powles, S. B. 2003. Simulating evolution of glyphosate resistance in *Lolium rigidum* II: past, present and future glyphosate use in Australian cropping. *Weed Res.* 43(6), 418-27. doi:10.1046/j.0043-1737.2003.00356.x.

Norsworthy, J. K., Ward, S. M., Shaw, D. R., Llewellyn, R. S., Nichols, R. L., Webster, T. M., Bradley, K. W., Frisvold, G., Powles, S. B., Burgos, N. R., et al. 2012. Reducing the risks of herbicide resistance: best management practices and recommendations. *Weed Sci.* 60(Special Issue), 31-62. doi:10.1614/WS-D-11-00155.1.

Norsworthy, J. K., Korres, N. E., Walsh, M. J. and Powles, S. B. 2016. Integrating herbicide programs with harvest weed seed control and other fall management practices for the control of glyphosate-resistant Palmer amaranth (*Amaranthus palmeri*). *Weed Sci.* 64(3), 540-50. doi:10.1614/WS-D-15-00210.1.

O'Donovan, J. T., Harker, K. N., Clayton, G. W., Newman, J. C., Robinson, D. and Hall, L. M. 2001. Barley seeding rate influences the effects of variable herbicide rates on wild oat. *Weed Sci.* 49(6), 746-54. doi:10.1614/0043-1745(2001)049[0746:BSRITE]2.0.CO;2.

Overland, L. 1966. The role of allelopathic substances in the 'smother crop' barley. *Amer. J. Bot.* 53(5), 423-32. doi:10.1002/j.1537-2197.1966.tb07355.x.

Paynter, B. H. 2010. Wide row spacing and rigid ryegrass (*Lolium rigidum*) competition can decrease barley yield. *Weed Technol.* 24(3), 310-8. doi:10.1614/WT-D-09-00062.1.

Paynter, B. H. and Hills, A. L. 2009. Barley and rigid ryegrass (*Lolium rigidum*) competition is influenced by crop cultivar and density. *Weed Technol.* 23(1), 40-8. doi:10.1614/WT-08-093.1.

Peltzer, S. 1999. Increased crop density reduces weed seed production without increasing screenings. In: *Proceedings of the 12th Australian Weeds Conference*, Hobart, Tasmania, pp. 510-12.

Preston, C. and Powles, S. B. 2002. Evolution of herbicide resistance in weeds: initial frequency of target site-based resistance to acetolactate synthase-inhibiting herbicides in *Lolium rigidum*. *Heredity* 88(1), 8-13. doi:10.1038/sj.hdy.6800004.

Sardana, V., Mahajan, G., Jabran, K. and Chauhan, B. S. 2017. Role of competition in managing weeds: an introduction to the special issue. *Crop Prot.* 95, 1-7. doi:10.1016/j.cropro.2016.09.011.

Seavers, G. P. and Wright, K. J. 1999. Crop canopy development and structure influence weed suppression. *Weed Res.* 39(4), 319-28. doi:10.1046/j.1365-3180.1999.00148.x.

Sikkema, P. H., Shropshire, C. and Soltani, N. 2008. Tolerance of spring barley (*Hordeum vulgare* L.) oats (*Avena sativa* L.) and wheat (*Triticum aestivum* L.) to saflufenacil. *Crop Prot.* 27(12), 1495-7. doi:10.1016/j.cropro.2008.07.009.

Soltani, N., Brown, L. R., Cowan, T. and Sikkema, P. H. 2014. Weed management in spring seeded barley, oats, and wheat with prosulfuron. *Int. J. Agron.* 2014, 1-5. doi:10.1155/2014/950923.

Somerville, G. J., Powles, S. B., Walsh, M. J. and Renton, M. 2018. Modeling the impact of harvest weed seed control on herbicide-resistance evolution. *Weed Sci.* 66(3), 395–403. doi:10.1017/wsc.2018.9.

Storrie, A. M. (Ed.). 2014. *Integrated Weed Management in Australian Cropping Systems*. Grains Research and Development Corporation.

Swanton, C. J., Harker, K. N. and Anderson, R. L. 1993. Crop losses due to weeds in Canada. *Weed Technol.* 7(2), 537–42. doi:10.1017/S0890037X00028049.

Swanton, C. J., Mahoney, K. J., Chandler, K. and Gulden, R. H. 2008. Integrated weed management: knowledge-based weed management systems. *Weed Sci.* 56(1), 168–72. doi:10.1614/WS-07-126.1.

Tidemann, B. D., Hall, L. M., Harker, K. N. and Alexander, B. C. S. 2016. Identifying critical control points in the wild oat (*Avena fatua*) life cycle and the potential effects of harvest weed-seed control. *Weed Sci.* 64(3), 463–73. doi:10.1614/WS-D-15-00200.1.

Tuesca, D., Puricelli, E. and Papa, J. C. 2001. A long-term study of weed flora shifts in different tillage systems. *Weed Res.* 41(4), 369–82. doi:10.1046/j.1365-3180.2001.00245.x.

Vencill, W. K., Nichols, R. L., Webster, T. M., Sotores, J. K., Mallory-Smith, C., Burgos, N. R., Johnson, W. G. and McClelland, M. R. 2012. Herbicide resistance: Toward an understanding of resistance development and the impact of herbicide-resistant crops. *Weed Sci.* 60(SP1), 2–30. doi:10.1614/WS-D-11-00206.1.

Walsh, M. J. and Powles, S. B. 2014. High seed retention at maturity of annual weeds infesting crop fields highlights the potential for harvest weed seed control. *Weed Technol.* 28(3), 486–93. doi:10.1614/WT-D-13-00183.1.

Walsh, M., Newman, P. and Powles, S. 2013. Targeting weed seeds in-crop: A new weed control paradigm for global agriculture. *Weed Technol.* 27(3), 431–6. doi:10.1614/WT-D-12-00181.1.

Walsh, M., Ouzman, J., Newman, P., Powles, S. and Llewellyn, R. 2017. High levels of adoption indicate that harvest weed seed control is now an established weed control practice in Australian cropping. *Weed Technol.* 31(3), 341–7. doi:10.1017/wet.2017.9.

Walsh, M. J., Broster, J. C., Aves, C. and Powles, S. B. 2018. Influence of crop competition and harvest weed seed control on rigid ryegrass (*Lolium rigidum*) seed retention height in wheat crop canopies. *Weed Sci.* 66(5), 627–33. doi:10.1017/wsc.2018.28.

Watson, P. R., Derksen, D. A. and Van Acker, R. C. 2006. The ability of 29 barley cultivars to compete and withstand competition. *Weed Sci.* 54(4), 783–92. doi:10.1614/WS-05-020R3.1.

Werth, J., Walker, S., Boucher, L. and Robinson, G. 2010. Applying the double knock technique to control *Conyza bonariensis*. *Weed Biol. Manag.* 10(1), 1–8. doi:10.1111/j.1445-6664.2010.00360.x.

Werth, J., Boucher, L., Thornby, D., Walker, S. and Charles, G. 2013. Changes in weed species since the introduction of glyphosate-resistant cotton. *Crop Pasture Sci.* 64(8), 791–8. doi:10.1071/CP13167.

Widderick, M. and McLean, A. 2018. Optimal intervals differ for double knock application of paraquat after glyphosate or haloxyfop for improved control of *Echinochloa colona*, *Chloris virgata* and *Chloris truncata*. *Crop Prot.* 113, 1–5. doi:10.1016/j.cropro.2018.07.003.

Widderick, M. J., Walker, S. R., Sindel, B. M. and Bell, K. L. 2010. Germination, emergence, and persistence of *Sonchus oleraceus*, a major crop weed in subtropical Australia. *Weed Biol. Manag.* 10(2), 102–12. doi:10.1111/j.1445-6664.2010.00370.x.

Wille, M. J., Thill, D. C. and Price, W. J. 1998. Wild oat (*Avena fatua*) seed production in spring barley (*Hordeum vulgare*) is affected by the interaction of wild oat density and herbicide rate. *Weed Sci*. 46, 336–43.

Wilson, B. J. and Peters, N. C. B. 1982. Some studies of competition between *Avena fatua* L. and spring barley I. The influence of *A. fatua* on yield of barley. *Weed Res*. 22(3), 143–8. doi:10.1111/j.1365-3180.1982.tb00157.x.

Wu, H., Walker, S., Rollin, M. J., Tan, D. K. Y., Robinson, G. and Werth, J. 2007. Germination, persistence, and emergence of flaxleaf fleabane (*Conyza bonariensis*. [L.] Cronquist). *Weed Biol. Manag*. 7(3), 192–9. doi:10.1111/j.1445-6664.2007.00256.x.

Part 4

Quality

Chapter 13

Developing barley crops for improved malt quality

Glen Fox, University of California-Davis, USA and The University of Queensland, Australia; and Reg Lance, Queensland Department of Agriculture and Fisheries, Australia

1 Introduction
2 Malting quality
3 Case study: modern varieties for twenty-first century brewing
4 A brief history of barley improvement in Australia
5 Requirements for successful programmes in malting quality improvement
6 Conclusion
7 Future trends
8 Where to look for further information
9 References

1 Introduction

Barley has been used for making alcoholic beverages for millennia and will continue to do so for millennia to come. However, the future is not without challenges for malting barley. Current challenges include (i) how and when genetic modified barley will be available and (ii) malt quality consistency in light of climate change. Plant breeding in the twenty-first century allows for more targeted trait selection using an array of molecular techniques and 'omics technologies. In the case of malting barley, molecular technology has developed from using RFLPs in the 1980s through to current use of SNP analysis to identify allelic variation in genes such as beta-amylase linked to barley quality. However, there is always a play-off between breeding targets such as increasing yield and improving disease resistance compared to targets such as improving malting quality. Yield is the first priority for barley farmers and thus for breeders, and this priority for production (and increased financial return for farmers) is understood by maltsters and brewers.

Not all barley varieties are suitable for use in brewing, and rigorous malting and brewing trials are undertaken over several years for a variety to

http://dx.doi.org/10.19103/AS.2019.0060.20

be accredited. The malting process is critical stage prior to brewing. There are numerous traits considered important for malted barley in brewing. However, there are only a handful of traits used for testing malt quality. These relate to the hydrolysis of the three major substrates, namely beta-glucan, endosperm storage proteins and starch. Starch-degrading enzymes (diastases enzymes), free amino nitrogen and soluble nitrogen, and wort beta-glucan are the main malt traits used by brewers to buy malt. Due to limitations in methodologies, costs or lack of relevance of current methods to modern brewing, other traits are not used to assess malt quality.

1.1 Craft brewing

The momentous increase in craft brewing industry in recent years may challenge the current standard malt specifications. The majority of craft brewers use all-malt and hence the malt quality specifications would be different to large breweries which may use solid adjuncts. For adjunct brewing, a higher level of diastase is required to hydrolyse the starchy endosperm of the adjunct, usually rice or corn. In contrast, all-malt brewing uses the endogenous malt enzymes to hydrolyse the malt starch.

The mashing conditions in all-malt craft brewing are more important than large-scale brewing for a number of reasons. Malt gelatinisation temperature is critical for the release of fermentable sugars as well as solubilisation of flavour and colour compounds. Modern brewing, with well-modified malt, mash at temperatures between 63°C and 68°C. At these temperatures, starch should quickly gelatinise and thus be available for diastatic enzymes to break down the starch, releasing primarily maltose but also maltotriose and glucose.

Let us briefly discuss the diastatic enzymes before going back to the significance of starch. There has been an enormous amount of research on starch-degrading enzymes, including possible variation in individual enzyme activities, thermostability of enzymes, levels of malt modification, correlating (and modelling) diastatic power (DP) in relation to fermentability and, importantly, breeding to achieve high levels of these enzymes (and collective DP). Generally these studies have been important to fully understand DP (and partial individual enzymes) as current varieties are required to have high levels of enzymes for solid adjunct brewing. But this is not the case for all-malt (and liquid adjunct) brewing. Most malt specifications will include DP and usually alpha-amylase, and these results generally inform the brewer about how the mash should perform, but there is no information on fermentable sugars which are paramount for fermentation.

As mentioned above, there is a range in gelatinisation from 58°C to 68°C in barley malt and these ranges are due, in part, to variation in the structure of starch. This variation can be in the degree of branching of amylopectin, and to a much lesser extent amylose, the average chains lengths of the C, B and A chains on amylopectin and, finally, the average chain length of amylose. Recent studies have highlighted the variation in structure as well as the impact on fermentable sugar profiles (Gous et al., 2015; Fox, 2016; Fox et al., 2019; Yu et al., 2019). Yu et al. (2018) showed the variation in amylose structure and fermentable sugars and also differences in the residual dextrins in wort which would contribute to flavour (Yu et al., 2019). However, no information on gelatinisation is provided in a malt specification though most list saccharification time (5–10 min). Longer saccharification times would usually suggest loss of enzyme activity due to the thermolability of enzymes but Evans and Fox (2017) showed the starch-degrading enzymes can survive longer in a hot mash.

For the craft industry, the malt also is considered as contributing to a positive flavour profile. However, the flavour can be influenced by both malting and mashing conditions. But is there a varietal effect? Malt is usually purchased on quality but also on the inclusion of one or more malting accredited varieties. It is important to note that lipoxygenase has been shown to induce staling in beer, and barley varieties have been shown to influence this off-flavour. In this context, there has been a recent re-emergence of what is termed 'heirloom varieties', such as Maris Otter and Golden Promise with improved flavour profiles. But what is malt flavour? Is it malt aroma and/or malt taste? Or are the flavours volatile or non-volatile?

Recent studies have shown sensory traits can be derived from the barley variety and growing environment as well as by processing (Bettenhausen et al., 2018; Herb et al., 2017a,b). The growing evidence of barley variety influencing malt characteristics and beer flavour raises new challenges for breeding programmes and the malting industry. The thousands of individual compounds of malt that survive into the final beer have now become potential targets for breeders, opening the door for craft brewers to request specific flavour specification in the malt they buy.

An additional challenge for the malting industry, which would flow back into breeding programmes, is the need to reduce costs (water and electricity) and reduce carbon footprint. But the maltster will still be required to produce the same malt quality in shorter malting conditions, usually with a reduced germination period.

The following sections review standard malting traits. We then discuss traits which are important for brewing but are not measured in malt.

2 Malting quality

2.1 Grain size

The first physical measure of malt quality is barley grain size which is a physical measure. Most countries have standards based on small grain sizes (screenings) and large grain size fractions (retention). A lower amount of screenings would indicate an on-average large grain size. Grain size distributions have been related to variation in malting and brewing performance (Yu et al., 2017) with both genetic and environment influencing grain size (Fox et al., 2006). The resultant variation in grain size influences components within the kernel such as starch, non-starch polysaccharides (beta-glucan and arabinoxylan) and protein. Starch is the most abundant component within the endosperm. Both genetic and environmental factors impact not just the amount of starch (and its two components, namely amylose and amylopectin) but also its structure (Yu et al., 2017; Gous et al., 2013, 2015, 2017; Gous and Fox, 2017). Hot and/or dry conditions during the critical stages of grain fill produce smaller kernels which in turns impacts starch synthesis, resulting in decreased starch content and ratios of amylose to amylopectin (Savin et al., 1997).

2.2 Protein

The general understanding is that larger grain size would have more starch and lower protein content. Protein content is the first chemical constituent measured and is related to both malting and brewing performance. Most barley-producing countries have similar ranges in protein for the malting industry with a minimum of 9.0% (dry basis) and up to 13.5% (dry basis), depending upon the commercial malting facility and brewing requirements. But the range in barley protein can vary between 8% and up to 20% depending upon environmental conditions (Corke, 1995).

For those countries with bulk grain handling systems, barley varieties will be segregated and there may be additional segregations for different levels of protein, for example 9.0–10.5% and 10.5–12.5%. However, grain handling systems cannot account for the fact that two or more samples of the same variety at the same protein content will have a slightly different protein compositional profiles of in-situ proteins, as well as the potential synthesis of proteins (enzymes) during malting. Furthermore, different malting conditions and levels of protein modification will result in different compositional profiles.

Total protein content is measured in barley and malt, but the difference is somewhat minor after malting, despite the internal changes from barley to malt. Numerous studies have explored changes in specific groups of proteins or expression of individual protein such as changes in hordeins during malting, or synthesis of major substrate degrading enzymes (beta-glucanase, xylanase, proteases and peptidases, alpha-amylase and limit dextrinase). These modifications have been linked to a range of beer quality traits. These traits include hordein peptides being foam negative protein, the final composition of amino acids for yeast nutrition during fermentation, poorly hydrolysed beta-glucan impacting on beer clarity, or starch hydrolysis enzymes to produce fermentable sugars. However, numerous small proteins survive the whole malting and brewing process intact or with minimal modification. These include lipid transfer protein (LTP), protein Z or the CM proteins (putative inhibitors) (Jones and Fontanini, 2003; Jones, 2005a).

The importance of protein content is universally accepted. However, there are better ways to profile protein composition and identify key in-situ barley proteins before malting, such as those proteins which have a positive or negative impact on brewing. The challenge is to have rapid and accurate measurement technologies at receival. The higher degree of segregation of varieties based on individual proteins also creates significant logistical issues.

2.3 Germination

The malting process is the conversion of grain components to simpler starch, amino acids and other water-soluble constituents which can then be utilised in brewing. This process is a controlled germination process where the enzyme system naturally convert the grain components to supply the growing embryo (Chu et al., 2014; Jones, 2005b; Briggs, 1992). Maltsters requires a minimum of 98% germination. Un-germinated seeds (whether dead or dormant) have a negative impact on brewing performance (Briggs, 2002).

Dormancy is a barley trait that varies in its importance depending upon the growing region. Excessive dormancy in barley can result in less than ideal germination rates and can delay the production of new season grain. However a lack of any dormancy risks pre-harvest sprouting in those regions with rainfall around harvest time.

Storage conditions also impact on germination efficiency. Clean, sound grain stored correctly will remain viable for around 12 months (Woonton et al., 2005a,b; Bueckert et al., 2007).

2.4 After malting

Since they need to produce malt to the expected quality, maltsters are highly skilled. As malting progresses, they need to know how to influence important traits such as colour, extract and enzyme development. Reading through current brewing industry methods' manuals (Institute of Brewing & Distilling (IoB), European Brewery Convention (EBC) and American Society of Brewing Chemists (ASBC)) highlights hundreds of methods to measure different aspects of malt and beer quality. However, the first specification on any malt sheet is the extract.

Malt extract, or hot water extract, is prepared by the solubilisation of malt compounds during a stirred and heat mash. The specification for the extract is usually expressed as a percentage (%). Malt is typically sold as extract prepared using the European Brewery Convention method which provides optimal temperatures and times for beta-glucanase, proteases, amylases and limit dextrinase to work. Mash-in temperature is at 45°C but finishes at 70°C over 80 minutes with protein and amylase rests at 55°C and 65°C, respectively. The American Society of Brewing Chemists method is very similar to the EBC method. However, whilst they specify the maximum potential extract, these methods no longer reflect contemporary industrial brewing processes. Conversely, the Institute of Brewing method follows a temperature profile (using more well modified malts) where malt is mashed at higher temperatures (around 65°C), reflecting modern brewing practice. Extract results are usually reported as percentage but can be converted to degree Plato (°Plato).

One of the major differences between the styles of mashing methods is the temperature that influences enzyme hydrolysis of substrates. The high-temperature mash severely reduces beta-glucanase and protease activity and can reduce beta-amylase and limit dextrinase activity, although a recent study suggest the starch-degrading enzyme may survive mashing longer than expected (Evans and Fox, 2017). Other mash condition also impact the final result including grist size and grist to liquor ratio (Fox, 2016).

Both the EBC and IoB methods are used in breeding programmes as the high-temperature mash allows the differentiation of potentially good breeding lines compared to poor breeding lines (Fox and Henry, 1993). The EBC method still provides an industry standard. However, to fully understand the potential performance of breeding lines, the IoB is preferred with a high grist to liquor ratio (1:3).

The amount of extract is important but more important is what is in the extract. The extract or wort contains many types of carbohydrates in the form of sugars, proteins in the form of peptides and amino acids, lipids, polyphenols as well as many minerals. Of all these different types of compounds, only free amino acids or free amino nitrogen (FAN) are measured routinely and provided on a malt specification. Amino acids have been correlated to wort colour (Coghe et al., 2005; Hellwig et al., 2016) and flavour profiles (Procopio et al., 2013; Bettenhausen et al., 2018). However, there is no easy test for total amino acids composition or for any individual amino acid in wort except for the use of infrared techniques (Gorinstein et al., 1999).

Other than amino acids as a form of nitrogen, there are hundreds of individual proteins present in wort which survive boiling and fermentation (Jin et al., 2013; Schulz et al., 2018). While specific small proteins such as LTP and protein Z are considered important for foam stability (Evans and Hejgaard, 1999; Evans et al., 1999; Nierop et al., 2004; Kauffman et al., 1994), chloroform-methanol (CM) proteins are reported to be foam negative (Kauffman et al., 1994), as well as being inhibitors (Fontanini and Jones, 2002; Jones and Fontanini, 2003; Jones, 2005a). Peptides from hordeins are foam negative but there is no standard measurement for total hordein or hordein composition in barley, nor for hordein peptides in malt, wort or beer. Proteomics offers the opportunity to identify the numerous foam positive and form negative proteins and peptides as well as numerous other proteins (Jin et al., 2013; Iimure and Sato, 2013; Schulz et al., 2018). However, there are no portable or simple systems that would allow such methods to be used in a routine malt laboratory.

The composition of the wort also includes a range of carbohydrates in the form of simple sugars, including maltose and glucose, and longer chain sugars such as maltotriose and maltotetrose and branched dextrins (Gous and Fox, 2017; De Rouck et al., 2013; Huuskonen et al., 2010; Lei et al., 2012). These simple sugars contribute to fermentation, specifically glucose and maltose, whilst other sugars contribute to flavour, especially dextrins (Duke and Henson, 2011). Additional sugars also come from adjuncts. Liquid adjuncts such as cane sugar or corn syrup provide sucrose as a ready form of fermentable sugar. Solid adjuncts including corn or rice undergo pre-gelatinisation and are added as a liquid to the mash, but still undergo starch hydrolysis and thus contribute fermentable sugars (Fox et al., 2019; Montanari et al., 2005; Zhu et al., 2017; He et al., 2018).

Fermentable sugars are produced from the hydrolysis of starch polymers, amylopectin and amylose (Koljonen et al., 1995; Yu et al., 2019) but also minor polymers of beta-glucan (Kanauchi and Bamforth, 2008) and arabinoxylan (Han, 2000) by specific enzymes activity of each of these polymers. Starch content and sometimes beta-glucan is measured in barley but less often in malt. Wort beta-glucan is an important trait in malt specifications. However, the structure of beta-glucan, as well as arabinoxylan and starch polymers is more important than the amount of all these polymers. It has been shown that, for each of these polymers, there is variation in structure which can influence malting and brewing qualities. The variation in amylose and amylopectin structure can influence the final amount of fermentable and non-fermentable sugars (Fox et al., 2019; Yu et al., 2019), whereas variation in beta-glucan structure, specifically molecular weight, can have a negative impact on filtration (Stewart et al., 2000). Non-fermentable sugars contribute to mouthfeel in beer. Recently a new phenomenon called 'hop creep' has been found to contribute to wort non-fermentable dextrins, where late hopping in fermentation may restart fermentation and increase alcohol formation (Kirkpatrick and Shellhammer, 2018).

The variation in starch structure in barley and malt (variation in average chain length of amylose) can result in different gelatinisation temperatures, sometimes higher than the mash temperature (Tester, 1997; Czuchajowska et al., 1998). However, no structural parameters are measured for starch, beta-glucan or arabinoxylan. The only trait measured is scarification time to indicate malt gelatinisation timing, usually between 5 to 10 minutes, which is associated with reduced starch enzyme activity. However, a recent study has suggested starch-degrading enzymes can survive longer than previously expected in the mash (Evans and Fox, 2017).

Both wort beta-glucan and arabinoxylan have been linked to wort viscosity, but only the former is usually tested along with wort viscosity. As mentioned previously, both content and structure are important in influencing viscosity. For breweries using a high-temperature mash, that is mash-in around 65°C, there is little to no time for beta-glucanase or xylanases to act on their respective substrates. Current malting practices with modern malting barleys typically modify the malt within four days of germination. However, wort beta-glucan is measured using wort produced using the EBC/ASBC ramping style methods which allow beta-glucanase to be active at the 45°C mash-in temperature. The difference between the method recommended in the EBC malt specification and commercial mashing conditions may misinform brewers as to the filtration efficiency of the wort. One major issue is that neither of these enzymes (beta-glucanase or xylanase) is measured at any stage. The focus is on the basic enzyme equation: substrate + enzyme = product i.e. mainly on the amount of product (wort viscosity), not the composition of the product nor the structure.

2.5 Starch-degrading enzymes

While little attention is paid to beta-glucanase and xylanase, hundreds of articles have been published on each of the individual starch-degrading enzymes, alpha-amylase, beta-amylase, limit dextrinase and alpha-glucosidase, as well as on their collective measurement. Typical malt quality tests are for DP and alpha-amylase. Beta-amylase is measured mostly by researchers to understand its thermostability as it is the main enzyme producing the most important fermentable sugar, maltose. Survival of starch-degrading enzymes during high-temperature mashing is important for the production of fermentable sugars. Current malt varieties have high levels of alpha-amylase and beta-amylase as these enzymes are required to hydrolysis the malt starch substrates as well as starch substrates from solid adjuncts. Breeding programmes select heavily to increase these enzymes so that new varieties meet the requirements for solid adjunct brewing. However, as noted earlier, there are missing components from the basic biochemical equation:

$$\text{Substrate} + \text{Enzymes} = \text{Products i.e. } [S] + [E] = [P].$$

These missing components have a significant effect on quality.

In general measurement includes DP and sometimes measurement of individual enzymes. However, neither the substrate (starch) nor the products (fermentable sugars) are measured. The increase in enzyme activity has been successfully measured via surrogate measurements of the apparent attenuation limit (AAL). The question is how much more efficient would the selection process be if even one of the substrate or fermentable sugars was measured. Although challenging, measurement technology is available (based on extract, DP and AAL) and can help to explain when fermentations do not go according to plan. A number of attempts have been made to predict fermentability using standard malt traits including the level of modification (KI) and individual enzymes (Edney et al., 2007; Evans et al., 2005, 2010; Koljonen et al., 1995), as well as fermentable sugars (Edney et al., 2007) . Evans et al. (2005) found no value from measuring starch gelatinisation temperature but suggested a gelatinisation temperature greater than 64°C might impact on fermentability. As mentioned above, if there were additional traits to be tested to better understand the interaction between starch-degrading enzymes and their impact on extract and AAL, then malt gelatinisation temperature would be of value.

2.6 Malt colour (wort colour)

Malt colour is not routinely measured whereas wort colour is an important specification and contributes to both beer colour and flavour. For a standard pale malt (base malt), Maillard reactions occur during kilning but the final drying temperatures and drying time are held at below 85°C for less than 3 to 4 hours. For more colour (which adds to flavour and aroma) higher drying temperatures are used. Most colour malts are added into the mash at around 10% with a significant contribution to colour and flavour (Coghe et al., 2004). Recently, variation in colour and flavour have been associated with genetic or environmental characteristics, thus making it possible to select for malt (wort) colour in breeding programmes (Herb et al., 2017a,b). Metabolomics analysis also suggests sugars and amino acids involved in producing Maillard products and other flavour compounds may be influenced by variety as well as by malting process (Bettenhausen et al., 2018). These recent findings suggest specific compounds could be selected for within breeding programmes, if there is access to metabolomics and proteomics analytical techniques.

2.7 Traits linked to malt quality but not routinely measured

Lipoxygenase (LOX) catalyses the hydroperoxidation of polyunsaturated fatty acids with a 1, 4-cis-cis-pentadien structure. Two isoforms of LOX (LOX-1 and LOX-2) have been identified during barley germination, with LOX-1 and LOX-2 catalysing the formation of 9-hydroxyperoxide (9-HPOD) and 13-hydroxyperoxide (13-HPOD), respectively (Van Mechelen et al., 1999).

LOX-1 is more important in brewing since 9-HPOD is a precursor of beer defects such as trans-2-nonenal (T2N) and trihydroxyoctadecenoic acid (THOD) (Kobayashi et al., 1993; Kuroda et al., 2002, 2003). T2N is known as a component of 'cardboard' flavour in stale beer and THOD is reported to have negative effect on mouthfeel of beer, and foam stability. Most of the interest in LOX-free barley has been in Japan with specific varieties breed to be LOX-free for markets such as China (Hoki et al., 2018a,b; Oozeki et al., 2017). Countries that export significant tonnages of malting barley to China, including Australia and Canada, grow these types of barleys.

2.8 Grain hardness (endosperm texture)

A trait that has been associated with barley and malt quality numerous times is grain hardness (related to endosperm texture). Harder barley grains generally have poorer malt quality (Psota et al., 2007). The internal grain texture is a physical measure related to higher protein content (Fox et al., 2007b), lower starch content (Fox et al., 2011), starch composition (waxy = high amylopectin content) (Czuchajowska et al., 1998) and beta-glucan content (Walker et al., 2011; Gamlath et al., 2008). All these produce harder kernels and are thus traits also associated with poor malt quality (Fox et al., 2009).

Both genetic and environmental factors influence barley grain texture (Fox et al., 2007b; Osborne et al., 2007; Walker et al., 2011). The genetics of barley hardness can be partially explained by hordoindolines (Beecher et al., 2002; Lee et al., 2006; Fox et al., 2007a). Other QTL linked to chemical traits such as proteins, starch and beta-glucan have also been identified and are also influenced by the growing environment.

Grain hardness is not routinely measured in barley breeding but friability is measured as part of the malt evaluation of breeding lines. Friability is a measure of endosperm modification during malting and positive correlated to the Kolbach Index (Fox et al., 2007a; Psota et al., 2007). Harder grain is generally poorly modified, with more storage protein and starch un-degraded, and thus a less friable malt (Psota et al., 2007; Vejrazka et al., 2008). Selecting for high friability would be a surrogate for grain hardness (or softer grain). Like a number of malt quality traits such as extract, there are a mix of characteristics that control friability but high friability is a good indication of malt quality.

Proteases (and peptidase) are perhaps one of the most important but least studied traits related to malt quality. This is partially due to difficulty in measuring the different classes of endoproteases (Osman et al., 2002; Jones, 2005b) . While a diverse set of protein substrates are targets for the different proteases, the most important in terms of malt quality are those responsible for degrading storage proteins. Currently, cysteine, metallo and aspartic proteases have been reported to degrade hordein (Davy et al., 1998, 2000; Terp et al., 2000; Osman et al., 2002; Jones, 2005b; Guerin et al., 1992). The physical degradation of hordein matrix allows the release of bound beta-amylase and also allows alpha-amylase, beta-amylase and limit dextrinase access to the starch granules. A high level of modification (the ratio of water-soluble proteins compared to the total amount of protein 40–48) indicates protease activity in degrading hordeins and other in-situ barley proteins.

Proteases are known to play a role during malting and mashing. Hordein peptides have been identified in wort, the final beer and in beer foam (Sheehan and Skerritt, 1997; Osman et al., 2003; Iimure and Sato, 2013; Schulz et al., 2018). The reduction of hordein during germination occurs one day after steeping, which coincides with the appearance of proteases.

The lack of routine testing for specific changes in hordein leaves the industry vulnerable to making untested assumptions about the protein composition of the wort. However, by testing for soluble nitrogen and free amino nitrogen or FAN (the total amount of amino acids in wort, including those amino acids cleaved from hordein peptides by exo-peptides), the industry has some control of the brewing and fermentation processes.

An additional dimension to the relevance and importance of proteases is their role in the release of beta-amylase during germination (malting). Beta-amylase is synthesised in the early stages of grain fill, as early as 5 to 10 days post anthesis, which coincides with initial synthesis of storage proteins. Around 60% of total beta-amylase is bound in the hordein matrix, with around 40% beta-amylase in a free form in the aleurone layer. Once germination commences, the race is on for the release of beta-amylase and cysteine protease (Guerin et al., 1992; Grime and Briggs, 1995, 1996). At the same time, a serine protease has been reported to degrade beta-amylase during germination with the final size of the beta-amylase protein being around 5000 daltons smaller than the bound form (Schmitt and Marinac, 2008).

There are very few reports on genetic and environmental effects on protease expression or activity, partially due to changes in measuring specific proteases. However, it has been reported that there were differences in protease activity in some Canadian barleys, although no information was provided on the growing conditions for these barleys (Rizvi et al., 2011). Genes have also been identified for proteases (Schmitt et al., 2013) which provides breeders with the opportunity to develop varieties with high expression of proteases. However, some brewers prefer lower levels and thus less hydrolysed hordein, whilst other brewers require modification over a shorter germination period, from 4 days to 3.5 days or shorter. There is a need for more protease research using new technologies such as proteomics.

3 Case study: modern varieties for twenty-first century brewing

Malting barley improvement in Australia has responded to the changing demands of both overseas and domestic markets. This section of the chapter will outline the principal changes from the release of the Clipper variety in the 1960s until the 2010s with a range of new varieties being produced for both domestic and export markets.

The domestic market within Australia is largely characterised by a brewing process which utilises liquid sugar as an adjunct, although some domestic brewing processes require 100% malt. More often breweries have adopted a high-temperature infusion mashing process which requires that there should be sufficient starch-degrading enzymes (diastase) to degrade the malt starch before being inactivated in the mash. The malting quality required for the '100% malt' or the 'malt plus adjunct' brewing methods requires well-modified malts with intermediate protein modification (Kolbach Index - KI), moderate extracts (80%) and AAL (90%) and moderate DP (220 WK plus). In comparison the export markets are largely characterised by a brewing process which utilises starch adjuncts from varying sources such as rice, wheat or maize. Requirements are for higher percentage malt extract and higher diastase.

Figure 1 shows series of micro-malts of about 100 samples of Schooner barley (the next major malting barley variety after Clipper) supplied from the Australian Barley Board in the mid-1980s. Malt extract was determined according to the EBC specification (fine-grind) and by DP using Windisch-Kolbach units. The relationships of percentage extract versus percentage protein, and that of DP verses percentage protein, are shown in Fig. 1. Targets for % Malt Extract (81%) and DP (300 WK) are a reflection of export market requirements, principally Japanese Brewers at the time. It should be noted that a mid-range of protein (for example between 10.0 and 11.0% (dry basis))

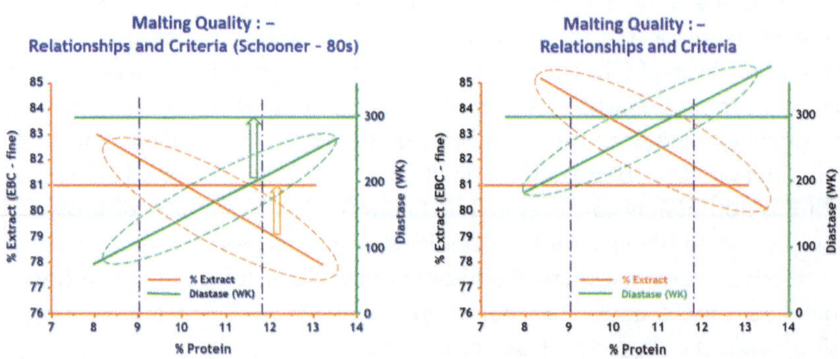

Figure 1 Malt quality relationships using Schooner, compared to modern varieties.

was adequate for the domestic market but not for export markets. In order to satisfy the requirements of expert markets (initially the Japanese malting barley market but later the Chinese, Vietnamese and other South-east Asian markets), improvements in both extract and diastase were required.

An example of the malting quality assessment from regional stage 4 variety testing trials is used to demonstrate the improvement in the quality traits. The results are taken from 2009 and represent varieties prior to the privatisation of the Western Australian Department of Agriculture's breeding programme. This programme was subsumed into InterGrain Pty. Ltd. which is now a major Australian Wheat and Barley breeding company. It should be stressed that these trials were principally designed to test or rank the varieties for relative comparisons and therefore do not represent their overall performance. The results are used to demonstrate basic principles in malting quality improvement and in the improvement in related quality traits over time (Table 1).

The fundamental quality relationships are between both DP and % Malt Extract (EBC) and Malt Protein (Fig. 1). The DP shows a strong and positive increase with protein with a significant number of varieties above the target of 300 WK units. Extract is classically negatively related to % protein. The example here indicate two groupings with varieties such as Bass, Flinders, Hindmarsh, Flagship and Baudin demonstrating significantly higher malt extract that are better suited to the export markets. The relationship between % Malt Extract and DP illustrates the joint improvements in both % Malt Extract and DP in a number of varieties. The variety Buloke also shows higher % Malt Extract with a DP of about 300 WK making it suitable for both export and domestic malting markets.

Even though the main foci were improvements in % Malt Extract and DP, other related traits have contributed to overall improvements in quality. There is a very strong positive relationship between Kolbach Index and FAN (Fig. 2). There is also a very strong decline in wort beta-glucan which is associated with an increase in malt beta-glucanase (Fig. 2). Such strong relationships suggest that there have been multiple improvements in both protein degrading enzymes and beta-glucanases concomitant with improvements in starch and beta-glucan degrading enzymes during the overall improvements in malting quality.

Up to and including the 1960s, varieties such as Clipper, Proctor and Weeah and Dampier were major domestic malting varieties (Fig. 3). In the 1970s and early 1980s varieties such as Schooner, Stirling, Shannon, Grimmett and Arapiles all had profiles better suited to domestic malting requirements. Franklin barley was bred in Tasmania by Wayne Vertigan and was a cross between Shannon (Proctor back-cross) and Triumph (a European malting variety). Franklin had improvements in both % Malt Extract and diastase, principally inherited from Triumph. It was readily adopted by the malting industry but was inherently late

Table 1 Summary of malt data for Australian commercial varieties

Variety	MPro (%)	MAA (U/g)	MBG (U/g)	KI (%)	DP (WK)	EFG (%)	SNit (%)	FAN (mg/L)	WVis (cP)	AAL (%)	WSN (mg/L)	WBG (mg/L)
Baudin	10.3	284.6	514	42.5	407	80.1	0.70	181	1.44	85.7	784	74
Buloke	9.3	230.0	497	38.5	299	80.6	0.57	138	1.47	83.5	641	118
Commander	9.5	224.4	537	39.3	258	78.8	0.60	152	1.42	82.7	671	80
Flagship	10.0	256.2	656	41.5	316	79.7	0.66	166	1.44	84.7	741	48
Gairdner	9.6	167.0	446	35.3	274	78.8	0.54	124	1.47	81.3	607	198
Hamelin	10.2	296.5	592	46.8	399	79.0	0.77	202	1.43	85.3	863	87
Hindmarsh	9.8	223.2	494	41.9	326	80.9	0.66	171	1.44	85.5	739	108
Vlamingh	10.3	226.8	513	38.9	298	79.3	0.64	152	1.44	83.1	713	88
Bass	10.2	307.7	639	45.0	392	80.3	0.73	192	1.40	85.7	814	70
Flinders	9.8	268.3	556	42.3	299	81.8	0.66	165	1.40	84.3	737	48

Data with permission from Intergrain.

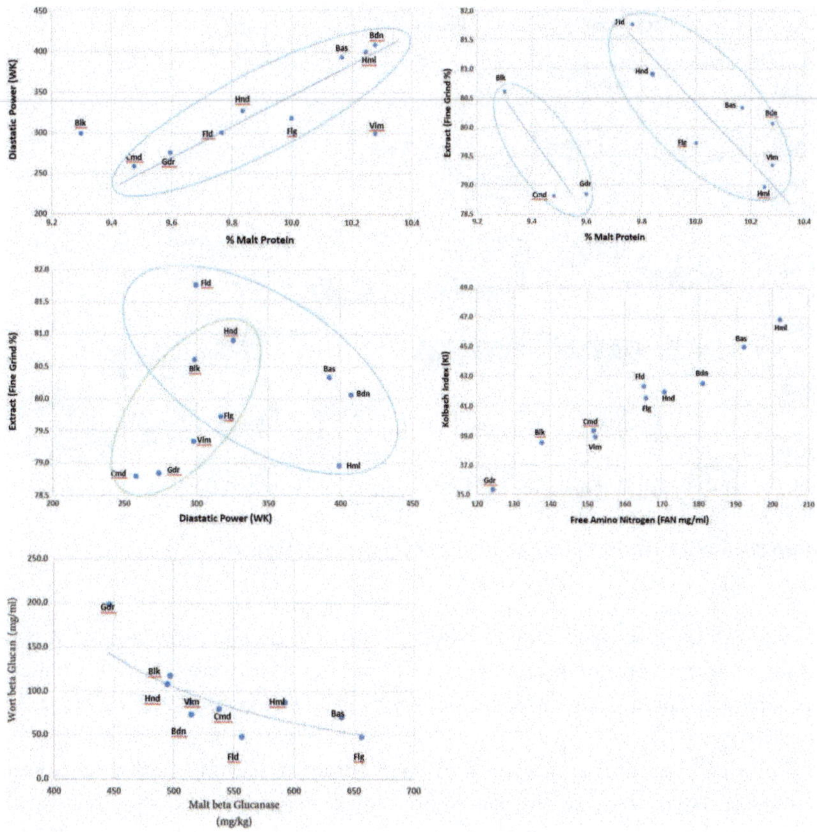

Figure 2 Performance of Australian varieties developed since 1990s for different malt traits.

maturing, often producing small grains. It did, however, represent a significant breakthrough in malting quality by utilising high quality overseas malting varieties as parents.

The next phase in malting quality improvement saw the use of varieties from Europe, North America and Japan. Varieties, released in the mid-1990s, such as Dhow (with quality traits derived from the Haruna Nijo variety, Japan), Sloop (and its derivatives, Sloop SA and Sloop Vic, with quality traits derived from Norbert, Canada) and Gairdner and Tallon (with quality traits from Franklin and Triumph, Europe).

Further improvements in quality and agronomic performance resulted from crosses between Australian barleys and those varieties with recent improvements in quality, for example Baudin (Stirling × Franklin) and Hamelin (Stirling × Harrington). Baudin demonstrated superior malting quality suitable for the Chinese and Japanese markets, good agronomic performance but

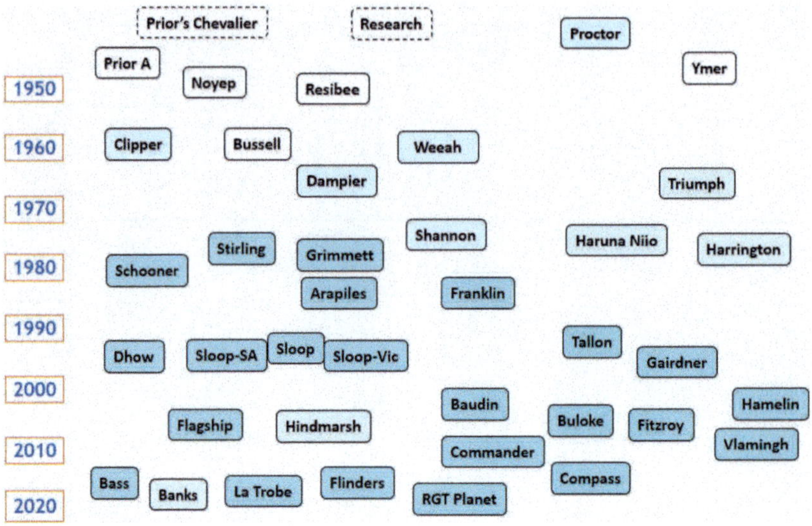

Figure 3 Australian malting barley varieties over the last 70 years.

was weak in disease resistance (powdery mildew and leaf rust). Hamelin was a significant improvement in quality over Stirling but its agronomic performance was similar to Stirling and was therefore never widely grown and adopted by the industry.

The most recent malting varieties have come from either Barley Breeding Australia (BBA) or the newer private companies utilising the better quality and agronomic performance of their predecessors. Buloke (Franklin × 2*VB9104), from the Victorian programme, again received its superior malting quality from Franklin and had been a mainstream malting barley. Hindmarsh (VB9409 × Dash) also from Victoria introduced a new semi-dwarf gene 'ari-e' to Australian barleys from the European variety Dash. The quality of Hindmarsh was good but it had a weakness with high malt beta-glucan and therefore wort viscosity. La Trobe is a reselection from Hindmarsh and does not have this problem and produces a high quality malt that can be used in both domestic and export markets.

The BBA-South programme produced several excellent malting quality varieties, Flagship ((Chieftain ×Barque) × (Manley × VB9104)), and Commander ((WI-2976 × Sloop) × Galaxy) sourcing malting quality from European and Canadian varieties.

The most recent malting varieties, Bass (B28719 × Alexis), La Trobe (Hindmarsh seln.), Compass (Commander × (County x Commander)) and Flinders (Baudin × Cooper) have made further improvements in yield, grain and malting quality, agronomic performance and disease resistances. La Trobe is a reselection from Hindmarsh and does not have the beta-glucan problem of Hindmarsh and produces a high quality malt that can be used in both domestic and export markets. RGT Planet (Tamtam × Concerto) is a newly accredited malting variety imported from the United Kingdom.

4 A brief history of barley improvement in Australia

The improvements in malting quality barley in Australia need to be put in the context of, firstly, the development of public breeding programmes, then the introduction of the Barley Breeding Australia (BBA) programme to transition from public to private barley breeding. The current situation is now that most barley improvement is undertaken by private breeding companies. The breeding activity is underpinned by substantial and comprehensive research and development and pre-breeding programmes co-funded by the Grains Research and Development Corporation (GRDC) and State and Federal Government departments, universities and barley industry partners.

Prior to the transition to private breeding companies, barley improvement was undertaken by State Departments of Agriculture or Primary Industries (vis-à-vis: Queensland, New South Wales, Victoria, Tasmania and Western Australia) and the university sectors. Historically, the South Australian Barley Improvement programme was located at the Waite Agricultural Research Institute, (now the University of Adelaide). There was a rationalisation in the state government programmes and the formation of a three node BBA programmes (BBA-North, BBA-South and BBA-West). BBA-North originated from the Queensland, BBA-South was developed from the University of Adelaide programme and BBA-West originated from the Western Australian Department of Agriculture. The New South Wales and Victorian programmes were abandoned and their germplasm dispersed. The Tasmanian programme continued but not formally as a part of BBA.

The BBA-North programme is now dispersed and the germplasm licenced to a private company. BBA-South has been disbanded with the germplasm dispersed and the breeders moving to private companies. BBA-West was subsumed in to InterGrain and operated nationally. There are now several key private companies, InterGrain, Australian Grain Technologies and Edstar Genetics, as well as smaller companies active in barley breeding in Australia such as Syngenta, SECOBRA Recherches, Malteurop and others.

Malting quality evaluation under the public system was serviced by State Government or university laboratories in Queensland, New South Wales, Victoria, Western Australia and South Australia. These still provide analyses under 'fee for service' arrangements with breeding companies or contract research.

4.1 Barley Australia

Barley Australia (www.barleyaustralia.com.au) is Australia's main industry body for the barley and malt industries. 'Barley Australia delivers leadership to increase the value and enhance the sustainability of Australia's barley industry.' It has representatives from the brewing, malting, barley marketing and breeding industries. Barley Australia's mandated responsibilities and key activities include:

- Providing timely and credible industry-specific information;
- Managing the malting barley accreditation process;
- Providing stewardship for the barley classification system;
- Providing barley industry relevant communication and education;
- Connecting with industry bodies on trade issues; and
- Guiding R&D investors and research providers on priorities.

A sub-committee of Barley Australia is the Malting and Brewing Industry Barley Technical Committee (MBIBTC). This committee is the national industry body responsible for the malting accreditation process for new barley varieties. The 'Malting Barley Evaluation Overview' and the 'Varietal Accreditation Pathway for Australian Malting Barley' can be accessed from the Barley Australia web site.

The criteria used in the accreditation process are carefully protected by the industry and are shared between Barley Australia and the breeding companies. A comprehensive list of accredited malting varieties as well as varieties/lines currently undergoing the accreditation process is available on the web site. Malting Barley Accreditation does not necessarily guarantee full commercial acceptance or success. This is essentially conducted by the owners of the variety, promoters and the barley or malt marketing companies.

5 Requirements for successful programmes in malting quality improvement

Improved malting varieties require traits encompassing, yield, grain and malting quality, agronomic traits, abiotic stress tolerance and disease and pest resistances. Figure 4 shows conceptual relationships between key quality traits for domestic and export markets. Developing new varieties requires, first and foremost, analytical laboratories capable of producing micro-malts and conducting the requisite industry assays for quality. Rapid, high throughput assays for estimating traits are also essential. This also involves having a capacity to conduct high throughput grain quality assessment and use appropriately calibrated near infra-red (NIR) analysis .

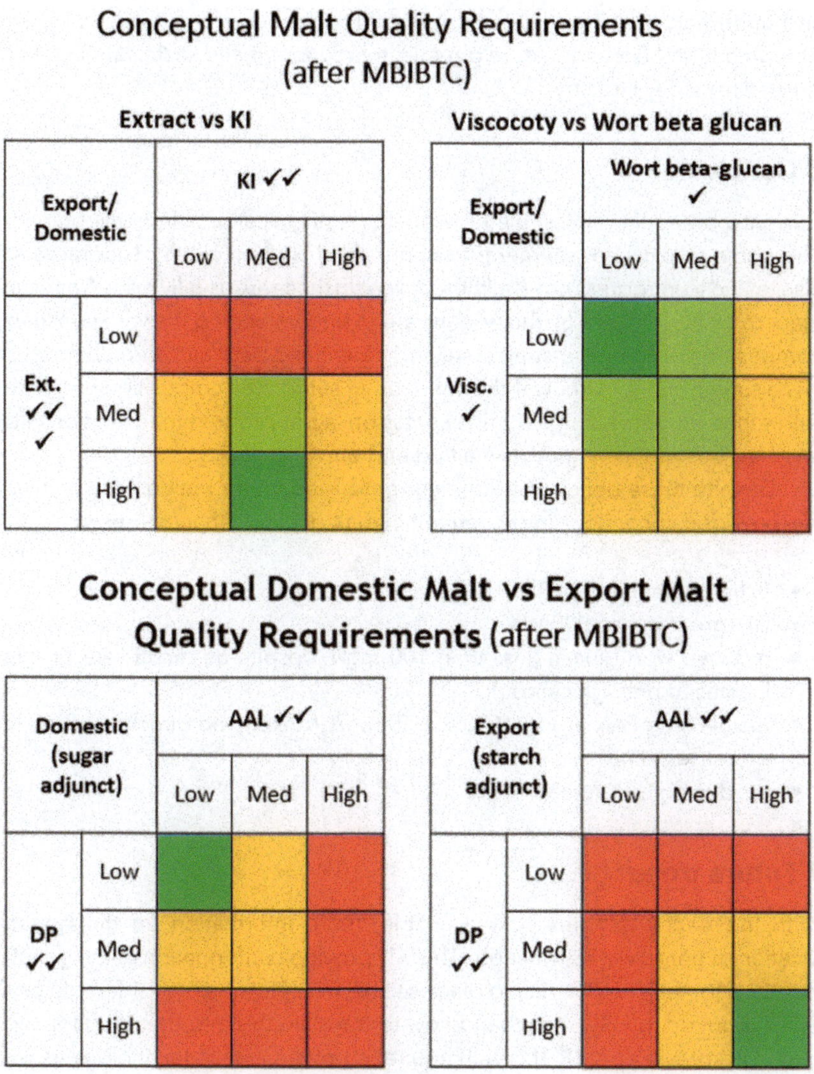

Figure 4 Desired levels and pair-wise relationships between key malt quality traits.

Germplasm introduction and evaluation is paramount for selecting the most appropriate parents for crossing. The introduction of new, high quality barleys has been the fundamental driver for success in improving the malting quality of Australian barley. This continues to be the case but increasingly new malting varieties are being produced from crosses between Australian varieties. The Australian Grains Genebank is the national centre for storing genetic material for plant breeding and research. It plays a significant role in sourcing and importing barleys which contribute through research, pre-breeding and breeding to enhance the future genetic improvement and sustainability of the Australian barley industry.

6 Conclusion

Currently breeders will test their material across typical growing environments plus some atypical environments to understand G×E on quality. Landraces are also tested to introgress novel alleles, but again these are usually tested for major (easy to test for) traits. The major gaps are in understanding the variation in the composition of major substrates, starch, protein and beta-glucan/arabinoxylan. It is also important to know if the variation in substrate composition correlates with substrate-degrading enzyme expression (a pre-programmed relationship), and how this variation impacts malting and brewing quality.

Despite these deficiencies, the current breeding for malt quality pathway has provided continuous improvement in malt quality with increases in:

- extract (greater than 82%);
- DP (greater than 500 WK);
- reduced wort-glucan (less than 100 mg/L despite no direct selection for increased beta-glucanase);
- maintaining FAN around 140 to 150 mg/L (despite no direct selection for proteases); and
- producing LOX-free varieties.

7 Future trends

With the barley genome now available, more information on the genetic location of genes will be available. This will provide conformation of the genetic locations for major traits such as diastase enzymes. However considerable work is still required to fully understand how to translate this into breeding for such traits as proteases or LTP. This will require improvements in technology as well as using specific populations to understand the effect of G×E and heritability. It will also be important to understand the effects of processing by testing for such traits during malting and brewing processes. Almost 50 years of research has unravelled diastase activity specifically in alpha-amylase, beta-amylase and limit dextrinase, although only alpha-amylase and beta-amylase are used in breeding programme. More attention needs to be paid to limiting dextrinase, especially as this enzyme is linked to the synthesis of amylopectin, the most abundant compound in barley, as well as its degradation.

Additional areas of research include the impact of starch structure on gelatinisation and fermentability as well as on potential malt flavour. Malt flavour will continue to attract the attention of craft maltsters and breweries but may not immediately give breeders new targets nor provide bulk handlers with new specifications for segregation.

8 Where to look for further information

There are numerous areas of research-involved aspects of barley quality related to malting and brewing research. This is especially so when developing new malting barley varieties in light of the climate change predicted in the next decades. Examples of such research can be found at the following universities:

- The University of Queensland, Australia (www.uq.edu.au).
- The University of California-Davis, USA (www.ucdavis.edu).

9 References

Beecher, B., Bowman, J., Martin, J. M., Bettge, A. D., Morris, C. F., Blake, T. K. and Giroux, M. J. 2002. Hordoindolines are associated with a major endosperm-texture QTL in barley (*Hordeum vulgare*). *Genome* 45(3), 584–91. doi:10.1139/g02-008.

Bettenhausen, H. M., Barr, L., Broeckling, C. D., Chaparro, J. M., Holbrook, C., Sedin, D. and Heuberger, A. L. 2018. Influence of malt source on beer chemistry, flavor, and flavor stability. *Food Research International* 113, 487-504. doi:10.1016/j.foodres.2018.07.024.

Briggs, D. E. 1992. Barley germination: biochemical changes and hormonal control. In: Shewry, P. R. (Ed.), *Barley: Genetics, Biochemistry, Molecular Biology and Biotechnology*. CAB International, Oxon, UK, 369–401.

Briggs, D. E. 2002. Malt modification – a century of evolving views. *Journal of the Institute of Brewing* 108(4), 395–405. doi:10.1002/j.2050-0416.2002.tb00567.x.

Bueckert, R. A., Lefol, E. B. and Harvey, B. L. 2007. Early detection of non-visible sprouting in barley seed using rapid viscosity analysis. *Canadian Journal of Plant Science* 87(1), 3–12. doi:10.4141/P05-068.

Chu, S., Hasjim, J., Hickey, L. T., Fox, G. and Gilbert, R. G. 2014. Structural changes of starch molecules in barley grains during germination. *Cereal Chemistry Journal* 91(5), 431–7. doi:10.1094/CCHEM-09-13-0174-R.

Coghe, S., Adriaenssens, B., Leonard, S. and Delvaux, F. R. 2004. Fractionation of colored Maillard reaction products from dark specialty malts. *Journal of the American Society of Brewing Chemists* 62(2), 79–86. doi:10.1094/ASBCJ-62-0079.

Coghe, S., D'Hollander, H., Verachtert, H. and Delvaux, F. R. 2005. Impact of dark specialty malts on extract composition and wort fermentation. *Journal of the Institute of Brewing* 111(1), 51–60. doi:10.1002/j.2050-0416.2005.tb00648.x.

Corke, H. 1995. Protein content and composition in crosses between wild and cultivated barley. *Cereal Research Communications* 23, 411–7.

Czuchajowska, Z., Klamczynski, A., Paszczynska, B. and Baik, B. -K. 1998. Structure and functionality of barley starches. *Cereal Chemistry Journal* 75(5), 747–54. doi:10.1094/CCHEM.1998.75.5.747.

Davy, A., Svendsen, I., Sorensen, S. O., Sorensen, M. B., Rouster, J., Meldal, M., Simpson, D. J. and Cameron-Mills, V. 1998. Substrate specificity of barley cysteine endoproteases EP-A and EP-B. *Plant Physiology* 117(1), 255–61. doi:10.1104/pp.117.1.255.

Davy, A., Sorensen, M. B., Svendsen, I., Cameron-Mills, V. and Simpson, D. J. 2000. Prediction of protein cleavage sites by the barley cysteine endoproteases EP-A and EP-B based on the kinetics of synthetic peptide hydrolysis. *Plant Physiology* 122(1), 137–46. doi:10.1104/pp.122.1.137.

De Rouck, G., Jaskula, B., De Causmaecker, B., Malfliet, S., Van Opstaele, F., De Clippeleer, J., De Brabanter, J., De Cooman, L. and Aerts, G. 2013. The influence of very thick and fast mashing conditions on wort composition. *Journal of the American Society of Brewing Chemists* 71(1), 1–14. doi:10.1094/ASBCJ-2013-0113-01.

Duke, S. H. and Henson, C. A. 2011. Tracking the progress of wort sugar production during Congress mashing with North American barley cultivars and comparisons to wort osmolyte concentrations and malt extract. *Journal of the American Society of Brewing Chemists* 69(4), 200–13. doi:10.1094/ASBCJ-2011-0829-01.

Edney, M. J., Eglinton, J. K., Collins, H. M., Barr, A. R., Legge, W. G. and Rossnagel, B. G. 2007. Importance of endosperm modification for malt wort fermentability. *Journal of the Institute of Brewing* 113(2), 228–38. doi:10.1002/j.2050-0416.2007.tb00280.x.

Evans, D. E. and Fox, G. P. 2017. Comparison of diastatic power enzyme release and persistence during modified institute of brewing 65°C and congress programmed mashes. *Journal of the American Society of Brewing Chemists* 75(4), 302–11. doi:10.1094/ASBCJ-2017-4707-01.

Evans, D. E. and Hejgaard, J. 1999. The impact of malt derived proteins on beer foam quality. Part I. The effect of germination and kilning on the level of protein Z4, protein Z7 and LTP1. *Journal of the Institute of Brewing* 105(3), 159–70. doi:10.1002/j.2050-0416.1999.tb00015.x.

Evans, D. E., Sheehan, M. C. and Stewart, D. C. 1999. The impact of malt derived proteins on beer foam quality. Part II: The influence of malt foam-positive proteins and non-starch polysaccharides on beer foam quality. *Journal of the Institute of Brewing* 105(3), 171–8. doi:10.1002/j.2050-0416.1999.tb00016.x.

Evans, D. E., Collins, H., Eglinton, J. and Wilhelmson, A. 2005. Assessing the impact of the level of diastatic power enzymes and their thermostability on the hydrolysis of starch during wort production to predict malt fermentability. *Journal of the American Society of Brewing Chemists* 63(4), 185–98. doi:10.1094/ASBCJ-63-0185.

Evans, D. E., Dambergs, R., Ratkowsky, D., Li, C., Harasymow, S., Roumeliotis, S. and Eglinton, J. K. 2010. Refining the prediction of potential malt fermentability by including an assessment of limit dextrinase thermostability and additional measures of malt modification, using two different methods for multivariate model development. *Journal of the Institute of Brewing* 116(1), 86–96. doi:10.1002/j.2050-0416.2010.tb00403.x.

Fontanini, D. and Jones, B. L. 2002. SEP-1 - a subtilisin-like serine endopeptidase from germinated seeds of *Hordeum vulgare* L. cv. Morex. *Planta* 215, 885–93.

Fox, G. P. 2016. Infrared spectral analysis of sugar profiles of worts from varying grist to liquor ratios using infusion and ramping mash styles. *Journal of the Institute of Brewing* 122(3), 437–45. doi:10.1002/jib.341.

Fox, G. P. and Henry, R. J. 1993. A rapid small-scale method for the determination of malt extract. *Journal of the Institute of Brewing* 99(1), 73–5. doi:10.1002/j.2050-0416.1993.tb01150.x.

Fox, G. P., Kelly, A., Poulsen, D., Inkerman, A. and Henry, R. 2006. Selecting for increased barley grain size. *Journal of Cereal Science* 43(2), 198–208. doi:10.1016/j.jcs.2005.08.004.

Fox, G. P., Nguyen, L., Bowman, J., Poulsen, D., Inkerman, A. and Henry, R. J. 2007a. Relationship between hardness genes and quality in barley (*Hordeum vulgare*). *Journal of the Institute of Brewing* 113(1), 87–95. doi:10.1002/j.2050-0416.2007.tb00261.x.

Fox, G. P., Osborne, B., Bowman, J., Kelly, A., Cakir, M., Poulsen, D., Inkerman, A. and Henry, R. 2007b. Measurement of genetic and environmental variation in barley (*Hordeum vulgare*) grain hardness. *Journal of Cereal Science* 46(1), 82–92. doi:10.1016/j.jcs.2006.12.003.

Fox, G. P., Kelly, A., Bowman, J., Inkerman, A., Poulsen, D. and Henry, R. 2009. Is malting barley better feed for cattle than feed barley? *Journal of the Institute of Brewing* 115(2), 95–104. doi:10.1002/j.2050-0416.2009.tb00352.x.

Fox, G. P., Kelly, A., Sweeney, N. and Hocroft, D. M. 2011. Development of a single kernel NIR barley protein calibration and assessment of variation in protein on grain quality. *Journal of the Institute of Brewing* 117(4), 582–6. doi:10.1002/j.2050-0416.2011.tb00507.x.

Fox, G. P., Yu, W., Nischwitz, R. and Harasymow, S. 2019. Variation in maltose in sweet wort from barley malt and rice adjuncts with differences in amylose structure. *Journal of the Institute of Brewing* 125(1), 18–27. doi:10.1002/jib.546.

Gamlath, J., Aldred, G. P. and Panozzo, J. F. 2008. Barley (1 -> 3; 1 -> 4)-beta-glucan and arabinoxylan content are related to kernel hardness and water uptake. *Journal of Cereal Science* 47(2), 365–71. doi:10.1016/j.jcs.2007.05.017.

Gorinstein, S., Zemser, M., Vargas-Albores, F., Ochoa, J. L., Paredes-Lopez, O., Scheler, C., Salnikow, J., Martin-Belloso, O. and Trakhtenberg, S. 1999. Proteins and amino acids in beers, their contents and relationships with other analytical data. *Food Chemistry* 67(1), 71–8. doi:10.1016/S0308-8146(99)00071-0.

Gous, P. W. and Fox, G. P. 2017. Amylopectin synthesis and hydrolysis - Understanding isoamylase and limit dextrinase and their impact on starch structure on barley (*Hordeum vulgare*) quality. *Trends in Food Science and Technology* 62, 23–32. doi:10.1016/j.tifs.2016.11.013.

Gous, P. W., Hasjim, J., Franckowiak, J., Fox, G. P. and Gilbert, R. G. 2013. Barley genotype expressing "stay-green"-like characteristics maintains starch quality of the grain during water stress condition. *Journal of Cereal Science* 58(3), 414–9. doi:10.1016/j.jcs.2013.08.002.

Gous, P. W., Warren, F., Mo, O. W., Gilbert, R. G. and Fox, G. P. 2015. The effects of variable nitrogen application on barley starch structure under drought stress. *Journal of the Institute of Brewing* 121(4), 502–9. doi:10.1002/jib.260.

Gous, P. W., Warren, F., Gilbert, R. and Fox, G. P. 2017. Drought-proofing barley (*Hordeum vulgare*): the effects of stay green on starch and amylose structure. *Cereal Chemistry Journal* 94(5), 873–80. doi:10.1094/CCHEM-02-17-0028-R.

Grime, K. H. and Briggs, D. E. 1995. Release and activation of barley beta-amylase. *Journal of the Institute of Brewing* 101(5), 337–43. doi:10.1002/j.2050-0416.1995.tb00870.x.

Grime, K. H. and Briggs, D. E. 1996. The release of bound beta-amylase by macromolecules. *Journal of the Institute of Brewing* 102(4), 261–70. doi:10.1002/j.2050-0416.1996.tb00911.x.

Guerin, J. R., Lance, R. C. M. and Wallace, W. 1992. Release and activation of barley beta-amylase by malt endopeptidases. *Journal of Cereal Science* 15(1), 5–14. doi:10.1016/S0733-5210(09)80052-0.

Han, J. Y. 2000. Structural characteristics of arabinoxylan in barley, malt, and beer. *Food Chemistry* 70(2), 131–8. doi:10.1016/S0308-8146(00)00075-3.

He, Y. Y., Cao, Y. F., Chen, S. F., Ma, C. Y., Zhang, D. L. and Li, H. J. 2018. Analysis of flavour compounds in beer with extruded corn starch as an adjunct. *Journal of the Institute of Brewing* 124(1), 9–15. doi:10.1002/jib.474.

Hellwig, M., Witte, S. and Henle, T. 2016. Free and protein-bound Maillard reaction products in beer: method development and a survey of different beer types. *Journal of Agricultural and Food Chemistry* 64(38), 7234–43. doi:10.1021/acs.jafc.6b02649.

Herb, D., Filichkin, T., Fisk, S., Helgerson, L., Hayes, P., Benson, A., Vega, V., Carey, D., Thiel, R., Cistue, L., Jennings, R., Monsour, R., Tynan, S., Vinkemeier, K., Li, Y., Nguygen, A., Onio, A., Meints, B., Moscou, M., Romagosa, I. and Thomas, W. 2017a. Malt modification and its effects on the contributions of barley genotype to beer flavor. *Journal of the Amercian Society of Brewing Chemists* 75, 354–62.

Herb, D., Filichkin, T., Fisk, S., Helgerson, L., Hayes, P., Meints, B., Jennings, R., Monsour, R., Tynan, S., Vinkemeier, K., Romagosa, I., Moscou, M., Carey, D., Thiel, R., Cistue, L., Martens, C. and Thomas, W. 2017b. Effects of barley (*Hordeum vulgare* L.) variety and growing environment on beer flavor. Journal of the Amercian Society of Brewing Chemists 75(4), 345–53. doi:10.1094/ASBCJ-2017-4860-01.

Hoki, T., Kanatani, R., Saito, W., Iimure, T., Zhou, T. S., Takoi, K., Tanigawa, A., Kihara, M. and Ogushi, K. 2018a. Breeding of lipoxygenase-1-less malting barley variety 'Satuiku 2 go'. *Journal of the Institute of Brewing* 124(2), 112–20. doi:10.1002/jib.483.

Hoki, T., Saito, W., Iimure, T., Hirota, N., Takoi, K., Shimase, M., Coventry, S., March, T., Box, A., Eglinton, J., Kihara, M. and Ogushi, K. 2018b. Breeding of lipoxygenase-1-less malting barley variety 'SouthernStar' and evaluation of malting and brewing quality. *Journal of Cereal Science* 83, 83–9. doi:10.1016/j.jcs.2018.07.019.

Huuskonen, A., Markkula, T., Vidgren, V., Lima, L., Mulder, L., Geurts, W., Walsh, M. and Londesborough, J. 2010. Selection from industrial lager yeast strains of variants with improved fermentation performance in very-high-gravity worts. *Applied and Environmental Microbiology* 76(5), 1563–73. doi:10.1128/AEM.03153-09.

Iimure, T. and Sato, K. 2013. Beer proteomics analysis for beer quality control and malting barley breeding. *Food Research International* 54(1), 1013–20. doi:10.1016/j.foodres.2012.11.028.

Jin, Z., Li, X. M., Gao, F., Sun, J. Y., Mu, Y. W. and Lu, J. 2013. Proteomic analysis of differences in barley (*Hardeum vulgare*) malts with distinct filterability by DIGE. *Journal of Proteomics* 93, 93–106. doi:10.1016/j.jprot.2013.05.038.

Jones, B. L. 2005a. The endogenous endoprotease inhibitors of barley and malt and their roles in malting and brewing. *Journal of Cereal Science* 42(3), 271–80. doi:10.1016/j.jcs.2005.06.002.

Jones, B. L. 2005b. Endoproteases of barley and malt. *Journal of Cereal Science* 42(2), 139-56. doi:10.1016/j.jcs.2005.03.007.

Jones, B. L. and Fontanini, D. 2003. Trypsin/alpha-amylase inhibitors inactivate the endogenous barley/malt serine endoproteinase SEP-1. *Journal of Agricultural and Food Chemistry* 51(19), 5803–14. doi:10.1021/jf030075x.

Kanauchi, M. and Bamforth, C. W. 2008. The relevance of different enzymes for the hydrolysis of beta-glucans in malting and mashing. *Journal of the Institute of Brewing* 114(3), 224–9. doi:10.1002/j.2050-0416.2008.tb00332.x.

Kauffman, J. A., Mills, E. N. C., Brett, G. M., Fido, R. J., Tatham, A. S., Shewry, P. R., Onishi, A., Proudlove, M. O. and Morgan, M. R. A. 1994. Immunological characterization of barley polypeptides in lager foam. *Journal of the Science of Food and Agriculture* 66(3), 345–55. doi:10.1002/jsfa.2740660312.

Kirkpatrick, K. R. and Shellhammer, T. H. 2018. A cultivar-based screening of hops for dextrin degrading enzymatic potential. *Journal of the American Society of Brewing Chemists* 76(4), 247–56. doi:10.1080/03610470.2018.1546091.

Kobayashi, N., Kaneda, H., Kano, Y. and Koshino, S. 1993. The production of linoleic and linolenic acid hydroperoxides during mashing. *Journal of Fermentation and Bioengineering* 76(5), 371–5. doi:10.1016/0922-338X(93)90024-3.

Koljonen, T., Hamalainen, J. J., Sjoholm, K. and Pietila, K. 1995. A model for the prediction of fermentable sugar concentrations during mashing. *Journal of Food Engineering* 26(3), 329–50. doi:10.1016/0260-8774(94)00061-D.

Kuroda, H., Kobayashi, N., Kaneda, H., Watari, J. and Takashio, M. 2002. Characterization of factors that transform linoleic acid into di- and trihydroxyoctadecenoic acids in mash. *Journal of Bioscience and Bioengineering* 93(1), 73–7. doi:10.1016/S1389-1723(02)80057-3.

Kuroda, H., Furusho, S., Maeba, H. and Takashio, M. 2003. Characterization of factors involved in the production of 2(E)-nonenal during mashing. *Bioscience, Biotechnology, and Biochemistry* 67(4), 691–7. doi:10.1271/bbb.67.691.

Lee, M. S., Jang, C. S., Lee, S. S., Kim, J. Y., Lee, B. M., Seong, R. C. and Seo, Y. W. 2006. Hordoindolines are predominantly expressed in the aleurone layer in late kernel development in barley. *Breeding Science* 56(1), 63–8. doi:10.1270/jsbbs.56.63.

Lei, H. J., Zhao, H. F., Yu, Z. M. and Zhao, M. M. 2012. Effects of wort gravity and nitrogen level on fermentation performance of brewer's yeast and the formation of flavor volatiles. *Applied Biochemistry and Biotechnology* 166(6), 1562–74. doi:10.1007/s12010-012-9560-8.

Montanari, L., Floridi, S., Marconi, O., Tironzelli, M. and Fantozzi, P. 2005. Effect of mashing procedures on brewing. *European Food Research and Technology* 221(1–2), 175–9. doi:10.1007/s00217-005-1166-8.

Nierop, S. N. E. V., Evans, D. E., Axcell, B. C., Cantrell, I. C. and Rautenbach, M. 2004. Impact of different wort boiling temperatures on the beer foam stabilizing properties of lipid transfer protein 1. *Journal of Agricultural and Food Chemistry* 52(10), 3120–9. doi:10.1021/jf035125c.

Oozeki, M., Sotome, T., Haruyama, N., Yamaguchi, M., Watanabe, H., Okiyama, T., Kato, T., Takayama, T., Oyama, M., Nagamine, T., Suzuki, Y., Toyoshima, T., Sekiwa, T., Oono, K., Saito, T., Usui, M., Arai, S., Kumekawa, T., Suzuki, E., Shirama, K., Kihara, M., Hoki, T., Matsubara, H. and Ohsawa, R. 2017. The two-row malting barley cultivar 'New Sachiho Golden' with null lipoxygenase-1 improves flavor stability in beer and was developed by marker assisted selection. *Breeding Science* 67(2), 165–71. doi:10.1270/jsbbs.16104.

Osborne, B. G., Fox, G. P., Kelly, A. M. and Henry, R. J. 2007. Measurement of barley grain rheology for the quality selection of breeding material. *Journal of the Institute of Brewing* 113(2), 135–41. doi:10.1002/j.2050-0416.2007.tb00269.x.

Osman, A. M., Coverdale, S. M., Cole, N., Hamilton, S. E., De Jersey, J. and Inkerman, P. A. 2002. Characterisation and assessment of the role of barley malt endoproteases during malting and mashing. *Journal of the Institute of Brewing* 108(1), 62–7. doi:10.1002/j.2050-0416.2002.tb00125.x.

Osman, A. M., Coverdale, S. M., Onley-Watson, K., Bell, D. and Healy, P. 2003. The gel filtration chromatographic-profiles of proteins and peptides of wort and beer: effects of processing - Malting, mashing, kettle boiling, fermentation and filtering. *Journal of the Institute of Brewing* 109(1), 41–50. doi:10.1002/j.2050-0416.2003.tb00592.x.

Procopio, S., Krause, D., Hofmann, T. and Becker, T. 2013. Significant amino acids in aroma compound profiling during yeast fermentation analyzed by PLS regression. *LWT - Food Science and Technology* 51(2), 423–32. doi:10.1016/j.lwt.2012.11.022.

Psota, V., Vejrazka, K., Famera, O. and Hrcka, M. 2007. Relationship between grain hardness and malting quality of barley (*Hordeum vulgare* L.). *Journal of the Institute of Brewing* 113(1), 80–6. doi:10.1002/j.2050-0416.2007.tb00260.x.

Rizvi, S. M. H., Beattie, A. D., Rossnagel, B. and Scoles, G. 2011. Thermostability of barley malt proteases in Western Canadian two-row malting barley. *Cereal Chemistry Journal* 88(6), 609–13. doi:10.1094/CCHEM-04-11-0043.

Savin, R., Stone, P. J., Nicolas, M. E. and Wardlaw, I. F. 1997. Grain growth and malting quality of barley .1. Effects of heat stress and moderately high temperature. *Australian Journal of Agricultural Research* 48(5), 615–24. doi:10.1071/A96064.

Schmitt, M. R. and Marinac, L. 2008. Beta-amylase degradation by serine endoproteinases from green barley malt. *Journal of Cereal Science* 47(3), 480–8. doi:10.1016/j.jcs.2007.06.002.

Schmitt, M. R., Skadsen, R. W. and Budde, A. D. 2013. Protein mobilization and malting-specific proteinase expression during barley germination. *Journal of Cereal Science* 58(2), 324–32. doi:10.1016/j.jcs.2013.05.007.

Schulz, B. L., Phung, T. K., Bruschi, M., Janusz, A., Stewart, J., Mehan, J., Healy, P., Nouwens, A. S., Fox, G. P. and Vickers, C. E. 2018. Process proteomics of beer reveals a dynamic proteome with extensive modifications. *Journal of Proteome Research* 17(4), 1647–53. doi:10.1021/acs.jproteome.7b00907.

Sheehan, M. C. and Skerritt, J. H. 1997. Identification and characterisation of beer polypeptides derived from barley hordeins. *Journal of the Institute of Brewing* 103(5), 297–306. doi:10.1002/j.2050-0416.1997.tb00960.x.

Stewart, D., Freeman, G. and Evans, E. 2000. Development and assessment of a small-scale wort filtration test for the prediction of beer filtration efficiency. *Journal of the Institute of Brewing* 106(6), 361–6. doi:10.1002/j.2050-0416.2000.tb00526.x.

Terp, N., Thomsen, K. K., Svendsen, I., Davy, A. and Simpson, D. J. 2000. Purification and characterization of hordolisin, a subtilisin-like serine endoprotease from barley. *Journal of Plant Physiology* 156(4), 468–76. doi:10.1016/S0176-1617(00)80161-3.

Tester, R. F. 1997. Influence of growth conditions on barley starch properties. *International Journal of Biological Macromolecules* 21(1-2), 37–45. doi:10.1016/S0141-8130(97)00039-1.

van Mechelen, J. R., Schuurink, R. C., Smits, M., Graner, A., Douma, A. C., Sedee, N. J. A., Schmitt, N. F. and Valk, B. E. 1999. Molecular characterization of two lipoxygenases from barley. *Plant Molecular Biology* 39(6), 1283–98. doi:10.1023/a:1006118003998.

Vejrazka, K., Psota, V., Ehrenbergerova, J. and Hrstkova, P. 2008. Relationship between grain milling energy and malting quality of barley. *Cereal Research Communications* 36(1), 97–105. doi:10.1556/CRC.36.2008.1.10.

Walker, C. K., Panozzo, J. F., Ford, R., Eckermann, P., Moody, D., Lehmensiek, A. and Appels, R. 2011. Chromosomal loci associated with endosperm hardness in a malting barley cross. *TAG. Theoretical and Applied Genetics. Theoretische und Angewandte Genetik* 122(1), 151–62. doi:10.1007/s00122-010-1431-5.

Woonton, B. W., Jacobsen, J. V., Sherkat, F. and Stuart, I. M. 2005a. Changes in germination and malting quality during storage of barley. *Journal of the Institute of Brewing* 111(1), 33–41. doi:10.1002/j.2050-0416.2005.tb00646.x.

Woonton, B. W., Sherkat, F. and Maharjan, P. 2005b. The influence of barley storage on respiration and glucose-6-phosphate dehydrogenase during malting. *Journal of the Institute of Brewing* 111(4), 388–95. doi:10.1002/j.2050-0416.2005.tb00224.x.

Yu, W. W., Tan, X. L., Zou, W., Hu, Z. X., Fox, G. P., Gidley, M. J. and Gilbert, R. G. 2017. Relationships between protein content, starch molecular structure and grain size in barley. *Carbohydrate Polymers* 155, 271–9. doi:10.1016/j.carbpol.2016.08.078.

Yu, W. W., Zou, W., Dhital, S., Wu, P., Gidley, M. J., Fox, G. P. and Gilbert, R. G. 2018. The adsorption of alpha-amylase on barley proteins affects the in vitro digestion of starch in barley flour. *Food Chemistry* 241, 493–501. doi:10.1016/j.foodchem.2017.09.021.

Yu, W., Tao, K., Gidley, M. J., Fox, G. P. and Gilbert, R. G. 2019. Molecular brewing: molecular structural effects involved in barley malting and mashing. *Carbohydrate Polymers* 206, 583–92. doi:10.1016/j.carbpol.2018.11.018.

Zhu, L. J., Ma, T., Mei, Y. M. and Li, Q. 2017. Enhancing the hydrolysis of corn starch using optimal amylases in a high-adjunct-ratio malt mashing process. *Food Science and Biotechnology* 26(5), 1227–33. doi:10.1007/s10068-017-0140-0.

Chapter 14

Developing barley crops for improved brewing quality

Søren Knudsen, Finn Lok and Ilka Braumann, Carlsberg Research Laboratory, Denmark

1 Introduction
2 Converting barley into beer
3 Breeding barley for the brewing process
4 Brewing traits related to the quality of the final product
5 Conclusion and future trends
6 Acknowledgements
7 Where to look for further information
8 References

1 Introduction

Domesticated barley has been grown in the Fertile Crescent as early as 9000 years ago (Kilian et al., 2009) and it is hypothesized that the main motivation for growing barley might not have been access to food but to fermented beverages (Braidwood et al., 1953). Recently, a 5000-year-old Chinese recipe for brewing beer that included barley was found (Wang et al., 2016) and today, barley remains the cereal of choice for the production of beer worldwide (Gupta et al., 2010).

In order to ensure an energy-efficient and economical industrial brewing process, new barley varieties need to meet a wide range of quality requirements, some of which have been actively bred for in the past decades, for example high grain starch content, optimal grain protein content and high hydrolytic enzyme potential. To illustrate where in the brewing process the different quality aspects specific for malting barley apply, we will commence this chapter with a short illustration of the most important steps in the conversion of barley into beer (see also Fig. 1). These include both the operations in the malthouse and in the brewhouse, as these processes are interdependent. In the following sections of this chapter, we will however focus on breeding malting barley for quality aspects relevant in the brewhouse only.

http://dx.doi.org/10.19103/AS.2019.0060.21

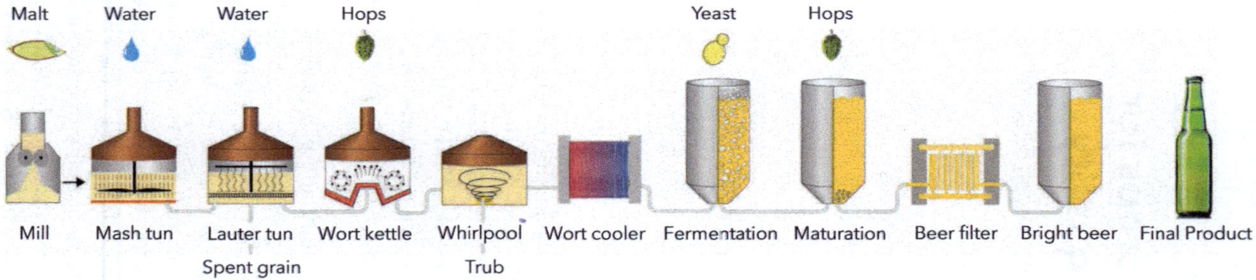

Figure 1 From barley to beer – an overview on the operations in the brewhouse. Barley malt (germinated and dried grain) is delivered to the brewhouse and milled. The milled malt is then transferred into the mash tun, where it is heated. After mashing, the mash is pumped into a lauter tun where the wort is separated from the spent grain. After transfer into the wort kettle, the wort is boiled and hops is added. Non-soluble material, the trub, is subsequently removed in the whirlpool and the wort is cooled down to reach fermentation temperature and filled into the fermentation tank, where the yeast is added. After fermentation, the yeast is harvested and the beer is either left in the fermentation tank or transferred to a new tank for maturation. After maturation, the beer is filtered, bottled and usually pasteurized. Source: adapted from Fangel et al. (2018).

2 Converting barley into beer

There is not one standard process to make beer, as brewing is very much influenced by regional traditions and preferences as well as specific local conditions such as the availability of certain raw materials. The basic principle however is common to most brewing operations, whether of an industrial scale or not. Readers interested in various technical aspects of brewing are referred to technical compendia such as the *Handbook of Brewing* (Priest and Stewart, 2006) or *Technology Brewing & Malting* (Kunze, 2010). The summary within this section is based on the latter if not stated otherwise.

The transition from barley to beer commences with the malting processes, during which barley grains are germinated, or malted, under controlled conditions. The malting process is typically initiated by immersing dry barley grains in water to raise their moisture content to about 40–45% (w/w). The steeped grains are then spread for germination under controlled temperature and airflow conditions. This allows for the activation and synthesis of endogenous enzymes necessary to degrade starch and protein reserves as well as the cell walls of the starchy endosperm. Accordingly, a partial mobilization of endosperm starch, proteins and cell walls occurs during malting. The process is stopped after a period of 4–6 days and the wet malt is dried in a process termed 'kilning'. The kiln-dried malt is subsequently transported from the malthouse to the brewery by road, rail or sea, where the kiln-dried malt is milled (summarized in Narziss, 1976). Milling is followed by mashing, during which the content of the milled malt is extracted with hot water. In simplified terms, mashing is designed to allow for the degradation of the starch stored in the barley endosperm into fermentable sugars, which are subsequently consumed by the yeast to produce alcohol in the latter stages of beer production.

During mashing, the temperature is increased to allow for proteases, cell wall-degrading enzymes and starch-degrading enzymes contained in the malt to act. As the temperature optima for these enzymes differ, step mashing is performed in modern breweries. An example is the classic mashing profile used by Carlsberg in the last century, starting with a protein rest at 35°C for 15 min, a glucan rest at 45°C for 15 min followed by a gelatinization step at 63°C for 60 min, a saccharification rest at 72°C for 20 min and terminated with mashing out at 78°C for 10 min. However, in experimental set-ups and to identify malt quality in the malt trading business the EBC standard mashing regime is used, which follows mashing in at 48°C for 20 min and ramping to 55°C followed by a rest for 30 min before ramping to 72°C for 30 min for mashing out. The extract obtained during mashing, which contains the fermentable sugars, is referred to as 'wort' (European Brewery Convention, 2004).

After mashing, spent grains need to be separated from the wort. Lauter tuns contain a perforated false bottom allowing for the formation of a filter bed from spent grains. Barley grains used in classical brewing set-ups therefore require largely intact husks, even though barley husks do not contribute to extract yield and are also not involved in the subsequent downstream processes that yield beer. Notably, many cereals used for human or animal consumption are hull-less with the exception of barley. So it is tempting to speculate that the tradition of growing hulled barley also for animal fodder originates from the widespread use of lauter tuns in the brewing industry. Today, most modern breweries are equipped with mash filters that generally extract 1–2% more of the fermentable material from the mash (Melis and Eyben, 1992).

In the next step, the wort is transferred to the wort kettle and boiled. Boiling serves two major purposes, it sterilizes the wort and maximizes the formation of bittering acids from the added hops. Boiling further results in the concentration of wort which is often accompanied by a change to a slightly darker coloured wort, and minor flavour refinement associated with the evaporation of certain off-flavour compounds during the boiling process. In addition, the brewer uses boiling to inactivate endogenous and potentially added exogenous enzymes that coagulate alongside polyphenols as hot trub. This effect is facilitated by the teacup effect in the whirlpool whereby hot and later on cold break trub are removed.

Depending on the beer style the wort is then cooled to the desired temperature that is optimal for the respective yeast used in fermentation. Following transfer to a fermenter, the temperature is kept at 6–18°C in order to get the optimal expression of flavours characterizing the beer style. Classically, pilsner-style beers are produced using bottom-fermented yeast at lower temperatures compared to ale-type fermentations which are often conducted at higher temperatures. Depending on the fermentation temperature, alcohol production can be completed in 5–12 days, and depending on fermentation cellar set-up, after yeast harvest the young beer is transferred into a maturation tank for maturation and stabilization. Finally, the beer is filtered, bottled and usually pasteurized before consumption.

3 Breeding barley for the brewing process

Barley bred for brewing purposes is commonly referred to as malting barley to illustrate that this type of barley carries certain traits that are useful for the malting and brewing operations, in contrast to those barley varieties used as animal fodder. Some traits characteristic of malting barley are relevant both during malting and brewing processes, while many play a role only in the brewing process. Nevertheless, all traits specific to malting barley are usually referred to as 'malting quality' traits even though many are, strictly speaking, 'brewing quality' traits. In the following sections we will highlight a number of such quality traits and explain where in the brewing process they apply. Further, we address relevant trait development and breeding efforts.

3.1 Efficient breakdown of starch during mashing

After malting, the first step in the brewing operation is the degradation of endosperm starch into fermentable sugars during mashing through the activity of malt-derived amylolytic enzymes. Accordingly, the efficiency of starch hydrolysis during mashing is mainly determined through two different factors: (1) the activity of the malt enzymes and (2) the structure of the starch itself.

3.1.1 Starch quality in brewing

Starch is a glucose homopolymer consisting of two distinct molecules: the predominantly linear amylose and the branched amylopectin. To form linear chains, glucose molecules are mostly bound together through alpha-1,4-glycosidic linkages, while branch points are introduced through alpha-1,6 linkages (Zeeman et al., 2010). Both amylose and amylopectin are tightly packed into semi-crystalline starch granules (Zeeman et al., 2010), which, though partially attacked, remain more or less intact during malting (Brennan et al., 1997) and are thus still present at the beginning of the mashing process.

Four malt-derived enzymes attack the complex three-dimensional starch structure at different and distinct locations – limit dextrinase cleaves the alpha-1,6 branch points in the amylopectin molecules while alpha- and beta-amylase as well as alpha-glucosidase attack the alpha-1,4 linkages (Bamforth, 2003). Alpha-amylase displays endo-hydrolytic activity, hydrolysing large starch molecules into smaller units, thereby producing substrate for beta-amylase, which catalyses the release of maltose units from the non-reducing ends of alpha-glucans (Zeeman et al., 2010). Malt alpha-glucosidase produces glucose units instead and only plays a minor role during mashing (Bamforth, 2003). Brewers use the term 'diastatic power' (DP) to describe the amount of these enzymes available in malted barley to degrade starch.

Despite the optimization of the mashing protocols to these enzymes, a complete breakdown of starch is usually not achieved. Branched dextrins which are complex, unfermentable carbohydrates that are produced by the partial hydrolysis of starch during mashing (Langstaff and Lewis, 1993) make up 20–25% of the carbohydrate content of wort (Park et al., 2014) and beer usually contains 10–50 g of dextrin per litre (Langstaff and Lewis, 1993). The occurrence of branched dextrins in wort suggests limited hydrolysis by limit dextrinase during mashing. It is further known that alpha-amylase activity is rarely limiting during mashing and that the enzyme remains active for over an hour at 65°C, while beta-amylase is rapidly denatured at temperatures above 55°C (summarized in Gupta et al., 2010). However, as dextrins are believed to add to the mouthfeel of beer, a complete breakdown of starch is usually not desired.

Nevertheless, fast and efficient hydrolysis of starch into fermentable sugars during mashing is certainly considered essential in an industrial brewing process (Gupta et al., 2010). However, the tight packing of starch granules and the fact that starch is a water-insoluble molecule render it inaccessible for enzymatic degradation. It is therefore necessary for the semi-crystalline structure of starch to be broken down in the presence of water and heat (Copeland et al., 2009), in a process referred to as gelatinization during mashing. The term 'starch gelatinization temperature' is then defined as the actual temperature at which the internal structure of starch is disrupted (Ratnayake and Jackson, 2009).

To gelatinize the starch in barley malt, mashing is performed either isothermally at approximately 65°C, or using ramped temperature profiles. Starch gelatinization temperature is however not fixed, but a variable trait dependant both on environmental and genetic factors. Depending on the crop year, the variability in gelatinization temperature can become of great relevance for the mashing process: as mentioned above, especially the beta-amylase present in barley malt has been found to be heat labile (Bamforth, 2003) so it is beneficial if the starch gelatinizes at a temperature significantly below the inactivation temperature of said enzymes. However, starch gelatinization temperature has been reported to increase in barley plants exposed to elevated temperatures during grain filling (Tester et al., 1991; Myllärinen et al., 1998). Especially in the context of higher ambient temperatures and the increased occurrence of extreme weather events like droughts or heat waves due to climate change (Forzieri et al., 2016), breeding of barley varieties with decreased gelatinization temperature is certainly desirable. However, grain starch content is a major factor determining grain yield in cereals like barley. Consequently, breeding for lower gelatinization temperature, which ultimately means breeding for altered starch structure, comes with the risk of yield penalties that most probably will not be accepted by growers. It has, for instance, been shown that loss-of-function mutants at the SSIIa locus in barley that harbours the gene encoding the starch biosynthetic enzymesoluble starch synthase (SS) IIa, produce starch with a gelatinization temperature decreased by 10°C. However, due to an overall reduced starch content the grain weight is reduced by more than 30% in these mutants (Morell et al., 2003). To the best of our knowledge, there is currently no barley variety on the market that produces starch with a gelatinization temperature consistently below the critical inactivation temperature of barley malt enzymes. Therefore, despite the obvious importance of starch quality, increased gelatinization temperature is compensated for by the addition of recombinant, heat-stable enzymes if needed for the specific brew stream.

An additional factor influencing starch gelatinization temperature is the starch granule size distribution. The starch found in barley grains displays a characteristic bimodal granule size distribution (Jaiswal et al., 2014; Chia et al., 2017) with smaller, rather spherical B-type granules (1–10 μm in diameter) and larger, more lenticular A-type granules (10–40 μm). B granules make up only 10% of the total starch by weight but account for 90% by number (Tillett and Bryce, 1993). It is known that B granules have a higher peak gelatinization temperature. They further gelatinize over a wider temperature range than A granules (MacGregor, 1991). In addition, B granules are more resistant to alpha-amylase attack during mashing than A granules (Bathgate and Palmer, 1973; Slack et al., 1979), which is of particular relevance when non-malted barley is added to the mash (Bathgate and Palmer, 1973). Insufficient degradation of B granules during malting and mashing can correspond to a loss of 5% of the total starch as it is not broken down to fermentable sugars and hence not available for conversion into alcohol (Tillett and Bryce, 1993) during fermentation. B granules remaining in the wort can cause starch haze and block filter beds in lauter tuns. This ultimately leads to filtration problems such as increased runtime in the brewhouse. Notably, in recent years, reports on wheat lines completely devoid of B granules have been published (Howard et al., 2011; Chia et al., 2017) and first results suggest that the lack of B granules does not correspond to a lower grain starch content (Chia et al., 2017). The identity of the gene controlling B granule content in wheat is however still unknown.

3.1.2 Enhancing brewing quality with heat-stable beta-amylase

As outlined above, it is specifically beta-amylase activity that is potentially impacted negatively by the temperatures applied during mashing necessary to gelatinize barley starch. The gene encoding beta-amylase, designated *Bmyl*, which is located on chromosome 4H in barley, has been cloned and sequenced by Kreis et al. (1987) and Yoshigi et al. (1994). Thacker et al. (1992) showed that the activity of isolated beta-amylase rapidly declines at temperatures above 55°C. However, the rate of thermal inactivation under true mashing conditions is somewhat slower due to the high protein content in the mash and the presence of stabilizing carbohydrates (Eglinton et al., 1998 and references therein).

The critical role played by beta-amylase during mashing has prompted breeders to improve beta-amylase thermostability in their breeding lines. Kihara et al. (1998) investigated the thermostability of beta-amylase activity in malt extracts from 75 different European, North American, Australian and Japanese varieties. The authors found that beta-amylase activity in Japanese varieties tends to be of high thermostability, while thermostability in barley from other geographic origin was intermediate or low. The authors speculated that the observed differences in beta-amylase thermostability could be due to sequence deviations between different barley lines as described by Kreis et al. (1987) and Yoshigi et al. (1994), especially as Yoshigi et al. (1995) have shown in *in vitro* studies that certain amino acid substitutions can alter the thermostability in barley. Eglinton et al. (1998) finally described three alleles of the beta-amylase encoding gene *Bmyl* in cultivated barley, *Sd1*, *Sd2L* and *Sd2H*, as well as the *Sd3* allele in *Hordeum spontaneum* and related the results to fermentability. Purified beta-amylase activity from extracts of barley lines carrying the *Sd2L*, the *Sd1* allele and the *Sd3* allele showed different rates of thermal inactivation with T50 values of 56.8°C, 58.5°C and 60.8°C, respectively. In the same study, it was further shown that increased beta-amylase thermostability resulted in more efficient starch degradation. Accordingly, malt from barley lines carrying the *Sd2L* allele showed the lowest fermentability, while malt from lines carrying the *Sd2* or *Sd2H* allele showed medium and malt from lines with the *Sd3* allele highest fermentability. In a subsequent study by Ma et al. (2001) it was shown that both amino acid substitutions, V233A and L347S, which were identified as the cause of altered thermostability (Eglinton et al., 1998), contribute to increased thermostability via different mechanisms: L347S slows down thermal unfolding while V233A accelerates the refolding after heat treatment.

The identification of polymorphism leading to increased beta-amylase stability enables both classical breeding strategies designed to introgress favourable alleles into elite germplasm and transgenic strategies. Kihara et al. (2000) for instance developed stable transgenic barley lines expressing a mutagenized thermostable beta-amylase gene under the control of the barley beta-amylase promoter. More recently, Xu et al. (2018) used marker-assisted selection to introgress the *Sd3 Bmyl* allele from *Hordeum spontaneum* into the commercial variety Gairdner. The results demonstrated that the *Sd3* allele not only increased enzyme thermostability but dramatically enhanced DP.

3.2 Beta-glucans and their impact on mashing and filterability

The efficient degradation of starch during mashing is further influenced by the content of 1-3,1-4 β-D-glucan in the mash, which is an important quality aspect in barley used for malting and brewing (Jin et al., 2004). 1-3,1-4 β-D-glucan constitutes up to 75% of the cell walls in the starchy endosperm thereby representing its main constituent (Burton and Fincher, 2014). The malting process is optimized to reduce the cell wall content within the starchy endosperm efficiently. This is because reduced or incomplete conversion of 1-3,1-4 β-D-glucan during malting renders parts of the starch and protein stored in the endosperm inaccessible for enzymatic degradation during mashing. Incomplete beta-glucan hydrolysis consequently leads to poor mash conversion leaving residual starch and protein in the spent grain. This results in lower extract yields in the brewhouse. Further, these residuals form a layer of fines that retards lautering (Bathgate and Palmer, 1975). 1-3,1-4 β-D-glucan remaining in the wort after lautering can also cause problems further down the line in the brewing process, as filtration of beer made with wort rich in high molecular weight 1-3,1-4 β-D-glucan lowers the filtration capacity of the filters (Leedham et al., 1975).

Classical barley breeding approaches for improved malting and brewing purposes have consequently focused on obtaining fast modification of the cell walls during malting. This can be achieved by increased hydrolytic activity enabling breakdown of the 1-3,1-4 β-D-glucan as well as by reducing the 1-3,1-4 β-D-glucan content of the secondary cell wall in the starchy endosperm. In fact, thinner cell walls have been suggested to correlate with good malting quality for a long time (Fincher, 1975), even though some researchers observed the opposite (Gram, 1982). Consequently, several experiments have been performed to identify barley lines with thinner cell walls and corresponding lower 1-3,1-4 β-D-glucan content. The first mutants displaying lower 1-3,1-4 β-D-glucan in their cell walls were however only characterized biochemically, while the causative mutations could not be identified (Aastrup, 1983; Molina-Cano et al., 1989).

Research within the past decades however has provided scientists and molecular breeders with insights into the genetics and biochemistry of cell wall biosynthesis. This allowed targeted screenings for barley breeding lines to control 1-3,1-4 β-D-glucan biosynthesis in particular (Schreiber et al., 2014 and references therein). Recently, mutants have been identified, which are devoid of CslF6, the key enzyme in 1-3,1-4 β-D-glucan biosynthesis (Tonooka et al., 2009; Taketa et al., 2012), or where ClsF6 enzyme activity was reduced due to an amino acid substitution in the enzyme (Hu et al., 2014). This mutant only has 50% of the wild-type 1-3,1-4 β-D-glucan content; however, the rate of broken grains is increased fourfold (Hu et al., 2014). Thus, further breeding is required to mitigate this off trait related to mutations in the *ClsF6* gene.

4 Brewing traits related to the quality of the final product

The malting quality aspects described in Section 3 mainly relate to technical requirements in the brewing process. In the following sections we will focus on malting traits that directly impact the quality of the final product and that are therefore highly relevant not only to the brewer but also to the end consumer.

4.1 Elimination of off-flavours through targeted breeding

In contrast to hops, and also yeast, the actual barley variety used for brewing beer has only very limited impact on the taste of the final product. Nevertheless, barley is the source of some very prominent off-flavours in beer that decrease the likability of the final product. Molecular and biochemical research on how these off-flavours evolve have therefore enabled breeders to develop barley lines, which, when used for malting and brewing, significantly reduce the development of undesired, barley-derived off-flavours in beer.

It is well known that storage of beer at temperatures above 30°C induces formation of a papery off-flavour. Jamieson and Van Gheluwe (1970) identified the molecule conferring this cardboard-like flavour as trans-2-nonenal (T2N), a volatile C9 alkenal. Since the taste-threshold level for T2N in humans is extremely low, previously determined to be around 0.7 nM or 0.1 ppb (Meilgaard, 1975), products with even minute levels of this aldehyde are perceived as aged. However, T2N levels are generally very low in fresh beer (Lermusieau et al., 1999). Therefore, it has been speculated that during beer storage, free T2N may be liberated from T2N adducts (Nyborg et al., 1999). This notion was supported by a subsequent observation that the T2N potential in wort correlates with the formation of T2N after product storage (Kuroda et al., 2005).

The formation of T2N during beer storage is related to the activity of a lipoxygenase (LOX) derived from barley. Several LOX enzymes are expressed in the barley grain, of which the LOX-1 enzyme is of particular relevance for T2N formation. The LOX-1 enzyme is synthesized in the embryo during both grain filling and germination (van Mechelen et al., 1999). It catalyses the formation of 9-hydroperoxy octadecadienoic acid (9-HPODE) – a precursor of both T2N and trihydroxy octadecenoic acids (abbreviated THAs) – from linoleic acid. To control the formation of T2N during beer maturation, barley populations carrying sodium azide-induced variations have been generated to identify lines characterized by abolished LOX-1 synthesis (Breddam et al., 2005; Oozeki et al., 2017). These barley mutants have been tested in brewing trials and the beers produced from them were characterized by better flavour stability as well as unexpected improved foam stability.

Notably, Hirota et al. (2006) identified a barley line devoid of LOX-1 activity in a collection of old Japanese landraces. By intensive crossing with modern cultivars a Canadian LOX-less line CDC PolarStar was developed and shown to perform similarly to the induced LOX mutants (Hoki et al., 2013; Yu et al., 2014) with respect to flavour stability.

Dimethyl sulphide (DMS) is another barley-derived off-flavour that occurs during the brewing process. DMS constitutes an undesirable cooked canned maize odour and is derived from the precursor *S*-methyl-L-methionine (SMM) (Dickenson, 1979). SMM is a compound unique to plants produced in the SMM cycle, a metabolic process potentially involved in the storage of methionine (Ranocha et al., 2001). The formation of SMM during malting is confined to the developing coleoptile. The Met-*S*-methyltransferase (MMT) enzyme catalyses the transfer of a methyl group from *S*-adenosyl methionine (AdoMet) to methionine (Met), forming SMM (Ranocha et al., 2001).

It is well known that beer production-related incubations at high temperatures, such as kiln drying of malt and the heating of wort after mashing, initiates the chemical conversion of SMM to DMS, thereby enhancing off-flavour formation. The boiling point of DMS is low at 37–38°C; however, at temperatures above 65–70°C, the volatility of DMS decreases considerably (Scheuren et al., 2015), while conditions for further oxidation to dimethyl sulphoxide (DMSO) are established. In beer, DMS is present typically in concentrations up to 144 mg/L depending on the respective beer style; the sensory threshold is around 30–45 mg/L (Meilgaard, 1982). At low levels, DMS contributes to palate fullness and overall beer aroma, while higher levels lead to the characteristic canned corn-like off-flavour. The MMT enzyme is encoded by a single gene in barley consisting of 12 exons and 11 introns (Knudsen et al., 2010). A mutant with aberrant splicing at the 5′ splice site of intron 5 was identified from a barley population carrying induced variation. Malting and brewing experiments showed very low levels of SMM and DMS in malt and beer produced from this barley line (Knudsen et al., 2010).

4.2 Colloidal haze formation

Beyond off-flavour formation, barley can also contribute to other undesired post-production effects that decrease consumer acceptance of the final product. Among these is the formation of precipitates in the beer referred to as colloidal haze.

It is well documented that polyphenols derived from barley play a major role in the development of haze. Proline-rich proteins displaying high affinity for di- or trimeric proanthocyanidins (PAs) have been identified to be critical for haze formation (Outtrup et al., 1987). PAs are present in beer as monomers, dimers, trimers and higher polymers of catechin, epicatechin and gallocatechin. The two most prominent dimeric proanthocyanidins, procyanidin B3 (catechin-catechin) and prodelphinidin B3 (gallocatechin-catechin), are of particular relevance in haze formation (McMurrough et al., 1996).

To produce beer which remains clear during storage at low temperatures the beer is stabilized by either removing part of the PA or degrading the proline-rich peptides. The PA can be removed by polyvinylpolypyrrolidone (PVPP) filtration (Leiper et al., 2005) and the proline-rich proteins (prolamins) deriving from the barley grain can be reduced in the boiled wort by treatment with commercial enzyme preparations for example Brewers Clarex® containing proline-specific proteinases derived from *Aspergillus niger* (Lopez and Edens,

2005). This stabilization process adds complexity to beer filtration and increases the environmental impact of the brewing process.

However, since most of the PAs derive from barley with little contribution from hops (Erdal, 1986), an alternative to the beer-stabilizing processes is the use of PA-free barley. The biosynthesis of PA is complex and involves many catalytic enzymes (Nakano et al., 2011). Several genes are therefore involved in the biosynthesis of PA, and breeders have struggled to develop PA-free barley cultivars with good agronomic performance and acceptable malting quality. The first barley lines low in PA were identified more than 40 years ago (von Wettstein et al., 1977). In the subsequent years more than 400 mutants have been identified in at least ten different genes (Jende-Strid, 1993), but only few PA-free cultivars have been successfully commercialized during these years. CRISP Malting Group is currently delivering PA-free malt specifically for craft brewers in the United Kingdom (http://www.crispmalt.co).

5 Conclusion and future trends

When breeding barley that is optimized for brewing, the breeder has to consider the needs of several major stakeholders: the grower, the maltster, the brewer and the beer consumer. Still, their needs will usually go hand in hand, as growers and maltsters will only be able to sell raw materials that allow the brewer to run his plant efficiently while producing beer that meets the preferences and demands of the end consumer.

Therefore, taking a bird's-eye view, breeders actually face two challenges when developing novel barley lines: (1) In times of climate change, water scarcity in many parts of the world and increasing energy costs, the crop must deliver high malting quality as well as high yield even under suboptimal growing conditions. However, breeding for stable agronomic performance under challenging climatic conditions is not the matter of this chapter and suitable approaches as well as relevant physiological pathways and candidate genes have been summarized elsewhere (see, for example Fita et al., 2015; Krannich et al., 2015; Ismail and Horie, 2017). (2) Further, the raw material must enable the brewer to maintain and even increase the quality of both fresh and shelved products. One route to achieve this quality improvement certainly is to optimize the protein composition of the barley grain. Three examples of relevant breeding targets linked to grain protein content are listed below:

1 During beer storage, the chemical composition of the product changes, which alters its sensory properties. Ageing in beer is, in contrast to wine, usually considered to have a negative impact on flavour quality (Vanderhaegen et al., 2006). Off-flavour formation during beer storage is, among other phenomena, related to Strecker degradation (Guedes

De Pinho and Silva Ferreira, 2006), a conversion of amino acids into aldehydes. The aldehydes form especially when the product is not refrigerated. In principle, a wide range of different amino acids can give rise to Strecker aldehydes. Still, when one considers both the concentration of individual amino acids in beer and the specific flavour thresholds for certain Strecker aldehydes, only methionine, leucine, valine, isoleucine and phenylalanine are of relevance. These are precursors for the corresponding Strecker aldehydes methional, 3-methylbutanal, 2-methylpropanal, 2-methylbutanal and phenylacetaldehyde, respectively (summarized in Baert et al., 2012). The low protein content of 9.5–11.5% that is usually targeted in the production of malting barley ensures that the formation of Strecker aldehydes is somewhat confined, as amino acid content in the final beer is limited. However, depending on geographic and climatic conditions, this target cannot always be met. Therefore, fine-tuning the grain storage protein composition to increase the proportion of proteins that have a low content of those amino acids mentioned above could pave the way to improve beer flavour stability during storage. This would benefit the consumer and reduce the carbon footprint of beer as distribution and storage under cool conditions would not be a necessity anymore.

2 A recent example for the feasibility of modifying grain protein composition by means of classical breeding, though not related to off-flavours, is the development of ultra-low gluten barley (Tanner et al., 2016). Gluten is known to be the cause of coeliac disease, a condition that forces those affected to live on a life-long gluten-free diet. Gluten-like proteins in barley are termed hordeins (Biesiekierski, 2017) and these, as they are also present in the final beer (Grochalová et al., 2017), can trigger the symptoms of coeliac disease upon consumption in affected individuals. Tanner et al. (2016) combined three recessive alleles, which all lowered the hordein content independently from each other. Subsequent breeding significantly increased the yield of the derived barley line as well as its malting quality, so that the ultra-low gluten barley produces grains with normal appearance and malting and brewing properties.

3 Protein composition is further relevant in the context of haze formation. As mentioned above, haze forms through the interaction of proline-rich proteins and PAs. Breeding efforts have mainly focused on eliminating biosynthesis of PAs (Jende-Strid, 1993), while tailoring protein composition appears promising as well. Robinson et al. (2004) for instance were able to identify a 12 kDa haze-active protein and could show that beer brewed from varieties lacking this specific protein formed significantly less haze. Genetic analysis of more than 200 varieties could identify a recessive allele at a single locus on chromosome 3H, the

barley trypsin inhibitor-CMe precursor (BTI-CMe), as the genetic cause for the reduced haze formation (Robinson et al., 2007a,b), enabling marker-assisted breeding for this trait. Iimure and Sato (2013) have summarized a large number of proteins in barley, including many hordeins, which are known to be haze active and could be targeted in breeding programmes. The same report also lists foam-active proteins, which are relevant when improving foam quality in beer.

These examples, though certainly not an exhaustive list of all protein-related traits relevant in barley used for brewing, illustrate how focusing on grain protein can enable the development of novel, significantly improved malting barley quality to the benefit of the brewer and their consumers. However, grain protein is also of relevance when breeding barley varieties with improved performance under suboptimal future climate conditions, as those are predicted to increase the grain protein content (Ingvordsen et al., 2016) with adverse impact on malting quality. Therefore, future malting barley breeding programmes could benefit from focusing on grain protein.

6 Acknowledgements

We thank Drs. Toni Wendt, Kenneth Fredlund and Olayide Oladokun for their critical review of the manuscript and Ms. Anne Sofie Lok for her contributions to Fig. 1.

7 Where to look for further information

Readers who wish to learn more about brewing-related quality traits in malting barley can find some suggestions for additional reading in Table 1. The books listed there are recommended as they give a detailed overview of barley in general as well as on malting and brewing operations. Reading these books will help to understand the traits discussed in this chapter in the wider context of the brewing process.

The newest scientific achievements in breeding for malting and brewing relevant traits in barley, as described in this chapter, are published in journals with a clear focus on plant breeding. These are certainly well known to the reader and therefore not listed here. Journals more focused on scientific advancements related to brewing chemistry are the *Journal of the American Society of Brewing Chemists* (*ASBC*) and the *Journal of the Institute of Brewing* (Table 1). Both journals publish papers - among others - concerning raw materials such as barley used in brewing.

The interested reader's attention is further drawn to the European Brewery Convention (EBC, see Table 1), that is active in dissemination of

Table 1 Where to find further information on barley and brewing

Books	Shewry, P. R. and Ullrich, S. E. 2014. *Barley: Chemistry and Technology*. American Association of Cereal Chemists.
	Briggs, D. E. 1998. *Malts and Malting*. Springer.
	Kunze, W. 2019. *Technology Brewing and Malting* (6th revised English edn.). VLB Berlin.
Journals and conferences	*Journal of the Institute of Brewing*
	Journal of the American Society of Brewing Chemists
	European Brewery Convention (EBC) - https://europeanbreweryconvention.eu/
Research and educational institutions	VLB Berlin (Versuchs-und Lehranstalt für Brauerei in Berlin e.V., Berlin, Germany) - www.vlb-berlin.org
	IFBM (Institut Francais des Boissons, de la Brasserie & de la Malterie, Vandoeuvre-lès-Nancy, France) - www.ifbm.fr/fr
	Lehrstuhl für Brau- und Getränketechnologie, Wissenschaftszentrum Weihenstephan für Ernährung, Landnutzung und Umwelt, Technische Universität München, Freising, Germany - http://lbgt.wzw.tum.de
	IBD (Institute of Brewing and Distilling, London, UK) - www.ibd.org.uk
	ICBD (The International Centre for Brewing and Distilling, Edinburgh, Scotland) - https://icbd.hw.ac.uk/

brewing knowledge and research via congresses and symposia. In addition the webpages of research and educational institutions dealing with multiple aspects of brewing beer including barley as the key raw material can serve as a starting point when searching for background information on the topic of this chapter. A non-comprehensive list of such institutions can be found in Table 1 as well.

8 References

Aastrup, S. 1983. Selection and characterization of low β-glucan mutants from barley. *Carlsberg Research Communications* 48(4), 307–16. doi:10.1007/BF02911913.

Baert, J. J., De Clippeleer, J., Hughes, P. S., De Cooman, L. and Aerts, G. 2012. On the origin of free and bound staling aldehydes in beer. *Journal of Agricultural and Food Chemistry* 60(46), 11449–72. doi:10.1021/jf303670z.

Bamforth, C. W. 2003. Barley and malt starch in brewing: a general review. *MBAA Technical Quarterly* 40(2), 89–97.

Bathgate, G. N. and Palmer, G. H. 1973. The *in vivo* and *in vitro* degradation of barley and malt starch granules. *Journal of the Institute of Brewing* 79(5), 402–6. doi:10.1002/j.2050-0416.1973.tb03557.x.

Bathgate, G. N. and Palmer, G. H. 1975. The prediction of brewing behaviour of malts from bench scale trials. *Proceedings of the Congress European Brewery Convention*, Nice, pp. 1561–81.

Biesiekierski, J. R. 2017. What is gluten? *Journal of Gastroenterology and Hepatology* 32, 78–81. doi:10.1111/jgh.13703.

Braidwood, R. J., Sauer, J. D., Helbaek, H., Mangelsdorf, P. C., Cutler, H. C., Coon, C. S., Linton, R., Steward, J. and Oppenheim, A. L. 1953. Symposium: did man once live by beer alone? *American Anthropologist* 55(4), 515–26. doi:10.1525/aa.1953.55.4.02a00050.

Breddam, K., Olsen, O., Skadhauge, B., Lok, F., Knudsen, S. and Bech, L. M. 2005. Patent WO2005087934A2: Barley for production of flavor-stable beverage.

Brennan, C. S., Amor, M. A., Harris, N., Smith, D., Cantrell, I., Griggs, D. and Shewry, P. R. 1997. Cultivar differences in modification patterns of protein and carbohydrate reserves during malting of barley. *Journal of Cereal Science* 26(1), 83–93. doi:10.1006/jcrs.1996.0103.

Burton, R. A. and Fincher, G. B. 2014. Evolution and development of cell walls in cereal grains. *Frontiers in Plant Science* 5, 456. doi:10.3389/fpls.2014.00456.

Chia, T., Adamski, N. M., Saccomanno, B., Greenland, A., Nash, A., Uauy, C. and Trafford, K. 2017. Transfer of a starch phenotype from wild wheat to bread wheat by deletion of a locus controlling B-type starch granule content. *Journal of Experimental Botany* 68(20), 5497–509. doi:10.1093/jxb/erx349.

Copeland, L., Blazek, J., Salman, H. and Tang, M. C. 2009. Form and functionality of starch. *Food Hydrocolloids* 23(6), 1527–34. doi:10.1016/j.foodhyd.2008.09.016.

Dickenson, C. J. 1979. The relationship of dimethyl sulphide levels in malt, wort and beer. *Journal of the Institute of Brewing* 85(4), 235–9. doi:10.1002/j.2050-0416.1979.tb03914.x.

Eglinton, J. K., Langridge, P. and Evans, D. E. 1998. Thermostability variation in alleles of barley beta-amylase. *Journal of Cereal Science* 28(3), 301–9. doi:10.1016/S0733-5210(98)90010-8.

Erdal, K. 1986. Proanthocyanidin-free barley – malting and brewing. *Journal of the Institute of Brewing* 92, 220–4.

European Brewery Convention. 2004. *EBC Anlytica: 4.5.1 - Extract of Malt: Congress Mash*. Fachverlag Hans Carl, Nuremberg, Germany.

Fangel, J. U., Eiken, J., Sierksma, A., Schols, H. A., Willats, W. G. T. and Harholt, J. 2018. Tracking polysaccharides through the brewing process. *Carbohydrate Polymers* 196, 465–73. doi:10.1016/j.carbpol.2018.05.053.

Fincher, G. B. 1975. Morphology and chemical composition of barley endosperm cell walls. *Journal of the Institute of Brewing* 81(2), 116–22. doi:10.1002/j.2050-0416.1975.tb03672.x.

Fita, A., Rodríguez-Burruezo, A., Boscaiu, M., Prohens, J. and Vicente, O. 2015. Breeding and domesticating crops adapted to drought and salinity: a new paradigm for increasing food production. *Frontiers in Plant Science* 6, 978. doi:10.3389/fpls.2015.00978.

Forzieri, G., Feyen, L., Russo, S., Vousdoukas, M., Alfieri, L., Outten, S., Migliavacca, M., Bianchi, A., Rojas, R. and Cid, A. 2016. Multi-hazard assessment in Europe under climate change. *Climatic Change* 137(1–2), 105–19. doi:10.1007/s10584-016-1661-x.

Gram, N. H. 1982. The ultrastructure of germinating barley seeds. II. Breakdown of starch granules and cell walls of the endosperm in three barley varieties. *Carlsberg Research Communications* 47(3), 173–85. doi:10.1007/BF02904380.

Grochalová, M., Konečná, H., Stejskal, K., Potěšil, D., Fridrichová, D., Srbová, E., Ornerová, K. and Zdráhal, Z. 2017. Deep coverage of the beer proteome. *Journal of Proteomics* 162, 119–24. doi:10.1016/j.jprot.2017.05.001.

Guedes De Pinho, P. and Silva Ferreira, A. C. 2006. Role of strecker aldehydes on beer flavour stability. *Developments in Food Science* 43, 529–32. doi:10.1016/S0167-4501(06)80125-1.

Gupta, M., Abu-Ghannam, N. and Gallaghar, E. 2010. Barley for brewing: characteristic changes during malting, brewing and applications of its by-products. *Comprehensive Reviews in Food Science and Food Safety* 9(3), 318–28. doi:10.1111/j.1541-4337.2010.00112.x.

Hirota, N., Kuroda, H., Takoi, K., Kaneko, T., Kaneda, H., Yoshida, I., Takashio, M., Ito, K. and Takeda, K. 2006. Development of novel barley with improved beer foam and flavor stability – the impact of lipoxygenase-1-less barley in the brewing industry. *MBAA Technical Quarterly* 43(2), 131–5.

Hoki, T., Saito, W., Hirota, N., Shirai, M., Takoi, K., Yoshida, S., Shimase, M., Saito, T., Takaoka, T., Kihara, M., et al. 2013. Breeding of lipoxygenase-1-less malting barley variety CDC PolarStar and effect of lipoxygenase-1-less trait on beer quality at pilot and commercial scale brewing. *BrewingScience* 66, 37–45.

Howard, T., Rejab, N. A., Griffiths, S., Leigh, F., Leverington-Waite, M., Simmonds, J., Uauy, C. and Trafford, K. 2011. Identification of a major QTL controlling the content of B-type starch granules in Aegilops. *Journal of Experimental Botany* 62(6), 2217–28. doi:10.1093/jxb/erq423.

Hu, G., Burton, C., Hong, Z. and Jackson, E. 2014. A mutation of the cellulose-synthase-like (CslF6) gene in barley (*Hordeum vulgare* L.) partially affects the β-glucan content in grains. *Journal of Cereal Science* 59(2), 189–95. doi:10.1016/j.jcs.2013.12.009.

Iimure, T. and Sato, K. 2013. Beer proteomics analysis for beer quality control and malting barley breeding. *Food Research International* 54(1), 1013–20. doi:10.1016/j.foodres.2012.11.028.

Ingvordsen, C. H., Gislum, R., Jørgensen, J. R., Mikkelsen, T. N., Stockmarr, A. and Jørgensen, R. B. 2016. Grain protein concentration and harvestable protein under future climate conditions. A study of 108 spring barley accessions. *Journal of Experimental Botany* 67(8), 2151–8. doi:10.1093/jxb/erw033.

Ismail, A. M. and Horie, T. 2017. Genomics, physiology, and molecular breeding approaches for improving salt tolerance. *Annual Review of Plant Biology* 68, 405–34. doi:10.1146/annurev-arplant-042916-040936.

Jaiswal, S., Båga, M., Ahuja, G., Rossnagel, B. G. and Chibbar, R. N. 2014. Development of barley (*Hordeum vulgare* L.) lines with altered starch granule size distribution. *Journal of Agricultural and Food Chemistry* 62(10), 2289–96. doi:10.1021/jf405424x.

Jamieson, A. M. and Van Gheluwe, J. E. A. 1970. Identification of a compound responsible for cardboard flavor in beer. *Proceedings of the American Society of Brewing Chemists* 29, 192–7. doi:10.1080/00960845.1970.12006980.

Jende-Strid, B. 1993. Genetic control of flavonoid biosynthesis in barley. *Hereditas* 119(2), 187–204. doi:10.1111/j.1601-5223.1993.00187.x.

Jin, Y., Speers, R. A., Paulson, A. T. and Stewart, R. J. 2004. Barley β-glucans and their degradation during malting and brewing. *MBAA Technical Quarterly* 41, 231–40.

Kihara, M., Kaneko, T. and Ito, K. 1998. Genetic variation of beta-amylase thermostability among varieties of barley, *Hordeum vulgare* L., and relation to malting quality. *Plant Breeding* 117(5), 425–8. doi:10.1111/j.1439-0523.1998.tb01967.x.

Kihara, M., Okada, Y., Kuroda, H., Saeki, K., Yoshigi, N. and Ito, K. 2000. Improvement of β-amylase thermostability in transgenic barley seeds and transgene stability in progeny. *Molecular Breeding* 6(5), 511–7. doi:10.1023/A:1026535407570.

Kilian, B., Özkan, H., Pozzi, C. and Salamini, F. 2009. Domestication of the Triticeae in the fertile Crescent. In: Feuillet, C. and Muehlbauer, G. (Eds), *Genetics and Genomics of the Triticeae*. Springer-Verlag, New York, NYpp. 81–119. doi:10.1007/978-0-387-77489-3.

Knudsen, S., Hambraeus, G., Bech, L. M., Sørensen, S. B., Skadhauge, B., Breddam, K. and Olsen, O. 2010. Patent WO 2010/063288 A2: Barley and malt-derived beverages with low DMS-level.

Krannich, C. T., Maletzki, L., Kurowsky, C. and Horn, R. 2015. Network candidate genes in breeding for drought tolerant crops. *International Journal of Molecular Sciences* 16(7), 16378–400. doi:10.3390/ijms160716378.

Kreis, M., Williamson, M., Buxton, B., Pywell, J., Hejgaard, J. and Svendsen, I. 1987. Primary structure and differential expression of β-amylase in normal and mutant barleys. *European Journal of Biochemistry* 169(3), 517–25. doi:10.1111/j.1432-1033.1987.tb13640.x.

Kunze, W. 2010. *Technology Brewing & Malting* (4th edn.). VLB Berling, Berlin.

Kuroda, H., Kojima, H., Kaneda, H. and Takashio, M. 2005. Characterization of 9-fatty acid hydroperoxide lyase-like activity in germinating barley seeds that transforms 9(S)-hydroperoxy-10(E),12(Z)-octadecadienoic acid into 2(E)-nonenal. *Bioscience, Biotechnology, and Biochemistry* 69(9), 1661–8. doi:10.1271/bbb.69.1661.

Langstaff, S. A. and Lewis, M. J. 1993. The mouthfeel of beer – a review. *Journal of the Institute of Brewing* 99(1), 31–7. doi:10.1002/j.2050-0416.1993.tb01143.x.

Leedham, P., Savage, D. J., Crabb, D. and Morgan, G. T. 1975. Materials and methods of wort production that influence beer filtration. *Proceedings of the Congress European Brewery Convention*, Nice, pp. 201–16.

Leiper, K. A., Stewart, G. G., McKeown, I. P., Nock, T. and Thompson, M. J. 2005. Optimising beer stabilisation by the selective removal of tannoids and sensitive proteins. *Journal of the Institute of Brewing* 111(2), 118–27. doi:10.1002/j.2050-0416.2005.tb00657.x.

Lermusieau, G., Noël, S., Liégeois, C. and Collin, S. 1999. Nonoxidative mechanism for development of trans-2-nonenal in beer. *Journal of the American Society of Brewing Chemists* 57, 29–33.

Lopez, M. and Edens, L. 2005. Effective prevention of chill-haze in beer using an acid proline-specific endoprotease from *Aspergillus niger*. *Journal of Agricultural and Food Chemistry* 53(20), 7944–9. doi:10.1021/jf0506535.

Ma, Y. F., Evans, D. E., Logue, S. J. and Langridge, P. 2001. Mutations of barley β-amylase that improve substrate-binding affinity and thermostability. *Molecular Genetics and Genomics: MGG* 266(3), 345–52. doi:10.1007/s004380100566.

MacGregor, A. 1991. The effect of barley structure and composition on malt quality. In: *European Brewery Convention Proceedings of the 23rd Congress*, Lisbon, pp. 37–50.

McMurrough, I., Madigan, D., Kelly, R. J. and Smyth, M. R. 1996. The role of flavanoid polyphenols in beer stability. *Journal of the American Society of Brewing Chemists* 54(3), 141–8. doi:10.1094/ASBCJ-54-0141.

Meilgaard, M. C. 1975. Flavor chemistry of beer: part II: flavor and threshold of 239 aroma volatiles. *MBAA Technical Quarterly* 12, 151–68.

Meilgaard, M. C. 1982. Prediction of flavor differences between beers from their chemical composition. *Journal of Agricultural and Food Chemistry* 30(6), 1009–17. doi:10.1021/jf00114a002.

Melis, M. and Eyben, D. 1992. The mash filter 2001: latest industrial results. *MBAA Technical Quarterly* 29, 18–9.

Molina-Cano, J. L., de Togores, F. R., Royo, C. and Pérez, A. 1989. Fast-germinating low beta-glucan mutants induced in barley with improved malting quality and yield. *Theoretical and Applied Genetics* 78(5), 748–54. doi:10.1007/BF00262573.

Morell, M. K., Kosar-Hashemi, B., Cmiel, M., Samuel, M. S., Chandler, P., Rahman, S., Buleon, A., Batey, I. L. and Li, Z. 2003. Barley sex6 mutants lack starch synthase IIa activity and contain a starch with novel properties. *Plant Journal* 34, 173–85. doi:10.1046/j.1365-313X.2003.01712.x.

Myllärinen, P., Schulman, A. H., Salovaara, H. and Poutanen, K. 1998. The effect of growth temperature on gelatinization properties of barley starch. *Acta Agriculturae Scandinavica, Section B – Soil and Plant Science* 48(2), 85–90. doi:10.1080/09064719809362484.

Nakano, H., Kawada, N., Yoshida, M., Ono, H., Iwaura, R. and Tonooka, T. 2011. Isolation and identification of flavonoids accumulated in proanthocyanidin-free barley. *Journal of Agricultural and Food Chemistry* 59(17), 9581–7. doi:10.1021/jf2019819.

Narziss, L. 1976. *Die Technologie der Malzbereitung* (6th edition). Enke Verlag, Stuttgart, Germany.

Nyborg, M., Outrrup, H. and Dreyer, T. 1999. Investigations of the protective mechanism of sulfite against beer staling and formation of adducts with trans-2-nonenal. *Journal of the American Society of Brewing Chemists* 57, 24–8.

Oozeki, M., Sotome, T., Haruyama, N., Yamaguchi, M., Watanabe, H., Okiyama, T., Kato, T., Takayama, T., Oyama, M., Nagamine, T., et al. 2017. The two-row malting barley cultivar 'New Sachiho Golden' with null lipoxygenase-1 improves flavor stability in beer and was developed by marker-assisted selection. *Breeding Science* 67(2), 165–71. doi:10.1270/jsbbs.16104.

Outtrup, H., Fogh, R. and Schaumburg, K. 1987. The interaction between proanthocyanidins and peptides. *European Brewery Convention, Proceedings of the 21st Congress*, Madrid, pp. 586-90.

Park, J. Y., Lee, J. Y., Choi, S. H., Ko, H. M., Kim, I. C., Lee, H. B. and Bai, S. 2014. Construction of dextrin and isomaltose-assimilating brewer's yeasts for production of low-carbohydrate beer. *Biotechnology Letters* 36(8), 1693-9. doi:10.1007/s10529-014-1530-5.

Priest, F. and Stewart, G. (Eds.). 2006. *Handbook of Brewing* (2nd edn.). Taylor & Francis.

Ranocha, P., McNeil, S. D., Ziemak, M. J., Li, C., Tarczynski, M. C. and Hanson, A. D. 2001. The S-methylmethionine cycle in angiosperms: ubiquity, antiquity and activity. *The Plant Journal: for Cell and Molecular Biology* 25(5), 575–84. doi:10.1046/j.1365-313X.2001.00988.x.

Ratnayake, W. S. and Jackson, D. S. 2009. Starch gelatinization. *Advances in Food and Nutrition Research* 55, 221–68. doi:10.1016/S1043-4526(08)00405-1.

Robinson, L. H., Evans, D. E., Kaukovirta-Norja, A., Vilpola, A., Aldred, P. and Home, S. 2004. The interaction Between malt protein quality and brewing conditions and their impact on beer colloidal stability. *MBAA Technical Quarterly* 41, 353–62.

Robinson, L. H., Healy, P., Stewart, D. C., Eglinton, J. K., Ford, C. M. and Evans, D. E. 2007a. The identification of a barley haze active protein that influences beer haze stability: the genetic basis of a barley malt haze active protein. *Journal of Cereal Science* 45(3), 335–42. doi:10.1016/j.jcs.2006.08.013.

Robinson, L. H., Juttner, J., Milligan, A., Lahnstein, J., Eglinton, J. K. and Evans, D. E. 2007b. The identification of a barley haze active protein that influences beer haze stability: cloning and characterisation of the barley SE protein as a barley trypsin inhibitor of the chloroform/methanol type. *Journal of Cereal Science* 45(3), 343–52. doi:10.1016/j.jcs.2006.08.012.

Scheuren, H., Sommer, K. and Dillenburger, M. 2015. Explanation for the increase in free dimethyl sulphide during mashing. *Journal of the Institute of Brewing* 121(3), 418–20. doi:10.1002/jib.234.

Schreiber, M., Wright, F., MacKenzie, K., Hedley, P. E., Schwerdt, J. G., Little, A., Burton, R. A., Fincher, G. B., Marshall, D., Waugh, R., et al. 2014. The barley genome sequence assembly reveals three additional members of the CslF (1,3;1,4)-β-glucan synthase gene family. *PLoS ONE* 9(3), e90888. doi:10.1371/journal.pone.0090888.

Slack, P. T., Baxter, E. D. and Wainwright, T. 1979. Inhibition by hordein of starch degradation. *Journal of the Institute of Brewing* 85(2), 112–4. doi:10.1002/j.2050-0416.1979.tb06837.x.

Taketa, S., Yuo, T., Tonooka, T., Tsumuraya, Y., Inagaki, Y., Haruyama, N., Larroque, O. and Jobling, S. A. 2012. Functional characterization of barley betaglucanless mutants demonstrates a unique role for CslF6 in (1,3;1,4)-β-D-glucan biosynthesis. *Journal of Experimental Botany* 63(1), 381–92. doi:10.1093/jxb/err285.

Tanner, G. J., Blundell, M. J., Colgrave, M. L. and Howitt, C. A. 2016. Creation of the first ultra-low gluten barley (*Hordeum vulgare* L.) for coeliac and gluten-intolerant populations. *Plant Biotechnology Journal* 14(4), 1139–50. doi:10.1111/pbi.12482.

Tester, R. F., South, J. B., Morrison, W. R. and Ellis, R. P. 1991. The effects of ambient temperature during the grain-filling period on the composition and properties of starch from four barley genotypes. *Journal of Cereal Science* 13(2), 113–27. doi:10.1016/S0733-5210(09)80029-5.

Thacker, S. P., Ramamurthy, V. and Kothari, R. M. 1992. Characterisation of barley p-amylase for application in maltose production. *Starch – Stärke* 44(9), 339–41. doi:10.1002/star.19920440906.

Tillett, I. J. and Bryce, J. H. 1993. The regulation of starch granule size in endosperm of developing barley grains. *European Brewery Convention Proceedings of the 24th Congress Oslo*, pp. 45–52.

Tonooka, T., Aoki, E., Yoshioka, T. and Taketa, S. 2009. A novel mutant gene for (1–3, 1–4)-beta-D-glucanless grain on barley (*Hordeum vulgare* L.) chromosome 7H. *Breeding Science* 54, 47–54.

van Mechelen, J. R., Schuurink, R. C., Smits, M., Graner, A., Douma, A. C., Sedee, N. J., Schmitt, N. F. and Valk, B. E. 1999. Molecular characterization of two lipoxygenases from barley. *Plant Molecular Biology* 39(6), 1283–98. doi:10.1023/A:1006118003998.

Vanderhaegen, B., Neven, H., Verachtert, H. and Derdelinckx, G. 2006. The chemistry of beer aging – a critical review. *Food Chemistry* 95(3), 357–81. doi:10.1016/j.foodchem.2005.01.006.

von Wettstein, D., Jende-Strid, B., Ahrenst-Larsen, B. and Sørensen, J. A. 1977. Biochemical mutant in barley renders chemical stabilization of beer superfluous. *Carlsberg Research Communications* 42(5), 341–51. doi:10.1007/BF02906119.

Wang, J., Liu, L., Ball, T., Yu, L., Li, Y. and Xing, F. 2016. Revealing a 5,000-y-old beer recipe in China. *Proceedings of the National Academy of Sciences of the United States of America* 113(23), 6444–8. doi:10.1073/pnas.1601465113.

Xu, Y., Zhang, X., Harasymow, S., Westcott, S., Zhang, W. and Li, C. 2018. Molecular marker-assisted backcrossing breeding: an example to transfer a thermostable β-amylase gene from wild barley. *Molecular Breeding* 38(5), 63. doi:10.1007/s11032-018-0828-8.

Yoshigi, N., Okada, Y., Sahara, H. and Koshino, S. 1994. PCR cloning and sequencing of the β-amylase cDNA from barley. *Journal of Biochemistry* 115, 47–51.

Yoshigi, N., Okada, Y., Maeba, H., Sahara, H. and Tamaki, T. 1995. Construction of a plasmid used for the expression of a sevenfold-mutant barley β-amylase with increased thermostability in Escherichia coli and properties of the sevenfold-mutant β-amylase. *Journal of Biochemistry* 118(3), 562–7. doi:10.1093/oxfordjournals.jbchem.a124946.

Yu, J., Huang, S., Dong, J., Fan, W., Huang, S., Liu, J., Chang, Z., Tian, Y., Hao, J. and Hu, S. 2014. The influence of LOX-less barley malt on the flavour stability of wort and beer. *Journal of the Institute of Brewing* 120(2), 93–8. doi:10.1002/jib.122.

Zeeman, S. C., Kossmann, J. and Smith, A. M. 2010. Starch: its metabolism, evolution, and biotechnological modification in plants. *Annual Review of Plant Biology* 61, 209–34. doi:10.1146/annurev-arplant-042809-112301.

Chapter 15

Optimising the use of barley as an animal feed

David M. E. Poulsen, Queensland University of Technology, Australia

1 Introduction
2 What is 'feed barley'?
3 What do we want from 'feed barley'?
4 Optimising feed barley use
5 Understanding and optimising feed barley quality for different livestock species
6 Future trends and research opportunities
7 Conclusion
8 Where to look for further information
9 References

1 Introduction

Barley (*H. vulgare* sp. *vulgare*) is one of the six most globally significant feed grains. Wheat (*Triticum aestivum*), corn (*Zea mays*), rye (*Secale cereale*), oats (*Avena sativa*) and sorghum (*Sorghum bicolour*) round out the sextet. Triticale (× *Triticosecale*) is a relatively minor crop which adds to the pool of grain used as feed in some countries. Pulse and oilseed meals are often incorporated into livestock rations with the above-mentioned grains for balancing protein, fats and other nutritional factors.

More than 80% of the world's barley crop is used as feed (Blake et al., 2011) in both grain-fed and grain-supplemented livestock production systems. Barley is arguably the most adaptable of the feed grains. It is the most important feed grain in cooler climates where it can be more successfully grown than corn and the other feed grains (Hunt, 1996). Barley is used to feed ruminant animal species (beef cattle, dairy cattle, sheep and goats), monogastric animal species (pigs and horses), poultry and fish. Because of its outer husk, barley has a higher fibre and lower starch content than the other feed grains. However, barley starch is more degradable in the rumen than sorghum or corn starch (Hunt, 1996). While the percentage of the barley starch intake which is fermented in the rumen or

http://dx.doi.org/10.19103/AS.2019.0060.22

hind-gut is similar to wheat and oats, it has a similar whole tract digestibility to corn starch and is intermediate between sorghum (low) and wheat and oats (high; Rowe et al., 1999). These, and other, properties of the barley grain make it a highly useful ingredient in feeding regimes as it can be strategically used to temper and moderate animal performance. Varying the structure and composition of barley grain, through exploiting genetic variation and processing opportunities, means that a great deal of flexibility can be applied to the use of barley in feeding systems. This chapter will add to the growing body of work demonstrating that barley has been undervalued as a feed grain.

The second significant use for barley is for making beer by the malting and brewing industry. Malting barley selection is driven by highly defined and narrow grain specifications which are required to achieve optimal malting and brewing performance. To encourage production of grain which meets these specifications, malting barley is purchased at a market-driven premium price. While making little or no difference to the price premiums, malting barley and malt specifications can be further split into unofficial categories to meet the requirements of different brewing processes; namely, malt only, malt plus solid adjuncts and malt plus liquid adjuncts. It is a general rule that malting barley can only be produced using specific barley varieties which have been approved by industry for that purpose.

The remaining two mainstream uses for barley occur in the distillation industries (e.g. whisky, shochu) and for use as human food. Selling barley into the distilling industry works similarly to the malting/brewing sector, in that grain attracts premium prices to meet the demand under a strictly defined set of specifications. Food barley specifications and prices are set according to consumer and/or processor demands. Given that consumer markets for food barley span first-, second- and third-world economies, the specifications and prices for food grade barley are very wide ranging.

The premium prices for malting barley tend to drive the worldwide barley industry, despite the heavy use of the crop for feed. Consequently, most of the global crop is grown using accredited malting varieties. However, the strict requirements to achieve malting grade classification in most markets combined with very high Genotype × Environment (G × E) and Genotype × Environment × Management (G×E×M) effects on grain quality (Eagles et al., 1995; Molina-Cano et al., 1997; O'Brien, 1999; Fox et al., 2008) mean that a significant proportion of the annual crop fails to meet malting specifications. Grain which does not achieve the malting, distilling or food specifications, along with production from designated feed varieties, then tends to be amalgamated into the bulk commodity known as 'feed barley'. This has unfortunately created a generalised perception in most markets that 'feed barley' is essentially failed malting barley. However, the real situation is highly complex and feed barley is much, much more.

2 What is 'feed barley'?

The common use of the words 'feed barley' is a generic term used by industry to classify any form of the grain which is fed to livestock, poultry or fish in a managed production system. It is, in effect, a catch-all term. Different classification names are used for feed barley segregation around the world; these include 'feed barley' in Australia, 'general-purpose barley' in Canada and 'barley' in the United States. Most jurisdictions are also likely to have two or more sub-categories of 'feed barley' which are segregated on the basis of physical attributes which are believed to be related to the energy content and feed value of the grain.

Malting barley is classified as 'malting barley' worldwide. It is mostly stored and traded in its unprocessed form as single variety lots. Malting is a managed bulk grain germination process and is best undertaken in single variety batches to reduce the risk of processing complications from uneven germination and grain modification rates. Finished batches of malted barley, otherwise known as malt, can be blended to brewers' specifications with less risk of affecting the end-product quality.

The added costs of segregating into single variety or style loads mean that most of the globally traded feed barley consists of blended bulk grain. In general, loads of feed barley will usually consist of a combination of grain types and varieties. Multi-variety bulks may be traded on an as-is basis, or, blended further to meet specific market requirements (e.g. blending to achieve a mean grain protein and/or test weight specification). It should be noted that there are exceptions in countries with highly sophisticated infrastructure such as Canada, where barley can be segregated to the 'boxcar' level. Where such segregation is available, it allows for some trading of more specific feed barley batches with particular attributes; for instance, classified to specific varieties or grain types including hulless, waxy, low-phytate and ruminant diet specific. Another situation where more specific batches of barley are traded as feed is when direct grain farm to livestock producer contracts are in use. Even malting barley has been traded for feed use under such contracts (Kym McIntyre [Queensland Barley Industry Development Officer], pers. comm.).

For the most part, the selection and development of varieties released as feed barley has been less constrained by specific quality parameters and more focussed on factors influencing grain production and crop protection. Yield has been the primary drive feed variety selection, with the attention to feed quality mostly limited to choosing genotypes with plump and heavy grain. Grain size and density has been said to lead to better feed quality by indicating more starch per grain, and better malting quality through enabling a high extract when the enzymes which drive good modification are also present (Burger and Laberge, 1985). This has led to the overall situation where feed barley varieties worldwide tend to outyield malting varieties (Blake et al., 2011). However, there is no doubt that yields of both feed and malting varieties have continued to increase. Friedt (2011) indicates that overall barley productivity has risen by an annual rate of 1–2% per annum due to both genetic and production system gains, and, that this has included significant gains in malting quality as well as improvements in yield stability through enhanced agronomic traits and genetic resistances to diseases and pests.

The sources of feed barley varieties have been both releases from dedicated high-yielding feed breeding programmes and as spin-off releases of high-yielding but lower-quality material from dedicated malting barley breeding programmes. Because malting barley commands a premium over feed barley in most countries, farmers have a perceived incentive to continue to plant malting varieties. Variety choice for sowing becomes a calculated risk, with farmers needing to consider the risks of not achieving malt grade against the potential of achieving the maximum yield advantage of the competing feed varieties. Blake et al. (2011) provided an elegant summary showing how the malting premiums drive the profitability of malting over feed barley feed in the United States. The data indicated that feed barley varieties need to achieve a minimum of 10% in yield advantage over malting varieties to become financially competitive. Our own experience in eastern Australia confirmed this information (McIntyre, pers. comm.), and even suggested that the yield advantage alone required to encourage farmers to switch to growing feed types was at least 15–20%, this being more than the offered premium for malt. The net effect of this market-led perception is that malting barley varieties continue to dominate the worlds production. With increasing economic pressures and funding shortfalls, the number of feed barley programmes focussing on yield improvement alone appears to have significantly decreased.

3 What do we want from 'feed barley'?

The aim of livestock production systems is to optimise the balance of feeding efficiency with average daily output of produce from the species in question, regardless of whether the end-product is meat, eggs, wool, milk, breeding capacity or performance. The feeding regime is, therefore, critical as it can substantially impact feeding efficiency, animal production and the produce end quality. As well as considering the grain, there needs to be an understanding of the animal as this is a complex biological system with multiple layers of genetic, environmental and management effects in play. Animal species is obviously important, as is genetic variation for a multitude of traits within the species. The age of the stock will also have an impact on how a diet is digested and so specific feed formulations will usually be applied at the various growth and development stages for each species.

Intensive livestock production systems tend to use blended feed rations to provide diet consistency over long time periods, as it is understood that changes of diet in those systems can set back productivity while the stock adapt to the new feed source. Free range and mixed production systems will typically use whole or processed grain as a supplementary feedstock to make up for gaps in the natural foraging of the stock or to 'finish off' stock growth prior to sale.

Barley is used as a source of energy and protein when fed in livestock production, with energy being the most important of the two traits. It may be fed as whole grain, rolled or flaked, processed and blended into a mash type ration, or, other styles of product such as extruded pellets. To work out what is required from a feed barley, it is best to start by defining the characteristics of a malting barley as these are more precisely known.

Classification of a barley lot as either malting or feed grade first depends on its variety of origin. Malting grade barley can only come from an industry approved and certified malting variety. Modern malting barley varieties have been bred for the physical and biochemical attributes required for use in industrial malting and brewing industries. To be classified as malting, they have passed the stringent quality testing regimes of both the breeding programmes and the industry.

Malting barley varieties have largely been selected for their genetic capacity to efficiently provide sugars and starches to feed yeasts in the brewing, or distilling, processes. For most of the 10 000 years or so that barley has been in cultivation and used to make beer, this has typically been understood as being lines with plump and uniform grain, high malt extract and quick and even germination when malted (Sparrow and Doolette, 1975). In recent history, however, the inherent attributes malting varieties must possess are far more defined and include high starch levels, appropriate levels of germination enzyme activity to modify starch and protein as required in the brewing process, moderate protein levels (e.g. 8–11%), large (but not too large) grain size, low-to-moderate grain dormancy and low levels of residual biochemicals such as β-glucan and unattractive flavour compounds which interfere with the brewing process and products. Some variation in the ranges of these traits is required to meet the needs of different brewing styles; however, all modern malting varieties will have been selected for these attributes.

The grain produced from malting barley varieties can be rejected from the premium classification for many reasons. The variety identity must be declared; noting that the grain accumulator usually has access to physical or biochemical tests which can prove identity in cases where there is doubt of authenticity. The barley must then meet the physical malting classification specifications as demanded by the market and regulated within the country of production. Reasons for rejecting barley from malt classification include, but are not limited to, protein outside of the specified range for malting, small grain, low test weight, weather damage, grain colour, presence of black point, the presence of fungal contaminants, contamination with other barley varieties, weed seeds or soil. If rejected as malt at a grain receival depot, the barley will usually be added to the appropriate bulk feed grade storage according to its sub-classification. It is also noteworthy that some of the above-mentioned reasons for rejecting malt barley can also have negative impacts on the value of the grain for feed purposes.

As malting barley specifications have become more defined, there have also been efforts to more clearly define attributes contributing to better barley feeding quality. Bhatty et al. (1975) stated that data from a study with 16 barley cultivars showed that good malting quality in barley is not incompatible feed value, specifically with high digestibility of energy (D) and digestible energy (DE), as determined by mouse feeding. Other studies with livestock of various kinds continue to demonstrate that malting barley quality is largely consistent with feed quality (Hart et al., 2008; Fox et al., 2009), with some exceptions that will be discussed later.

Feed grade barley is classified by relatively few characteristics, when compared to malting barley. Bulk density measurements, test weight or volume weight, have been the main assessment standard in many markets. However, most evidence from studies indicates that bulk density does not correlate well with animal performance from varying sources of fed barley (Grimson et al., 1987; Hunt, 1996). The ability to undertake research into feed barley quality has been constrained by several factors, including the complexity of dealing with multiple target species, the cost and ethical considerations involved in live animal feed-quality assays and a significant lack of funding support. As a consequence, most so-called feed barleys released prior to the 1990s were subjected to any form of selection for feed quality apart from grain size or bulk density measurements. Since then, however, there has been a trend towards the release of some varieties of feed barley which have been shown to address specific animal feed requirements.

Substantial genetic variation has been reported in barley for a range of attributes which may have impact on the dietary value of the grain for different species. One of the key characteristics identified for feed barley breeding aimed at monogastric species has been the hulless trait, controlled by the recessive *nud* gene. In hulless barley, the hull is less firmly attached to the kernel than in the conventional genotypes and consequently becomes detached during threshing. This results in a barley which has a higher starch content and DE per unit of volume or weight. Reports of other notable feed-related attributes have included high amylose or low amylopectin (Merritt, 1967), high-lysine (Munck et al., 1970; Oram and Doll, 1981) and high-lipid content (Parsons and Price, 1974). Each of these have been investigated further as breeding targets. An extensive study of the USDA barley core collection demonstrated the existence of a substantial amount of exploitable genetic variation for barley feed quality (Bowman et al., 2001).

Apart from yield and specific quality attributes, attention must be given to the disease and pest resistance traits which are required for crop protection. These have been given equally high priority in selection of both feed and malting varieties. This is warranted by the combination of negative effects which diseases and pests may have on yield, production stability, the starch and other quality attributed of starch content of grain from affected plants. In some cases, resistances are critical to protect from other negative biotic effects of pathogens, such as the potential presence of toxins in grain affected by Fusarium Head Blight (FHB).

To summarise this section, when considering what is required from a feed grain, in this case barley, it is important to consider the entire system from sowing the seed through to the end-product from the livestock production. In the modern world where input resources such as water, land and fertiliser are increasingly limited and precious, this means that optimisation of the entire supply and value chain is the only logical way forward to achieving an outcome for sustainable production and consumption. What we are really looking for is the feed grain system which gives us the best and most efficient animal production per hectare of sown ground.

4 Optimising feed barley use

4.1 Optimising feed barley production

G × E × M interactions have a significant effect on the quantity and quality of barley produced in a crop, with yield, starch content, grain protein concentration and grain size distribution being highly vulnerable to heat and moisture stresses (Eagles et al., 1995). The impact of heat and moisture stress on starch accumulation in the grain is most severe in the early stages of grain filling, while protein is affected by stress applied from early through to late grain fill (Savin and Nicolas, 1999). Nitrogen application can also have significant impacts on these factors, including yield, protein and starch concentration (Perrott et al., 2018b). Genotype interactions with both the seasonal environment and management practices can be highly significant, and the relative maturity class of the variety can be a major co-factor in influencing the genotypic responses (Eagles et al., 1995). These G × E × M effects create a large amount of variation in the feed value of barley. As the variation in barley grain protein and energy content is demonstrably large, and arguably greater than seen in other feed grains, it becomes essential to analyse barley which is to be used for either feed or malting purposes (Hunt, 1996). A study of over 1600 feed barley samples from the northwest United States demonstrated highly variable chemical composition and ruminal digestibility, with source, location and variety accounting for approximately equal variability in two data sets (Reynolds et al., 1992). Therefore, to be able to produce feed barley, it becomes essential to optimise the relevant production system for every growing area. A simple 'one-size-fits-all recipe' does not work due to broad suite of environments in which barley is produced. Regional studies to determine optimal growing conditions need to be routinely conducted in order to correctly sample the relevant genotypes, soil and environmental conditions. All of these production inputs vary over time, with changing environmental conditions being foremost in many minds. Barley production is discussed elsewhere in this book, and for the purposes of this chapter a recent sample study (Perrott et al., 2018a,b) is referenced as an example of the ongoing agronomic research required to achieve optimal feed grain production.

4.2 Breeding feed barley varieties

For reasons described previously, continuing to pursue the 'yield alone' (incorporating disease resistance) strategy for the development of feed barley varieties is commercially questionable in the current world market. High-yielding feed barleys have generally only succeeded at commercial levels in areas where the production of malting barley was somehow constrained. For instance, the variety Galleon was highly successful in certain South Australian production zones for many years due to its resistance to Cereal Cyst Nematode (CCN) and the variety Grout was popular in Queensland for several years due to its ability to produce a decent crop under drought conditions which had significantly impacted the available malting varieties. In both case, as new malting varieties with equivalent agronomic traits and effective genetic resistances were released, the area sown to the feed varieties rapidly decreased and the malting varieties became dominant.

Nevertheless, the breeding for 'just yield' strategy has been proven as useful in improving the overall barley gene pool, as genetics from the high-yielding material has made its way back into malting barley development and helped to increase the yields of malting barley varieties. Researchers have investigated the strategy of using molecular marker techniques to introduce malting quality genes and quantitative trait loci (QTL) into feed barley backgrounds for the express purpose of rapid yield improvement of malting types (Vasos et al., 2004).

The availability of a substantial level of genetic variation in barley for improvement of the crops feed quality has already been mentioned. This resource has led to the development of breeding programmes in which specific feed-quality traits have been targeted as part of the overall variety improvement strategy. Breeding programmes of this type have been established at least in North America, Europe, Australia and Asia. The programmes tend to either focus on developing feed types for ruminant species or for monogastric species.

Barley hull content is known to be a major contributor to the lower digestibility of barley grain when compared to wheat, oats and maize. In the monogastric industries, this is being addressed through the increasing use of hulless barley varieties. For hulled barley, the hull content varies from 7% to 25% among the full gene pool including two-row and six-row barley genotypes, whereas two-row barley as a subclass commonly displays lower hull content with a mean of 10% (Evers et al., 1999). Genetic variation within the modern cultivated two-row barley gene pool certainly exists; however, it is relatively narrow as demonstrated by the hull content range of 9–11% observed in six North American two-row varieties (Du et al., 2009).

The pork industry seems to have been the most proactive in supporting the development of specific feed varieties. This is reflected in results as, arguably, the most successful quality-specific feed barley varieties have been those produced for the North American pork industry by the breeding programmes operated by Alberta Agriculture and the University of Saskatchewan. These have included hulless, waxy, waxy-hulless and low-phytate varieties. The success of these barleys has been driven in part by both the investment of the pig producers in the breeding programme and the preparedness of the pig producers to pay a premium for grain that they want (Jim Helm, former Alberta Agriculture barley breeder, pers. comm.).

Another well-publicised and moderately successful specific purpose feed barley variety has been Valier (Blake et al., 2002), released by the Montana State University specifically to improve the performance of feedlot cattle. Valier was selected for the combination of excellent agronomic characteristics with improved cattle feeding performance, the latter being demonstrated in calf feeding trials comparing the variety as a breeding line against both of its parents (Boss et al., 1999). A comparative analysis of the nutritional value for cattle of Valier versus the predominant malting variety Harrington was conducted (Yu et al., 2003). The authors concluded that the nutritional advantage of Valier was highly associated with the characteristics of the feed after the barley was dry rolled. Despite chemical and nutritional composition of grain from each variety, there was little difference in rumen degradation characteristics between the two barleys.

The potential for developing even better barley varieties with improved generic and/or specific feed barley qualities is very high. While there is still a lot of research and development work to do, increased knowledge about animal dietary requirements and combined new breeding and selection technologies such as near-infrared spectroscopy (NIR) and molecular markers, have given us the capability to be highly focussed on making the necessary improvements to optimise feed barley performance.

4.3 The Premium Grains for Livestock Program: a model for developing the knowledge required to optimise the use of grains in feeding livestock

Arguably, knowledge is the most powerful weapon in a plant breeders arsenal. It enables the breeding team to make decisions which significantly increase the probabilities of finding the best genetic combinations for each situation. The breeding programmes must, of course, be market driven and working closely with the industry which represents their target market. Industries must be similarly driven and work together to achieve the best possible mutual outcomes, instead of competing against each other to squeeze the last dollar out of the value chain for themselves. This approach has worked for the Canadian pork industry, as discussed previously. The Australian grains and livestock industries have also been working together to gain a better understanding of the requirements for optimal feed grain performance in animal production. The Premium Grains for Livestock Program (PGLP) was a milestone project conducted in Australia, with the aim of characterising all available feed grain sources along with livestock feeding, digestion and growth factors to generate a major source of knowledge for use by the industry in optimising the national grain-fed animal value chain (Black, 2018). Barley, as Australia's third most commonly produced feed grain, featured throughout the PGLP studies and publications.

For ease of access, a series of papers from the PGLP Stage I were collated and published in a special edition of the *Australian Journal of Agricultural Research* (AJAR, 1999, Vol 50, Issue 5). This volume is a seminal source of information for researchers and others seeking information on the topic of optimising the utilisation of feed grains in the animal industries. The journal contains a number of useful articles, including important reviews of the various *in-vitro* and *in-vivo* feed-quality testing protocols in use for ruminants (Hogan and Flinn, 1999; Kitessa et al., 1999; White and Ashes, 1999), pigs (Moughan, 1999), poultry (Farrell, 1999; Ravindran and Brydan, 1999) and general (Peterson et al., 1999; Wrigley, 1999) assessment activities.

Kaiser (1999) presents arguments in favour of reducing grain processing costs and improving animal performance by increasing the digestibility of whole grains in cattle through the dual strategies: (1) breeding grains are more readily damaged during mastication and with more digestible seed coats, and (2) treatment of the grains with fibrolytic enzymes or chemicals such as NaOH.

A review of the effects of the common practice of combining grain with forage in ruminant feeds to improve animal performance (Dixon and Stockdale, 1999) concluded that both positive and negative effects were possible. The positive effects were mostly related to the grain providing additional nutrients which were deficient in the forage, or by stimulating forage digestion when added at low percentages of net intake. Negative effects were associated with lower intake and digestion of forage components, but could be minimised by careful selection of compatible grain and forage combinations. Barley, along with wheat and oats, is more likely to contribute to negative intake and digestion than sorghum or maize because of the risk of lower rumen pH from the highly fermentable starch. On the other hand, the authors suggest that a satisfactory compromise forage mixture using grain processed only to the extent necessary to be adequately digested in the entire gastrointestinal tract (GIT) will usually be easier to achieve with rapidly fermented grains such as barley than with slowly fermented grains such as sorghum.

In grain-fed production systems, the types of grain and processing need to be balanced to achieve optimal production and minimise risks of low pH issues in the rumen and/or intestines. Manipulating feed rations by adjusting mixes to take account for the relative digestibility of wheat, oats, barley, maize and sorghum and how this can be managed by the amount of rolling and steam flaking can be used to manipulate the main sites of starch digestion within the GIT of various species (Rowe et al., 1999).

The PFG Program highlighted that all feeding and breeding strategies which rely on grain quality as a key piece of knowledge must have rapid, cost-effective and accurate analytical tools in order to succeed (Wrigley, 1999). Past means of assessing have included visual inspection, physical characterisation and detailed laboratory chemical analysis. Visual and physical assessment tools provide an indirect, rudimentary and largely inaccurate means of estimating feed quality. On the other hand, laboratory analyses provide detailed information but can be expensive and take a long time to produce results for comparison of multiple samples. One of the most game-changing technologies for rapid and accurate grain-quality assessment is NIR, an imaging system which is now being used in grain receival depots, feed manufacturers, plant breeding programmes and research laboratories. One of the outputs from Stages I and II of the PFG was the release of a suite of NIR calibrations developed to give feed-quality predictions across the full range of grains being utilised in Australia (Black, 2018). Stage I of the PGLP produced a set of calibrations useful for characterising the DE of cereals to be used for pig feed (van Barneveld et al., 1999). By the end of the second stage of the Program, calibrations had been released to industry for use with whole and milled grains, on a dry matter or as fed basis and for cattle, sheep, pigs, broiler chickens and layer chickens. These calibrations have been useful to grain accumulators, feed manufacturers and livestock producers. However, they are of relatively little use in barley genetic improvement since, due to the nature of NIR technology and the physical and chemical variation between the grain of crop species, only the use of barley specific algorithms is likely to achieve significant feed-quality gains through plant breeding (Glen Fox, pers. comm.).

4.4 Near-infrared spectroscopy (NIR) and molecular markers: modern selection tools for improving feed quality

Fast, cost-effective and reliable screening techniques which allow the assessment of grain from large numbers of breeding lines is a critical need for any breeding programme targeting quality improvement for any purpose. The non-invasive predictive NIR technology is now widely adopted by crop breeding programmes as a means of rapidly screening for a diverse range of attributes, from predicting relatively simple traits such as protein and starch content, through to the significantly more complex traits such as DE and ADG in different livestock species which would usually rely on both *in-vitro* and *in-vivo* trials to confirm (Wrigley, 1999; van Barneveld et al., 1999; Fox et al., 2011).

One of the earliest uses of NIR to make feed-quality selections in barley occurred in the Alberta Agriculture breeding programme, with pigs as the primary target species (Jim Helm, Alberta Agriculture barley breeder, pers. comm.). The Queensland barley programme in Australia was also a relatively early adopter of NIR to select for feed quality, developing calibrations and using the technology to predict traits which had been identified from studies at Montana State University (Surber et al., 2000) as pertinent to developing better-quality barley varieties for feedlot cattle. By 2006, the Queensland barley breeding programme was testing grain samples harvested from over 10 000 plots within a 10-week period (Poulsen, 2006). This capability was expanded further as the breeding evolved over the next 5 years (Franckowiak, 2011). Like the physical and chemical traits which it predicts, significant variability can be observed in barley NIR trait measurements between genotypes, environments, crop management systems and the various combinations of interactions, particularly G × E (Fox et al., 2011). This reinforces the need for breeding programmes to constantly update and refine calibration sets to account for the gene pool/s in use and the test environments. As well as its use for selection per se, studies have shown that NIR can be used as a reliable means of developing data sets for the identification of molecular markers linked to single genes and QTL conferring barley malting and feed-quality attributes (Gous et al., 2012b).

Plant breeders also have access to molecular marker technologies to assist with both speeding up and improving selection efficiency in crop breeding programmes. The first two molecular marker maps of barley were developed using European and wild barley (*Hordeum spontaneum*) germplasm (Heun et al., 1991; Graner et al., 1991). These were a precursor of what was to come, and major steps with the development of barley molecular marker technologies occurred with the two major multi-organisational initiatives of the

North American Barley Genome Mapping Project (NABGMP) and the Australian Winter Cereals Molecular Marker Program (AWCMMP), the latter which featured genome mapping for both barley and wheat. The significance of these initiatives, in which the barley research between the two programmes was also linked, was that they used the combined resources of breeding programmes, research teams and laboratories from multiple organisations, growing regions and countries to produce an unsurpassed series of plant genetic maps using multiple marker types and linking these to genes, the expression of traits and QTL through the use of a globally collected phenotypic dataset.

The first NABGMP high-density map from the cross between the six-row feed variety Steptoe and 6-row malting variety Morex (Kleinhofs et al., 1993) was one of the most significant steps in modern barley improvement, in that it provided a framework to study, identify and map genetic variation for a broad set of traits. Grain and malting quality traits were quickly added to the Steptoe × Morex map (Hayes et al., 1993), and subsequently to the second major NABGMP population from the two-row malting barley cross Harrington × TR306 (Mather et al., 1997; Igartua et al., 2000).

The Australian-led barley mapping project was somewhat different from the NABMMP, in that multiple barley populations were mapped to represent a broad selection of the gene pool being used by the several state-based breeding programmes which were in operation during the period that the AWCMMP was initiated (Karakousis et al., 2003a,b; Barr et al., 2003a,b; Read et al., 2003; Cakir et al., 2003; Pallotta et al., 2003; Long et al., 2003). The individual populations were mapped and then used to construct the first published consensus map of the barley genome (Karakousis et al., 2003c) based on five of the AWCMMP barley maps and using Singe Sequence Repeat (SSR), Random Fragmented Polymorphic DNA (RFLP) and Amplified Fragmented Polymorphic Marker (AFLP) marker types.

The AWCMMP mapping populations included feed and malting barley parents to ensure that phenotypic data for a broad range of grain, malting and feed-quality attributes could be captured and mapped. The 'feed' parents were not speciality feed varieties, but were bred either for high yield in specific target environments or had been released as potential malting varieties but later classed as feed after being rejected by failing to meet malting and brewing industry testing protocols. Therefore, most of the grain-quality traits mapped through the AWCMMP barley populations were either the general quality traits required to meet receival specifications (grain size, protein content) or more specific malting attributes.

A comprehensive study conducted as part of the AWCMMP and using a combination of Australian and International barley maps indicated that the detectable QTL for grain weight and size were, at that time, all associated with barley developmental genes (Coventry et al., 2003). These genes included the *Ppd-H1* locus (long day photoperiod response) in European mapping populations and studies, the *eps2* locus (developmental rate determination) in Australian mapping populations and studies, the *sg1*, *sg2* and *sg3* vernalisation loci and the allelic *denso* and *sdw1* semi-dwarfing loci found in two- and six-row barleys, respectively. Because of the high level of influence of those genes on grain size, and the significant G × E effects which affect their expression, the mapping of non-developmental genes affecting grain size requires a specific mapping population where the major developmental loci are not segregating. However, the use of molecular markers to select for grain size may well be unnecessary, as it has been demonstrated that grain size is highly heritable (Fox et al., 2006) and that the combination of modern statistical analysis with traditional breeding and selection methods can be highly effective in selecting large-grained genotypes.

Genes contributing to the variation in grain protein have also been mapped (See et al., 2002; Emebiri et al., 2003). While both of the studies were intended to develop marker systems for selecting low protein varieties to increase the chances of growing malting grade barley in less favourable conditions, they could also be used to enhance feed barley breeding strategies.

The hulless barley gene *nud* has been mapped to the long arm of chromosome 7H (Fedak et al., 1972), and linked to microsatellite markers (Becker and Heun, 1995). Nevertheless, it took some time for additional specific feed-quality traits to be investigated by either the NABMMP or AWCMMP. A population derived from a cross between the two-row hulled feed barley Valier and six-row hulless Swiss landrace PI370970 has been used to map and identify multiple QTL for in-situ dry-matter digestibility (ISDMD) and post dry rolling particle size (PS), with 35–62% of the variation in ISDMD and 26–53% of the variation in each trait, respectively, linked to the vrs1 locus which controls the two-row/six-row trait (Abdel-Haleem et al., 2010b). Thirty main effect and four epistatic QTL were also mapped in the Valier × PI370970 population as related to the expression of ADF, starch content and protein content (Abdel-Haleem et al., 2010a). Again, there were strong associations with the feed-quality traits and a major gene. In this second study of the population, major QTL for ADF and starch content were associated with the *nud* gene which confers the hulless trait. This association between major developmental traits and QTLs reflected information from previous studies of the impact of the genes on physical and chemical grain characteristics.

Thirty-two main effect and five epistatic QTL for cattle feed-quality characteristics were added to the Steptoe × Morex mapping data; namely, QTL for crude protein, ADF, starch content, PS, ISDMD for cracked and ground grain and the starch digestibility of ground grain (Abdel-Haleem et al., 2012). An Australian study, using the Tallon × Scarlett AWCMMP mapping population and fistulated cattle, identified one QTL associated with digested protein and two QTL associated with DMD (Gous et al., 2012a). The genetic associations as described previously indicate that selection for improvement of specific feed barley quality attributes can be effectively managed through a combination of traditional section for major developmental genes and marker-assisted selection for more subtle genetic effects. However, an advantage of marker-assisted selection is that it can be applied in seedlings as a tool to select or eliminate heterozygotes with otherwise undetectable recessive genes and homozygotes before the plants need to mature to the age where they are expressed.

4.5 Processing

Whole grains can be readily fed to sheep and chickens; however, as whole grain can be difficult for most other animals to digest and the barley hull has been shown to reduce the grains digestibility when compared to wheat, some form of grain processing can be used to improve the accessibility of barley starch and protein within the GIT (Rowe et al., 1999). The best type of processing to use is generally determined by the animal species, and particularly, adjusting the feed mix so that the main sites of grain digestion within each species GIT are optimised to make the most of the nutritional value of the grain.

One of the simplest processing methods to improve digestibility is dry rolling of the whole grain, which can then be fed as is or blended into a ration mixture with forage material and/or seed meals. Dry rolling has been shown to significantly increase grain digestibility, with the effect of dry rolling in barley being significantly greater than seen in most other grains. For example, an Australian experiment using 2-year-old steers demonstrated an improvement in digestibility after dry rolling in barley, wheat and oats of 41.9%, 27.7% and 4.0%, respectively (Toland, 1976). An alternative to dry rolling is hammer milling, which similarly uses physical force to break the grain, thereby exposing the starch in the endosperm by both reducing the effect of husk and increasing the access to starch by the larger surface area-to-volume ratio of the rolled or milled grain particles. Milling has a significantly more disruptive effect on the grain, in particular resulting in small PS. Care must be taken not to over process grain, because rapid digestion as a result of excessive small particles or flour in the feed is likely to lead to low rumen pH and subsequent acidosis and low-fibre digestibility in ruminant species (Rowe et al., 1999). It is important, therefore, to adjust the rolling or milling equipment to achieve optimal PS to match the characteristics of each separate load of grain.

Steam flaking can also significantly improve barley digestibility. Most studies show that steam flaking offers little advantage over dry rolling in barley (Rowe et al., 1999). There appears to be little difference in digestibility in cattle between rolled and steam flaked barley, apart from the first month of feeding where steam flaking may offer a small advantage (Grimson et al., 1987). The downside of such treatments to markedly increase the digestibility of starch is that they can contribute to a rapid breakdown of starch in the rumen and the consequent problem of acidosis, leading to poor health and performance in cattle (Blake et al., 2011). Sheep are also vulnerable to acidosis, which will affect both weight gain and wool production (Hynd and Allden, 1985).

In general, the optimal way to process barley for ruminants is dry rolling (Rowe et al., 1999). Steam flaking will improve overall digestibility; however, it comes with a higher risk of high fermentation and acidosis. Additionally, the economics of the two systems means that dry rolling is likely to be more cost effective due to the additional processing costs of steam flaking for relatively little, if any, gain. The opposite holds for pigs, where steam flaking results in higher intestinal and overall digestibility of grains, and gut fermentation is not as significant a risk to the animal. As indicated previously, poultry can be fed whole grain; however, steam flaking will improve the overall digestibility when compared to either whole or cracked grain feed preparation methods.

A minor use of barley in stock feeding is the addition of small amounts of whole or rolled barley to silage as a means of increasing the digestibility of the forage component. This practice has been shown to significantly improve cattle feeding performance when compared to a silage-only diet (Jacobs et al., 1995), and is particularly useful in dairy cattle production. A 2011 study showed that barley, corn-dried distillers grain, wheat-dried distillers grain and wheat middlings when fed with corn silage were interchangeable in such a feeding system without adverse impacts on production (Holtshausen et al., 2011).

Another form of barley modification for feed rations, and one which is not available to most other grains, is the use of spent brewers grain. The malting and brewing process significantly modifies the barley, with the resulting material having a lower carbohydrate content and elevated protein levels. Because the spent grain is usually wet, it is not suitable for long-distance transport and this suggests that such material is more likely to be useful to industries closer to the urban centres where the breweries are located. Studies have shown that spent brewers grain is a very useful feed source in dairy production, and can be used to replace rolled barley in diets incorporating a mixture of hay and grain. Daily yields of milk, protein and non-fat milk solids have been shown as significantly higher from diets incorporating brewers grain over rolled barley (Valentine and Wicks, 1982).

5 Understanding and optimising feed barley quality for different livestock species

As has already been discussed in this chapter, barley is fed to diverse species. Cereal grains are primarily fed to animals as an energy source

(Rowe et al., 1999), with the protein content and type being an important secondary condition. Hulled barley is also recognised as a source of dietary fibre, particularly for ruminants. The different dietary needs of each species, together with the varying dietary needs of animals and birds in different age categories, means that it becomes very complicated to define a specific set of criteria for the ideal and generic 'feed barley' (Bleidere and Gaile, 2012). For example, hulled barley is arguably better for ruminant species as it is easier to regulate gut fermentation due to the increased fibre, while hulless barley has become preferred for feeding monogastric species for a range of reasons including higher energy value and lower faecal output. Many grain-quality attributes have been identified, including physical, biochemical and nutritional characteristics. It thus becomes necessary that to optimise feed barley use, we need to understand what each type of animal requires from the grain.

5.1 Ruminants: cattle

Barley is the most significant feed grain used for cattle production systems in North America, and competes region by region with wheat and sorghum because of seasonal availability and crop distributions in Australia. The barley quality requirements for beef and dairy cattle are essentially the same, noting that there are significant differences between the feeding systems used for beef production, milk production and breeding. Most breeder cattle are fed on pasture and forage diets, with grain used for supplementation when the natural feed supplies are running short.

Beef cattle are either pasture or grain fed, with grain feeding being used mostly to 'finish off' animals which have been pasture fed through to a desired age before being introduced to a high-grain diet. This system has been shown to result in faster weight gain and higher quality meat. For example, 12-month-old Friesian steers showed a 31% faster growing rate on a predominantly barley diet than similar cattle fed predominantly on pasture, reaching a target fasted liveweight of 457 kg on average 69 days faster (Purchas and Lloyd Davies, 1974). Testing of meat from the slaughtered animals showed that the grain-fed steers also had significant higher fat content and greater tenderness measurements in specified cuts. A laboratory taste panel also indicated a more acceptable flavour from topside roasts of the barley-fed animals.

Numerous studies have been undertaken on the preferred barley attributes to optimise the performance of beef cattle. The optimal use of barley for ruminant feed can be related to both physical and chemical aspects of the grain. As previously discussed, large and plump grains are preferred because they contain a high percentage of starch compared to husk. Because of the presence of the husk, hulled barley contains 10–15% less starch per unit weight than maize, sorghum, wheat or oats (van Barneveld, 1999b). However, whole-tract digestibility of barley is superior to sorghum, similar to maize and slightly below wheat and oats. Anecdotal evidence suggests that the optimal barley grain protein levels for feeding cattle are believed to be in the range of 12–13%, as compared to the 8–12% protein range preferred for the malting process. The reasoning behind this being thought of as an optimal range for cattle is that the relatively indigestible hordein storage proteins are significantly higher in the high-protein barley, and as protein drops below 12% the nutrition available to support the rumen bacteria begins to decline. The endosperm protein matrix is also important in starch digestion, as it supports the starch granules, and as it degrades, it enables the ruminal bacteria to access the surface area of the starch. This is a primary factor in the lower digestibility of sorghum and corn as the protein matrices in those grains are less digestible than that found in barley, wheat and oats (van Barneveld, 1999b).

A high level of starch fermentation in the rumen is undesirable, as it can lead to a low pH and the consequent problem of acidosis. Acidosis can have significant negative effects on animal growth, milk production and overall animal health. In extreme cases, animal deaths will occur. Feedlots have learned how to manage acidosis, by using appropriate feed blends incorporating grain, silage or dry forage and pulse and oilseed meals. Systems vary between production areas and feedlot dieticians and managers learn how to manage acidosis risk accordingly, depending on the availability of produce at a cost-effective price. Anecdotal evidence suggests that this appears to correlate with the preference to use feed wheat as compared to feed barley, respectively, in Australian versus North American feedlots. A Canadian study investigating mixture proportions of silage and feed barley determined that optimum feed utilisation was obtained using more extensively processed barley combined with 3–6% of the dietary DM as barley silage (Koenig and Beauchemin, 2011). Fine-tuning of the diet formulations to decrease the risk of acidosis were shown as linked to decreased feed-conversion efficiency, and it was concluded that a decrease in the cost of animal mortality or sickness was necessary to justify the increased cost and use of low-risk acidosis.

Studies have demonstrated that there is a very high level of genetic variability for rumen digestibility between feed barley cultivars and as much as 70% of that can be attributed to variation in bulk density, starch content and kernel weight (Khorasani et al., 2000). Surber et al. (2000) identified a number of more specific feed-quality traits as important to beef cattle production, namely, low ADF, low ruminal DMD and large PS after dry rolling. In contrast with wheat, grain hardness in barley as conferred by the *hin b1* and *hin b2* genes appears not to be associated with feed quality, including PS (Fox et al., 2007c). Genetic variation in the above-mentioned traits is available (Bowman et al., 2001), and some breeding programmes adopted selection for these traits as part of their breeding strategy (Blake et al., 2002; Poulsen, 2006). The Canadian varieties

CDC Dolly and CDC Helgason were identified as being good candidates as feed for cattle over four other Canadian varieties, on the basis of lower hull content, FA, PCA, fibres and moderate mean/median PS after dry rolling (Du et al., 2009).

5.2 Ruminants: sheep

As ruminants, sheep face similar issues as cattle when fed grains with readily accessible starch. One difference, however, is that sheep can be fed whole grain readily (REF). There appears to be little to no advantage in feeding processed barley to lambs (Sormunen-Cristian, 2013). Sheep are more likely to be pasture fed with supplemented grain, and are not typically housed in feedlot-like situations. Like cattle, sheep are vulnerable to low rumen pH and acidosis as a consequence of rapid digestion of starch. Acidosis can affect both animal growth and wool production. One way of managing the transition to barley-fed diets so as to minimise the risk of acidosis include a gradual introduction of the barley into the diet, administration of a course of antibiotics such as virginiamycin and the transfer of rumen fluid from well-adapted animals (Godfrey et al., 1992). An arguably better way to manage acidosis in sheep is simply to use a mixed-grain diet. The addition of slowly fermenting grains such as sorghum or corn into a barley diet has been shown to improve rumen pH, and by association, animal performance (Yahaghi et al., 2012).

Lambs appear to adapt faster to diets supplemented with barley as opposed to oats (Kenney, 1986; Sormunen-Cristian, 2013). Kenney (1986) found that liveweight was greater, but dressing percentage was lower on oat diets as compared to barley or wheat diets. In comparison, growth rate, carcass weight, slaughter percentage, kidney fat and carcass fat were shown to be higher with barley than oat dietary supplementation in the study conducted by Sormunen-Cristian (2013).

5.3 Pigs

Wheat and barley are commonly used in pig diets, with barley known to have lower average DE than wheat. In general, early studies demonstrated that pigs fed barley as compared to wheat tended to produce a similar to slightly higher liveweight gain despite a poorer efficiency of feed conversion (Lawrence, 1970). Barley, and other cereals, are most effective in pig diets when they have been processed, as all forms of processing significantly increase DE over that of grains (van Barneveld, 1999a).

A Canadian study assessing the dietary performance in recently weaned pigs of several wheat and one barley variety found that the barley-based diet had both lower DE and metabolizable energy; however, the difference was not reflected in live pig performance in an ad libitum feeding regime (Bowland, 1974). It was theorised that the higher lysine and threonine levels in the barley when compared to the wheat may have contributed to the good growth performance of the pigs on the barley diet, despite the lower DE than the wheat diets. More recently, a large feeding study using three feed mixes of soybean meal combined with either wheat, 'low-quality' hulled barley and 'high-quality' hulled barley echoed the aforementioned 1974 study and concluded that barley can full replace wheat in pig diets and achieve equivalent or better growth performance, despite the lower nutrient digestibility and energy value of the barley (Nasir et al., 2015). The addition of enzymes to pig diets can help to improve the digestibility of lower-quality barley (Clarke et al., 2018). By adding a combined β-glucanase and β-xylanase supplement to both low- and high-quality barley feed mixes, ADG, ADFI and nutrient digestibility were improved in the low-quality barley mix.

Data from analysis of over 125 barley cultivars has produced a range of DE estimates from 11.7 to 16.0 MJ/kg DM (van Barneveld, 1999a), this being an economically significant range when it comes to pig producers being able to optimise their productivity. The data suggested that energy digestibility of barley does not reflect gross energy (GE) content. The analysis of this, and data from other grains, indicated that cultural conditions and agronomic practices such as fertiliser application appear to have a greater influence on the availability of amino acids and energy than do the growing region or year.

Bhatty et al. (1975) recommended the desirability of developing hulless feed barley cultivars, especially for pigs and poultry, from mouse feeding trial results showing that the proportion of hull content being the major factor influencing the D and DE of barley. In their study, the hulless genotypes showed an approximately 8% increase in DE over the hulled types. Given that the average hull content of hulled barley is approximately 10% (Evers et al., 1999), this strongly suggests a direct effect from removing the hull and not as a by-product of the biochemistry of husk adhesion. Bhatty et al. (1975) also suggested that lipid content of barley may have a major influence on DE, while kernel size and lysine content had no effect on DE and were therefore viable breeding targets for improving DE, grain appearance and amino balance, respectively. Their follow-up study on pigs confirmed the above-mentioned recommendations with the DE of hulless barley being equal to or superior than corn in the new feeding trials (Bhatty et al., 1979). Additionally, the trials indicated that hulless barley could be a better feed for pigs than corn, because of the higher and superior protein content of the barley. The dietary advantage of hulless over hulled barley in the trial was shown to be 11.1% for D and 14.7% for DE. In another study, Mitchall et al. (1976) confirmed the advantages of hulless versus hulled barley by providing increased DE and total dietary protein content, even though there appeared to be a decrease in the protein digestibility of the hulless barley tested relative to its hulled isogenic type. By the 1990s, several hulless barley varieties had been released in North America and were in regular use as feed by the pork industry (Aherne, 1990; Darroch et al., 1996). Much of this work had been made possible by funding contributions made by the pig producers to breeding and research programmes, such as those run by Alberta Agriculture (Jim Helm, pers. comm.).

As well as being a source of energy for pigs, barley also contributes a substantial proportion of their protein requirements. Thacker et al. (1988) concluded that by combining an improvement in feed efficiency of hulless over hulled barley with the lower levels of soya-bean meal required with barley diets in order to meet the pigs' requirements for essential amino acids, sufficient incentive may be provided to encourage the incorporation of hulless barley in pig rations. However, the deficiency of some essential amino acids in barley protein, particularly lysine and threonine, necessitate the addition of protein supplements such as fish, meat or oilseed meal with or without pure methionine and lysine (Oram and Doll, 1981). Consequently, there have been plant breeding efforts to change the balance of amino acids in barley to render it more suitable for the porcine diet. Multiple high-lysine mutants have been discovered in barley; however, all are known to be associated with yield depression and lower-energy content due to reduced endosperm production. Genetic studies conducted in Australia and Europe, as reported by Oram and Doll (1981), identified potential biological pathways which could be exploited in breeding programmes to increase the yields of high-lysine variants and bring them closer to commercially viable yields. Exploitation of the material has been successful, with high-lysine varieties released and additional breeding lines showing greater commercial potential (Gabert et al., 1995; Darroch et al., 1996). This material has been shown to have a greater biological value for monogastric diets, as well as substantially reducing N excretion into the environment due to a reduction in the prolamin protein fraction (Gabert et al., 1996; Darroch et al., 1996). Transgenic approaches have been suggested to further increase the levels of lysine in feed and food barley (Shewry et al., 1994). The added attraction of the use of hulless varieties with higher levels of lysine, such as the variety Condor released in 1988, is that they may provide an economic benefit to pig producers because of their higher DE energy content when compared to hulled barley and wheat, reducing the need for costly fat supplements (Darroch et al., 1996).

Due to the cost, regulation and large amounts of grain required for conducting animal feeding trials, predictive tools had to be developed so that a greater degree of sophistication than just selecting for a hulless barley could be employed in pig-focussed breeding programmes. NIR is now the predictive tool of choice; however, earlier studies identified that apparent energy digestibility (AED) and total protein digestibility (TPD) assessment data from rats could effectively predict barley performance in pigs. *In-vitro* organic matter digestibility and mobile nylon bag dry-matter digestibility could also be used for AED. Gabert et al. (1996) used rats and young pigs to prove that high-lysine barleys would be beneficial to both monogastric animal health and performance and help to reduce N excretion into the environment.

The excretion of phosphorous into the environment from piggeries also has the potential to be a significant environmental problem (Pierce, 2000). This occurs because about 70% of the phosphorous found in barley is bound as phytate (Lott et al., 2000) and the necessary digestive enzyme phytase does not occur in pigs (Pointillart et al., 1984). There are two primary solutions to this problem, these being either addition of phytase to pig diets or development of barley varieties with lower phytate content. After the identification of low-phytate barley mutants (Larson et al., 1998), studies have concluded that the addition of this trait into feed barley varieties could offer significant advantages to the pig industry through lowering ration costs by reducing the need to purchase the inorganic phosphorous currently needed to provide a balanced diet (Thacker et al., 2003).

Shewry (2007) argued in favour of the use of genetic engineering to develop and release a competitively yielding low-phytate barley, given the yield disadvantages demonstrated by material developed from the two barley lines of this type, namely, Hiproly and RisØ (1508). A low-phytate variety named Piggy and released by Carlsberg in 1987 failed to achieve commercial success, possibly because the increased feed value did not compensate sufficiently for a small-yield penalty. Another conventionally bred low-phytate hulless variety, CDC Lophy-I, was released by the University of Saskatchewan's Crop Development Centre in 2008 (Rossnagel et al., 2008). Feeding trials conducted with the variety showed that pig phosphorous utilisation increased and that the grain had a similar DE by comparison to regular hulless barley (Ige et al., 2010). Lophy-I appears to produce reasonably competitive yields when compared with other hulless varieties grown in Canada (Gray et al., 2012). A second Canadian feeding study demonstrated higher ADG and nutrient digestibility than conventional hulless barley, and that addition of dietary phytase to the low-phytate barley diet improved feed conversion and nutrient digestibility (Woyengo et al., 2012).

5.4 Poultry

Although barley is a potential major source in energy in poultry feed, it has tended to be passed by in favour of other grains such as wheat, corn and sorghum. One of the reasons for this is that barley and wheat are also a source of anti-nutritive components which can have an impact on how effectively the complete diet is used by poultry (Hughes and Choct, 1999). Barley has been linked to a growth-depressing effect in broiler chicks (Anderson et al., 1960; Classen et al., 1985). Following multiple observations of the positive impact of using enzyme mixtures with barley-based diets in chickens, it has been determined that the β-glucan polymer which occurs at high levels in barley and is associated with viscous intestinal contents that reduce gut digestion is largely responsible for the effect (Burnett, 1966; Mannion, 1981; Classen et al., 1988a). Additionally, this leads to sticky droppings which can lead to significant waste management problems in egg and poultry production systems.

Australian tests of commercial enzyme admixture to poultry feed in Australia (Mannion, 1981) confirmed that the inclusion of β-glucanases and α-amylases increased chicken feed consumption, digestion and weight gain and appeared to eliminate the occurrence of sticky droppings when compared to the control birds. Classen et al. (1988a) confirmed that the addition of β-glucanases with barley grain rations increased growth and feed conversion in young birds and reduced the variability in performance created from feeding the chicks a range of barley samples with varying grain quality. The benefits from inclusion enzymes, such as β-glucanase, have continued to be demonstrated in other studies (Campbell et al., 1993), leading to the regular inclusion of exogenous glycanases in broiler feeds for many years (Bedford and Morgan, 1996).

There is some evidence which suggests that the effects of barley β-glucan/viscosity levels on growth rate may be age related. In Classens' et al. (1988a) experiments, the growth of chicks was slower in the first 21 days of feeding on the barley rations as compared to those on a wheat/corn diet; however, both diets then produced similar growth rates from day 21 on. Additionally, it has been observed that the negative effects of barley on broiler chick growth are not evident in older laying hens, meaning that barley can indeed have a place in poultry rations (Classen et al., 1988b).

Dunstan (1973) reported findings that replacing mixed-grain diets of wheat and oats with barley in laying hens did not significantly affect the number of eggs produced or hen body weights. They did, however, note that the feed consumption was greater for the barley diets as opposed to the mixed-grain diets used for the experiments. This was adjudged to be due to the lower metabolizable energy of the barley diets, which were estimated to be 91% of the mixed-grain diet.

The use of hulless instead of hulled barley in chicken feed rations has been shown to have significant advantages. Removal of the hull results in lower crude fibre and higher levels of valuable nutrients, including crude protein. Hulless barley has been shown as comparable to wheat in the rations of egg laying hens, and contributes to the production of larger eggs and increased body weight over the use of conventional hulled barley (Classen et al., 1988b). In broiler chicks, hulless barley led to higher weight gain and lower feed-conversion rates when compared to hulled barley feed, with β-glucanase added to both rations (Campbell et al., 1993).

As with pigs, the locking up of phosphorous in barley as phytate presents both a nutritional and waste problem for poultry producers. It seems likely that low-phytate, hulless barley varieties may be useful to the industry. However, as phytases are relatively inexpensive, their use to reduce the need for expensive phosphorous supplements to address both issues has been very popular in Europe and North America. It seems likely that Australian chicken farmers will follow the same strategy, with lower feed costs being of most interest to broiler chicken producers while the egg industry will have a greater motivation to address waste management issues (Hughes and Choct, 1999).

6 Future trends and research opportunities

There seems little doubt that the largest challenge facing future feed barley production will be the need to develop adapted varieties and production systems which address the threats imposed by climate change. To summarise the available literature, it appears that the forecasted changes to the global climate will have significant impacts on both the quantity and quality of the current sources of all feed grains, including barley. This conjecture has been supported by a range of studies using both *in-situ* and modelling research techniques.

The projected impact of climate change on grain yield is variable, with many areas likely to suffer from production decline (Doltra et al., 2014; Ghahramani and Moore, 2016; Liu et al., 2017; Cammarano et al., 2019; Pirttioja et al., 2019), while some others may benefit from increased yield potential (Dubey and Sharma, 2018; Masud et al., 2018). A few production areas may see little change in yield and may even benefit from needing to apply less N fertiliser to barley crops (Holden and Brereton, 2006).

The key factor implicated in causing climate change is the increase of atmospheric CO_2. This is a complicating factor in trying to predict the impact of our changing climate as, apart from the impact of changing weather patterns on crop production, yield and quality, higher levels of atmospheric CO_2 are anticipated to affect all aspects of plant growth. As early as the 1990s, researchers were beginning to investigate the potential impact of higher CO_2 levels. In C_3 cereal species such as barley, wheat and rice, increases in atmospheric CO_2 have been shown to result in increased plant growth and grain yield. This is due to the greater sensitivity of the C_3 biological mechanisms to CO_2, as compared to the C_4 species, such as sorghum and maize, which have a tenfold greater CO_2 assimilation rate with low water demand and a better CO_2 capture system (Hashiguchi et al., 2010). A 1974 controlled environment pot study of barley and wheat found that barley responded to increased CO_2 by raising grain yield, largely through increased grain number per plant (Thompson and Woodward, 1994). However, grain nitrogen concentration was significantly reduced as the atmospheric CO_2 was increased. Significantly though, the overall loss of grain nitrogen was proportionately greater than the increase in yield. Another CO_2 study including two spring barley cultivars resulted in lower grain nitrogen, magnesium and calcium concentrations, and lower grain protein concentration with some variation observed in the relative proportion of amino acids between the ambient and increased CO_2 treatments (Manderscheid et al., 1995). Erbs et al. (2010) demonstrated significant reductions in crude protein concentration and sulphur in grain from a winter barley variety grown under elevated CO_2 levels. The data from the study indicated that at least part of the effect was caused by dilution due to increased starch content under the high CO_2 growth conditions.

These trends have been confirmed in a meta-analysis of data from 228 studies in barley, rice, wheat, soybean and potato, which indicated that reduction of grain nitrogen and protein is highly likely to occur across crop species with the increasing atmospheric CO_2 (Taub et al., 2008). In barley, wheat and rice, the impact on grain protein concentration of elevated CO_2 (540–958 $\mu mol\ mol^{-1}$) was a production of 10–15% less protein versus the concentration obtained under the control ambient CO_2 (315–400 $\mu mol\ mol^{-1}$) conditions. Higher rates of N fertiliser can be used to maintain the protein content of cereals under higher atmospheric CO_2 levels (Kimball et al., 2001), however, may be neither economical nor practical under the broad-scale dryland production systems used for most cereals.

Hogy and Fangmeier (2008) identified statistically significant CO_2-induced effects on the grain content of essential amino acids such as threonine, valine, isoleucine, leucine and phenylalanine. The relative globulin and B-hordein protein fractions of barley have been demonstrated to be significantly increased due to an elevated atmospheric CO_2 concentration (Wroblewitz et al., 2014). A meta-analysis which included wheat, rice and corn but not barley indicated that zinc (Zn) and iron (Fe) concentrations also decrease in grain under elevated atmospheric CO_2 levels, noting that both crop hydration and nitrogen fertiliser modified the crop responses (Al-Hadeethi et al., 2019).

However, looking at elevated CO_2 alone presents an unrealistic view of the potential impacts of climate change on grain crop production and quality. Increased temperatures and changed rainfall patterns will also have a significant impact on crop performance and grain quality (Högy et al., 2013), and interactions of all three environmental attributes will come into play in overall crop performance. While higher temperature stress and moisture stress have negative effects on vegetative growth and grain filling, higher CO_2 levels appear to have positive effects on the same traits (Jagadish et al., 2014). However, all three environmental factors have negative effects on grain physical and nutritional quality. A key question, therefore, becomes whether the positive impacts of higher CO_2 on vegetative growth and grain fill are enough to overcome the negative growth effects of heat and moisture stresses, as well as the reduction of grain nutritional quality.

In studies with multiple barley genotypes, the combination of elevated CO_2 and higher temperature was found to have a profound impact on grain yield and protein levels (Ingvordsen et al., 2015; Ingvordsen et al., 2016). While the elevated CO_2 decreased grain protein concentration by 5%, elevated temperatures resulted in a 29% increase in protein content. The net effect of the two treatments applied in combination was an increase of 8% protein content across the full set of genotypes. However, the combined treatments also resulted in a significant yield reduction, with the overall impact of a 23% reduction in harvestable protein. Given the significant effects of temperature and moisture stress on grain size (Eagles et al., 1995), it would seem likely that the overall impact of climate change on the global barley crop will be a large increase in the rejection of grain from the malting receival specifications. This should lead to a proportionate increase in bulk feed barley, which may help to offset some of the tonnages lost from droughted growing regions. However, the feed value of the bulk grain is likely to be less than the current situation, as a direct consequence of the likelihood of reduced grain protein content due to the higher atmospheric CO_2.

There is evidence that concerted research and development into adaptation of farming systems, changes to crop management and plant breeding can offset at least some of the impacts of increased temperatures, rainfall variability and other notable factors (Nendel et al., 2014; Liu et al., 2017; Ghahramani and Bowran, 2018). However, much more research is required.

Barley seems likely to be more amenable to breeding efforts to minimise the impact of climate change on bulk grain production, because its proven broad adaptability when compared to other cereal crops means that the necessary genetic variation is likely to exist to allow plant breeders to develop new varieties. Barley is grown in climatic zones which range from cold through to subtropical. We know that grain size in barley is highly heritable (Fox et al., 2006). Relatively untapped traits in barley such as stay-green (Gous et al., 2013) and root architecture (Abdel-Ghani et al., 2019) are also likely to assist the development of new varieties with improved adaptation to yielding as well as possible under environmental stress.

It seems highly likely that the issue of reduced grain nutritional quality because of climate change will be more difficult to address than plant adaptation. Breeders are in a position where they have access to many modern tools which can be used to speed up and refine the selection cycles to address the multiple stressors (Jagadish et al., 2014). These include the availability of multiple finely mapped QTLs for stress tolerances, technology platforms for high-throughput molecular marker analysis and steadily improving understanding of the genome, proteome and critical developmental pathways. Much of the investigations to date have focussed on individual stressors; however, to successfully address the potential impacts of climate change there will need to be more work conducted on the understanding of, and selection for, tolerance to multiple abiotic stresses.

One scientific discipline which appears to offer substantial promise in understanding plant physiological requirements to overcome the plant growth and grain-quality challenges posed by climate change is that of proteomics.

Proteomics is the study of the full range of proteins extracted from specific tissues, structures or organs and is becoming a powerful tool to understand and enable biotechnological improvement of crops. This includes developing the abilities to understand and specifically modify mechanisms of plant growth and development, responses to biotic and abiotic stresses and processes of seed or fruit development (Tan et al., 2017). Used in combination with fieldwork, the other 'omics' and molecular marker technologies, breeders will be able to direct their selection processes towards the enhanced or reduced expression of particular proteins at specific times in plant growth and development. For example, proteomic studies of leaf senescence and grain development rates will hopefully lead to the selection of cereal varieties which are more tolerant to heat and moisture stresses occurring during grain fill, and thereby, reduce yield loss from heat wave conditions or in warmer environments (Dupont and Altenbach, 2003). To be of most use to plant breeders, the proteome information should to be connected to other useful genetic tools. Such work has commenced, for example, the proteome and molecular marker maps for barley chromosomes 1H, 2H, 3H, 5H and 7H developed and published by Finnie et al. (2009).

In barley, proteomic studies are likely to create opportunities to further improve crop adaptation in response to climate change and for improving grain-quality attributes for malting, feed and food purposes. A substantial amount of work has been done to date and can be tracked down in the literature. As a starting point related to grain quality and the possibilities for use of proteomics in variety improvement, excellent reviews of cereal grain proteomics have been written by Finnie and Svensson, 2009; Finnie et al. (2011) and Labuschagne (2018).

Aqueous extracts from barley indicate the presence of at least 1000 individual proteins in developed grains, with the larger proportion in both concentration and number being found in the aleurone layer and embryo (Finnie and Svensson, 2003). The starchy endosperm contains less than 50% of the grain protein content by weight, despite being the largest component of the grain. Examination of the endosperm proteome indicates that starch synthesis enzymes are found throughout the internal structure of the starch granules, while hordein and serpin storage proteins and two non-starch synthesising enzymes are located on the granule surfaces (Borén et al., 2004). The aleurone and seedling proteomes appear to contain many of the same proteins as the endosperm, plus many others which include the enzymes involved in germination and endosperm breakdown (Finnie and Svensson, 2003; Finnie et al., 2011).

Most proteome studies of barley grain and related to barley end-use quality have been looking at factors which contribute to the making and quality of beer (Colgrave et al., 2013; Lastovicková and Bobálová, 2012; Iimure and Sato, 2013.) In a study including 18 feed and malting cultivars, Finnie et al. (2009) found that the subset of malting cultivars clustered together more closely through a pattern analysis of their proteomes than by a similar analysis of their SSR marker profiles. This supported the selection history of malting varieties by stringent quality assessment in laboratory and processing situations, and indicated that selection by proteome may be a more effective tool to develop high-quality barley varieties. Notably, the feed variety proteomes in the study did not cluster in the same manner as the malting varieties, a likely consequence of the less selective processes and lower emphasis on quality used to develop 'feed' varieties, as discussed earlier in this chapter. However, the clustering of the malting varieties was an indication that proteome selection for feed barley quality characteristics could be a highly effective way to improve the grains nutritional content.

Proteome studies have identified grain quality-related pathways which can be enhanced and/or selected through the DNA manipulation and tracking technologies now available to plant breeders. Examples include grain lysine content (Forsyth et al., 2005), chitinase (for protection from grain attacking pathogens) heat shock and putative nitrogen accumulation-related proteins (Finnie et al., 2004). Kaspar-Shoenefeld et al. (2016) conducted a major study into the proteomic systems in developing barley grain and identified and described 226 proteins expressed during grain fill. Protein activity was at its greatest in the earlier stages of grain development, especially from cellularisation through to the switch to the storage phase or during storage product accumulation. Transgenic C-hordein antisense barley lines were developed and found in a proteomic study to possess an improved balance of beneficial amino acids, as a result of alterations in the non-storage protein profiles in the grain (Schmidt et al., 2016). In particular, the expression of lysine-rich proteins was found to have been upregulated. The increase in lysine was not the product of a single change, but rather a consequence of multiple changes to production of storage and non-storage proteins as well as changes in the lysine mechanism itself. Further research is certainly warranted.

Another topic under investigation through proteomic research is in the identification, monitoring and regulation of toxins in food and feed products

(Giacometti et al., 2013). Current and future research opportunities include both detection and monitoring of toxins, and the potential to understand and manipulate the relationship between host and pathogen through which the toxins are produced and enter the grain or other feed material. There are several host pathogen relationships in barley, for example *Fusarium graminearum*, which would warrant investigation to help protect humans and animals from fungal toxins which can be present in or on the grain.

The genetic diversity of barley features many genes and attributes which can be exploited for improving the health, performance and productivity of all grain-fed animal and bird species. Conventional plant breeding methods have already produced successful varieties with a range of useful feed traits, and with the vast amount of information being gathered through the application of new technology platforms, significant additional improvements are highly achievable. Genetic gain is maximised when rapid and convenient genotype testing is available to the plant breeders to tilt the selection probabilities in their favour. This requires knowledge of which traits are the most efficient in both improving performance and heritability. Therefore, there is considerable room for researchers to improve the understanding of the interaction of barley with animal and bird digestive systems and identify additional traits or specifications which can be used to optimise feed performance. Animal proteomics will no doubt feature highly in the research undertaken to explore these possibilities. Alongside the trait discovery phase of this research, is the need for development of rapid, inexpensive and high-volume diagnostic tests to select the best genotypes for advancement in the breeding programmes.

There is also potential to further expand the genetic base in barley. Investigation of the world barley collections and wild populations is ongoing. Mutation, whether spontaneous or induced, has proven to be highly effective in developing new genetic variation for feed-quality traits. Finally, genetic engineering technologies are also likely to contribute significantly to the protein, compositional quality and adaptability of future barley varieties (Forsyth et al., 2005; Shewry, 2007), especially as the demand for feed grain production under increasing environmental and resource pressures continues.

However, very little of this work will be able to proceed without changes in the funding paradigms for barley research and the overall perceptions of feed barley as a bulk commodity created from downgraded malting barley. Significant changes in the approach of both the livestock and grain industries will be essential, in order to achieve a realisation that the best result for all will be to develop a system where both the production gains and profits are shared by all producers and industries from sowing to plate. Increasing pressures from population growth, climate change and the demands for food security are likely to be key drivers to stimulate the required changes to fall into place.

7 Conclusion

This chapter has explored the global barley markets and argued that feed barley should be viewed as comprising a series of different and potentially valuable categories, rather than the current perception as feed being a bulk commodity comprised of failed malting barley. Past and current opportunities for developing and enhancing different styles of barley have been explored, along with traits suited to the production of different species including ruminants, monogastric mammals, poultry and fish. Future challenges such as climate change have been explored, together with the potential of more recent technologies such as proteomics to help address the developing issues.

While primarily exploring the opportunities for optimising the use of barley as a feed through processing and genetic improvement, we have aimed to show that the ultimate solution to get the best feed performance from the crop is to take a higher-level viewpoint. To continue to succeed in the future world, the barley production and utilisation industries must seek to improve the whole system from crop production and stability, through grain handling, processing and ration formulation through to the enhancement of animal management systems.

8 Where to look for further information

There is quite a reasonable amount of additional information available to the researcher with an interest in feed barley. Feed barley specifications can be obtained from the relevant grain-marketing bodies in the various jurisdictions where barley is grown and/or used. Barley research and development is ongoing as a worldwide activity. Topics such as barley genetics, production agronomics, malting and brewing quality, and, the various 'omics are studied and published around the world. Notable output on specific quality attributes of feed barley has mostly been produced over the past decade from Australia, the United States of America and Canada.

9 References

Abdel-Ghani, A. H., Sharma, R., Wabila, C., Dhanagond, S., Owais, S. J., Duwayri, M. A., Al-Dalain, S. A., Klukas, C., Chen, D., Lübberstedt, T., von Wirén, N., Graner, A., Kilian, B. and Neumann, K. (2019) Genome-wide association mapping in a diverse spring barley collection reveals the presence of QTL hotspots and candidate genes for root and shoot architecture traits at seedling stage. *BMC Plant Biol.* 19:216. https://bmcplantbiol.biomedcentral.com/track/pdf/10.1186/s12870-019-1828-5.

Abdel-Haleem, H., Bowman, J., Kanazin V et al. (2010b) Quantitative trait loci of dry matter digestibility and particle size traits in two-rowed 9 six-rowed barley population. *Euphytica*. 172:419–33.

Abdel-Haleem, H., Bowman, J., Giroux, M., Kanazin, V., Talbert, H., Surber, L. and Blake, T. (2010a) Quantitative trait loci of acid detergent fiber and grain chemical composition in hulled × hull-less barley population. *Euphytica*. 172:405–18.

Abdel-Haleem, H., Bowman, J. G. P., Surber, L. and Blake, T. (2012) Variation in feed quality traits for beef cattle in Steptoe × Morex barley population. *Mol. Breed.* 29:503–14.

Aherne, F. X. (1990) Barley: Hulless. In *Nontraditional Feed Sources for Use in Swine Production*. Ed. Thacker, P. A. and Kirkwood, R. N., Butterworth, Toronto, pp. 23–31.

Al-Hadeethi, I., Seneweera, S., Li, Y., Odhafa, A. K. H., Al-Hadeethi, H. and Lam, S. K. (2019) Assessment of grain quality in terms of functional group response to elevated [CO_2], water, and nitrogen using a meta-analysis: Grain protein, zinc, and iron under future climate. *Ecol. Evol.* 9:7425–37.

Anderson, J. O., Wagstaff, N. K. and Dobson, O. C. (1960) Value of barley and hulless barley in rations for laying hens. *Poult. Sci.* 39:1230 (Abstr.).

Barr, A. R., Jefferies, S. P., Broughton, S., Chalmers, K. J., Kretschmer, J. M., Boyd, W. J. R., Collins, H. M., Roumeliotis, S., Logue, S. J., Coventry, S. J., Moody, D. B., Read, B. J., Poulsen, D., Lance, R. C. M., Platz, G. J., Park, R. F., Panozzo, J. F., Karakousis, A., Lim, P., Verbyla, A. P. and Eckermann, P. J. (2003a) Mapping and QTL analysis of the barley population Alexis × Sloop. *Aust. J. Agric. Res.* 54:1117–23.

Barr, A. R., Karakousis, A., Lance, R. C. M., Logue, S. J., Manning, S., Chalmers, K. J., Kretschmer, J. M., Boyd, W. J. R., Collins, H. M., Roumeliotis, S., Coventry, S. J., Moody, D. B., Read, B. J., Poulsen, D., Li, C. D., Platz, G. J., Inkerman, P. A., Panozzo, J. F., Cullis, B. R., Smith, A. B., Lim, P. and Langridge, P. (2003b) Mapping and QTL analysis of the barley population Chebec × Harrington. *Aust. J. Agric. Res.* 54:1125–30.

Becker, J. and Heun, M. (1995) Barley microsatellites: Allele variation and mapping. *Plant. Mol. Biol.* 27:835–45.

Bedford, M. R. and Morgan, A. J. (1996) The use of enzymes in poultry diets. *World's Poultry Science Journal*. 52:61–8.

Bhatty, R. S., Berdahl, J. D. and Cristison, G. I. (1975) Chemical composition and digestible energy of barley. *Can. J. Anim. Sci.* 55:759–64.

Bhatty, R. S., Christison, G. I. and Rossnagel, B. G. (1979) Energy and protein digestibilities of hulled and hulless barley determined by swine-feeding. *Can. J. Anim. Sci.* 59:585–8.

Black, J. (2018) Premium Grains for Livestock Program, Component 1: Co-ordination, An overview of outcomes from PGLP 1 & 2, Final Report, September 2008. https://ses.library.usyd.edu.au/bitstream/2123/5441/1/PGLPfinalreport.pdf

Blake, T., Bowman, J. P. G., Hensleigh, P., Kushnak, G. et al. (2002) Registration of 'Valier' barley. *Crop. Sci.* 42:1748.

Blake, T., Blake, V. C., Bowman, J. P. G and Abdel-Haleem, H. (2011) Barley feed uses and quality improvement. In *Barley: Production Improvement and Uses*. Ed. Ullrich, S. E., Wiley-Blackwell, pp. 522–31.

Bleidere, M. and Gaile, Z. (2012) Grain quality traits important in feed barley. *Proc. Latvian Acad. Sci., Section B*. 66: No. 1/2.

Borén, M., Larsson, H., Falk, A. and Jansson, C. (2004) The barley starch granule proteome – internalized granule polypeptides of the mature endosperm. *Plant Sci. J.* 166:617–26.

Boss, D. L., Bowman, J. G. P., Surber, L. M. M. Anderson, D. C. and Blake, T. K. (1999) Improving the feed value of barley for finishing steers. *Proc. West. Sec. Am. Soc. Anim. Sci.* 50:293–6.

Bowland, J. P. (1974) Comparison of several wheat cultivars and a barley cultivar in diets for young pigs. *Can. J. Anim. Sci.* 54:629–38.

Bowman, J. G. P., Blake, T. K., Surber, L. M. M., Habernicht, D. K. and Bockelman, H. (2001) Feed-quality variation in the barley core collection of the USDA National Small Grains Collection. *Crop. Sci.* 41:863–70.

Burger, W. C. and Laberge, D. E. (1985) Malting and brewing quality. In *Barley*. Ed. Rasmussen, D. C. American Society of Agronomy: Madison WI, pp. 367–401.

Burnett, G. S. (1966) Studies of viscosity as the probable factor involved in the improvement of certain barleys for chickens by enzyme supplementation. *Brit. Poult. Sci.* 7:55–75.

Cakir, M., Poulsen, D., Galwey, N. W., Ablett, G. A., Chalmers, K. J., Platz, G. J., Park, R. F., Lance, R. C. M., Panozzo, J. F., Read, B. J., Moody, D. B., Barr, A. R., Johnston, P., Li, C. D., Boyd, W. J. R., Grime, C. R., Appels, R., Jones, M. G. K. and Langridge, P. (2003) Mapping and QTL analysis of the barley population Tallon × Kaputar. *Aust. J. Agric. Res.* 54:1155–62.

Cammarano, D., Ceccarelli, S., Grando, S., Romagosa, I., Benbelkacem, A., Akar, T., Al-Yassin, A., Pecchioni, N., Francia, E. and Ronga, D. (2019) The impact of climate change on barley yield in the Mediterranean basin. *Eur. J. Agron.* 106:1–11.

Campbell, G. L., Rossnagel, B. G. and Bhatty, R. (1993) Evaluation of hull-less barley genotypes varying in extract viscosity in broiler chick diets. *Anim. Feed Sci. Technol.* 41:191–7.

Clarke, L. C., Sweeney, T., Curley, E., Gath, V., Duffy, S. K., Vigors, S., Rajauria, G. and O'Doherty, J. V. (2018) Effect of β-glucanase and β-xylanase enzyme supplemented barley diets on digestibility, growth performance and expression of intestinal nutrient transporter genes in finisher pigs. *Anim. Feed Sci. Technol.* 238:98–110.

Classen, H. L., Campbell, G. J., Rossnagel, B. G., Bhatty, R. S. and Reichert, R. D. (1985) Studies on the use of hulless barley in chick diets: Deleterious effects and methods of alleviation. *Can. J. Anim. Sci.* 65:725–33.

Classen, H. L., Campbell, G. L. and Grootwassink, J. W. D. (1988a) Improved feeding value of Saskatchewan-grown barley for broiler chickens with dietary enzyme supplementation. *Can. J. Anim. Sci.* 68:1253–9.

Classen, H. L., Campbell, G. L., Rossnagel, B. G. and Bhatty, R. S. (1988b) Evaluation of hulless barley as replacement for wheat or conventional barley in laying hen diets. *Can. J. Anim. Sci.* 68:1261–6.

Colgrave, M. L., Goswami, H., Howitt, C. A. and Tanner, G. J. (2013) Proteomics as a tool to understand the complexity of beer. *Food. Res. Int.* 54: 1001–12.

Coventry, S. J., Barr, A. R., Eglinton, J. K. and McDonald, G. K. (2003) The determinants and genome locations influencing grain weight and size in barley (Hordeum vulgare L.). *Aust. J. Agric. Res.* 54:1103–15.

Darroch, C. S., Aherne, F. X., Helm, J., Sauer, W. C. and Jaikaran, S. (1996) Effects of dietary level of barley hulls and fiber type on protein and energy digestibilities of Condor hulless barley in growing swine. *Anim. Feed Sci. Technol.* 61:173–82.

Dixon, R. M. and Stockdale, C. R. (1999) Associative effects between forages and grains: Consequences for feed utilisation. *Aust J. Agric. Res.* 50:757–73.

Doltra, J., Laegdsmand, M. and Olesen, J. E. (2014) Impacts of projected climate change on productivity and nitrogen leaching of crop rotations in arable and pig farming systems in Denmark. *J. Agric. Sci.* 152:75–92.

Du, L., Yu, P., Rossnagel, B. G., Christensen, D. A. and McKinnon, J. J. (2009) Physiological characteristics, hydroxycinnamic acids (ferulic acid, p-coumaric acid) and their ratio, and in situ biodegradability: Comparison of genotypic differences among six barley varieties. *J. Agric. Food. Chem.* 57:4777–83.

Dubey, S. K. and Sharma, D. (2018) Assessment of climate change impact on yield of major crops in the Banas River Basin, India. *Sci. Total. Environ.* 635:10–19.

Dunstan, E. A. (1973) The performance of laying hens on diets using barley as the main energy source. *Aust. J. Exp Agr. Anim. Husb.* 13:251-6.

Dupont, F. M. and Altenbach, S. B. (2003) Molecular and biochemical impacts of environmental factors on wheat grain development and protein synthesis. *J. Cereal Sci.* 38:133–46.

Eagles, H. A., Bedggood, A. G., Panozza, J. F. and Martin, P. J. (1995) Cultivar and environmental effects on malting quality in barley. *Aust. J. Agric. Res.* 46:831–44.

Emebiri, L. C., Moody, D. B., Panozzo, J. F., Chalmers, K. J., Kretschmer, J. M. and Ablett, G. A. (2003) Identification of QTLs associated with variations in grain protein concentration in two-row barley. *Aust. J. Agric. Res.* 54:1211–21.

Erbs, M., Manderscheid, R., Jansen, G., Seddig, S., Pacholski, A. and Weigel, H. J. (2010) Effects of free-air CO_2 enrichment and nitrogen supply on grain quality parameters and elemental composition of wheat and barley grown in a crop rotation. *Agr. Ecosyst. Environ.* 136:59–68.

Evers, A. D., Blakeney, A. B. and O'Brien, L. (1999) Cereal structure and composition. *Aust. J. Agric. Res.* 50:629–50.

Farrell, D. J. (1999) In vivo and in vitro techniques for the assessment of the energy content of feed grains for poultry: A review. *Aust. J. Agric. Res.* 50:881–8.

Fedak, G., Tsuchiya, T. and Helgason, S. B. (1972) Use of monotelotrisomics for linkage mapping in barley. *Can. J. Genet. Cytol.* 14:949–57.

Finnie, C. and Svensson, B. (2003) Feasibility study of a tissue-specific approach to barley proteome analysis: Aleurone layer, endosperm, embryo and single seeds. *J. Cereal. Sci.* 38:217-27.

Finnie, C. and Svensson, B. (2009) Barley seed proteomics from spots to structures. *J. Proteom.* 72:315–24.

Finnie, C., Steenholdt, T., Noguera, O. R., Knudsen, S., Larsen, J., Brinch-Pedersen, H., Holm, P. B., Olsen, O. and Svensson, B. (2004) Environmental and transgene expression effects on the barley seed proteome. *Phytochemistry*. 65:1619–27.

Finnie, C., Bagge, M., Steenholdt, T., Østergaard, O., Bak-Jensen, K. S., Backes, G., Jensen, A., Giese, H., Larsen, J., Roepstorff, P. and Svensson, B. (2009) Integration of the barley genetic and seed proteome maps for chromosome 1H, 2H, 3H, 5H and 7H. *Funct. Integr. Genomics*. 9:135–43.

Finnie, C., Sultan, A. and Grasser, K. D. (2011) From protein catalogues towards targeted proteomics approaches in cereal grains. *Phytochemistry*. 72:1145–53.

Forsyth, J. L., Beaudoin, F., Halford, N. G., Sessions, R. B., Clarke, A. R. and Shewry, P. R. (2005) Design, expression and characterisation of lysine-rich forms of the barley seed protein CI-2. *Biochimica et Biophysica Acta*. 1747:221–7.

Fox, G. P., Kelly, A., Poulsen, D., Inkerman, A. and Henry, R. (2006) Selecting for increased barley grain size. *J. Cereal. Sci.* 43:198–208.

Fox, G. P., Nguyen, L., Bowman, J., Poulsen, D., Inkerman, A. and Henry, R. J. (2007) Relationship between hardness genes and quality in barley (Hordeum vulgare). *J. Inst. Brew.* 113:87–95.

Fox, G. P., Bowman, J., Kelly, A., Inkerman, A., Poulsen, D. and Henry, R. (2008) Assessing for genetic and environmental effects on ruminant feed quality in barley (*Hordeum vulgare*). *Euphytica*. 163:249–57.

Fox, G., Kelly, A., Bowman, J., Inkerman, A., Poulsen, D. and Henry, R. (2009) Is malting barley better feed for cattle than feed barley? *J. Inst. Brew.* 115:95–104.

Fox, G., Borgogne, M. G., Flinn, P. and Poulsen, D. (2011) Genetic and environmental analysis of NIR feed quality predictions on genotypes of barley (Hordeum vulgare L.) *Field. Crops. Res.* 20:380–6.

Franckowiak, J. (2011) Barley Breeding Australia - Northern Node DAQ00110. Final Report. http://era.daf.qld.gov.au/id/eprint/3003/1/GRDC_final_report_DAQ00110.pdf.

Friedt, W. (2011) Barley Breeding History, Progress, Objectives, and Technology - Europe. In Barley feed uses and quality improvement. In *Barley: Production Improvement and Uses*. Ed. Ullrich, S. E. Wiley-Blackwell, pp. 160–71.

Gabert, V. M., Brunsgaard, G., Eggum, B. O. and Jensen, J. (1995) Protein quality and digestibility of new high-lysine barley varieties in growing rats. *Plant. Food. Hum. Nutr.* 48:169–79.

Gabert, V. M., Jorgensen, H., Brunsgaard, G., Eggum, B. O. and Jensen, J. (1996) The nutritional value of new high-lysine barley varieties determined with rats and young pigs. *Can. J. Anim. Sci.* 76:443–50.

Ghahramani, A. and Bowran, D. (2018) Transformative and systemic climate change adaptations in mixed crop-livestock farming systems. *Agric. Sys.* 164:236–51.

Ghahramani, A. and Moore, A. D. (2016) Impact of climate changes on existing crop-livestock farming systems. *Agric. Sys.* 146:142–55.

Giacometti, J., Tomljanović, A. B. and Josić, D. (2013) Application of proteomics and metabolomics for investigation of food toxins. *Food. Res. Int.* 54:1042-51.

Godfrey, S. I., Boyce, M. D., Rowe, J. B. and Speijers, E. J. (1992) Changes within the digestive tract of sheep following engorgement with barley. *Aust. J. Agric. Res.* 44:1093–101.

Gous, P. W., Martin, A., Lawson, W., Kelly, A., Fox, G. P. and Sutherland, M. W. (2012a) QTL associated with barley (Hordeum vulgare) feed quality traits measured through in situ digestion. *Euphytica*. 185:37–45.

Gous, P. W., Martin, A., Lawson, W., Kelly, A., Fox, G. P. and Sutherland, M. W. (2012b) Correlation between NIRS generated and chemically measured feed quality data in barley (Hordeum vulgare), and potential use in QTL analysis identification. *Euphytica*. 188:325–32.

Gous, P. W., Hasjim, J., Franckowiak, J., Fox, G. P. and Gilbert, R. G. (2013) Barley genotype expressing "stay-green"-like characteristics maintains starch quality of the grain during water stress condition. *J. Cereal. Sci*. 58:414–19.

Graner, A., Jahoor, A., Schondelmaier, J., Siedler, H., Pillen, K., Fischbeck, G., Wenzel, G. and Herrmann, R. G. (1991) Construction of an RFLP map of barley. *Theor. Appl. Genet.* 83:250-6.

Gray, R., Nagy, C. and Guzel, A. (2012) Returns to Research, Western Grains Research Foundation, Wheat and Barley Varietal Development, 31 October 2012. http://westerngrains.com/wp-content/uploads/2017/01/Final-WGRF-ROR-STUDY2.pdf.

Grimson, R. E., Weisenburger, R. D., Basarab, J. A. and Stilborn, R. P. (1987) Effects of barley volume-weight and processing method on feedlot performance of finishing steers. *Can. J. Anim. Sci.* 67:43-53.

Hart, K. J., Rossnagel, B. G. and Yu, P. (2008) Chemical characteristics and in-situ ruminal parameters of barley for cattle: Comparison of the malting cultivar AC Metcalfe and five feed cultivars. *Can. J. Anim. Sci.* 88:711-9.

Hashiguchi, A., Ahsan, N. and Komatsu, S. (2010) Proteomics application of crops in the context of climatic changes. *Food. Res. Int.* 43:1803-13.

Hayes, P. M., Liu, B. H., Knapp, S. J., Chen, F., Jones, B., Blake, T., Franckowiak, J., Rasmusson, D., Sorrells, M., Ullrich, S. E., Wesenberg, D. and Kleinhofs, A. (1993) Quantitative trait locus effects and environmental interaction in a sample of North American barley germ plasm. *Theor. Appl. Genet.* 87:392-401.

Heun, M., Kennedy, A. E., Anderson, J. A., Lapitan, N. L. V., Sorrells, M. E. and Tanksley, S. D. (1991) Construction of a restriction fragment length polymorphism map for barley (*Hordeum vulgare*). *Genome*. 34:437-47.

Hogan, J. P. and Flinn, P. C. (1999) An assessment by *in vivo* methods of grain quality for ruminants. *Aust J. Agric. Res.* 50:843-54.

Högy, P. and Fangmeier, A. (2008) Effects of elevated atmospheric CO_2 on grain quality of wheat. *J. Cereal. Sci.* 48:580-91.

Högy, P., Poll, C., Marhan, S., Kandeler, E. and Fangmeier, A. (2013) Impacts of temperature increase and change in precipitation pattern on crop yield and yield quality of barley. *Food. Chem.* 136:1470-7.

Holden, N. M. and Brereton, A. J. (2006) Adaptation of water and nitrogen management of spring barley and potato as a response to possible climate change in Ireland. *Agric. Water. Manag.* 82:297-17.

Holtshausen, L., Beauchemin, K. A., Schwartzkoph-Genswein, K. S., Gonzalez, L. A., McAllister, T. A. and Gibb, D. J. (2011) Performance, feeding behaviour and rumen pH profile of beef cattle fed corn silage in combination with barley grain, corn or wheat distillers' grain or wheat middlings. *Can. J. Anim. Sci.* 91:703-10.

Hughes, R. J. and Choct, M. (1999) Chemical and physical characteristics of grains related to variability in energy and amino acid availability in poultry. *Aust. J. Agric. Res.* 50:689-01.

Hunt, C. W. (1996) Factors affecting the feeding quality of barley for ruminants. *Anim. Feed. Sci. Technol.* 62:37-48.

Hynd, P. J. and Allden, W. G. (1985) Rumen fermentation pattern, postruminal protein flow and wool growth rate of sheep on a high barley diet. *Aust. J. Agric. Res.* 36:451-60.

Igartua, E., Edney, M., Rossnagel, B. G., Spaner, D., Legge, W. G., Scoles, G. J., Eckstein, P. E., Penner, G. A., Tinkier, N. A., Briggs, K. G., Falk, D. E. and Mather, D. E. (2000) Marker-based selection of QTL affecting grain and malt quality in two-row barley. *Crop. Sci.* 40:1426-33.

Ige, D. V., Kiarie, E., Akinremi, O. O., Rossnagel, B., Flatten, D. and Nyachoti, C. M. (2010) Energy and nutrient digestibility in a hulless low-phytate barley fed to finishing pigs. *Can. J. Anim. Sci.* 90:393-9.

Iimure, T. and Sato, K. (2013) Beer proteomics analysis for beer quality control and malting barley breeding. *Food. Res. Int.* 54:1013-20.

Ingvordsen, C. H., Lyngkjær, M. F., Peltonen-Sainio, P., Mikkelsen, T. N and Jørgensen, R. B. (2015) A 10-days heatwave around flowering superimposed on climate change conditions significantly affects production of 22 barley accessions. *Procedia. Environ. Sci.* 29:160–1.

Ingvordsen, C. H., Gislum, R., Jørgensen, J. R., Mikkelsen, T. N., Stockmarr, A. and Jørgensen, R. B. (2016) Grain protein concentration and harvestable protein under future climate conditions. A study of 108 spring barley accessions. *J. Exp. Bot.* 67:2151–8.

Jacobs, J. L., Morris, R. J. and Zorilla-Rios, J. (1995) Effect of ensiling whole barley grain with pasture on silage quality and effluent production and the performance of growing cattle. *Aust. J. Exp Agric.* 35:731–8.

Jagadish, K. S. V., Kadam, N. N., Xiao, G., Melgar, R. J., Bahuguna, R. N., Quinones, C., Tamilselvan, A. and Prasad, P. V. V. (2014) Agronomic and physiological responses to high temperature, drought, and elevated CO_2. *Adv. Agron.* 127:111–56.

Kaiser, A. G. (1999) Increasing the utilisation of grain when fed whole to ruminants. *Aust. J. Agric. Res.* 50:737–56.

Karakousis, A., Barr, A. R., Kretschmer, J. M., Manning, S., Logue, S. J., Roumeliotis, S., Collins, H. M., Li, C. D., Lance, R. C. M. and Langridge, P. (2003a) Mapping and QTL analysis of the barley population Galleon × Haruna Nijo. *Aust. J Agric. Res.* 54:1131–5.

Karakousis, A., Barr, A. R., Kretschmer, J. M., Manning, S., Jefferies, S. P., Chalmers, K. J., Islam, A. K. M. and Langridge, P. (2003b) Mapping and QTL analysis of the barley population Clipper × Sahara. *Aust. J. Agric. Res.* 54:1137–40.

Karakousis, A., Gustafson, J. P., Chalmers, K. J., Barr, A. R. and Langridge, P. (2003c) A consensus map of barley integrating SSR, RFLP and AFLP markers. *Aust J. Agric. Res.* 54:1173–85.

Kaspar-Schoenefeld, S., Merx, K., Jozefowicz, A. M., Hartmann, A., Seiffert, U., Weschke, W., Matros, A. and Mock, H. P. (2016) Label-free proteome profiling reveals developmental-dependent patterns in young barley grains. *J. Proteom.* 143:106–21.

Kenney, P. A. (1986) Productivity of early-weaned lambs fed high-grain diets of wheat, oats or barley with or without lupin grain. *Aust. J. Exp. Agric.* 26:279–84.

Khorasani, G. R., Helm, J. and Kennelly, J. J. (2000) In situ rumen degradation characteristics of sixty cultivars of barley grain. *Can. J. Anim. Sci.* 80:691–701.

Kimball, B. A., Morris, C. F., Pinter Jr, P. J., Wall, G. W., Hunsaker, D. J., Adamsen, F. J., LaMorte, R. L., Leavitt, S. W., Thompson, T. L., Matthias, A. D. and Brooks, T. J. (2001) Elevated CO_2, drought and soil nitrogen effects on wheat grain quality. *New. Phytol.* 150: 295–303.

Kitessa, S., Flinn, P. C. and Irish, G. G. (1999) Comparison of methods used to predict the in vivo digestibility of feeds in ruminants. *Aust. J Agric. Res.* 50:825–41.

Kleinhofs, A., Kilian, A., Saghai Maroof, M. A., Biyashev, R. M., Hayes, P., Chen, F. Q., Lapitan, N., Fenwich, A., Blake, T. K., Kanazin, V., Ananiev, E., Dahleen, L., Kudrna, D., Bollinger, J., Knapp, S. J., Liu, B., Sorrells, M., Heun, M., Franckowiak, J. D., Hoffman, D., Skadsen, R. and Steffenson, B. J. (1993) A molecular, isozyme and morphological map of the barley (*Hordeum vulgare*) genome. *Theor. Appl. Genet.* 86:705–12.

Koenig, K. M. and Beauchemin, K. A. (2011) Optimum extent of barley grain processing and barley silage proportion in feedlot cattle diets: Growth, feed efficiency and faecal characteristics. *Can. J. Anim. Sci.* 91:411–22.

Labuschagne, M. T. (2018) A review of cereal grain proteomics and its potential for sorghum improvement. *J. Cereal. Sci.* 84:151–8.

Larson, S. R., Young, K. A., Cook, A., Blake, T. K. and Raboy, V. (1998) Linkage mapping of two mutations that reduce phytic acid content of barley grain. *Theor. Appl. Genet.* 97: 141-6.

Lastovicková, M. and Bobálová, J. (2012) MS based proteomic approaches for analysis of barley malt. *J. Cereal. Sci.* 56: 519-30.

Lawrence, T. L. J. (1970) High level cereal diets for the growing-finishing pig. IV. Comparison of two slaughter weights (120 and 200lb.) of diets containing high levels of maize, sorghum, wheat and barley. *J. Agr. Sci.* 74:539-48.

Liu, D. L., Zeleke, K. T., Wang, B., Macadam, I., Scott, F. and Martin, R. J. (2017) Crop residue incorporation can mitigate negative climate change impacts on crop yield and improve water use efficiency in a semiarid environment. *Eur. J. Agron.* 85:51-68.

Long, N. R., Jefferies, S. P., Warner, P., Karakousis, A., Kretschmer, J. M., Hunt, C., Lim, P., Eckermann, P. J. and Barr, A. R. (2003) Mapping and QTL analysis of the barley population Mundah × Keel. *Aust. J. Agric. Res.* 54:1163-71.

Lott, J. N., Ockenden, I., Raboy, V. and Batten, G. D. (2000) Phytic acid and phosphorus in crop seeds and fruits: A global estimate. *Seed. Sci. Res.* 10: 11-33.

Manderscheid, R., Bender, J., Jager, H. J. and Weigel, H. J. (1995) Effects of season long CO_2 enrichment on cereals. II. Nutrient concentrations and grain quality. *Agr. Ecosyst. Environ.* 54:175-85.

Mannion, P. F. (1981) Enzyme supplementation of barley based diets for broiler chickens, *Aust. J. Exp. Agric. Anim. Husb.* 21:296-302.

Masud, M. B., McAllister, T., Cordeiro, M. R. C. and Faramarzi, M. (2018) Modeling future water footprint of barley production in Alberta, Canada: Implications for water use and yields to 2064. *Sci. Total. Environ.* 616-17:208-22.

Mather, D. E, Tinker, N. A., LaBerge, D. E., Edney, M., Jones, B. L., Rossnagel, B. G., Legge, W. G., Briggs, K. G., Irvine, R. B., Falk, D. E. and Kasha, K. J. (1997) Regions of the genome that affect grain and malt quality in a North American two-row barley cross. *Crop. Sci.* 37:544-54.

Merritt, N. R. (1967) A new strain of barley with starch of high amylose content. *J. Inst. Brewing.* 73:583-6.

Mitchall, K. G., Bell, J. M. and Sosulski, F. W. (1976) Digestibility and feeding value of hulless barley for pigs. *Can. J. Anim. Sci.* 56:505-11.

Molina-Cano, J. L, Francesh, M., Perez-Vendrell, A. M., Ramo, T., Voltas, J. and Brufau, J. (1997) Genetic and environmental variation in malting and feed quality of barley. *J. Cereal. Sci.* 25:37-47.

Moughan, P. J. (1999) In vitro techniques for the assessment of the nutritive value of feed grains for pigs: A review. *Aust J. Agric. Res.* 50:871-79.

Munck, L., Karlsson, K. E., Hagberg, A. and Eggum, B. P. (1970) Gene for improved nutritional value in barley seed protein. *Science* 168:985-7.

Nasir, Z., Wang, L. F., Young, M. G., Swift, M. L., Beltranena, E. and Zijlstra, R. T. (2015) The effect of feeding barley on diet nutrient digestibility and growth performance of starter pigs. *Anim. Feed. Sci. Technol.* 210:287-94.

Nendel, C., Kersebaum, K. C., Mirschel, W. and Wenkel, K. O. (2014) Testing farm management options as climate change adaptation strategies using the MONICA model. *Eur. J. Agron.* 52:47-56.

O'Brien, L. (1999) Genotype and environmental effects on feed grain quality. *Aust. J. Agric. Res.* 50:703-19.

Oram, R. N. and Doll, H. (1981) Yield improvement in high lysine barley. *Aus. J. Agric. Res.* 32:425–34.
Pallotta, M. A., Asayama, S., Reinheimer, J. M., Davies, P. A., Barr, A. R., Jefferies, S. P., Chalmers, K. J., Lewis, J., Collins, H. M., Roumeliotis, S., Logue, S. J., Coventry, S. J., Lance, R. C. M., Karakousis, A., Lim, P., Verbyla, A. P. and Eckermann, P. J. (2003) Mapping and QTL analysis of the barley population Amagi Nijo × WI2585. *Aust. J. Agric. Res.* 54:1141–4.
Parsons, J. G. and Price, D. B. (1974) Search for barley (*Hordeum vulgare*) with high lipid content. *Lipids*. 9:804–8.
Perrott, L. A., Strydhorst, S. M., Hall, L. M., Yang, R. C., Pauly, D., Gill, K. S. and Bowness, R. (2018a) Advanced agronomic practices to maximise feed barley yield, quality and standability in Alberta, Canada. I. Responses to plant density, a plant growth regulator, and foliar fungicides. *Agron J.* 110:1447-57.
Perrott, L. A., Strydhorst, S. M., Hall, L. M., Yang, R. C., Pauly, D., Gill, K. S. and Bowness, R. (2018b) Advanced agronomic practices to maximise feed barley yield, quality, and standability in Alberta, Canada. II. Responses to supplemental post-emergence nitrogen. *Agron J.* 110:1458–66.
Peterson, D. S., Harris, D. J., Rayner, C. J., Blakeney, A. B and Choct, M. (1999) Methods for the analysis of premium livestock grains. *Aust. J. Agric. Res.* 50:775–87.
Pierce, J. (2000) Phytase, production and pollution. In *Concepts in Pig Science 2000*. Ed. Lyons, T. P. and Cole, D. J. A. Nottingham University Press, Nottingham, UK, pp. 97–116.
Pirttioja, N., Palosuo, T., Fronzeka, S., Räisänen, J., Rötter, R. P. and Carter, T. R. (2019) Using impact response surfaces to analyse the likelihood of impacts on crop yield under probabilistic climate change. *Agr. Forest. Meteorol.* 264:213–24.
Pointillart, A., Fontaine, N. and Thomasset, M. (1984) Phytate phosphorus utilization and intestinal phytases in pigs fed low phosphorus wheat or corn diets. *Nutr. Rept. Int.* 29: 473–83.
Poulsen, D. (2006) DAQ00038 - Barley Improvement for the GRDC Northern Region. Final Report. https://grdc.com.au/research/reports/report?id=224.
Purchas, R. W. and Lloyd Davies, H. (1974) Carcass and meat quality of Friesian steers fed on either pasture or barley. *Aust. J. Agric Res.* 25:183-92.
Ravindran, V. and Bryden, W. L. (1999) Amino acid availability in poultry - *in vitro* and *in vivo* measurements. *Aust. J. Agric. Res.* 50:889-908.
Read, B. J., Raman, H., McMichael, G., Chalmers, K. J., Ablett, G. A., Platz, G. J., Raman, H., Genger, R. K., Boyd, W. J. R., Li, C. D., Grime, C. R., Park, R. F., Wallwork, H., Prangnell, R. and Lance, R. C. M. (2003) Mapping and QTL analysis of the barley population Sloop × Halcyon. *Aust. J. Agric. Res.* 54:1145–53.
Reynolds, W. K., Hunt, C. W., Eckert, J. W. and Hall, M. H. (1992) Evaluation of the feeding value of barley as affected by variety and location using near infrared reflectance spectroscopy. *Proc. West. Sect. Am. Soc. Anim. Sci*. 43:498-501.
Rossnagel, B. G., Zatorski, T., Arganosa, G. and Beattie, A. D. (2008) Registration of 'CDC Lophy-I' Barley. *J. Plant. Registr*. 2: 169–73.
Rowe, J. B., Choct, M. and Pethick, D. W. (1999) Processing cereal grains for animal feeding. *Aust. J. Agric. Res.* 50:721-36.
Savin, R. and Nicolas, M. E. (1999) Effects of timing of heat stress and drought on growth and quality of barley grains. *Aust. J. Agric. Res.* 50:357–64.

Schmidt, D., Gaziola, S. A., Boaretto, L. F. and Azevedo, R. A. (2016) Proteomic analysis of mature barley grains from C-hordein antisense lines. *Phytochemistry*. 125:14–26.

See, D., Kanazin, V., Kephart, K. and Blake, T. (2002) Mapping genes controlling variation in barley grain protein concentration. *Crop. Sci.* 42:680-5.

Shewry, P. R. (2007) Improving the protein content and composition of cereal grain. *J. Cereal. Sci.* 46:239–50.

Shewry, P. R., Tatham, A. S., Halford, N. G., Barker, J. H. A., Hannappel, U., Gallois, P., Thomas M and Kreis, M. (1994) Opportunities for manipulating the seed protein composition of wheat and barley in order to improve quality. *Transgenic. Res.* 3:3–12.

Sormunen-Cristian, R. (2013) Effect of barley and oats on feed intake, live weight gain and some carcass characteristics of fattening lambs. *Small. Rumin. Res.* 9:22–7.

Sparrow, D. H. B and Doolette, J. B. (1975) Barley. In *Australian Field Crops: Wheat and Other Temperate Cereals*. Ed. Lazenby, A. and Matheson, E. M. Vol. 1. 2nd edn. Angus and Robinson Publishers, Australia.

Surber, L. M. M., Bowman, J. G. P., Blake, T. K., Hinman, D. D., Boss, D. L. and Blackhurst, T. C. (2000) Prediction of barley feed quality for beef cattle from laboratory analyses. *Proc. West. Sec. Am. Soc. Anim. Sci.* 51:454–7.

Tan, B. C., Lim, Y. S. and Lau, S. (2017) Proteomics in commercial crops: An overview. *J. Proteom.* 169:176–88.

Taub, D. R., Miller, B. and Allen, H. (2008) Effects of elevated CO_2 on the protein concentration of food crops: A meta-analysis. *Glob. Chang. Biol.* 14:565–75.

Thacker, P. A., Bell, J. M., Classen, H. L., Campbell, G. L. and Rossnagel, B. G. (1988) The nutritive value of hulless barley for swine. *Anim. Feed Sci. Technol.* 19:191–6.

Thacker, P. A., Rossnagel, B. G. and Raboy, V. (2003) Phosphorus digestibility in low-phytate barley fed to finishing pigs. *Can. J. Anim. Sci.* 83:101–4.

Thompson, G. B. and Woodward, F. I. (1994) Some influences of CO_2 enrichment, nitrogen nutrition and competition on grain yield and quality in spring wheat and barley. *J. Exp. Bot.* 45:937–42.

Toland, P. C. (1976) The digestibility of wheat, barley or oat grain fed wither whole or rolled at restricted levels with hay to steers. *Aust. J. Exp. Agr. Anim. Husb.* 16:71–5.

Valentine, S. C and Wicks, R. B. (1982) The production and composition of milk from dairy cows fed hay and supplemented with either brewers grains or rolled barley grain. *Aust. J. Exp. Agric. Anim. Husb.* 22:155–8.

Van Barneveld, R. J. (1999a) Chemical and physical characteristics of grains related to variability in energy and amino acid availability in pigs: A review. *Aust. J. Agric. Res.* 50:667–87.

Van Barneveld, S. L. (1999b) Chemical and physical characteristics of grains related to variability in energy and amino acid availability in ruminants: A review. *Aust. J. Agric. Res.* 50:651–66.

Van Barneveld, R. J., Nuttall, J. D., Flinn, P. C. and Osborne, B. G. (1999) Near infrared reflectance measurement of the digestible energy content of cereals for growing pigs. *J. Near. Infrared. Spec.* 7, 1–7.

Vasos, E. J., Barr, A. R. and Eglinton, J. K. (2004) Genetic conversion of feed barley varieties to malting types. *Proceedings of the 9th International Barley Genetics Symposium. Czech, June*: 20–6.

White, C. L. and Ashes, J. R. (1999) A review of methods for assessing the protein value of grain fed to ruminants. *Aust. J. Agric. Res.* 50:855–69.

Woyengo, T. A., Akinremi, O. O., Rossnagel, B. G. and Nyachoti, C. M. (2012) Performance and total tract nutrient digestibility of growing pigs fed hulless low phytate barley. *Can. J. Anim. Sci.* 92:505–11.

Wrigley, C. W. (1999) Potential methodologies and strategies for the rapid assessment of feed-grain quality. *Aust. J. Agric. Res.* 50:789–805.

Wroblewitz, S., Hüther, L., Manderscheid, R., Weigel, H. J., Wätzig, H. and Dänicke, S. (2014) Effect of rising atmospheric carbon dioxide concentration on the protein composition of Cereal Grain. *J. Agric. Food. Chem.* 62:6616–25.

Yahaghi, M., Liang, J. B., Balcells, J., Valizadeh, R., Alimon, A. R. and Ho, Y. W. (2012) Effect of replacing barley with corn or sorghum grain on rumen fermentation characteristics and performance of Iranian Baluchi lamb fed high concentrate rations. *Anim. Prod. Sci.* 52:263–8.

Yu, P., Meier, J. A., Christensen, D. A., Rossnagel, B. G. and McKinnon, J. J. (2003) Using the NRC-2001 model and the DVE/OEB system to evaluate nutritive values of Harrington (malting-type) and Valier (feed-type) barley for ruminants. *Anim. Feed Sci. Technol.* 107:45–60.

Chapter 16

Nutritional and bioactive compounds in barley

Nancy Ames, Joanne Storsley, Lovemore Malunga and Sijo Joseph Thandapilly, Agriculture and Agri-Food Canada, Canada

1 Introduction
2 Key issues and challenges
3 Barley bioactives
4 Health benefits of barley foods
5 Enhancing barley bioactivity
6 Summary
7 Future trends
8 Where to look for further information
9 References

1 Introduction

Historical and archaeological evidence points to barley as one of the most commonly consumed food grains prior to the start of the twentieth century; cultivation of barley began nearly 10 000 years ago in the Middle East. Barley is still a significant food staple in places such as North Africa, the Near East and Central Asia. Countries such as Canada and the United States use barley mainly for malting and for animal feed, but with a growing awareness worldwide of its many health benefits and modern consumer interest in nutrition and health, demand for food barley is increasing. To meet the potential market need for barley with improved nutrition and quality, many newer Canadian hull-less food barley cultivars, including waxy and high amylose, have been bred specifically for food use (Izydorczyk and Dexter, 2008).

Barley has much to offer consumers, not only as a rich source of basic nutrients, but also as a source of bioactive components that exert positive physiological health effects. For example, barley contains high amounts of beta-glucan dietary fibre, which has exhibited physicochemical properties and health benefits both *in vitro* and *in vivo*, including the ability to lower

http://dx.doi.org/10.19103/AS.2019.0060.23
Published by Burleigh Dodds Science Publishing Limited, 2020.

cholesterol, glycaemic response and blood pressure, as well as the ability to positively impact the gut microbiome. Human feeding trials and meta-analyses examining the effect of barley consumption on human health have largely focussed on beta-glucan as the bioactive. However, beta-glucan is only one of a multitude of compounds in barley that contribute to health benefits. Barley can be utilized as a whole grain which offers additional opportunities in product development and consumer benefits. The ability to utilize barley as a whole grain food product contributes to its sustainable food production.

This chapter will review the known and potential bioactive components in barley, research that has been carried out on barley and health and the mechanisms behind the observed health effects, case studies examining the effect of bioactive dosage on acute health effects (cholesterol and blood glucose lowering) and the research that has been published examining potential influences on bioactivity, such as genotype, environment and processing.

2 Key issues and challenges

The greatest challenge faced by the barley industry is to create accessible, acceptable and high-quality products for consumers, in order to position barley as a convenient, healthy and delicious food option. It has been established in numerous studies that barley beta-glucan can significantly lower low-density lipoprotein (LDL) cholesterol and blood glucose levels; as a result, health claims for barley and cholesterol and/or glycaemic response have been approved in many jurisdictions worldwide, including Europe, Canada and the United States (FDA, 1996; USFDA, 2006; EFSA, 2011; Health Canada, 2012). This has been a positive step in promoting increased consumption of barley and barley foods. However, food products currently on the market which contain barley or beta-glucan from barley in high enough amounts to influence cholesterol or blood glucose[1] are limited (EFSA, 2011; Health Canada, 2012). Pearled barley or partially pearled barley (known as pot barley) is readily available in grocery stores and is commonly used as a soup or stew ingredient but may not provide a large enough serving size and beta-glucan quantity to meet health claim criteria. Incorporating pearled barley or barley flour into staple food products such as hot porridges, extruded snacks, bread and baked products may allow larger volumes of barley and barley beta-glucan to be consumed (Kirjoranta et al., 2012; Lin et al., 2012; Rieder et al., 2012; Thondre et al., 2012). Growing novel barley varieties for industry with increased levels of beta-glucan would enable manufacturers to readily meet the minimum 3 g of beta-glucan per serving in the food products they are incorporated into, and would assist

1 Minimum of 3 g beta-glucan per day for cholesterol lowering; 4 g per 30 g available carbohydrate for glycaemic response lowering.

Published by Burleigh Dodds Science Publishing Limited, 2020.

in getting the health message to consumers via health claims on barley food packaging. There are varieties already developed (through traditional plant breeding) containing very high levels of beta-glucan, an extreme example of which is the variety Prowashonupana, which contains 30% dietary fibre and 17.5% beta-glucan (Liljeberg et al., 1996).

Another major challenge currently faced by barley growers and food manufacturers is ensuring that health benefits of barley bioactives are retained throughout processing, storage, food preparation and within the food matrix. For example, barley beta-glucan bioactivity partly depends on maintaining physiochemical properties such as molecular weight and viscosity throughout processing and during human digestion. Since barley and other grains commonly contain endogenous beta-glucanase enzymes, which can degrade beta-glucan polymers and in this way reduce viscosity and bioactivity, reducing enzyme degradation is an important processing step. Beta-glucan may also be susceptible to degradation by endogenous enzymes in wheat, and other ingredients such as ascorbic acid (Kivelä et al., 2009a), within the food matrix. Steps must be taken to ensure that barley functional compounds such as beta-glucan remain biologically active (e.g. heat treatment to inactivate beta-glucanase enzymes) to achieve health benefits.

The protein in barley may prove to be a challenge with regard to bread making and consumer acceptability. Unlike wheat, the gluten in barley does not contribute significantly to bread loaf volume, which is why traditional bread made with barley is often blended with wheat flour. It has been reported in the literature that barley can replace wheat flour up to a 10–20% level without compromising loaf volume and quality attributes such as crumb structure and porosity (Al-Attabi et al., 2017). However, barley bread made without the addition of wheat flour has been successful; a group of researchers found that bread made with 100% barley flour with 5% hydroxypropyl methylcellulose (HPMC)[2] had acceptable loaf volume and rated highest in consumer acceptability compared to 100% oat and 100% whole wheat breads with and without HPMC (Kim and Yokoyama, 2011). Beta-glucan fibre-rich milling fractions and beta-glucan isolates from barley have also been added to wheat breads to improve fibre levels, and have demonstrated acceptable quality (Izydorczyk et al., 2008; Skendi et al., 2010).

In addition to providing health benefits and meeting approved health claims, barley foods must also meet consumer taste, texture and quality preferences. Working with culinary food experts and conducting sensory panels may help in the development and promotion of barley foods that the consumer is likely to buy and ideally consume on a regular basis.

2 HPMC, a soluble fibre often added to gluten-free breads.

Another challenge in consumer acceptability is that barley contains gluten and it is not suitable for those with celiac disease. Although celiac disease only affects a small portion of the population, the recent trend in non-celiac individuals choosing gluten-free diets may deter a significant segment of the population from consuming barley foods.

Beyond the challenge of consumer acceptability, is that barley food products must be easy to identify, prepare and incorporate into a consumer's typical diet. Both product development and marketing up to this point have not been successful in making barley consumption mainstream, and that has made it difficult for consumers to understand how, when and why to use barley.

3 Barley bioactives

3.1 Soluble dietary fibre: beta-glucan

Cereal-derived beta-glucan, otherwise known as (1,3) (1,4) beta(β)-D-glucan, is known for its contribution as a component of soluble dietary fibre (SDF). It is a heterogeneous linear (unbranched) non-starch polysaccharide containing beta (1-4) linked glucose residues with beta (1-3) linked glucose residues every two to three units (Zielke et al., 2017). The beta (1-3) linkages found within the polymers makes them partially water soluble (Theander et al., 1993). Relative to other cereal grains, barley has the highest beta-glucan content at 2–20 g per 100 g dry weight; approximately 65% is water soluble (El Khoury et al., 2012). In comparison, oats contain 3–8 g beta-glucan per 100 g dry weight, and 82% is water soluble (El Khoury et al., 2012). Beta-glucans are located in the endosperm cell walls of barley.

Barley is a genetically diverse cereal grain (Baik and Ullrich, 2008), and therefore it is no surprise that beta-glucan content and characteristics of beta-glucan such as solubility, viscosity and molecular weight distribution may vary considerably depending on barley cultivar. A study characterizing extracted non-starch polysaccharides in eight barley cultivars with four different starch types (normal, high amylose, waxy and zero amylose waxy) found that the normal starch varieties had water extractable beta-glucan with lower weight average molecular weight and viscosity compared to the cultivars from the other starch types (Storsley et al., 2003). Physical features such as the length of cellulosic regions (regions of the beta-glucan chain with continuous beta 1-4 linkages) varied considerably, depending on both the barley genotype as well as the temperature of water extraction; beta-glucans extracted at 45°C had fewer cellulosic regions than beta-glucans isolated at 95°C (Storsley et al., 2003).

In addition to genotypic differences, studies on oat and barley have shown beta-glucan content to vary with growing location or environment, and in some

Published by Burleigh Dodds Science Publishing Limited, 2020.

cases, interaction between genotype and environment has also been observed. These are significant considerations in sourcing barley for optimum nutritional and health benefits.

The health benefits of soluble beta-glucan from barley are the health benefits normally associated with barley consumption and those studied most extensively, namely lowering of LDL cholesterol and improved glycaemic response. These will be discussed later in the chapter.

3.2 Insoluble dietary fibre

As alluded to earlier in the chapter, beta-glucan in barley is only partially soluble; the remaining beta-glucan is part of insoluble dietary fibre (IDF) in barley. Dietary fibres are classified as either SDF or IDF based on their solubility in water. Cereal grain structural or matrix fibres (lignins, cellulose and some hemicelluloses) are insoluble, whereas the natural gel-forming fibres (pectins, gums, mucilages and the remainder of the hemicelluloses including beta-glucan) are soluble (Brown et al., 1999). Constituents of IDF in barley in addition to a portion of beta-glucan include arabinoxylans, cellulose, lignins and resistant starch (RS). IDF content in barley is very high, ranging from 11% to 14% in one study (Fastnaught et al., 1996) to 17% in others (USDA National Nutrient Database April 2018).

There have been many positive health outcomes associated with the consumption of IDF, including enhanced weight loss in individuals with obesity, and reduction in gastrointestinal disorders such as duodenal ulcers, diverticulitis, gastroesophageal reflux disease, constipation and haemorrhoids (Anderson et al., 2009). Reduced gastric transit time is thought to be the underlying mechanism behind these effects.

There has also been research to show that IDF increases satiety, or a sense of fullness. A specific example of this in barley was demonstrated in a recent study, where consumption of the high-IDF barley treatment by healthy human subjects elicited the highest glucagon such as peptide -1, or GLP-1[3] response (Ames et al., 2015a).

3.2.1 Arabinoxylans

Arabinoxylans are found in barley in relatively high amounts compared to oats, but less than that of rye and sometimes wheat. Levels range from approximately 4.9% to 5.2% in barley, 1.6% to 2% in oats, 2.3% to 10.8% in wheat (with up to 22% in wheat bran) and 8.9% in rye (Pritchard et al., 2011; Andersson et al., 2013; Frølich et al., 2013; Izydorczyk et al., 2014). A recent

3 A hormone which promotes slower gastric emptying, and ultimately, increased satiety.

review paper indicated that approximately 10–20% of arabinoxylans are soluble in water at 40°C (Izydorczyk and Dexter, 2008), while the remaining 80–90% are insoluble.

Like beta-glucans, arabinoxylans in barley are partially soluble in water, are found within the cell walls of the grain and the soluble portion may contribute to viscosity along with beta-glucan. It is believed that arabinoxylans may have similar health benefits to beta-glucans but more research is needed to support this theory. Research on arabinoxylans from wheat show evidence for reducing serum glucose, insulin, triglycerides and total plasma ghrelin (a hunger-signalling hormone) (Lu et al., 2000; Garcia et al., 2007).

3.2.2 Resistant starch

Resistant starch (RS) is defined as part of starch that is not digestible with human gastrointestinal digestive enzymes. RS exists in four classes, namely RS1, RS2, RS3 and RS4 (Englyst et al., 1992; Sajilata et al., 2006). RS1 is the type of starch that is physically inaccessible to digestive enzymes, whereas RS2 represents native starch which is in crystalline or granular form that is resistant to digestion. On the other hand, RS3, which is the most abundant, is retrograded starch formed as a consequence of cooling of gelatinized starch. Finally, RS4 includes chemically modified starches which may not be relevant to this chapter.

RS is considered to be part of IDF, as it resists digestion and is fermented in the large intestine, producing short-chain fatty acids (SCFA) (Wong et al., 2006). Different barley genotypes can have different inherent starch characteristics, and according to the literature, high amylose barley cultivars may have the greatest potential for increasing RS levels. In fact, a novel high amylose barley variety, Himalaya 292, was found to improve markers of gut health in a clinical trial where participants ate foods made from that particular barley variety (Bird et al., 2008).

RS levels in barley tend to increase with processing (Lahouar, 2016), such as application of heating and cooling cycles, which promote starch retrogradation, rendering the starch resistant to digestion. Not only does retrograded starch (amylose) form RS, increasing IDF levels, it lowers glycaemic response by remaining inaccessible to digestive enzymes that would otherwise break down the starch into glucose.

RS has also been shown to have a positive effect on gut health. A meta-analysis examining the effect of RS from barley (Himalaya 292) and other sources, such as high amylose corn, on bowel function in healthy individuals demonstrated higher faecal weight, higher levels of butyrate and a lowering of faecal pH (Shen et al., 2017) in individuals consuming RS, compared to those on control diets without RS.

3.3 Phenolics

Barley is a rich source of phenolic compounds, which are secondary metabolites usually associated with plant defence systems (Todd et al., 1971) and contribute to their sustainability (Bourdon et al., 1999). Because of their bitterness, they deter disease causing agents, foraging animals and insects. Phenolic compounds can impart a colour in plants which attract predators; due to the bitterness of the phenolics the predators prey on surrounding invading species, protecting the plant. Highly disease resistant plants and cultivars tend to be high in phenolic compounds (Wise et al., 2008).

It has been observed that increasing the nutrient density of the soil, through fertilization, (Nørbæk et al., 2003), and application of pesticides and/or insecticides tend to lower the amount of phenolic compounds in plants. Considering that pesticides and fertilization usage contribute significantly to the carbon footprint of barley, breeding strategies to increase phenolic compound concentration may reduce reliance on pesticides. Like most secondary metabolites, phenolic compound concentration is dependent on genotype, environment and their interaction.

Phenolic antioxidants in cereals exist as either readily extractable (free) or as less extractable (bound); bound phenolics are covalently linked to macromolecules such as arabinoxylans (Holtekjølen et al., 2008). In barley, the majority of bound phenolics exist as phenolic acids (ferulic acid and p-coumaric acid), whereas the free phenolics are mainly flavanols and tocopherols (Holtekjølen et al., 2006). All of these are known to have antioxidant activity and, therefore, possible health benefits, such as improved vascular health, reduced inflammation, reduced sensitivity of LDL to oxidization and improved glucose metabolism (Andreasen, 2001; Beecher, 2004). Cereal grains in general contain high levels of bound phenolic components (Pérez Jiménez and Saura Calixto, 2005). For this reason, cereals may potentially be good sources of natural antioxidants, and industrial processes that increase the bioaccessibility or bioavailability of these phenolic compounds continues to be a priority for researchers (Ragaee et al., 2014).

Alkylresorcinols, or resorcinolic lipids, are found in barley (26–103 μg/g on a dry basis). They are composed of a phenolic ring with two hydroxyl groups 1, 3 and an alkyl chain (Kozubek and Tyman, 1999). These alkylresorcinols are distinguished based on their chain length, C_{15} to C_{27} (Zarnowski and Suzuki, 2004; Gómez-Caravaca et al., 2015) and degree of saturation. The C25:0 and C21:0 homologues are the most abundant, accounting for 60–70% of the total alkylresorcinols in barley. Because they are amphiphilic, they have a large scope of bioactivities, including antioxidant, antimicrobial and antiparasitic, and contribute to the many health benefits associated with intake of whole grains. Although found in larger amounts in wheat (1000 μg/g) and rye (>3200 μg/g), a recent study by Bordiga et al. found that levels could be increased through pearling, as alkylresorcinols are distributed unevenly in the grain (Bordiga et al., 2016). Considering their localization in the outer layers, they are being explored to be used as markers for whole grain (wheat, rye, triticale and barley) (Ross et al., 2004).

The total phenolic acid content in barley ranges from approximately 969 to 1132 μg/g (Ndolo and Beta, 2014). These phenolic acids include hydroxycinnamic acid derivatives (p-coumaric, caffeic, ferulic and sinapic acid) and hydroxybenzoic acid derivatives (p-hydroxybenzoic, syringic and vanillic acid) (Andersson et al., 2008). Ferulic acid constitutes 70–90% of the total phenolic acids in barley (Andersson et al., 2008; Ndolo and Beta, 2014; Gómez-Caravaca et al., 2015). The phenolic acids present in barley are mostly bound to arabinoxylan (451–585 μg/g) with a small proportion either free (1.2–23 μg/g) or conjugated (86–198 μg/g) (Gómez-Caravaca et al., 2015; Idehen et al., 2017). Consequently, most of the phenolic acids are allocated in grain outer layers (aleurone and pericarp) (Ndolo and Beta, 2014). Phenolic acid derivatives have been shown, *in vitro*, to possess antidiabetic properties by inhibiting the brush border's carbohydrases and glucose uptake proteins in the gut (Malunga and Eck, 2016); they also have been found to exhibit antioxidant properties by quenching oxidized lipids and inhibiting lipid hydroperoxidation during gastric digestion (Kanner and Lapidot, 2001; Malunga et al., 2017). Hydroxycinnamic acid derivatives have proved to be more potent alpha-glucosidase inhibitors (Malunga et al., 2018) and antioxidants (Teixeira et al., 2013) compared to hydroxybenzoic acid derivatives having the same number and position of hydroxyl or methoxy groups.

Flavonoids are phenolic compounds biosynthesized through the phenylpropanoid-acetogenic pathway and are identified as having two phenyl rings (commonly assigned ring A and B) joined by a heterocyclic 3 carbon ring (Ring C). Depending on the level of oxidation, connection position of the B ring and the annularity of the C ring, flavonoids are classified into flavanols, anthocyanidins, flavones, flavonols, flavanones and isoflavonoids (Kühnau, 1976). They are mostly allocated in the testa of mature grain as catechin monomers or their condensed dimers and trimers (tannins). While flavonoids are constitutively synthesized as antimicrobial agents in plants, some are considered phytoalexins, produced in response to pathogens. Consequently, the flavonoid concentration is high in fusarium-resistant barley cultivars (Skadhauge et al., 1997). Because of flavonoid involvement in haze formation in beer, flavonoid-free cultivars have been developed. In barley, flavonoids have mostly been studied for their role in beer haze formation. However, flavonoids have been shown to have anti-diabetic, anti-inflammatory, anti-allergic, antioxidant and antiarthritic properties (Prior and Cao, 2000; Yao et al., 2004). Also, as reviewed by Spencer (2008), flavonoids are thought to play a significant role in human memory and neurocognitive function.

The three classes of flavonoids present in barley are flavonols (3-hydroxy-2-phenylchromen-4-one), flavanols (2-phenyl-3,4-dihydro-2*H*-chromen-3-ol) and anthocynidins. Together, flavonoids represent more than 95% of the total free phenolic compounds in barley. The flavanol family includes (+)-catechin (C), (-)-epicatechin and their gallic acid esters (gallocatechin (GC) or epigallocatechin). Catechin (30–43 μg/g) is the common flavanol found in barley grain followed by glycosides (catechin-3-glucose) (8–12 μg/g) (Gómez-Caravaca et al., 2015). Flavonols reported in barley include quercetin, kaempferol and myricetin. These flavonols, such as most flavonoids, are mostly found in the pericarp of barley. Hulled purple barley contains higher amounts of flavonols (107 μg/g) compared to unhulled purple barley (50 μg/g) (Yang et al., 2013). Common barley on the other hand has very low amounts of flavonols (26 μg/g). Regardless of their differences in whole grain flavonol concentration, barley endosperm flour has a negligible amount of flavonols (Yang et al., 2013).

Anthocyanins are oligomeric dimers or trimers of flavanol. In mature barley grains, three major flavanol dimers have been reported, namely prodelphinidin B-3 (PD B3), procyanidin B-3 (PC B3), and propelargonidin. The concentration of PD B3 and PC B3 ranges from 100 μg/g to 800 μg/g in barley whole grain (Zieliński and Kozłowska, 2000; Gómez-Caravaca et al., 2015). Five proanthocyanidin trimers have been reported in barley grains, namely GC-GC-C, GC-C-C, C-GC-C, C-C-GC and C-C-C (Gómez-Caravaca et al., 2015). Total concentration of flavanol trimers in barley is about 130–180 μg/g, of which 45–55% is C-GC-C (Gómez-Caravaca et al., 2015).

There are several important functional properties of cereal phenolics besides antioxidant activity. Phytochemicals such as phenolics in grains influence quality attributes of the food product, such as colour, flavour and texture. For example, anthocyanins contribute purple colours in barley (Martinez et al., 2018), and phenolic acids and the flavanol polymers have distinct flavours such as sour or bitter (Dimberg et al., 1996; Lesschaeve and Noble, 2005). Finally, phenolics may influence bread quality by interfering with dough formation (Okada et al., 1987; Piber and Koehler, 2005; Wang et al., 2004).

3.4 Other bioactives

3.4.1 Plant sterols

Phytosterols are naturally occurring compounds in plants that are structurally very similar to mammalian cell-derived cholesterol. The best dietary sources of phytosterols are unrefined vegetable oils, seeds, cereals, nuts and legumes. Phytosterols are mainly accumulated in plant cell walls and membranes. It has been well documented in animal and human trials that dietary phytosterols can lower serum cholesterol levels by reducing intestinal cholesterol absorption from the diet (Hendriks et al., 1999), and they may also be beneficial in reducing the risk of colon cancer (Jones and AbuMweis, 2009).

Barley is considered to be a good source of phytosterols, although the concentration of phytosterols is less compared to other major grains (Frølich et al., 2013). The content of phytosterols in ten barley varieties studied showed a range between 820 μg/g and 1153 μg/g (Andersson et al., 2008). Phytosterols in barley can be found in both free and bound states. Similar to other cereal grains, sitosterol is the most abundant form of phytosterol present in barley, which contributes about 60% of the total sterols present (Lampi, 2004). Most of the phytosterols are located in the outer layers of the kernel, and because of this, research has been undertaken to use pearling of barley as a means of increasing phytosterol content in the portion removed (Lampi, 2004).

3.4.2 Tocols

Barley is unique to other cereal grains in that it contains high levels of tocols, or vitamin E: all eight tocol vitamers in the vitamin E family are present in barley (Andersson et al., 2008), and of these, a significant percentage, up to 76.8%, are tocotrienols (Ward et al., 2008). A study by Martinez et al. (2018) on tocol levels in barley found levels ranging from 39.9 μg/g to 81.6 μg/g. The positive health effects of tocols, which may include modulation of degenerative diseases such as cancer and cardiovascular disease (CVD) (Tiwari and Cummins, 2009b), are thought to arise from their antioxidant action, whereby they impede lipid peroxidation (Martínez et al., 2018).

3.4.3 Protein/peptides

Plant-based protein isolates and concentrates are important in the food industry because of the trend towards increasing protein consumption to build and maintain lean muscle mass, control appetite and the trend towards eating more plant-based foods. Barley represents a valuable source of plant proteins with levels ranging from 11% to 15%. Protein levels in barley can be increased further through air classification creating high-protein products and low-protein content fractions which could be used for malting and brewing (Mohamed et al., 2007; Yalçln et al., 2008; Cho and Rhee, 2009; Alu'datt et al., 2012; Houde et al., 2018). Solubility and emulsifying properties such as emulsion stability or capacity are important for their use in different applications, and therefore more research should be dedicated to exploring the use of barley protein fractions in this capacity.

4 Health benefits of barley foods

4.1 Barley and cholesterol

The relationship between barley consumption and cholesterol lowering is well established. Research has shown that a minimum daily intake of 3 g beta-glucan soluble fibre from barley can significantly reduce blood cholesterol concentrations. The United States Food and Drug Administration (FDA) and Health Canada have approved the health claim that daily consumption of 3 g of soluble beta-glucan from whole-grain barley or certain dry-milled barley products would have a physiologically relevant cholesterol-lowering effect, similar to oat products. There have been two meta-analyses that have evaluated the cholesterol-lowering effect of beta-glucan derived from barley. A meta-analysis of eight studies conducted in 2009 by Talati et al. showed an LDL-C reduction of −10 mg/dL or −0.26 mmol/L with barley beta-glucan treatment (Talati et al., 2009). Later, a meta-analysis published in 2010 based on 11 studies showed a reduction in LDL-C of −0.27 mmol/L (AbuMweis et al., 2010).

Several mechanisms have been proposed to explain the cholesterol-lowering effects of beta-glucan. Cholesterol homeostasis in humans results from a balance between cholesterol absorption and elimination, as well as endogenous cholesterol synthesis (Ames et al., 2018). The body gets its cholesterol directly from the diet, but also synthesizes it in the liver. Cholesterol removal from the body can occur by its elimination from the gastrointestinal tract, its conversion to bile acids, the synthesis of steroid hormones and by its incorporation into cell membranes (Dietschy, 1984; Russell, 1992). Beta-glucan from barley may therefore influence more than one of these biological pathways.

Beta-glucan from barley can form a thick (viscous) solution during digestion in the small intestine. The viscosity of barley beta-glucan will vary since it is dependent on both the content and molecular weight distribution (which is influenced by physicochemical structure). The viscosity of beta-glucan formed in the small intestine can interfere with micelle formation and cholesterol transport to the intestinal brush border by increasing the thickness of unstirred water layer (UWL) located at the intestinal lumen-membrane interface (Martlett et al., 1994). Furthermore, fermentation of beta-glucan by bacteria in the gastrointestinal tract produces SCFA such as propionate (Wong et al., 2006), which may inhibit endogenous cholesterol synthesis. Acetate and propionate, two of the SCFA produced by this fermentation process, have also been reported to lower circulating lipid levels. The relative amounts of propionate and acetate appear to be important; acetate increases hepatic cholesterol synthesis, whereas propionate may reverse this effect by decreasing the activity of HMG-CoA reductase, a rate-limiting enzyme in cholesterol biosynthesis. Hence, a beta-glucan rich diet could increase the amount of propionate relative to acetate, thereby improving the blood lipid profile. More research is needed to back the idea of cholesterol synthesis reduction as a possible mechanism for cholesterol-lowering via beta-glucan consumption.

Bile acids serve as detergents in the intestine to help digest and absorb fats and fat-soluble molecules. Beta-glucan may bind with bile acids, and through this mechanism decrease re-absorption of bile acids and in turn lower cholesterol. Decreased re-absorption of bile acids due to binding with beta-glucan will lower circulating cholesterol levels by increasing the synthesis of more bile acids to restore the bile acid pool in the body. Reduced cholesterol and bile acid levels in the liver activates the LDL receptors, causing increased removal of LDL and other lipoproteins from the blood (Ames et al., 2018).

4.1.1 Cholesterol case study

In a recent study, researchers conducted a human feeding trial to determine the physiological mechanism(s) by which beta-glucan lowers blood cholesterol. Thirty participants consumed breakfast meals containing barley beta-glucan or a non-barley control food for 5 weeks in a randomized crossover design. The results of the trial demonstrated that the reduction in circulating cholesterol levels was mediated through the increased elimination of bile acids through the faeces. The increased excretion of bile acids resulted in higher bile acid synthesis from cholesterol to restore the bile acid pool (Wang et al., 2016b). An additional finding of the study was that depending on genetics of the individual participants, some had a greater increase in bile acid synthesis, resulting in a greater decrease in circulating cholesterol than others; individuals carrying the CYP7A1 SNPs rs3808607-G allele were more responsive to the cholesterol-lowering effect of beta-glucan with HMW than TT carriers (Wang et al., 2016b). Contrary to the aforementioned mechanisms of cholesterol lowering by beta-glucan, this study did not find any changes in intestinal cholesterol absorption or endogenous cholesterol synthesis.

4.2 Barley and glycaemic response

The ability of beta-glucan containing food products to attenuate postprandial glycaemic response (PPGR), or changes in blood glucose following a meal, is related to the content and viscosity of beta-glucan (Regand et al., 2011). In general, consumption of beta-glucan prevents or delays the action of digestive enzymes on starch and other food components due to the formation of a highly viscous meal bolus in the gut, which can result in delayed gastric emptying (Ames et al., 2015b). The increased viscosity produced by beta-glucan has been shown to impede intestinal glucose absorption. *In vivo* studies, and several *in vitro* digestion experiments, have supported the idea that beta-glucan can slow down the rate of starch digestion. Beta-glucan can undergo fermentation by the gut microbiota to produce SCFA including acetate, propionate and butyrate. The acute effects of beta-glucan are delayed enzyme activity and gastric emptying and delayed glucose absorption, due to the development of high viscosity in the gut. SCFA produced via fermentation in the large intestine may be attenuating the postprandial glucose response at subsequent meals.

A review on the effect of beta-glucan from both oat and barley on glycaemic response looked at human feeding studies, 16 of which utilized barley foods (Tosh, 2013). When only the studies using products which contained at least 4 g beta-glucan were compared, 76% of the treatments gave a reduction in glycaemic response, indicating that in order for beta-glucan to have a significant effect, a certain minimum dosage, in this case 4 g, is required. Moreover, barley treatments produced using processing methods that enhanced beta-glucan solubility and that prevented degradation by enzymes or acid had the greatest glycaemic impact (Tosh, 2013).

The lack of consistency in glycaemic response to barley consumption between trials is due in part to the wide variations in the composition and matrix of the foods used in these studies. For example, processing methods such as pearling, as well as food form (e.g. wet pasta versus dried), seem to influence the extent of blood glucose and/or insulin lowering (Aldughpassi et al., 2012). Lack of consistency between barley glycaemic response studies may also be related to treatments made from differing genotypes of barley, differing composition (e.g. different ingredients added) and/or differing processing techniques (e.g. differences in particle size of milled barley). These factors and others may affect digestion and absorption of carbohydrate.

4.2.1 Glycaemic response case study

A study by Ames et al. examined human PPGR to consumption of barley test foods in the form of a fresh tortilla. Test foods were designed using two barley cultivars, but varying the ratios of different milling fractions to make specific compositional comparisons. Results indicated that the type of dietary fibre in barley flour and its milling fractions can influence postprandial glucose and insulin response. Choosing high beta-glucan barley genotypes and specialized milling fractions can be a way to optimize barley's functionality with respect to postprandial glycaemia (Ames et al., 2015a). The data also supported the hypothesis that increasing beta-glucan intake specifically is beneficial in the control of PPGR (Ames et al., 2015a).

A meta-analysis supports the notion that incorporating barley and/or beta-glucan from barley into one's diet may improve glycaemic response after a meal, although the results were not conclusive (AbuMweis et al., 2016). There was a significantly reduced postprandial area under the glucose curve (G iAUC), a widely used parameter for calculating the glycaemic response of a food (Rozendaal et al., 2018), and glycaemic index (GI) observed across studies; insulin area under the curve (AUC) was reduced, and a significant drop in insulin index was seen. The meta-regression analysis did not show a statistically significant relationship between beta-glucan dose and G iAUC in spite of the trend towards higher beta-glucan dose and lower G iAUC, because of inconsistent results between clinical trials. As mentioned in Section 4.2, varying results in clinical trials can be due to inconsistencies in barley composition and food matrix; these factors and others contributed to the high heterogeneity in the current meta-analysis, and point to the need for further clinical studies that control for some of these variables.

4.3 Barley and gut health

Researchers are becoming more aware of the integral role of microbial diversity in intestinal health and also in preventing certain diseases, and how diet can influence the microbiome of the large intestine. Research has shown that bioactives found in cereal grains can positively impact the gut microbiota (Angelov et al., 2006; Costabile et al., 2008; Lamsal and Faubion, 2009; Morowitz et al., 2011; Parnell and Reimer, 2012). Gut microbiota are known to influence host nutrition and metabolism through the production of microbial fermentation products such as SCFA, amino acids and vitamins (Morowitz et al., 2011). These products may produce a substantial physiological effect and influence host health and disease. The make-up of the gut microbiota has been linked to obesity, diabetes (Karlsson et al., 2013; Shen et al., 2013) and CVD. The gut microbiome is highly influenced by diet and the fermentability of the fibres consumed. Beta-glucan, a fermentable fibre, has been previously reported to actively impact gut microbiota in *in vitro* fermentation settings and *in vivo* intervention studies. A clinical trial conducted by Wang et al. discovered that consuming 3 g beta-glucan with high molecular weight (1349 kDa) daily for 35 days results in increased *Bacteroidetes* and decreased *Firmicutes* compared to consuming the wheat and rice controls in moderately hypercholesterolaemic adults (Wang et al., 2016a). The change in bacterial phyla composition observed in this study was opposite to the microbiota pattern observed in people who are obese. In addition, a greater abundance of bifidobacteria and butyrate producing bacteria is considered as metabolically beneficial. The authors speculated that high MW beta-glucan may have the potential to improve weight control in overweight or obese adults through gut-mediated changes.

Insights into how dietary fibre bioactives such as barley beta-glucan impact the gastrointestinal microbiota are emerging; however, well-designed clinical trials are needed to validate if changes in the composition and subsequent functioning of the human gastrointestinal microbiota translates to improvement in human health. Future studies will require examination not only of the effect of dietary fibre on the gastrointestinal bacterial taxa, but also of the effects of other bioactives such as phenolic compounds on the gut microbiome, and of how microbial metabolites affect physiological outcomes, such as body fat composition, circulating cholesterol, glucose metabolism, blood pressure and inflammation.

4.4 Other health benefits of barley

4.4.1 Antioxidant activity

Oxidative stress, which is an imbalance of reactive oxygen species and antioxidants, is considered to be one of the main factors involved in ageing, as well as in the development of diseases including arteriosclerosis, CVD, diabetes and cancer. Moreover, it is thought that in order to prevent such diseases from occurring, scavenging of the reactive oxygen species is imperative (Kofuji et al., 2012).

There have been a number of studies examining free radical scavenging activity in grains in various forms, including flour and bread. In one such study, comparing flours from wheat, rye, sorghum and barley, barley was found to have moderate scavenging activity relative to the other grains, measured using a common colorimetric test (Ragaee et al., 2006). In another study, incorporation of 40% barley from three different hulled varieties with different starch types (normal, waxy and high amylose) increased the antioxidant properties of the breads compared to the control bread. These antioxidant properties were dependent on both barley variety and extraction rate of the flour (Holtekjølen et al., 2008).

Not surprisingly, methanolic extracts from barley, which contain high levels of phenolics, have been shown to have strong antioxidant properties, and since phenolic compounds are located mainly within the bran fraction, barley bran extracts may possess greater antioxidant capacity than whole-grain barley or flour. For example, methanol extracts from seven pearling fractions of two hull-less barley varieties (Falcon and Metcalfe) were investigated for free radical scavenging activity. In both barley varieties, the fractions coming from the outermost layers of the kernel (F1) yielded the highest phenolic content and scavenging activity (Madhujith et al., 2006).

Although beta-glucan is not typically thought of as an antioxidant, some research has shown that it may also have antioxidant capability. Kofuji et al. found that barley beta-glucan extracts exhibited significant hydroxyl radical scavenging activity, greater than the other non-starch polysaccharides tested, including pectin, gelatin, pullulan and xanthan gum (Kofuji et al., 2012).

4.4.2 Blood pressure lowering

Hypertension is known to be a risk factor for heart disease, stroke and renal disease (Chobanian et al., 2003; El Khoury et al., 2012), and consumption of whole grains has been recommended for managing blood pressure and in turn reducing the risk of these diseases. A recent meta-analysis showed that higher consumption of beta-glucan fibre from barley and oats was associated with lower systolic and diastolic blood pressure compared to other fibre types (Evans et al., 2015). Although the precise mechanism by which beta-glucan lowers blood pressure is not clear, it is possible that because insulin resistance is an underlying mechanism contributing to the development of hypertension, beta-glucan might be lowering blood pressure by regulating insulin metabolism (Ferrannini et al., 1987; Ferri et al., 1999). Beta-glucan-mediated cholesterol lowering may also play a role in improving endothelial function, thereby improving vasodilation, and in this way lowering blood pressure. Beta-glucan has also been shown to have an effect on anti-hypertensive signalling molecules such as nitric oxide, which may influence blood pressure (Tabesh et al., 2014). The blood pressure-lowering effect of barley beta-glucan may be greater in hypertensive individuals compared to normotensive. The precise way in which beta-glucan lowers blood pressure requires further investigation.

5 Enhancing barley bioactivity

5.1 Processing

It is important to consider the effect that barley processing may have on nutritional or nutraceutical value, specifically beta-glucan content and its functionality (Ames et al., 2015b), which can be modified (either positively or negatively) during processing (Gansmann and Vorwerck, 1995; Johansson et al., 2007; Tiwari and Cummins, 2009a). For example, heat treatment of barley through boiling, autoclaving, pressure cooking or micronization has resulted in significantly higher beta-glucan viscosity, which is likely due to the inactivation of endogenous beta-glucanases and subsequent preservation of beta-glucan molecular weight and viscosity (Izydorczyk et al., 2003; Ames et al., 2011). A study by Ames et al. looked at the effect of grain heat treatments on the viscosity and MW of beta-glucan in barley products. Pre-treating the whole-grain barley via boiling, autoclaving, pressure cooking and micronization produced barley flour with high beta-glucan viscosity and MW compared to untreated grain (Ames et al., 2015b). Moreover, this information was utilized to develop test meals for a clinical trial examining the effects of consuming high MW and low MW barley beta-glucan on human cholesterol metabolism. The various types of grain heat treatments produced barley flour for incorporation into test meal foods with high beta-glucan viscosity and MW. These high viscosity foods were shown to significantly lower blood cholesterol levels (Wang et al., 2016b).

Other types of extreme heat treatment have been shown to reduce the MW of beta-glucan in barley, while increasing solubility. In a study by Klamczynski et al., the MW of beta-glucan in barley was reduced by puffing (high heat) and jet cooking treatments (heat and pressure), but solubility of beta-glucan was dramatically increased (Klamczynski et al., 2004). Viscosity was not measured; the authors concluded that although greater amounts of beta-glucan in solution may contribute to an increase in viscosity, decreases in MW may be a moderating factor affecting potential nutritional properties.

5.2 Food matrix

Beta-glucan and its physicochemical properties can be highly influenced by the other components present in the food system. For example, when barley or oat is used as an ingredient in bread along with wheat or rye, beta-glucan viscosity is reduced due to endogenous beta-glucanase enzymes in wheat and/or rye (Cleary et al., 2007). In cases where barley (or oat) is blended with other flours, reduced dough resting time may be required to prevent beta-glucan degradation via enzymes present in the other flours. However, in dough made from 100% barley flour and water, beta-glucan viscosity and molecular weight will also decrease with increasing rest time (Ames et al., 2011), due to the activity of endogenous enzymes in barley. Oats are typically heat treated through kilning and flaking, inactivating most endogenous enzyme activity, including beta-glucanase. Heat treating barley, and perhaps wheat and/or rye, at least in certain applications, might be considerations in helping to inactivate enzymes and retain beta-glucan health benefits.

Application of heat to the barley kernel, with or without added moisture (e.g. boiling, autoclaving, pressure cooking or micronization) has resulted in higher beta-glucan viscosity compared to untreated barley in several studies, because of the inactivation of beta-glucanases (and subsequent preservation of beta-glucan molecular weight), and by increasing beta-glucan solubility (Ames et al., 2015a; Boyd et al., 2017). In addition to maintaining or increasing beta-glucan functionality, heat/moisture treatment may have additional benefits, including enhanced food safety and shelf life (Boyd et al., 2017), and improvement in milling performance (Izydorczyk et al., 2003).

Addition of ascorbic acid, or vitamin C, to beta-glucan leads to breakdown of the beta-glucan molecule, as has been demonstrated by Kivelä et al. (Kivelä et al., 2009a,b, 2011). This is another important piece of information for processors to keep in mind when formulating foods or beverages made with barley beta-glucan, especially if high viscosity is desired for quality or efficacy. Knowing the effect ascorbic acid has on beta-glucan, Ames et al. intentionally added ascorbic acid to develop treatments with low beta-glucan viscosity and MW for a clinical trial where they wanted to compare high and low viscosity treatments. Incorporating apple juice containing high levels of vitamin C into food formulations such as barley porridge and crepes provided a safe and effective means to reduce beta-glucan viscosity and MW (Ames et al., 2011, 2015b).

5.3 Genotypic and environmental influences

As mentioned previously, levels and characteristics of bioactives in cereals are influenced by genotype and environment. A study by Dickin et al. examined the effect of genotype, environment and agronomic management on beta-glucan concentration of several hull-less barley cultivars grown in three different years and examined the effect of autumn sowing and fungicide use on beta-glucan levels; two barley samples with differing content of beta-glucan were then made into porridge and fed to humans to determine whether a difference in glycaemic response was present (Dickin et al., 2011). Results showed that genotype had the greatest influence on determining the beta-glucan content of the barley grain, but the observed variation in beta-glucan due to environment was great enough to influence postprandial blood glucose in humans. Furthermore, autumn-sown cultivars had more consistent beta-glucan content across years; this may be an advantage in the production of barley with more predictable and stable beta-glucan content (Dickin et al., 2011).

6 Summary

It is clear from the published literature that barley has numerous health benefits, many of which have been attributed to beta-glucan soluble fibre. Other bioactives such as arabinoxylans and phenolic compounds in barley have been studied to some extent and likely deserve further investigation into their role in human health.

Many human clinical trials have demonstrated that consuming beta-glucan from barley has positive physiological outcomes such as cholesterol lowering, attenuation of blood glucose levels and improvement in gut microbiota. Recent research has helped us understand the impact of various factors influencing the bioactivity of beta-glucan, and has identified a relationship between human genetics and the degree of cholesterol lowering. Emerging evidence of other bioactives in barley warrants further investigation into barley's potential to provide additional health benefits.

As the range of potentially beneficial physiological effects associated with barley is becoming better understood, the food industry is challenged to develop products with desirable sensory properties while ensuring that the physiological effects of the bioactives present are retained.

The research presented in this chapter may contribute to enhanced and sustainable barley production by helping breeders and farmers select/grow barley varieties not only with high yield or disease resistance, but with the added traits of higher bioactive levels and improved physicochemical characteristics of beta-glucans (e.g. high molecular weight); at the same time, research on processing influences on bioactivity will help processors understand how to utilize various treatments such as heat/moisture to improve health benefits. Increased production of high-quality barley will further elevate barley as a functional food and increase consumer demand for barley foods.

7 Future trends

Value-added processing is an area of increasing interest in the barley industry, in part due to increasing political and social pressure to reduce waste and pollution from industrial processes. Using residue from an industrial process as material for another process would be an example of value-added processing (Mussatto et al., 2006). For example, barley bran fractions produced as a by-product of the pearling process could be used as high-fibre ingredients.

Although most of this chapter has focussed on hull-less barley which is commonly the barley of choice for human consumption, an excellent example of a potential value-added product from malting barley for human consumption would be brewer's spent grain (BSG).

BSG is the major by-product of the brewing process, currently utilized primarily for low-value applications such as animal feed (Parekh et al., 2017). Researchers have examined its composition and found the residue to contain high amounts of fibre, protein and oil, as well as phenolics with antioxidant properties. One of the challenges in upcycling BSG is to halt microbial activity shortly after malting in order for it to be safe for human consumption. Robertson et al. (2010) found that BSG had significant levels of microbial growth within the first 24 h after malting (Robertson et al., 2010). BSG has a very favourable nutritional composition (arabinoxylan content 20–30%), and products made with BSG have been shown to have a low GI and high antioxidant activity (Reis and Abu-Ghannam, 2014).

8 Where to look for further information

Some excellent review papers to find further information include one by Baik et al. on barley for food (Baik and Ullrich, 2008), a review on studies examining the glycaemic response-lowering effect of beta-glucan from oat and barley (Tosh, 2013), two meta-analyses on barley beta-glucan and cholesterol lowering (Abumweis et al., 2010; Zou et al., 2015) and a meta-analysis on glycaemic response lowering (AbuMweis et al., 2016). Research and reviews on the effect of processing on barley bioactives and bioactivity include the following papers (Granfeldt et al., 2000; Vaikousi and Biliaderis, 2005; Ames et al., 2011, 2015a,b; Aldughpassi et al., 2012; Boyd et al., 2017).

9 References

AbuMweis, S. S., Jew, S. and Ames, N. P. 2010. B-glucan from barley and its lipid-lowering capacity: a meta-analysis of randomized, controlled trials. *European Journal of Clinical Nutrition* 64(12), 1472–80. doi:10.1038/ejcn.2010.178.

AbuMweis, S. S., Thandapilly, S. J., Storsley, J. and Ames, N. 2016. Effect of barley beta-glucan on postprandial glycaemic response in the healthy human population: a meta-analysis of randomized controlled trials. *Journal of Functional Foods* 27, 329–42. doi:10.1016/j.jff.2016.08.057.

Al-Attabi, Z. H., Merghani, T. M., Ali, A. and Rahman, M. S. 2017. Effect of barley flour addition on the physico-chemical properties of dough and structure of bread. *Journal of Cereal Science* 75, 61–8. doi:10.1016/j.jcs.2017.03.021.

Aldughpassi, A., Abdel-Aal, E. S. M. and Wolever, T. M. S. 2012. Barley cultivar, kernel composition, and processing affect the glycemic index. *The Journal of Nutrition* 142(9), 1666–71. doi:10.3945/jn.112.161372.

Alu'datt, M. H., Rababah, T., Ereifej, K., Alli, I., Alrababah, M. A., Almajwal, A., Masadeh, N. and Alhamad, M. N. 2012. Effects of barley flour and barley protein isolate on chemical, functional, nutritional and biological properties of Pita bread. *Food Hydrocolloids* 26(1), 135–43. doi:10.1016/j.foodhyd.2011.04.018.

Ames, N., Storsley, J., Gamel, T. and Tosh, S. 2011. Validating the health benefits of barley foods: effect of processing on physicochemical properties of beta-glucan in barley test foods. *Cereal Foods World* 56(4), A28.

Ames, N., Blewett, H., Storsley, J., Thandapilly, S. J., Zahradka, P. and Taylor, C. 2015a. A double-blind randomised controlled trial testing the effect of a barley product containing varying amounts and types of fibre on the post-prandial glucose response of healthy volunteers. *British Journal of Nutrition* 113(9), 1373–83. doi:10.1017/S0007114515000367.

Ames, N., Storsley, J. and Tosh, S. 2015b. Effects of processing on physicochemical and efficacy of beta-glucan from oats and barley. *Cereal Foods World* 60(1), 4–8. doi:10.1094/CFW-60-1-0004.

Ames, N., Storsley, J. and Thandapilly, S. J. 2019. Functionality of beta-glucan from oat and barley and its relation with human health. In: *Cereal Grain-based Functional Foods: Carbohydrate and Phytochemical Components*. The Royal Society of Chemistry, pp. 147–66. Chapter 8. doi: 10.1039/9781788012799-00147.

Anderson, J. W., Baird, P., Davis Jr., R. H., Ferreri, S., Knudtson, M., Koraym, A., Waters, V. and Williams, C. L. 2009. Health benefits of dietary fiber. *Nutrition Reviews* 67(4), 188–205. doi:10.1111/j.1753-4887.2009.00189.x.

Andersson, A. A. M., Lampi, A. M., Nyström, L., Piironen, V., Li, L., Ward, J. L., Gebruers, K., Courtin, C. M., Delcour, J. A., Boros, D., Fraś, A., Dynkowska, W., Rakszegi, M., Bedo, Z., Shewry, P. R. and Åman, P. 2008. Phytochemical and dietary fiber components in barley varieties in the HEALTHGRAIN diversity screen. *Journal of Agricultural and Food Chemistry* 56(21), 9767–76. doi:10.1021/jf802037f.

Andersson, A. A. M., Andersson, R., Piironen, V., Lampi, A. M., Nyström, L., Boros, D., Fraś, A., Gebruers, K., Courtin, C. M., Delcour, J. A., Rakszegi, M., Bedo, Z., Ward, J. L., Shewry, P. R. and Åman, P. 2013. Contents of dietary fibre components and their relation to associated bioactive components in whole grain wheat samples from the HEALTHGRAIN diversity screen. *Food Chemistry* 136(3–4), 1243–8. doi:10.1016/j.foodchem.2012.09.074.

Andreasen, M. F., Landbo, A. K., Christensen, L. P., Hansen, A. and Meyer, A. S. 2001. Antioxidant effects of *Phenolic rye* (L.) extracts, monomeric hydroxycinnamates, and ferulic acid dehydrodimers on human low-density lipoproteins. *Journal of Agricultural and Food Chemistry* 49(8), 4090–6, doi:10.1021/jf0101758.

Angelov, A., Gotcheva, V., Kuncheva, R. and Hristozova, T. 2006. Development of a new oat-based probiotic drink. *International Journal of Food Microbiology* 112(1), 75–80, doi:10.1016/j.ijfoodmicro.2006.05.015.

Baik, B.-K. and Ullrich, S. E. 2008. Barley for food: characteristics, improvement, and renewed interest. *Journal of Cereal Science* 48(2), 233–42, doi:10.1016/j.jcs.2008.02.002.

Beecher, G. R. 2004. Proanthocyanidins: biological activities associated with human health. *Pharmaceutical Biology* 42(Suppl. 1), 2–20. doi:10.3109/13880200490893474.

Bird, A. R., Vuaran, M. S., King, R. A., Noakes, M., Keogh, J., Morell, M. K. and Topping, D. L. 2008. Wholegrain foods made from a novel high-amylose barley variety (Himalaya 292) improve indices of bowel health in human subjects. *The British Journal of Nutrition* 99(5), 1032–40, doi:10.1017/S000711450783902X.

Bordiga, M., Locatelli, M., Travaglia, F., Arlorio, M., Reyneri, A., Blandino, M. and Coisson, J. D. 2016. Alkylresorcinol content in whole grains and pearled fractions of wheat and barley. *Journal of Cereal Science* 70, 38–46, doi:10.1016/j.jcs.2016.05.017.

Bourdon, I., Yokoyama, W., Davis, P., Hudson, C., Backus, R., Richter, D., Knuckles, B. and Schneeman, B. O. 1999. Postprandial lipid, glucose, insulin, and cholecystokinin responses in men fed barley pasta enriched with β-glucan. *The American Journal of Clinical Nutrition* 69(1), 55–63. doi:10.1093/ajcn/69.1.55.

Boyd, L., Holley, R., Storsley, J. and Ames, N. 2017. Effect of heat treatments on microbial load and associated changes to β-glucan physicochemical properties in whole grain barley. *Cereal Chemistry Journal* 94(2), 333–40. doi:10.1094/CCHEM-04-16-0099-R.

Brown, L., Rosner, B., Willett, W. W. and Sacks, F. M. 1999. Cholesterol-lowering effects of dietary fiber: a meta-analysis. *The American Journal of Clinical Nutrition* 69(1), 30–42. doi:10.1093/ajcn/69.1.30.

Cho, S. Y. and Rhee, C. 2009. Functional and film-forming properties of fractionated barley proteins. *Food Science and Biotechnology* 18(4), 889–94.

Chobanian, A. V., Bakris, G. L., Black, H. R., Cushman, W. C., Green, L. A., Izzo Jr., J. L., Jones, D. W., Materson, B. J., Oparil, S., Wright Jr., J. T., Roccella, E. J., Joint

National Committee on Prevention, Detection, Evaluation, and Treatment of High Blood Pressure. National Heart, Lung, and Blood Institute and National High Blood Pressure Education Program Coordinating Committee 2003. Seventh report of the Joint National Committee on prevention, detection, evaluation, and treatment of high blood pressure. *Hypertension* 42(6), 1206–52. doi:10.1161/01.HYP.0000107251.49515.c2.

Cleary, L. J., Andersson, R. and Brennan, C. S. 2007. The behaviour and susceptibility to degradation of high and low molecular weight barley β-glucan in wheat bread during baking and *in vitro* digestion. *Food Chemistry* 102(3), 889–97. doi:10.1016/j.foodchem.2006.06.027.

Costabile, A., Klinder, A., Fava, F., Napolitano, A., Fogliano, V., Leonard, C., Gibson, G. R. and Tuohy, K. M. 2008. Whole-grain wheat breakfast cereal has a prebiotic effect on the human gut microbiota: a double-blind, placebo-controlled, crossover study. *The British Journal of Nutrition* 99(1), 110–20, doi:10.1017/S0007114507793923.

Dickin, E., Steele, K., Frost, G., Edwards-Jones, G. and Wright, D. 2011. Effect of genotype, environment and agronomic management on β-glucan concentration of naked barley grain intended for health food use. *Journal of Cereal Science* 54(1), 44–52. doi:10.1016/j.jcs.2011.02.009.

Dietschy, J. M. 1984. Regulation of cholesterol metabolism in man and in other species. *Klinische Wochenschrift* 62(8), 338–45. doi:10.1007/bf01716251.

Dimberg, L. H., Molteberg, E. L., Solheim, R. and Frolich, W. 1996. Variation in oat groats due to variety, storage and heat treatment: 1. Phenolic compounds. *Journal of Cereal Science* 24, 263–72.

EFSA. 2011. Scientific Opinion on the substantiation of health claims related to beta-glucans from oats and barley and maintenance of normal blood LDL-cholesterol concentrations (ID 1236, 1299), increase in satiety leading to a reduction in energy intake (ID 851, 852), reduction of post-prandial glycaemic responses (ID 821, 824), and "digestive function" (ID 850) pursuant to Article 13(1) of Regulation (EC) No 1924/2006. *European Food Safety Authority* 9(6), 2207–28.

El Khoury, D., Cuda, C., Luhovyy, B. L. and Anderson, G. H. 2012. Beta glucan: health benefits in obesity and metabolic syndrome. *Journal of Nutrition and Metabolism* 2012, 851362. doi:10.1155/2012/851362.

Englyst, H. N., Kingman, S. M. and Cummings, J. H. 1992. Classification and measurement of nutritionally important starch fractions. *European Journal of Clinical Nutrition* 46, S33-50.

Evans, C. E., Greenwood, D. C., Threapleton, D. E., Cleghorn, C. L., Nykjaer, C., Woodhead, C. E., Gale, C. P. and Burley, V. J. 2015. Effects of dietary fibre type on blood pressure: a systematic review and meta-analysis of randomized controlled trials of healthy individuals. *Journal of Hypertension* 33(5), 897–911. doi:10.1097/HJH.0000000000000515.

Fastnaught, C. E., Berglund, P. T., Holm, E. T. and Fox, G. J. 1996. Genetic and environmental variation in β-glucan content and quality parameters of barley for food. *Crop Science* 36(4), 941-6. doi:10.2135/cropsci1996.0011183X003600040021x.

FDA. 1996. Food labeling: health claims; oats and coronary heart disease: proposed rule. *Federal Register* 61, 296–313.

Ferrannini, E., Buzzigoli, G., Bonadonna, R., Giorico, M. A., Oleggini, M., Graziadei, L., Pedrinelli, R., Brandi, L. and Bevilacqua, S. 1987. Insulin resistance in essential

hypertension. *The New England Journal of Medicine* 317(6), 350–7. doi:10.1056/NEJM198708063170605.
Ferri, C., Desideri, G., Valenti, M., Bellini, C., Pasin, M., Santucci, A. and De Mattia, G. 1999. Early upregulation of endothelial adhesion molecules in obese hypertensive men. *Hypertension* 34(4 Pt 1), 568–73. doi:10.1161/01.hyp.34.4.568.
Frølich, W., Åman, P. and Tetens, I. 2013. Whole grain foods and health – a Scandinavian perspective. *Food and Nutrition Research* 57(1). doi:10.3402/fnr.v57i0.18503.
Gansmann, W. and Vorwerck, K. 1995. Oat milling, processing and storage. In: Welch, R. W. (Ed.), *The Oat Crop*. Springer, Dordrecht, pp. 369–408.
Garcia, A. L., Otto, B., Reich, S. C., Weickert, M. O., Steiniger, J., Machowetz, A., Rudovich, N. N., Möhlig, M., Katz, N., Speth, M., Meuser, F., Doerfer, J., Zunft, H. J. F., Pfeiffer, A. H. F. and Koebnick, C. 2007. Arabinoxylan consumption decreases postprandial serum glucose, serum insulin and plasma total ghrelin response in subjects with impaired glucose tolerance. *European Journal of Clinical Nutrition* 61(3), 334–41. doi:10.1038/sj.ejcn.1602525.
Gómez-Caravaca, A. M., Verardo, V., Candigliota, T., Marconi, E., Segura-Carretero, A., Fernandez-Gutierrez, A. and Caboni, M. F. 2015. Use of air classification technology as green process to produce functional barley flours naturally enriched of alkylresorcinols, β-glucans and phenolic compounds. *Food Research International* 73, 88–96. doi:10.1016/j.foodres.2015.02.016.
Granfeldt, Y., Eliasson, A. C. and Björck, I. 2000. An examination of the possibility of lowering the glycemic index of oat and barley flakes by minimal processing. *The Journal of Nutrition* 130(9), 2207–14. doi:10.1093/jn/130.9.2207.
Health Canada. 2012. Summary of health Canada's assessment of a health claim about barley products and blood cholesterol lowering. https://www.canada.ca/en/health-canada/services/food-nutrition/food-labelling/health-claims/assessments/assessment-health-claim-about-barley-products-blood-cholesterol-lowering.html (Accessed on 26 August 2019).
Hendriks, H. F. J., Weststrate, J. A., Van Vliet, T. and Meijer, G. W. 1999. Spreads enriched with three different levels of vegetable oil sterols and the degree of cholesterol lowering in normocholesterolaemic and mildly hypercholesterolaemic subjects. *European Journal of Clinical Nutrition* 53(4), 319–27. doi:10.1038/sj.ejcn.1600728.
Holtekjølen, A. K., Kinitz, C. and Knutsen, S. H. 2006. Flavanol and bound phenolic acid contents in different barley varieties. *Journal of Agricultural and Food Chemistry* 54(6), 2253–60. doi:10.1021/jf052394p.
Holtekjølen, A. K., Bævre, A. B., Rødbotten, M., Berg, H. and Knutsen, S. H. 2008. Antioxidant properties and sensory profiles of breads containing barley flour. *Food Chemistry* 110(2), 414–21. doi:10.1016/j.foodchem.2008.02.054.
Houde, M., Khodaei, N., Benkerroum, N. and Karboune, S. 2018. Barley protein concentrates: extraction, structural and functional properties. *Food Chemistry* 254, 367–76. doi:10.1016/j.foodchem.2018.01.156.
Idehen, E., Tang, Y. and Sang, S. 2017. Bioactive phytochemicals in barley. *Journal of Food and Drug Analysis* 25(1), 148–61doi:10.1016/j.jfda.2016.08.002.
Izydorczyk, M. S. and Dexter, J. E. 2008. Barley β-glucans and arabinoxylans: molecular structure, physicochemical properties, and uses in food products–a review. *Food Research International* 41(9), 850–68. doi:10.1016/j.foodres.2008.04.001.
Izydorczyk, M. S., Dexter, J. E., Desjardins, R. G., Rossnagel, B. G., Lagasse, S. L. and Hatcher, D. W. 2003. Roller milling of Canadian hull-less barley: optimization of roller

milling conditions and composition of mill streams. *Cereal Chemistry Journal* 80(6), 637–44. doi:10.1094/CCHEM.2003.80.6.637.

Izydorczyk, M. S., Chornick, T. L., Paulley, F. G., Edwards, N. M. and Dexter, J. E. 2008. Physicochemical properties of hull-less barley fibre-rich fractions varying in particle size and their potential as functional ingredients in two-layer flat bread. *Food Chemistry* 108(2), 561–70. doi:10.1016/j.foodchem.2007.11.012.

Izydorczyk, M. S., McMillan, T., Bazin, S., Kletke, J., Dushnicky, L., Dexter, J., Chepurna, A. and Rossnagel, B. 2014. Milling of Canadian oats and barley for functional food ingredients: oat bran and barley fibre-rich fractions. *Canadian Journal of Plant Science* 94(3), 573–86. doi:10.4141/CJPS2013-229.

Johansson, L., Tuomainen, P., Anttila, H., Rita, H. and Virkki, L. 2007. Effect of processing on the extractability of oat β-glucan. *Food Chemistry* 105(4), 1439–45. doi:10.1016/j.foodchem.2007.05.021.

Jones, P. J. and AbuMweis, S. S. 2009. Phytosterols as functional food ingredients: linkages to cardiovascular disease and cancer. *Current Opinion in Clinical Nutrition and Metabolic Care* 12(2), 147–51. doi:10.1097/MCO.0b013e328326770f.

Kanner, J. and Lapidot, T. 2001. The stomach as a bioreactor: dietary lipid peroxidation in the gastric fluid and the effects of plant-derived antioxidants. *Free Radical Biology and Medicine* 31(11), 1388–95. doi:10.1016/S0891-5849(01)00718-3.

Karlsson, F. H., Tremaroli, V., Nookaew, I., Bergstrom, G., Behre, C. J., Fagerberg, B., Nielsen, J. and Backhed, F. 2013. Gut metagenome in European women with normal, impaired and diabetic glucose control. *Nature* 498(7452), 99–103. doi:10.1038/nature12198.

Kim, Y. and Yokoyama, W. H. 2011. Physical and sensory properties of all-barley and all-oat breads with additional hydroxypropyl methylcellulose (HPMC) β-glucan. *Journal of Agricultural and Food Chemistry* 59(2), 741–6. doi:10.1021/jf103817h.

Kirjoranta, S., Solala, K., Suuronen, J. P., Penttilä, P., Peura, M., Serimaa, R., Tenkanen, M. and Jouppila, K. 2012. Effects of process variables and addition of polydextrose and whey protein isolate on the properties of barley extrudates. *International Journal of Food Science and Technology* 47(6), 1165–75. doi:10.1111/j.1365-2621.2012.02956.x.

Kivelä, R., Gates, F. and Sontag-Strohm, T. 2009a. Degradation of cereal beta-glucan by ascorbic acid induced oxygen radicals. *Journal of Cereal Science* 49(1), 1–3. doi:10.1016/j.jcs.2008.09.003.

Kivelä, R., Nyström, L., Salovaara, H. and Sontag-Strohm, T. 2009b. Role of oxidative cleavage and acid hydrolysis of oat beta-glucan in modelled beverage conditions. *Journal of Cereal Science* 50(2), 190–7. doi:10.1016/j.jcs.2009.04.012.

Kivelä, R., Sontag-Strohm, T., Loponen, J., Tuomainen, P. and Nyström, L. 2011. Oxidative and radical mediated cleavage of β-glucan in thermal treatments. *Carbohydrate Polymers* 85(3), 645–52. doi:10.1016/j.carbpol.2011.03.026.

Klamczynski, A. P., Knuckles, B. E. and Glenn, G. M. 2004. Puffing and jet cooking affect solubility and molecular weight of barley β-glucans. *Journal of Food Processing and Preservation* 28(6), 433–41. doi:10.1111/j.1745-4549.2004.23053.x.

Kofuji, K., Aoki, A., Tsubaki, K., Konishi, M., Isobe, T. and Murata, Y. 2012. Antioxidant activity of beta-glucan. *International Scholarly Research Network* 2012, 5. doi:10.5402/2012/125864.

Kozubek, A. and Tyman, J. H. 1999. Resorcinolic lipids, the natural non-isoprenoid phenolic amphiphiles and their biological activity. *Chemical Reviews* 99(1), 1–26. doi:10.1021/cr970464o.

Published by Burleigh Dodds Science Publishing Limited, 2020.

Kühnau, J. 1976. The flavonoids. A class of semi-essential food components: their role in human nutrition. In: *World Review of Nutrition and Dietetics*. Karger Publishers, pp. 117–91.

Lahouar, L., Arem, A. and Achour, A. 2016. Bioactive compounds in whole grain barley: nutraceutical properties and health benefits. *Journal of Bioresources Valorization* 1(1), 5–15.

Lampi, A. M., Moreau, R. A., Piironen, V. and Hicks, K. B. 2004. Pearling barley and rye to produce phytosterol-rich fractions. *Lipids* 39(8), 783–7. doi:10.1007/s11745-004-1296-1.

Lamsal, B. P. and Faubion, J. M. 2009. The beneficial use of cereal and cereal components in probiotic foods. *Food Reviews International* 25(2), 103–14. doi:10.1080/87559120802682573.

Lesschaeve, I. and Noble, A. C. 2005. Polyphenols: factors influencing their sensory properties and their effects on food and beverage preferences. *The American Journal of Clinical Nutrition* 81(1), 330S–5S. doi:10.1093/ajcn/81.1.330S.

Liljeberg, H. G. M., Granfeldt, Y. E. and Björck, I. M. E. 1996. Products based on a high fiber barley genotype, but not on common barley or oats, lower postprandial glucose and insulin responses in healthy humans. *The Journal of Nutrition* 126(2), 458–66. doi:10.1093/jn/126.2.458.

Lin, S. Y., Chen, H. H., Lu, S. and Wang, P. C. 2012. Effects of blending of wheat flour with barley flour on dough and steamed bread properties. *Journal of Texture Studies* 43(6), 438–44. doi:10.1111/j.1745-4603.2012.00352.x.

Lu, Z. X., Walker, K. Z., Muir, J. G., Mascara, T. and O'Dea, K. 2000. Arabinoxylan fiber, a byproduct of wheat flour processing, reduces the postprandial glucose response in normoglycemic subjects. *The American Journal of Clinical Nutrition* 71(5), 1123–8. doi:10.1093/ajcn/71.5.1123.

Madhujith, T., Izydorczyk, M. and Shahidi, F. 2006. Antioxidant properties of pearled barley fractions. *Journal of Agricultural and Food Chemistry* 54(9), 3283–9. doi:10.1021/jf0527504.

Malunga, L. N., Eck, P. and Beta, T. 2016. Inhibition of intestinal α-glucosidase and glucose absorption by feruloylated arabinoxylan mono-and oligosaccharides from corn bran and wheat aleurone. *Journal of Nutrition and Metabolism* 2016, 1932532. doi:10.1155/2016/1932532.

Malunga, L. N., Izydorczyk, M. and Beta, T. 2017. Effect of water-extractable arabinoxylans from wheat aleurone and bran on lipid peroxidation and factors influencing their antioxidant capacity. *Bioactive Carbohydrates and Dietary Fibre* 10, 20–6. doi:10.1016/j.bcdf.2017.05.001.

Malunga, L. N., Joseph Thandapilly, S. and Ames, N. 2018. Cereal-derived phenolic acids and intestinal alpha glucosidase activity inhibition: structural activity relationship. *Journal of Food Biochemistry* 42(6), e12635. doi:10.1111/jfbc.12635.

Marlett, J. A., Hosig, K. B., Vollendorf, N. W., Shinnick, F. L., Haack, V. S. and Story, J. A. 1994. Mechanism of serum cholesterol reduction by oat bran. *Hepatology* 20(6), 1450–7. doi:10.1002/hep.1840200612.

Martínez, M., Motilva, M. J., López de las Hazas, M. C., Romero, M. P., Vaculova, K. and Ludwig, I. A. 2018. Phytochemical composition and β-glucan content of barley genotypes from two different geographic origins for human health food production. *Food Chemistry* 245, 61–70. doi:10.1016/j.foodchem.2017.09.026.

Mohamed, A., Hojilla-Evangelista, M. P., Peterson, S. C. and Biresaw, G. 2007. Barley protein isolate: thermal, functional, rheological, and surface properties. *Journal of the American Oil Chemists' Society* 84(3), 281–8. doi:10.1007/s11746-006-1027-z.

Morowitz, M. J., Carlisle, E. M. and Alverdy, J. C. 2011. Contributions of intestinal bacteria to nutrition and metabolism in the critically ill. *The Surgical Clinics of North America* 91(4), 771–85. doi:10.1016/j.suc.2011.05.001.

Mussatto, S. I., Dragone, G. and Roberto, I. C. 2006. Brewers' spent grain: generation, characteristics and potential applications. *Journal of Cereal Science* 43(1), 1–14. doi:10.1016/j.jcs.2005.06.001.

Ndolo, V. U. and Beta, T. 2014. Comparative studies on composition and distribution of phenolic acids in cereal grain botanical fractions. *Cereal Chemistry Journal* 91(5), 522–30. doi:10.1094/CCHEM-10-13-0225-R.

Nørbæk, R., Aaboer, D. B., Bleeg, I. S., Christensen, B. T., Kondo, T. and Brandt, K. 2003. Flavone C-glycoside, phenolic acid, and nitrogen contents in leaves of barley subject to organic fertilization treatments. *Journal of Agricultural and Food Chemistry* 51(3), 809–13. doi:10.1021/jf0258914.

Okada, K., Negishi, Y. and Nagao, S. 1987. Factors affecting dough breakdown during overmixing. *Cereal Chemistry* 64, 428–34.

Parekh, I., Khanvilkar, A. and Naik, A. 2017. Barley-wheat brewers' spent grain: a potential source of antioxidant rich lipids. *Journal of Food Processing and Preservation* 41(6), e13244. doi:10.1111/jfpp.13244.

Parnell, J. A. and Reimer, R. A. 2012. Prebiotic fiber modulation of the gut microbiota improves risk factors for obesity and the metabolic syndrome. *Gut Microbes* 3(1), 29–34. doi:10.4161/gmic.19246.

Pérez Jiménez, J. and Saura Calixto, F. D. 2005. Literature data may underestimate the actual antioxidant capacity of cereals. *Journal of Agricultural and Food Chemistry* 53(12), 5036–40. doi:10.1021/jf050049u.

Piber, M. and Koehler, P. 2005. Identification of dehydro-ferulic acid-tyrosine in rye and wheat: evidence for a covalent cross-link between arabinoxylans and proteins. *Journal of Agricultural and Food Chemistry* 53(13), 5276–84. doi:10.1021/jf050395b.

Prior, R. L. and Cao, G. 2000. Flavonoids: diet and health relationships. *Nutrition in Clinical Care* 3(5), 279–88. doi:10.1046/j.1523-5408.2000.00074.x.

Pritchard, J. R., Lawrence, G. J., Larroque, O., Li, Z., Laidlaw, H. K. C., Morell, M. K. and Rahman, S. 2011. A survey of β-glucan and arabinoxylan content in wheat. *Journal of the Science of Food and Agriculture* 91(7), 1298–303. doi:10.1002/jsfa.4316.

Ragaee, S., Abdel-Aal, E. S. M. and Noaman, M. 2006. Antioxidant activity and nutrient composition of selected cereals for food use. *Food Chemistry* 98(1), 32–8. doi:10.1016/j.foodchem.2005.04.039.

Ragaee, S., Seetharaman, K. and Abdel-Aal, E.-S. M. 2014. The impact of milling and thermal processing on phenolic compounds in cereal grains. *Critical Reviews in Food Science and Nutrition* 54(7), 837–49. doi:10.1080/10408398.2011.610906.

Regand, A., Chowdhury, Z., Tosh, S. M., Wolever, T. M. S. and Wood, P. 2011. The molecular weight, solubility and viscosity of oat beta-glucan affect human glycemic response by modifying starch digestibility. *Food Chemistry* 129(2), 297–304, doi:10.1016/j.foodchem.2011.04.053.

Reis, S. F. and Abu-Ghannam, N. 2014. Antioxidant capacity, arabinoxylans content and *in vitro* glycaemic index of cereal-based snacks incorporated with brewer's spent grain. *LWT - Food Science and Technology* 55(1), 269–77. doi:10.1016/j.lwt.2013.09.004.

Rieder, A., Holtekjølen, A. K., Sahlstrøm, S. and Moldestad, A. 2012. Effect of barley and oat flour types and sourdoughs on dough rheology and bread quality of composite wheat bread. *Journal of Cereal Science* 55(1), 44–52. doi:10.1016/j.jcs.2011.10.003.

Robertson, J. A., l'Anson, K. J. A., Treimo, J., Faulds, C. B., Brocklehurst, T. F., Eijsink, V. G. H. and Waldron, K. W. 2010. Profiling brewers' spent grain for composition and microbial ecology at the site of production. *LWT - Food Science and Technology* 43(6), 890–6. doi:10.1016/j.lwt.2010.01.019.

Ross, A. B., Kamal-Eldin, A. and Åman, P. 2004. Dietary alkylresorcinols: absorption, bioactivities, and possible use as biomarkers of whole-grain wheat- and rye-rich foods. *Nutrition Reviews* 62(3), 81–95. doi:10.1111/j.1753-4887.2004.tb00029.x.

Rozendaal, Y. J., Maas, A. H., van Pul, C., Cottaar, E. J., Haak, H. R., Hilbers, P. A. and van Riel, N. A. 2018. Model-based analysis of postprandial glycemic response dynamics for different types of food. *Clinical Nutrition Experimental* 19, 32–45. doi:10.1016/j.yclnex.2018.01.003.

Russell, D. W. 1992. Cholesterol biosynthesis and metabolism. *Cardiovascular Drugs and Therapy* 6(2), 103–10. doi:10.1007/BF00054556.

Sajilata, M. G., Singhal, R. S. and Kulkarni, P. R. 2006. Resistant starch–a review. *Comprehensive Reviews in Food Science and Food Safety* 5(1), 1–17. doi:10.1111/j.1541-4337.2006.tb00076.x.

Shen, J., Obin, M. S. and Zhao, L. 2013. The gut microbiota, obesity and insulin resistance. *Molecular Aspects of Medicine* 34(1), 39–58, doi:10.1016/j.mam.2012.11.001.

Shen, D., Bai, H., Li, Z., Yu, Y., Zhang, H. and Chen, L. 2017. Positive effects of resistant starch supplementation on bowel function in healthy adults: a systematic review and meta-analysis of randomized controlled trials. *International Journal of Food Sciences and Nutrition* 68(2), 149–57. doi:10.1080/09637486.2016.1226275.

Skadhauge, B., Thomsen, K. K. and Von Wettstein, D. 1997. The role of the barley testa layer and its flavonoid content in resistance to *Fusarium* infections. *Hereditas* 126(2), 147–60. doi:10.1111/j.1601-5223.1997.00147.x.

Skendi, A., Biliaderis, C. G., Papageorgiou, M. and Izydorczyk, M. S. 2010. Effects of two barley β-glucan isolates on wheat flour dough and bread properties. *Food Chemistry* 119(3), 1159–67, doi:10.1016/j.foodchem.2009.08.030.

Spencer, J. P. 2008. Food for thought: the role of dietary flavonoids in enhancing human memory, learning and neuro-cognitive performance. *The Proceedings of the Nutrition Society* 67(2), 238–52. doi:10.1017/S0029665108007088.

Storsley, J. M., Izydorczyk, M. S., You, S., Biliaderis, C. G. and Rossnagel, B. 2003. Structure and physicochemical properties of β-glucans and arabinoxylans isolated from hull-less barley. *Food Hydrocolloids* 17(6), 831–44. doi:10.1016/S0268-005X(03)00104-8.

Tabesh, F., Sanei, H., Jahangiri, M., Momenizadeh, A., Tabesh, E., Pourmohammadi, K. and Sadeghi, M. 2014. The effects of beta-glucan rich oat bread on serum nitric oxide and vascular endothelial function in patients with hypercholesterolemia. *BioMed Research International* 2014, 481904. doi:10.1155/2014/481904.

Talati, R., Baker, W. L., Pabilonia, M. S., White, C. M. and Coleman, C. I. 2009. The effects of barley-derived soluble fiber on serum lipids. *Annals of Family Medicine* 7(2), 157–63, doi:10.1370/afm.917.

Teixeira, J., Gaspar, A., Garrido, E. M., Garrido, J. and Borges, F. 2013. Hydroxycinnamic acid antioxidants: an electrochemical overview. *BioMed Research International* 2013, 251754. doi:10.1155/2013/251754.

Theander, O., Westerlund, E. and Aman, P. 1993. Structure and components of dietary fiber. *Cereal Foods World* 38(3), 135–41.

Thondre, P. S., Wang, K., Rosenthal, A. J. and Henry, C. J. K. 2012. Glycaemic response to barley porridge varying in dietary fibre content. *The British Journal of Nutrition* 107(5), 719–24. doi:10.1017/S0007114511003461.

Tiwari, U. and Cummins, E. 2009a. Factors influencing β-glucan levels and molecular weight in cereal-based products. *Cereal Chemistry Journal* 86(3), 290–301, doi:10.1094/CCHEM-86-3-0290.

Tiwari, U. and Cummins, E. 2009b. Nutritional importance and effect of processing on tocols in cereals. *Trends in Food Science and Technology* 20(11–12), 511–20. doi:10.1016/j.tifs.2009.06.001.

Todd, G. W., Getahun, A. and Cress, D. C. 1971. Resistance in barley to the greenbug, Schizaphis graminum.1 1. Toxicity of Phenolic and Flavonoid Compounds and Related Substances2. *Annals of the Entomological Society of America* 64(3), 718–22. doi:10.1093/aesa/64.3.718.

Tosh, S. M. 2013. Review of human studies investigating the post-prandial blood-glucose lowering ability of oat and barley food products. *European Journal of Clinical Nutrition* 67(4), 310–7. doi:10.1038/ejcn.2013.25.

USFDA, 2006. FDA finalizes health claim associating consumption of barley products with reduction of risk of coronary heart disease. https://www.newhope.com/supply-news-amp-analysis/fda-finalizes-health-claim-associating-consumption-barley-products-reductio (Accessed on 26 August 2019).

Vaikousi, H. and Biliaderis, C. G. 2005. Processing and formulation effects on rheological behavior of barley β-glucan aqueous dispersions. *Food Chemistry* 91(3), 505–16. doi:10.1016/j.foodchem.2004.04.042.

Wang, M., van Vliet, T. and Hamer, R. J. 2004. How gluten properties are affected by pentosans. *Journal of Cereal Science* 39(3), 395–402. doi:10.1016/j.jcs.2004.02.002.

Wang, Y., Ames, N. P., Tun, H. M., Tosh, S. M., Jones, P. J. and Khafipour, E. 2016a. High molecular weight barley β-glucan alters gut microbiota toward reduced cardiovascular disease risk. *Frontiers in Microbiology* 7, 129. doi:10.3389/fmicb.2016.00129.

Wang, Y., Harding, S. V., Eck, P., Thandapilly, S. J., Gamel, T. H., Abdel-Aal, el-S. M., Crow, G. H., Tosh, S. M., Jones, P. J. and Ames, N. P. 2016b. High-molecular-weight β-glucan decreases serum cholesterol differentially based on the CYP7A1 rs3808607 polymorphism in mildly hypercholesterolemic adults. *The Journal of Nutrition* 146(4), 720–7. doi:10.3945/jn.115.223206.

Ward, J. L., Poutanen, K., Gebruers, K., Piironen, V., Lampi, A. M., Nyström, L., Andersson, A. A. M., Åman, P., Boros, D., Rakszegi, M., Bedő, Z. and Shewry, P. R. 2008. The HEALTHGRAIN cereal diversity screen: concept, results, and prospects. *Journal of Agricultural and Food Chemistry* 56(21), 9699–709. doi:10.1021/jf8009574.

Wise, M. L., Doehlert, D. C. and McMullen, M. S. 2008. Association of avenanthramide concentration in oat (*Avena sativa* L.) grain with crown rust incidence and genetic resistance. *Cereal Chemistry Journal* 85(5), 639–41. doi:10.1094/CCHEM-85-5-0639.

Wong, J. M., de Souza, R., Kendall, C. W., Emam, A. and Jenkins, D. J. 2006. Colonic health: fermentation and short chain fatty acids. *Journal of Clinical Gastroenterology* 40(3), 235–43. doi:10.1097/00004836-200603000-00015.

Yalçın, E., Çelik, S. and İbanoğlu, E. 2008. Foaming properties of barley protein isolates and hydrolysates. *European Food Research and Technology* 226(5), 967–74. doi:10.1007/s00217-007-0618-8.

Yang, T., Duan, C. L., Zeng, Y. W., Du, J., Yang, S. M., Pu, X. Y. and Yang, S. C. 2013. HPLC analysis of flavonoids compounds of purple, normal barley grain. *Advanced Materials Research* 634–638, 1486–90. doi:10.4028/www.scientific.net/AMR.634-638.1486.

Yao, L. H., Jiang, Y. M., Shi, J., Tomas-Barberan, F. A., Datta, N., Singanusong, R. and Chen, S. S. 2004. Flavonoids in food and their health benefits. *Plant Foods for Human Nutrition* 59(3), 113–22. doi:10.1007/s11130-004-0049-7.

Zarnowski, R. and Suzuki, Y. 2004. 5-n-alkylresorcinols from grains of winter barley (*Hordeum vulgare* L.). *Zeitschrift Fur Naturforschung. C* 59(5–6), 315–7. doi:10.1515/znc-2004-5-603.

Zieliński, H. and Kozłowska, H. 2000. Antioxidant activity and total phenolics in selected cereal grains and their different morphological fractions. *Journal of Agricultural and Food Chemistry* 48(6), 2008–16. doi:10.1021/jf990619o.

Zielke, C., Kosik, O., Ainalem, M. L., Lovegrove, A., Stradner, A. and Nilsson, L. 2017. Characterization of cereal β-glucan extracts from oat and barley and quantification of proteinaceous matter. *PLoS ONE* 12(2), e0172034, doi:10.1371/journal.pone.0172034.

Zou, Y., Liao, D., Huang, H., Li, T. and Chi, H. 2015. A systematic review and meta-analysis of beta-glucan consumption on glycemic control in hypercholesterolemic individuals. *International Journal of Food Sciences and Nutrition* 66(4), 355–62. doi:10.3109/09637486.2015.1034250.

Index

AA. *see* Amino acid (AA)
AAL. *see* Apparent attenuation limit (AAL)
Abiotic stress. *see* Physiology
Adopt-a-Crop online platform 340
AED. *see* Apparent energy digestibility (AED)
Afghanistan 100
Aflatoxins (AFs) 238, 243
AGG. *see* Australian Grains Genebank (AGG)
AGOUEB. *see* Association Genetics of UK Elite Barley project (AGOUEB)
Agricultural and Horticultural Development Board (AHDB) 184
  Barley Disease Management Guide for the United Kingdom 340
AGT. *see* Appressorial germ tube (AGT)
AHDB. *see* Agricultural and Horticultural Development Board (AHDB)
Alberta Agriculture breeding programme 438
Algeria 100
Alkylresorcinols 473–474
Alternative oxidases (AOX) 74
American Society of Brewing Chemists method 382
Amino acid (AA) 72–73
Amplified Fragment Length Polymorphisms and Simple Sequence Repeats 184
α-Amylases 449
Amylose-to-amylopectin ratio 69
Animal feed 427
  feed barley and 429–433
    breeding varieties 434–436
    cattle 443–445
    modern selection tools for improving feed quality 438–441
    pigs 445–448
    poultry 448–450
    Premium Grains for Livestock Program 436–438
    processing 441–442
    production, optimising 433–434
    sheep 445
  future trends 450–455
Anthocyanins 475
A-optimality criterion 165
AOX. *see* Alternative oxidases (AOX)
Apparent attenuation limit (AAL) 385
Apparent energy digestibility (AED) 447, 448
Appressorial germ tube (AGT) 279
Arabinoxylan 383, 384
Argentina 100, 233
Association Genetics of UK Elite Barley project (AGOUEB) 192, 193
Australia 101, 233, 274, 282, 355, 364, 365
Australian Barley Board 388
Australian Grains Genebank (AGG) 100, 395
Australian Grain Technologies 393
Australian Journal of Agricultural Research 436
Australian Winter Cereals Molecular Marker Program (AWCMMP) 439
Autoregressive process 165
Avena fatua 358, 367, 368
AWCMMP. *see* Australian Winter Cereals Molecular Marker Program (AWCMMP)
Awn 13
Axillary meristems (AXMs) 5

Barley Breeding Australia (BBA) 392, 393
Barley Oligo Pooled Arrays 185
  mapping, and yield traits 187–188
Barley stripe 304
Bass variety 392
BBA. *see* Barley Breeding Australia (BBA)
Beauvericin (BEA) 246

Beta-amylase 387, 411–412
Bile acids 478
Bin dryer 232
Biocontrol 254
approaches 252
Biostimulant 334–335
Biotrophic foliar diseases 270–274
Biotrophs 268
Bi-parental mapping 188
Bipolaris sorokiniana 282–285
Black point. *see* Kernel blight
Blocking 162–163
Blumeria graminis f. sp. *hordei* 278, 281
Brassinosteroids 10, 16
Breeding experiments advanced designs 159–161
early generation field trials designs 169–172
field trials experimental design 161–164
genetic relationship matrix 172–173
partially replicated design incorporating genetic relatedness 173–175
late-generation field trials designs 164–165
case study 166–168
multi-phase designs for laboratory experiments 176–178
multi-phase design in laboratory 178
Brewer's spent grain (BSG) 485
Brewers Clarex® 415
Brewing quality improvement 405–406
barley conversion to beer and 407–408
beta-glucans and impact on mashing and filterability and 412–413
breeding barley for brewing process and 408–412
final product quality traits 413
colloidal haze formation 415–416
off-flavours elimination through targeted breeding 413–415
future trends 416–418
Brown rust 275
BSA. *see* Bulk segregant analysis (BSA)
BSG. *see* Brewer's spent grain (BSG)
Bulk segregant analysis (BSA) 134, 135, 191
Buloke variety 392

Canada 365, 429, 467
Canadian Grain Commission 235
Canola 114
Canopy senescence 30
CAP. *see* Co-ordinated Agricultural Project (CAP)
Carbon dioxide 237
Caryopsis 56, 63, 76
*CBF*genes 27
CDC PolarStar 414
Chasmothecia 280
Chloroplast 26
CIM. *see* Composite interval mapping (CIM)
Clustered Regularly Interspaced Short Palindromic Repeats (CRISPR) 212, 213
Cly1 gene 13
Commander variety 392
Compass variety 392
Composite interval mapping (CIM) 129–130
Conidia 287–288, 298, 299
Conservation Agriculture systems 355
Co-ordinated Agricultural Project (CAP) 190
COR genes 26
cp TILLING 209
CRISPR. *see* Clustered Regularly Interspaced Short Palindromic Repeats (CRISPR)
Crop Monitor online platform 340
Cutlass variety 108

DArT. *see* Diversity Arrays Technology (DArT)
DE. *see* Digestible energy (DE)
Demethylation inhibitors (DMIs) 299
Denmark 337, 364
Deoxynivalenol (DON) 240, 241, 249, 254, 301, 303–304
DH. *see* Doubled haploid (DH) lines
Diastatic power (DP) 378, 388, 409
relationship with % Malt Extract 389
Diatomaceous earth-based grain protectants 237
Digestible energy (DE) 432, 445–446
Dimethyl sulphide (DMS) 414, 415
Disease escape 333
Disease tolerance 332–333
Distinctness, Uniformity and Stability (DUS) tests 184, 193, 333
Diversity Arrays Technology (DArT) 174
DMIs. *see* Demethylation inhibitors (DMIs)
DMS. *see* Dimethyl sulphide (DMS)
DON. *see* Deoxynivalenol (DON)
Doubled haploid (DH) lines 129
Double knock approach 357
Dovetail Chicago™ 138

DP. *see* Diastatic power (DP)
Drought 24
resilience to 28–33
Dry rolling 441
DUS. *see* Distinctness, Uniformity and Stability (DUS) tests

Economic Commission for Europe 103
Edstar Genetics 393
EFSA. *see* European Food Safety Authority (EFSA)
EMC *see* Equilibrium moisture content (EMC)
EMS. *see* Ethyl methanesulfonate (EMS)
Endosperm transfer cells (ETC) 58, 59, 62, 63, 64
Enniatins (ENNs) 246
Ensembl Plants database 140
eQTL. *see* Expression QTL (eQTL)
Equilibrium moisture content (EMC) 233
Ergot 238, 243
ETC. *see* Endosperm transfer cells (ETC)
Ethiopia 100, 101
Ethiopian Biodiversity Institute 100
Ethyl methanesulfonate (EMS) 205
mutation 135
European Brewery Convention method 382
European Food Safety Authority (EFSA) 241
Exome capture 127
Expression QTL (eQTL) 143

FDA. *see* US Food and Drug Administration (FDA)
Fermentable sugars 383
FHB. *see* Fusarium Head Blight (FHB)
Field fungi 247
Flagship variety 174, 392
Flavonoids 474–475
Flinders variety 392
Florets 13
Foliar fungicides 278, 299
Fragment Analyzer™ 135
FRAG-UK. *see* Fungicide Resistance Action Group UK (FRAG-UK)
Friability 386
Frost tolerance 25–28, 38
FT1 11
FTIR microspectroscopy 64
Fumigation 237
Fumonisins 240, 243
Fungal diseases 265–267
fusarium head blight 300–304
leaf rust 274–276
necrotrophic diseases 282–286
net blotch 286–291
plant genetic resistance to fungal pathogens and 267–269
powdery mildew 278–281
Ramularia leaf spot 291–294
scald 297–300
seed-borne disease 304
Septoria speckled leaf blotch 294–297
stem rust 270–274
stripe dust 276–278
Fungicide Resistance Action Group UK (FRAG-UK) 340
Fungicides 278, 280–281, 334–335
application methods 336
modes of action 337
stewardship 337–339
Fusarium graminearum 301–302
Fusarium Head Blight (FHB) 249, 433

G × E × M effects 433
γ-aminobutyric acid (GABA) 79
GBSS. *see* Granule-bound starch synthase (GBSS)
Genebank genomics 196
Gene editing 333
Gene-for-gene hypothesis 268
GENEVESTIGATOR database 139
Genome mapping and exploitation 123–124
association mapping boom 130–131
case study 140–142
classical strategies and improvement in genomics era 129–130
fast NGS-enabled technologies 133–138
future trends 142–144
genetic and genomic resources 139–140
germplasm conservation 138–139
high-resolution mapping and gene cloning 132–133
multiparental populations 131–132
new possibilities in genomics era 124–128
Genome-wide association studies (GWAS) 8, 10, 11, 14, 130, 183–184
barley breeding progress 184–187
in barley improvements, application of results from 192–195

future trends 195-197
mapping, in barley 188-192
Germplasm for yield improvement 97-98
barley origins 98-99
genetic diversity in barley 99-101
in breeding 101-102
see also Sulphur
β-Glucanases 449
Gluten 417
GrainGenes database 139
Grains Research and Development Corporation (GRDC) 393
Gramene 140
Granule-bound starch synthase (GBSS) 68
GRDC. *see* Grains Research and Development Corporation (GRDC)

Halophytes plants 34
Haustoria 271
Heated drying 233
Heat Shock Factors 32
Heirloom varieties 379
Hemoglobins 77
Herbicide resistance 355, 359-362
HGCA. *see* Home-Grown Cereals Authority (HGCA)
Hindmarsh variety 392
Home-Grown Cereals Authority (HGCA) 229
Hooded phenotypes 13
Hop creep 383
Hordeins 111
HR. *see* Hypersensitive response (HR)
Hulled barley 56, 435, 443, 475
Hulless barley 285, 432, 435, 440, 443, 446-450
HvDWARF4 10
Hypersensitive response (HR) 268

IB. *see* Incomplete block (IB)
ICARDA 100
ICP. *see* Integrated crop protection (ICP)
IDF. *see* Insoluble dietary fibre (IDF)
IMPROMALT project 193
Incomplete block (IB) 163
In-crop herbicides 358
India 100, 277
Inflorescence development 6
In-situ dry-matter digestibility (ISDMD) 440
Insoluble dietary fibre (IDF) 471-472
Institute of Brewing method 382
INTEGRA project 193
Integrated crop protection (ICP) 299
Integrated pest management (IPM) 323-324
agronomy 339
barley production context 324, 326
crop protectants 334-339
diseases overview 326-329
farming systems, soil and research platforms 342-344
future trends 344
inoculum management 330-331
knowledge sources and tools 339-340
uptake and communication of 340-342
varietal resistance 331-333
Integrated weed management 353-356
examples, in barley 367-368
in practice 366-367
weed control tactics
crop competition 364-365
crop rotation 363-364
harvest weed seed control 362-363
herbicides 356-362
on-farm hygiene 366
Intercropping farming system 113, 339, 344
InterGrain Pty. Ltd. 389, 393
Intermediate microflora 247
International Barley Core Collection 138
International Rice Research Institute (IRRI) 14
International Survey of Herbicide Resistant Weeds 359
Intracellular wash fluids (IWFs) 289
IPK Gatersleben (Germany) 100
IPM. *see* Integrated pest management (IPM)
Irradiation 251
IRRI. *see* International Rice Research Institute (IRRI)
ISDMD. *see* In-situ dry-matter digestibility (ISDMD)
IWFs. *see* Intracellular wash fluids (IWFs)

JEKYLL gene 61, 63

KASP™ 191, 193
Kernel blight 282
Kinship matrix 174

Landrace Collection LRC1485 139
Latinised row-column designs 165

La Trobe variety 392-393
Lattices 164
Lauter tuns 408
LD. *see* Linkage disequilibrium (LD)
Leaf angle 10
Leaf area 10-11
Leaf spot disease severity scale 282-283
Leaves 5-6
Lemma 6
Linear mixed models (LMM) 106
Linkage disequilibrium (LD) 130
Lipoxygenase (LOX) 385, 414
LMM. *see* Linear mixed models (LMM)
LOX. *see* Lipoxygenase (LOX)

MA. *see* Modified atmospheres (MA)
Mace variety 108
MAGB. *see* Maltsters Association of Great Britain (MAGB)
MAGIC. *see* Multi-parent advanced generation inter-cross (MAGIC)
Malteurop 393
Malting and Brewing Industry Barley Technical Committee (MBIBTC) 394
Malting Barley Accreditation 394
Malting Barley Committee 186
Malting quality 186
Malt quality improvement 377-378
  barley improvement history in Australia 393
    Barley Australia 394
  craft brewing 378-379
  future trends 396-397
  malting quality
    after malting 381-384
    germination 381
    grain hardness (endosperm texture) 386-387
    grain size 379-380
    linked traits but not routinely measured 385-386
    malt colour (wort colour) 385
    protein 380-381
    starch-degrading enzymes 384-385
  successful programmes requirements 394-396
  twenty-first century brewing modern varieties 388-393
Maltsters Association of Great Britain (MAGB) 186
MAMPs. *see* Microbe-associated molecular patterns (MAMPs)
Mapping-by-sequencing 136
Marker-Assisted Selection (MAS) 188
Mashing 407, 412-413
  efficient breakdown of starch during 409-412
MBCs *see* Methyl benzimidazole carbamates (MBCs)
MBIBTC. *see* Malting and Brewing Industry Barley Technical Committee (MBIBTC)
Mendel's law 132
Methanol extracts 481
Methyl benzimidazole carbamates (MBCs) 299
N-Methyl-N-nitroso-urea (MNU) 205
METs. *see* Multi-environment trials (METs)
Met-*S*-methyltransferase (MMT) 414-415
Mexico 277
Microbe-associated molecular patterns (MAMPs) 267
Micro-malting 186
Mildew-resistance locus O (MLO) 211
Mississippi Crop Situation online platform 340
Mla gene 281
MLO. *see* Mildew-resistance locus O (MLO)
mlo gene 281, 294
MMT. *see* Met-*S*-methyltransferase (MMT)
MNU *see* N-Methyl-N-nitroso-urea (MNU)
Mode of action (MOA) 334, 337, 359, 360
Modified atmospheres (MA) 237
Monocropping system 113
Morex BAC library 141
morexGenes database 139
Multi-environment trials (METs) 159
Multi-parent advanced generation inter-cross (MAGIC) 131, 132
MutChromSeq 137, 138
MutMap 135
MutMap-Gap 136
MutRenSeq 136, 137
Mycotoxins 228, 237-241
  and fungi 241, 243-246
  prevention and decontamination, in barley storage 248
    mycotoxin-contaminated barley removal 249-250
    stored barley treatment 250-252
  toxigenic fungi stability and development during barley storage and 246-248

NABGMP. *see* North American Barley Genome Mapping Project (NABGMP)
Naked barley 13, 56
NAM. *see* Nested association mapping (NAM)
NaN_3 *see* Sodium azide (NaN_3)
National Institute of Agricultural Botany (United Kingdom) 340
National Variety Trials 358
ND24260 174
Near infra-red (NIR) 435, 437
  and molecular markers 438–441
Near isogenic lines (NILs) 132
Necrotrophic effectors (NEs) 268, 288–289
Necrotrophic-effector-triggered susceptibility (NETS) 268
NEs. *see* Necrotrophic effectors (NEs)
Nested association mapping (NAM) 131, 132
Net form net blotch (NFNB) 286, 289
NETS. *see* Necrotrophic-effector-triggered susceptibility (NETS)
Next-generation sequencing (NGS) 124, 127, 207
NFNB. *see* Net form net blotch (NFNB)
NGS. *see* Next-generation sequencing (NGS)
NILs. *see* Near isogenic lines (NILs)
Nitrogen-use efficiency (NUE) 36, 37
NLR. *see* Nucleotide-binding-leucine-rich repeat (NLR)
Non-fermentable sugars 383
North American Barley Genome Mapping Project (NABGMP) 439
NSGC Barley Core 138, 139
Nuclear projection 59
Nucleotide-binding-leucine-rich repeat (NLR) 268
  genes 136, 137
Nud gene 13
NUE. *see* Nitrogen-use efficiency (NUE)
N-uptake efficiency (NUpE) 36
N-utilization efficiency (NUtE) 36
Nutritional and bioactive compounds 467–468
  arabinoxylans 471–472
  barley activity enhancement
    food matrix 483–484
    genotypic and environmental influences 484
    processing 482–483
  beta-glucan 470–471
  future trends 485
  health benefits
    antioxidant activity 481–482
    blood pressure lowering 482
    cholesterol 477–478
    glycaemic response 478–480
    gut health 480–481
  key issues and challenges 468–470
  phenolics 473–475
  plant sterols 475–476
  protein/peptides 476
  resistant starch 472
  tocols 476

Ochratoxins 240
od package 175
Okayama University Barley Germplasm Resources (Japan) collection 100
Open reading frames (ORFs) 141, 142
OsDWARF4 10
Osmotic adjustment 29, 38
Ozone 250–251

PAs. *see* Proanthocyanidins (PAs)
Pattern-triggered immunity (PTI) 267–268
PCD. *see* Programed cell death (PCD)
Pearling process 250
Pesticides 237
PGSB PlantsDB database 140
PGT. *see* Primary germ tube (PGT)
Pgt race QCCJB 272–273
Pgt race TTKSK 273
Ph. see Puccinia hordei (*Ph*)
Phenotyping malting quality 190
Photoperiod 25, 27
PHS. *see* Preharvest sprouting (PHS)
Physiology 23–25
  adaptation to environment 38–40
  cold acclimation 25–28
  low nitrogen 36–37
  new methodologies 28–33
  soil salinity adaptation 33–36
Phytohormones 7, 9, 10
Phytosterols 475–476
Pinoxaden 358
Plant development and architecture 3–4
  breeding implications 14–16
  reproductive development molecular control 11–13
  structure/morphology and growth habit 4–7
  vegetative development molecular control 7–11

Plants evolved cell-surface pattern recognition receptors (PRRs) 267
Plasma 251
PLEXdb database 139
Polyvinylpolypyrrolidone (PVPP) 415
Postharvest storage and handling 227-228
  future trends 255-256
  operations 228-229
    general considerations 229-233
    insects management in stored barley 236-237
    storage considerations for dormant and preharvest sprouted barley 235-236
    stored barely management 233-235
  post-storage treatment 252
    feed barley 253
    malting barley 253-255
  see also Mycotoxins
Pre-emergent herbicides 357-358
Preharvest sprouting (PHS) 235
Primary germ tube (PGT) 279
Proanthocyanidins (PAs) 415, 416
Programed cell death (PCD) 58, 59-60, 61, 268
Proteases 386-387
Proteomics 383, 453-454
PRRs. *see* Plants evolved cell-surface pattern recognition receptors (PRRs)
PTI. *see* Pattern-triggered immunity (PTI)
Puccinia graminis 270, 271, 273
Puccinia hordei (*Ph*) 274-275
Puccinia striiformis f. sp. *hordei* 276-277
PVPP. *see* Polyvinylpolypyrrolidone (PVPP)
Pyrenophora teres f. *maculata* 286, 287, 291
Pyrenophora teres f. *teres* 286, 287, 291

Quantitative trait loci (QTL) 8, 9, 10, 14, 30, 36, 131, 132, 188-192, 300, 440-441
Queensland barley programme 438

Rachis 12
RAM. *see* Root apical meristems (RAM)
Ramularia collo-cygni (*Rcc*) 291-294, 340
Randomised complete block (RCB) 162
Rapid Visco-Analyzer (RVA) 235
RCB. *see* Randomised complete block (RCB)
Rcc. see Ramularia collo-cygni (*Rcc*)
rcs5 gene 285
Recombinant inbred lines (RILs) 129, 132
Recommended List Trials (RLT) 184
Reduced height-1 (*Rht1*) genes 213
Residual maximum likelihood (REML) 106
Resistance elicitors 334, 339
Resorcinolic lipids. *see* Alkylresorcinols
Restriction fragment length polymorphisms (RFLPs) 125, 184
R-genes 281, 296
RGT Planet variety 392
Rht1. *see* Reduced height-1 (*Rht1*) genes
Rhynchosporium commune 297, 298, 299, 300
RLT. *see* Recommended List Trials (RLT)
RMRL. *see rpg4/Rpg5*-mediated resistance locus (RMRL)
Root apical meristems (RAM) 4
Row-column alpha designs 165
Rpg1 gene 272
rpg4/Rpg5-mediated resistance locus (RMRL) 273
Rph genes 275
rrBLUP package 174
Rumen 444
RVA. *see* Rapid Visco-Analyzer (RVA)
Rym1 gene 140
rym4/rym5 gene 140, 141
rym11 gene 140, 141, 142

Salt tolerance, in shoot tissues 35
SAM. *see* Shoot apical meristems (SAM)
Scepter variety 108
SCFA. *see* Short-chain fatty acids (SCFA)
SDF. *see* Soluble dietary fibre (SDF)
SECOBRA Recherches 393
Seed storage proteins (SSP) 71, 73
Semi-dwarf phenotypes 9
Septoria passerini 295, 296
SFNB. *see* Spot form net blotch (SFNB)
sgRNA. *see* Single guide RNA (sgRNA)
Shoot apical meristems (SAM) 4
SHOREmap 135
Short-chain fatty acids (SCFA) 472, 477, 479
Single guide RNA (sgRNA) 212, 213
Single nucleotide polymorphism (SNP) 127, 191
S-methyl-l-methionine (SMM) 414
SNP. *see* Single nucleotide polymorphism (SNP)
SNP iSelect platform 100
Sodium azide (NaN_3) 205
Soluble dietary fibre (SDF) 470-471
South Australian Barley Improvement programme 393
Speed-breeding 143

Spike growth, influence on grain traits 54
barley grain in botanical terms 55-56
cell death role in grain development 57
cellular mechanisms and main actors in cell death 60-61
death events arrangement in grain 57-60
implications 61
energy metabolism particularities 73-74
barley grain photosynthesis and respiration 74-76
hypoxia management at structural and metabolic levels 78-80
main energy metabolism components and pathways 74
O_2 gradients and indications of hypoxia 76-77
O_2 sensing mechanisms and balancing in seeds 77
functional orchestration of barley grain 80-83
grain growth temporal sequence 56-57
proteins and barley grain quality 71
properties and deposition 71-72
protein storage regulation and manipulation 72-73
starch in caryopsis development 67
starch storage variability and quality in mature grain 69-70
transient starch physiological role 68-69
sucrose allocation during grain-filling stage 62
cellular sucrose allocation pathways within grain 62-64
membrane transport events and active transporters 64, 66-67
sucrose concentration gradients in grain 64, 65
Spot blotch 282-286
Spot form net blotch (SFNB) 286, 290-291
Spring barleys 28
SSP. *see* Seed storage proteins (SSP)
Stagonospora avenae f. sp. *triticea* 295, 296
Starch gelatinization temperature 410
Starch quality, in brewing 409-411
Stay-green phenotype 30
Steam flaking 441-442
Stem elongation 5
Storage fungi 247
SUE. *see* Sulphur-use efficiency (SUE)
Sugars Will Eventually be Exported Transporters (SWEET) 66-67
Sulphur
effects on yield, quality, and response to stress 109-110
adaptation to abiotic and biotic stress factors 112-113
quality 111-112
yield 110-111
nutrition, and farming systems 113-114
nutrition effects assessment, on barley and wheat grain yield
data analysis 106-107
materials and methods 104-106
results 107-109
role, in barley growth 102-103
use, genotypic differences in
associated genes 116-117
varietal differences 114-116
Sulphur-use efficiency (SUE) 114, 115, 116
Svalbard Global Seed Vault (Norway) 100
SWEET. *see* Sugars Will Eventually be Exported Transporters (SWEET)
Syngenta 393

T2N. *see* Trans-2-nonenal (T2N)
TACCA. *see* Targeted chromosome-based cloning via long-range assembly (TACCA)
TALENS 133
Targeted chromosome-based cloning via long-range assembly (TACCA) 137, 138
Targeted induced local lesions in genomes (TILLING) 135, 141, 142, 203-204
artificial mutagenesis and mutant discovery
mutagen choice 204-206
mutation density 206-207
searching for mutations 207-208
TILLING *in silico* 208
tissue and organ for mutagenizing 204
current and future trends 209-212
resources, in barley 209
and reverse genetics tools compared 212-213
by Sequencing (TbyS) 207
and wheat Tilling compared 212-214
TB1 11
TCA. *see* Tricarboxylic acid (TCA) cycle

TCAP *see* Triticeae Co-Ordinated Project (TCAP)
Teliospores 271
THOD. *see* Trihydroxyoctadecenoic acid (THOD)
Tillering 7
Tillers 5
TILLING. *see* Targeted induced local lesions in genomes (TILLING)
Tissue tolerance 35
Total protein digestibility (TPD) 447
ToxA NE 284
TPD. *see* Total protein digestibility (TPD)
Trans-2-nonenal (T2N) 386, 414
Tricarboxylic acid (TCA) cycle 78, 79
Trichothecenes 240, 241
Trihydroxyoctadecenoic acid (THOD) 386
Triplet spikelet meristem 12
Triticale 427
Triticeae Co-Ordinated Project (TCAP) 189

UK Cereal Pathogen Virulence Survey 330
United Kingdom 338, 340
United States 467
Urediniospores 271
US Food and Drug Administration (FDA) 241, 477
Uzu gene 9–10

Value for Cultivation and Use (VCU) 333
Vernalization 27
Voluntary initiatives (VI) 338
Vomitoxin. *see* DON
VPE genes 61
Vrn-H1 gene 27
Vrs 1 gene 9
Vrs3 12

Weed seeds 366
Western Australian Department of Agriculture 389
Whole-genome sequencing (WGS) 135, 136
Wild Barley Diversity Collection 139
Wort, meaning of 407
Wort beta-glucan 383, 384

Yield stability 40

Zearalenone (ZEA) 241, 243